AF352867

Water Management for Sustainable Food Production

Water Management for Sustainable Food Production

Special Issue Editors

Narayanan Kannan
Aavudai Anandhi

MDPI • Basel • Beijing • Wuhan • Barcelona • Belgrade • Manchester • Tokyo • Cluj • Tianjin

Special Issue Editors
Narayanan Kannan
Tarleton State University
USA

Aavudai Anandhi
Florida A&M University
USA

Editorial Office
MDPI
St. Alban-Anlage 66
4052 Basel, Switzerland

This is a reprint of articles from the Special Issue published online in the open access journal *Water* (ISSN 2073-4441) (available at: https://www.mdpi.com/journal/water/special_issues/Wate_Food_Production).

For citation purposes, cite each article independently as indicated on the article page online and as indicated below:

LastName, A.A.; LastName, B.B.; LastName, C.C. Article Title. *Journal Name* **Year**, *Article Number*, Page Range.

ISBN 978-3-03928-941-7 (Hbk)
ISBN 978-3-03928-942-4 (PDF)

Contents

About the Special Issue Editors

Narayanan Kannan, Dr., is an agricultural engineer with expertise in water resources and hydrology. He received his Ph.D. from Cranfield University in the UK. Presently, he is a research scientist with the Texas Institute for Applied Environmental Research-Tarleton State University at Stephenville, TX, USA. Dr. Kannan's research is focused on the development and application of computer models for solving pressing environmental problems. His expertise includes non-point source pollution modeling, water sustainability analysis, stream health, urban hydrologic modeling, and life cycle analysis. He developed the "Animal Production Life Cycle Analysis Tool" (APLCAT) to estimate the water footprint, energy use, and greenhouse gas emissions associated with beef cattle production. To date, he has published 34 research articles in peer-reviewed scholarly journals and presented more than 50 oral and poster presentations at conferences. He is also an associate editor of the *Journal of Soil and Water Conservation*. He has traveled extensively to many countries in connection with international conventions and training people on the use of computer models.

Aavudai Anandhi, Dr., is an Assistant Professor in the Biological and Agricultural Systems Engineering (BASE) program in the College of Agriculture and Food Sciences at Florida A&M University. In Dr. Anandhi's group, during research, teaching, and outreach, they often use this concept: "There are known knowns. These are things we know that we know. There are known unknowns. That is to say, there are things that we now know we don't know. But, there are also unknown unknowns. These are things we do not know we don't know". This aids in understanding complex hydrological and ecological processes and their interactions at the food–water–energy (FEW) nexus. We then apply this understanding to improve predictions of the processes and of ecosystem services responses to environmental changes (extreme and mean: climate, water use, and land use). This supports us in developing novel solutions to increase our decision support capability of these critical systems using data, network, and computing resources for synergizing FEW systems to improve efficiency, productivity, and sustainability. We use a mixture of systems thinking approaches, conceptual and structural models, AI, spatial statistics, and machine learning for analysis of the ecosystem services of interconnected FEW systems in affecting human well-being by exploiting new data streams to advance system adaptation, resilience, and stress mitigation. These are well documented in publications (50 peer-reviewed articles; 100 oral and poster presentations; 25 invited talks; 4 book chapters in the past decade) and supported by funding (~$12 Million, as PI and co-PI) from various organizations (IUSSTF, NSF, USDA-NIFA, SoTL, USDA-OAP, Tilford, DoE). They have received multiple awards and recognitions, namely: "Emerging Researcher Award" at Florida A&M University, "Teacher of the year" award from the Florida section of ASABE, ASCE's ExCEEd fellow, and the "Blue Ribbon Award" for innovative educational materials from ASABE. Additionally, Dr. Anandhi's work experiences and collaborations with the software industry and state, federal, and non-governmental agencies in India, NY, KS, and FL have provided the skills, expertise, and knowledge to develop a successful program in complex systems engineering in FEW.

Editorial

Water Management for Sustainable Food Production

Narayanan Kannan [1] and **Aavudai Anandhi** [2,*]

[1] Texas Institute for Applied Environmental Research, Tarleton State University, Stephenville, TX 76402, USA; kannan@tarleton.edu

[2] Biological Systems Engineering, Florida Agricultural and Mechanical University, Tallahassee, FL 32307, USA

* Correspondence: anandhi@famu.edu

Received: 5 March 2020; Accepted: 9 March 2020; Published: 11 March 2020

Abstract: The agricultural community has a challenge of increasing food production by more than 70% to meet demand from the global population increase by the mid-21st century. Sustainable food production involves the sustained availability of resources, such as water and energy, to agriculture. The key challenges to sustainable food production are population increase, increasing demands for food, climate change, and climate variability, decreasing per capita land and water resources. To discuss more details on (a) the challenges for sustainable food production and (b) mitigation options available, a special issue on "Water Management for Sustainable Food Production" was assembled. The special issue focused on issues such as irrigation using brackish water, virtual water trade, allocation of water resources, consequences of excess precipitation on crop yields, strategies to increase water productivity, rainwater harvesting, irrigation water management, deficit irrigation, and fertilization, environmental and socio-economic impacts, and irrigation water quality. Articles covered several water-related issues across the U.S., Asia, Middle-East, Africa, and Pakistan for sustainable food production. The articles in the special issue highlight the substantial impacts on agricultural production, water availability, and water quality in the face of increasing demands for food and energy.

Keywords: deficit irrigation; excess precipitation; irrigation water quality; virtual water; water productivity; brackish groundwater; rainwater harvesting; socio-economic impacts

---

1. Introduction

Sustainable food production involves the sustained availability of resources, such as water and energy, to agriculture. The key challenges to sustainable food production are population increase, availability of resources at the right time and place, threats posed by climate variability and extremes to land and water which are often exacerbated by other biophysical limits such as declining per-capita land and water scarcity, as well as rising demand for agricultural products [1,2]. The emerging consensus is that the world likely will exceed nine billion people by 2050, requiring 70%, 80%, and 55%, more food, water, and energy, respectively [3–5]. Increasing agricultural productivity and developing sustainable water management techniques are needed to feed the ever-increasing population [1,6]. Projected climate change is expected to affect crop and livestock production substantially, and water availability and quality. Climate change and variability, as well as extremes, including floods and droughts, further aggravate the challenges to sustainable food production. Innovative strategies are needed to mitigate these negative impacts while meeting the increasing demands in a sustainable manner [7].

Efficient and smart use of resources, and the adoption of less water-intensive crop production systems, are the present requirements to achieve sustainable food production. Advanced crop production methods such as precision farming, access to low-cost data, advancement in electronic gadgets, and smart instruments have opened up plenty of opportunities for agricultural producers to gear up towards sustainable food production. However, the knowledge of managing water in agriculture with existing technology has not reached many parts of the world. Therefore, this special issue is developed to bring out the knowledge on water management towards sustainable food production in an open-access platform.

This Special Issue aims to bring forth the challenges and discuss the mitigation options on the availability of water to both rain-fed and irrigated agricultural production (including animal production) to sustain food production at local, regional, national, and global scales.

In particular, the Special Issue focused on:

1. Use of Smart technology (electronic gadgets, low-cost data sources, local technology) to manage water to obtain more crop per drop.
2. Agricultural production under shrinking land and water resources.
3. Availability of water to agricultural production under historical past and projected future climate change (including floods, droughts, and extremes of precipitation and temperature).
4. Sustaining agricultural production under population increase with existing water resources.

In this introductory article, we highlight the major findings from papers published in this collection. A summary of the articles in the special issue is presented in Table 1. This Edited Collection, "Water Management for Sustainable Food Production," includes fifteen articles (thirteen are research articles, two review articles). These studies cover a wide range of topics related to water management for various crop production systems in different parts of the world.

Table 1. Summary of articles included in the Special Issue on Water Management for Sustainable Food Production.

Authors	Citation	Food Production Systems	Water Management	Focus Region
Katuri et al., 2019 [8]	*Water* 2019, *11*(12), 2556 [8]	Olive	Drip irrigation systems using brackish groundwater	Israel
Ali et al., 2019 [9]	*Water* 2019, *11*(11), 2259 [9]	15 major agricultural commodities	Virtual water trade to analyze water use sustainability	Pakistan
Gedefaw et al., 2019 [10]	*Water* 2019, *11*(10), 1966 [10]	Rainfed agriculture (specific crop N/A)	Allocate water supplies to maximize economic benefits	Ethiopia
Sharma et al., 2019 [11]	*Water* 2019, *11*(9), 1920 [11]	Sorghum	Consequences of Excess precipitation on crop yields	Texas, USA
Lampayan et al., 2019 [12]	*Water* 2019, *11*(9), 1816 [12]	Rice	Crop management strategies to increase water productivity and decrease irrigation	Lao
Kaioglu et al., 2019 [13]	*Water* 2019, *11*(8), 1730 [13]	Canola	Crop management techniques and water productivity	North Dakota, USA
Silungwe et al., 2019 [14]	*Water* 2019, *11*(3), 578 [14]	Pearl Millet Yield	Rainwater harvesting techniques	Central Tanzania
Gadédjisso-Tosssou et al., 2019 [15]	*Water* 2018, *10*(12), 1803 [15]	Maize	Five irrigation management techniques	West Africa
Fan et al., 2019 [16]	*Water* 2018, *10*(11), 1637 [16]	Studied 17 crops Focus: corn, soybean	Farmers' decision making, economics, irrigation water use efficiency	48 states in the USA
Materu et al., 2019 [17]	*Water* 2018, *10*(8), 1018 [17]	Rice	Three irrigation management alternatives	Tanzania
Trifnov et al., 2019 [18]	*Water* 2018, *10*(8), 970 [18]	Potato	Reduced water (drip irrigation) and fertilization	Areal desert, Israel
Ket et al., 2019 [19]	*Water* 2018, *10*(5), 666 [19]	Lettuce	Water-saving irrigation strategies	Kampong Chhnang, Cambodia
Sekvi-Annan et al., 2019 [20]	*Water* 2018, *10*(5), 624 [20]	Tomato-Maize Rotation-System	Reservoir-Based Irrigation Scheme	Ghana
Velasco-Muñoz et al., 2019 [21]	*Water* 2019, *11*(9), 1758 [21]	Agricultural Production Systems	Environmental, economic, social impacts	Ghana
Malakar et al., 2019 [22]	*Water* 2019, *11*(7), 1482 [22]	Food crops	Irrigation water quality impacts on crop/soil: Source contaminants	Ghana

2. Content of the Special Issue

Two review articles in the special collection bring out the global perspective on irrigation, which is considered the highest consumptive use of freshwater [21,22]. Velasco-Muñoz et al. (2019) [21] reviewed

713 articles on sustainable irrigation in agriculture over the last twenty years (1999–2018) through a bibliometric analysis (quantitatively) and a systematic review based on keyword analysis (qualitatively). Their results show the study of sustainable irrigation has grown in recent years, which is higher than that of general research in irrigation. The study observed that the environmental dimension dominates far more than the social or economic perspectives in the research of sustainable irrigation. Substantial differences in specific approaches and preferred research topics varied with countries. The review brought out the need for integrating environmental, social, and economic dimensions in sustainable irrigation, and aimed to communicate the results of the research to society, as well as to provide greater knowledge of the environmental impacts of irrigation-related practices on different levels (plot, district, basin, region). The second review article [22] in this special collection addressed some of the environmental impacts of irrigation water quality from multiple sources (conventional sources like surface or groundwater, or nonconventional sources like reclaimed water) [22]. The study highlights the vulnerability of crop and soil quality, as well as the complexities of the composition of irrigation water to various emerging contaminants from various water sources. They focused on contaminants such as organic pollutants (e.g., pharmaceuticals, antibiotics, steroids, agrochemicals, cyanotoxins, and mycotoxins); biological contaminants such as bacteria, viruses, and antibiotic resistance, as well as inorganic contaminants such as geogenic source and nanomaterials. The study brought out the need for establishing regulations and clear guidelines for irrigation water quality to ensure healthy food production for human consumption. They emphasized the need to question the existing recommendations of contaminants when higher than recommended concentrations were bio-accumulated in crops. This understanding will help ensure adequate crop production to meet increased demand as well as to maintain proper food and soil quality.

Optimal allocation of water resources is important in basins facing water scarcity due to the increasing demands caused by population growth, urbanization, industrialization and agricultural intensification, and poor water resources management. The impacts due to these demands can have important implications in many developing countries. Three studies have demonstrated improved irrigation management strategies in Africa and Asia. These studies focus on increasing variability in rainfall and overcome deficits in current irrigation schemes. Gedefaw et al.'s (2019) [10] study formulated the water allocation networks under an irrigation expansion and climate change scenario using the calibrated water evaluation and planning (WEAP) model for the Awash River Basin, Ethiopia, facing water scarcity. In that study, water demands, water shortages, and supply alternatives were analyzed using three scenarios, namely: the reference scenario (1981–2016), the medium-term development (2017–2030), and the long-term development (2031–2050) as well as an economic parameter to maximize the economic benefits of water allocation. Their results showed that future water consumption would greatly increase in the Awash River Basin. Their results also highlight the requirement of water-saving measures to prevent future water shortages. Rice is a major food crop for more than half of the global population, with more than 90% of global rice production and consumption occurring in Asia. Efficient water use in rice production is a prerequisite to sustaining the world's food security. The objective of Lampayan et al.'s (2019) [12] study of rice production systems is to test their hypothesis that a delayed transplanting strategy reduces irrigation requirements and increases irrigation water productivity. They tested the hypothesis using a field experiment conducted in two wet seasons in Lao People's Democratic Republic by evaluating the interactive effects of seedling age, seedling density, and variety on post-transplanted rice crop development (e.g., tillering propensity), grain yield, and water productivity. Materu et al. (2019) [17] compared three variants of irrigation management alternatives (system of rice intensification (SRI)) using the conventional, continuously flooding system in Tanzania experimentally for both wet and dry seasons. SRI is a new production practice, which is a combination of the agronomic practices adopted by many farmers in Asia and Sub-Saharan Africa. The study observed that SRI resulted in water saving, an increased rice yield, and improved the economic productivity of water.

Three studies have used multiple water sources (e.g., reservoir-based, groundwater-based, brackish irrigation) to research improved irrigation management strategies in multiple crops (e.g., tomato, maize, canola, and olive) in the USA, Africa, and Asia. Sekyi-Annan et al.'s (2019) [20] study aimed to improve the traditional dry season irrigation practices in reservoir-based irrigation schemes for the tomato–maize rotation system in Ghana (Upper East region), and assessed the potential for introducing supplemental irrigation in the rainy season as an adaptation to climate change using the AquaCrop model under different climate scenarios. The improved irrigation schedule for dry season tomato cultivation resulted in water-saving (130–1325 mm) compared to traditional irrigation practices, and an increase in tomato yield (4–14%). Maize would require 107–126 mm of water in periods of low rainfall and frequent dry spells, and 88–105 mm in periods of high rainfall and rare dry spells. These water management techniques could make year-round irrigated crop production feasible, using water saved during dry season tomato cultivation for supplemental irrigation of maize in the rainy season. The main scope of Kaioglu et al.'s (2019) [13] study was to determine an optimum shallow groundwater depth to achieve a high yield, growth, and water use in canola plants using lysimetric experiments in greenhouse conditions in Fargo, North Dakota, USA. Canola plant characteristics (e.g., height, water use, total biomass and grain yield water use efficiency, root mass, root–shoot ratio, and harvesting results (total biomass, pod, and seed weight) were determined and compared for four different water table scenarios. The results suggested that the canola plant characteristics were affected by different water table levels and showed inverse linear relationships. The decrease in yields in olive orchards and replacing olive trees due to unprofitability in orchards irrigated with brackish irrigation water in the central Negev Desert, Israel, motivated Katuri et al.'s (2019) [9] study. The results of this study demonstrate that, following twenty years of irrigation with brackish irrigation water, salinization and sodification took place in the soil profile (0–60 cm, the active root zone of the olive trees). This study fills the knowledge gap regarding the spatial distribution of salinity and sodicity in long-term, sub-surface, drip-irrigated soils with brackish irrigation water. They concluded that, in the long term, utilizing marginal irrigation water sources, such as brackish water, may be fundamentally unsustainable, particularly in arid lands where precipitation is too low to leach the accumulated salts from the active root zone. The benefits of water-saving due to drip irrigation are masked by soil salinization and sodification. Reclamation of these soils with gypsum, for example, is essential. Any alternative practices, such as replacing olive trees and the further introduction of plants with even higher salinity tolerance (e.g., jojoba) in this region, will intensify the salt buildup without leaving any option for soil reclamation in the future.

Two studies addressed the availability of water to agricultural production under the historical past and projected future climate change. By documenting the relationships between reductions in rainfed crop yield and excess precipitation, Sharma et al. (2019) [11] address an existing knowledge gap. They used the historical crop yield data for Texas by county for grain sorghum from 1973 to 2000 and the corresponding daily precipitation data from weather stations within the counties, estimating the total precipitation of the growing season and the maximum four-day precipitation. Using the two parameters as independent variables, and the crop yield of sorghum as the dependent variable, they established graphical and mathematical relationships between excess precipitation and decreases in crop yields. Their results show a decrease in rainfed sorghum yields in Texas in the range of 18% and 38% due to excess precipitation. In another part of the world, Silungwe et al. (2019) [14] attempted to understand the rainfed pearl millet yield variability due to variations in seasonal rainfall in a 1500 ha area of a semiarid region in central Tanzania with and without rainwater harvesting management practice (flat and tied ridge). Rainfall data were collected from 38 rain gauge stations from November 2016 to May 2018, which includes two growing seasons. Yield data from plots near these locations were collected for both the practices (20 for the tied ridge, 18 for flat). The yield data showed the correlation with both the rainfall amounts and the number of events in a season. The use of tied ridges (an infield rainwater harvesting system) increased the pearl millet yield significantly.

Currently, the production of crops poses a greater challenge in managing effective inputs (e.g., water irrigation and fertilization) due to their sensitivity to water shortage and increasing costs. Deficit irrigation and fertilization are practices whereby a crop is irrigated and fertilized with an amount below the full requirement for optimal plant growth, thereby saving inputs and minimizing the economic impact on the harvest. Three studies have addressed this for vegetables (e.g., lettuce, potato) and cereal (e.g., maize) in different parts of the world (e.g., Cambodia, Israel, West Africa). The main objective of Ket et al.'s (2019) [19] study was to improve the water productivity of lettuce by assessing the impact of multiple water-saving scenarios (full, deficit irrigation) by developing the crop model, AquaCrop, for lettuce (currently not available in catalog) and calibrating it with observations from field experiments for Cambodia. The results suggested that a deficit irrigation strategy can save 20–60% of water compared to full irrigation scenarios. Potatoes are a high-value vegetable crop with a shallow, inefficient root system and high fertilizer rate requirements with a high risk of leaching below the root zone when grown often in sandy soils. Trifonov et al. (2019) [18] attempted to save water and fertilizer with reduced nutrient leaching in potato production. Their objective was to optimize potato growth under a low discharge drip irrigation (40%, 60%, 80%, and 100%) and fertigation (0%, 50%, and 100%) doses in Arava Desert, Israel. They used field experiments during the 2014–2015 time period. They found that water productivity was affected by water dose and nitrogen level with an 80% (438.6 mm) irrigation dose, and a 50% (50 mg N L^{-1}) fertigation dose showing optimal potato yield (about 40 ton ha^{-1}) without qualitative changes in the potato tuber. In another study, Gadédjisso-Tossou et al. (2019) [15] investigated alternative irrigation strategies in the dry savannah area of Togo, in the West African region, for a maize crop. They characterized the climate of a water-scarce region and evaluated five irrigation management strategies for combinations of no irrigation, conventional and supplemental irrigation, limited and full supply. They used the OCCASION framework (weather generator-LARS-WG, AquaCrop model, optimal irrigation scheduling algorithm) for their research. They observed (a) satisfactory performance of the LARS Weather Generator in predicting the climate of northern Togo, and (b) irrigation practice (0 to 600 mm) in agriculture lowered crop yield variability as well as crop failure.

Two studies provide additional perspectives on sustainable water use practices using multilevel models and virtual water trade assessments. Using multilevel linear regression models (MLMs), Fan et al.'s (2019) [16] study analyzed the effects of multiple factors on farmers' decision making and economical irrigation water use efficiency (EIWUE) in a multi-crop production system. The study was conducted across five regions, encompassing 48 states in the USA (Western, Plains, Midwestern, Southern, and Atlantic states) with different cropping patterns and climatic conditions. Originally, multiple factors (e.g., water sources, input costs, farming area, land characteristics, adoption of various irrigation systems, climate perceptions) at multiple levels (farm, state, regional) were identified from multiple sources (e.g., review, survey, observation) for 17 crops. However, the results in the paper focused on water source (surface, groundwater) cost and water use, as well as the adoption of pressure irrigation systems, adoption of enhanced irrigation systems, higher temperatures, and precipitation on corn and soybean yield and water use. This study could help farmers and policymakers adapt to potential climate risks, better manage the irrigation water application, and achieve the sustainable use of limited water resources. Their results show higher costs of surface water are not effective in reducing water use, while groundwater costs show a positive association with water use on both corn and soybean farms. In the second study of this category, the research by Ali et al. (2019) [9] on virtual water trade is one of the first studies to concentrate on a water-stressed, net virtual, water-exporting country (Pakistan), while most of the existing country-level studies on the virtual water trade focused on net virtual water importers, which are usually water-scarce countries as well. This paper assessed the trade and savings/losses of blue and green virtual water through 15 agricultural commodities, over the period of 1990–2016 and in 2030. The results of the study show that, in most of the studied commodities, blue VW is the major component in total water use. Pakistan has been a net exporter of blue VW, mostly through rice export to Asian and African countries. In terms of green VW, Pakistan has been a net importer (marginally), mainly through the import of palm oil from Indonesia and Malaysia. In the future (2030), both Pakistan's domestic savings of green and losses of virtual blue water will increase by more than 200%. Their results also suggest that Pakistan has been exporting more expensive (with high opportunity cost) blue VW through its agricultural trade to the rest of the world. However, there are opportunities for improving the water use efficiency (in the export-oriented crops) and adjustment in its export portfolio of agricultural commodities by promoting the export of commodities with higher value and lower water use intensity.

3. Conclusions

This special issue was organized to initiate an interdisciplinary dialogue with stakeholder groups about mitigation and adaptation strategies for sustainable agricultural production in the face of increasing demands due to population growth, urbanization, industrialization and agricultural intensification, and inadequate water resources management. Given the scope and extent of the impacts from these increasing demands, variety in crop production systems, water sources, climate variability, as well as changes in crop production systems, differences in water resources, and ecosystem health, there is a need to form global partnerships for developing sustainable agricultural production strategies. For example, developing holistic strategies for adaptation and mitigation (e.g., improving irrigation efficiency, virtual water assessment) for sustaining crop production during resource-scarce conditions. This Special Collection highlights the fact that agricultural production, water availability, and water quality are going to be substantially impacted by increasing demands in food, energy, and water sectors. There are technological options available to mitigate the adverse impacts of climate change, but adaptation strategies will require rethinking agricultural management practices to maintain crop and livestock production while protecting environmental quality.

Author Contributions: Both the authors N.K. and A.A. summarized the contents of the special issue and contributed to the production of this editorial. All authors have read and agreed to the published version of the manuscript.

Funding: This research received no external funding.

Acknowledgments: The guest editors of this special issue express their thanks to the MDPI team for their support to bring out this special issue.

Conflicts of Interest: The authors declare no conflict of interest.

References

1. Anandhi, A. CISTA: Conceptual model using indicators selected by systems thinking for adaptation strategies in a changing climate: Case study in agro-ecosystems. *Ecol. Model.* **2017**, *345*, 41–55. [CrossRef]
2. Liu, H.; Zhan, J.; Hussain, S.; Nie, L. Grain Yield and Resource Use Efficiencies of Upland and Lowland Rice Cultivars under Aerobic Cultivation. *Agronomy* **2019**, *9*, 591. [CrossRef]
3. Albrecht, T.R.; Crootof, A.; Scott, C.A. The Water-Energy-Food Nexus: A systematic review of methods for nexus assessment. *Environ. Res. Lett.* **2018**, *13*, 043002. [CrossRef]
4. Gragg Iii, S.; David, R.; Anandhi, A.; Jiru, M.; Usher, K. A Conceptualization of the Urbanizing Food-Energy-Water Nexus Sustainability Paradigm: Modeling from Theory to Practice. *Front. Environ. Sci.* **2018**, *6*, 133. [CrossRef]
5. Malekpour, S.; Caball, R.; Brown, R.R.; Georges, N.; Jasieniak, J. *Food-Energy-Water Nexus: Ideas for Monash University Clayton Campus*; Monash University: Melbourne, Australia, 2017.
6. Anandhi, A.; Kannan, N. Vulnerability assessment of water resources—Translating a theoretical concept to an operational framework using systems thinking approach in a changing climate: Case study in Ogallala Aquifer. *J. Hydrol.* **2018**, *557*, 460–474. [CrossRef]
7. Chaubey, I.; Bosch, D.; Muñoz-Carpena, R.; Harmel, R.D.; Douglas-Mankin, K.R.; Nejadhashemi, A.; Srivastava, P.; Shirmohammadi, A. Climate change: A call for adaptation and mitigation strategies. *Trans. ASABE* **2016**, *59*, 1709–1713.
8. Katuri, R.J.; Trifonov, P.; Arye, G. Spatial Distribution of Salinity and Sodicity in Arid Climate Following Long Term Brackish Water Drip Irrigated Olive Orchard. *Water* **2019**, *11*, 2556. [CrossRef]
9. Ali, T.; Nadeem, A.M.; Riaz, M.F.; Xie, W. Sustainable Water Use for International Agricultural Trade: The Case of Pakistan. *Water* **2019**, *11*, 2259. [CrossRef]

10. Gedefaw, M.; Wang, H.; Yan, D.; Qin, T.; Wang, K.; Girma, A.; Batsuren, D.; Abiyu, A. Water Resources Allocation Systems under Irrigation Expansion and Climate Change Scenario in Awash River Basin of Ethiopia. *Water* **2019**, *11*, 1966. [CrossRef]
11. Sharma, O.; Kannan, N.; Cook, S.; Pokhrel, B.; McKenzie, C. Analysis of the effects of high precipitation in Texas on rainfed sorghum yields. *Water* **2019**, *11*, 1920. [CrossRef]
12. Lampayan, R.; Xangsayasane, P.; Bueno, C. Crop Performance and Water Productivity of Transplanted Rice as Affected by Seedling Age and Seedling Density under Alternate Wetting and Drying Conditions in Lao PDR. *Water* **2019**, *11*, 1816. [CrossRef]
13. Kaioglu, H.; Hatterman-Valenti, H.; Jia, X.; Chu XAslan, H.; Simsek, H. Groundwater Table Effects on the Yield, Growth, and Water Use of Canola (*Brassica napus* L.) Plant. *Water* **2019**, *11*, 1730. [CrossRef]
14. Silungwe, F.R.; Graef, F.; Bellingrath-Kimura, S.D.; Tumbo, S.D.; Kahimba, F.C.; Lana, M.A. Analysis of Intra and Interseasonal Rainfall Variability and Its Effects on Pearl Millet Yield in a Semiarid Agroclimate: Significance of Scattered Fields and Tied Ridges. *Water* **2019**, *11*, 578. [CrossRef]
15. Gadédjisso-Tossou, A.; Avellan, T.; Schutze, N. Potential of Deficit and Supplemental Irrigation under Climate Variability in Northern Togo, West Africa. *Water* **2018**, *10*, 1803. [CrossRef]
16. Fan, Y.; Massey, R.; Park, S.C. Multi-Crop Production Decisions and Economic Irrigation Water Use Efficiency: The Effects of Water Costs, Pressure Irrigation Adoption, and Climatic Determinants. *Water* **2018**, *10*, 1637. [CrossRef]
17. Materu, S.T.; Shukla, S.; Sishodia, R.P.; Tarimo, A.; Tumbo, S.D. Water Use and Rice Productivity for Irrigation Management Alternatives in Tanzania. *Water* **2018**, *10*, 1018. [CrossRef]
18. Trifonov, P.; Lazarovitch, N.; Arye, G. Water and Nitrogen Productivity of Potato Growth in Desert Areas under Low-Discharge Drip Irrigation. *Water* **2018**, *10*, 970. [CrossRef]
19. Ket, P.; Garre, S.; Oeurng, C.; Hok, L.; Degre, A. Simulation of Crop Growth and Water-Saving Irrigation Scenarios for Lettuce: A Monsoon-Climate Case Study in Kampong Chhnang, Cambodia. *Water* **2018**, *10*, 1018. [CrossRef]
20. Sekyi-Annan, E.; Tischbein, B.; Diekkruger, B.; Khamzina, A. Year-Round Irrigation Schedule for a Tomato-Maize Rotation System in Reservoir-Based Irrigation Schemes in Ghana. *Water* **2018**, *10*, 624. [CrossRef]
21. Velasco-Muñoz, J.F.; Aznar-Sánchez, J.A.; Batlles-delaFuenta, A.; Fidelibus, M.D. Sustainable Irrigation in Agriculture: An Analysis of Global Research. *Water* **2019**, *11*, 1758. [CrossRef]
22. Malakar, A.; Snow, D.D.; Ray, C. Irrigation Water Quality—A Contemporary Perspective. *Water* **2019**, *11*, 1482. [CrossRef]

 *water*

Review

Irrigation Water Quality—A Contemporary Perspective

Arindam Malakar [1] , **Daniel D. Snow** [2,*] **and Chittaranjan Ray** [3]

[1]   Nebraska Water Center, part of the Robert B. Daugherty Water for Food Global Institute, 109 Water Sciences
     Laboratory, University of Nebraska, Lincoln, NE 68583-0844, USA
[2]   School of Natural Resources and Nebraska Water Center, part of the Robert B. Daugherty Water for Food
     Global Institute, 202 Water Sciences Laboratory, University of Nebraska, Lincoln, NE 68583-0844, USA
[3]   Nebraska Water Center, part of the Robert B. Daugherty Water for Food Global Institute 2021 Transformation
     Drive, University of Nebraska, Lincoln, NE 68588-6204, USA
*   Correspondence: dsnow1@unl.edu; Tel.: +01-402-472-7539

Received: 29 May 2019; Accepted: 13 July 2019; Published: 17 July 2019

Abstract: In the race to enhance agricultural productivity, irrigation will become more dependent on poorly characterized and virtually unmonitored sources of water. Increased use of irrigation water has led to impaired water and soil quality in many areas. Historically, soil salinization and reduced crop productivity have been the primary focus of irrigation water quality. Recently, there is increasing evidence for the occurrence of geogenic contaminants in water. The appearance of trace elements and an increase in the use of wastewater has highlighted the vulnerability and complexities of the composition of irrigation water and its role in ensuring proper crop growth, and long-term food quality. Analytical capabilities of measuring vanishingly small concentrations of biologically-active organic contaminants, including steroid hormones, plasticizers, pharmaceuticals, and personal care products, in a variety of irrigation water sources provide the means to evaluate uptake and occurrence in crops but do not resolve questions related to food safety or human health effects. Natural and synthetic nanoparticles are now known to occur in many water sources, potentially altering plant growth and food standard. The rapidly changing quality of irrigation water urgently needs closer attention to understand and predict long-term effects on soils and food crops in an increasingly fresh-water stressed world.

Keywords: crop uptake; food quality; geogenic; emerging contaminants; nanomaterials

---

1. Introduction

Irrigation is the controlled use of multiple water sources in a timely manner for increased or sustained crop production. Irrigation comprises of the water that is applied by an irrigation system during the growing season and also includes water applied during field preparation, pre-irrigation, weed control, harvesting, and for leaching salts from the root zone [1]. In 2015 it was estimated that in the United States irrigation alone accounted for 62% of water usage [1]. Globally, irrigation is the highest consumptive use of freshwater [2]. As the world's population grows, the risk increases that more people will be deprived of adequate food supplies in impoverished areas, particularly those subject to water scarcity [3]. Agricultural production of food needs to increase by an estimated 60% by 2050 to ensure global food security [3] and irrigation will increasingly be called upon to help meet this demand. In the race to enhance agricultural productivity, irrigation will become even more dependent on substandard sources of water. Therefore, it is of utmost importance to access our current state of knowledge and explore the effects of irrigation water quality on crops. This understanding will help ensure adequate crop production to meet increased demand as well as to maintain proper food and soil quality.

Groundwater exploitation (withdrawal for irrigation) can release naturally occurring geogenic contaminants, such as arsenic, from the solid phase to groundwater, while wastewater reuse can concentrate pesticides, pharmaceuticals and other emerging contaminants in irrigation water [4,5]. Use of untreated wastewater is becoming prevalent in developing countries where around 80–90% of wastewater remains untreated [6]. Polluted municipal, industrial or agricultural water used for irrigation significantly changes soil quality, increases the amount of trace elements in soil and plants, and acts as a source of various pathogens which affects food quality and safety [7,8]. Water of inadequate quality is a potential source of both direct and indirect contamination to food crops [9], and leads to increased contamination of soil and water [10,11]. In addition, the presence of synthetic and natural nanomaterials is beginning to be identified in crops [12–14]. In locations where excess irrigation is practiced, contaminants in soils are leached to the vadose zone, where they can contribute to geogenic contaminant mobilization and potentially increase contaminant levels in local groundwater [15]. Many aspects of water composition, such as hardness and iron content, also affect the suitability of a water source for newer, more efficient spray or drip irrigation techniques. Runoff, return flow, and leaching of irrigation water also contribute to local surface and groundwater contamination [16]. Increased usage of irrigation water has already led to impaired irrigation water and soil quality. Considering the presence of new contaminant types in different water sources (see Figure 1), it is essential to evaluate the impact of these contaminants within the context of modern agriculture. To date, very little research and regulatory attention has been paid to contaminants in irrigation water. Contamination of irrigation water supplies is likely to worsen unless additional efforts (research, guidelines, regulations, treatment methods) are brought to bear on this problem.

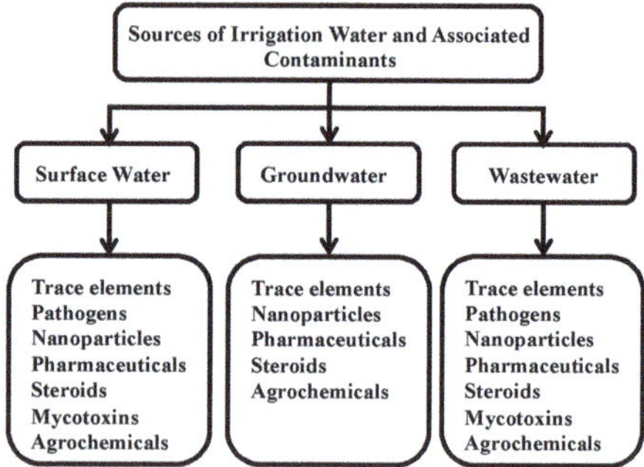

Figure 1. Main sources of irrigation water and different types of contaminants present in those sources impacting food, soil, and water quality. Note that surface water and wastewater are subject to similar types of contamination.

This review article looks at previous approaches to define irrigation water quality and compares to a current perspective with respect to impacts on human health. It evaluates the long-term effect and influence of the changing quality of water sources used in agricultural production. Although the article discusses traditional irrigation water quality concerns, such as salinization, it mainly emphasizes contemporary water quality issues like new or emerging contaminants, pathogens, geogenic trace elements and engineered nanomaterials. These contaminants are now widespread in various conventional and unconventional water sources used for modern-day irrigation. The article is organized as a short summary of conventional measures of irrigation water quality followed by a more detailed evaluation of the impact of contemporary irrigation water quality issues on soil and crop quality. Contemporary topics include emerging contaminants with separate sections on pharmaceuticals, antibiotics, steroid hormones, pesticides, cyanotoxins and mycotoxins, biological contaminants bacteria, virus and antibiotic resistance genes, modern inorganic contaminants, such as geogenic trace elements and nanomaterials. The review is summarized by considering the changing quality of water sources used for irrigation, and the need for additional work and improved regulation of irrigation water, especially for food production. The primary focus of this review is to recognize water quality issues that have a direct or indirect influence on surface soil contamination, crop uptake of contaminants and their potential to impact human health. This article shows a need for modern guidelines, regulations and research to understand the complex nature of irrigation water. Though it is a critically important topic from a human health standpoint, this review does not include an exhaustive discussion of contamination of irrigation water by human pathogens. While wastewater treatment technologies are constantly evolving and can address some of the issues presented here, a review of wastewater quality as a function of treatment technology is beyond the scope of this article. Moreover, treatment approaches are likely to be tailored to sources, and irrigation water sources are highly varied depending on climate, population, industry, crop, and livestock density.

2. Conventional Measures of Irrigation Water Quality

The effect of irrigation water composition on soil properties for crop production has been a focus for the past half century. Previous studies of water quality issues, and the suitability of freshwater sources for irrigation, have primarily been directed toward an understanding of potential problems to soil salinity, fertility and crop growth. For example, early work by the United States Geological Survey (USGS) [17] evaluated groundwater quality in Texas for irrigation and other potentially competing uses. A subsequent report by Schwennesen and Forbes characterized groundwater in San Simon Valley, Arizona and New Mexico, for domestic use and irrigation [18]. Clark reported on the chemical composition of groundwater in the Morgan Hill area of California [19], while Scofield and Headly [20] evaluated water composition with respect to irrigation potential. Most of these early works focused on understanding the impact of water quality on long-term viability of irrigation in arid regions of the United States.

Globally, irrigation water quality was described in Tanzania, Africa, with respect to pH and alkaline and alkaline-earth elements [21]. Taylor et al. reported that irrigation water pH was one of the main factors for wheat growth in Punjab, India [22]. A subsequent work reiterated that alkaline elements such as sodium play a crucial role in continued use of water for irrigation of cropland and quantified the maximum amount that may be tolerated [23]. The effects of soil salinization and trace element composition on crop growth have become more apparent over time. Eaton et al. reported that boron present in water around Hollister, California affected the growth of apricots and prunes [24]. In the subsequent years, the United States Department of Agriculture (USDA) conducted further studies and reported that sodium, boron and electrical conductivity are the best general measures for judging the suitability of water for irrigation [25]. From these studies, it was evident that continuous irrigation with water of marginal quality impacted soil and also affected crop growth [26–31]. In 1967, the American Society for Testing Materials (ASTM) developed a quantitative assessment of irrigation water quality, including new formulas for maximum permissible quantity of chloride and electrical conductivity based on infiltration rate, evapotranspiration rate, irrigation frequency and duration [32]. Traditionally, discussion on irrigation water quality has mainly focused on its effect on soil quality, and how soil quality was predicted to affect crop growth and yield. Color, turbidity, total dissolved solids (TDS), pH, specific conductance, odor and foam characterized the quality of water. Colorless, odorless, foamless water with minimum turbidity, TDS below 1000 mg L^{-1} at circumneutral pH and specific conductance below 1.5 mmhos/m is generally considered to be of good quality for irrigation purposes [33,34]. A higher TDS is not recommended for most crops as it can impact the salinity of soil and pore water will become highly concentrated when taken up by roots via osmosis. Excessive dissolved solids content, or salinity of irrigation water, has historically been the primary characteristic determining water suitability for irrigation. Salt accumulation in the crop root zone impedes water uptake and can eventually prevent plant growth altogether [34]. Excess salinity from sodium can affect soil structure and water infiltration. The proportion of sodium to calcium and magnesium is the primary factor controlling the hydraulic conductivity of water in soil [33–35]. Sodium is generally expressed as a sodium absorption ratio (SAR) [9]. Long-term irrigation of soils with elevated sodium concentrations relative to calcium and magnesium, bicarbonate, carbonate, and TDS will be limiting soil aggregate formation, which reduces infiltration and makes less water available to crops [34].

Seiler et al., under the National Irrigation Water Quality Program (NIWQP) of the U.S. Department of the Interior (DOI), studied the effect of irrigation-induced contamination of water, soil and biota in the western United States. NIWQP data from the 26 areas under study suggested that degradation of groundwater quality due to irrigation is a common occurrence [11,36]. The study indicated that selenium was the most common contaminant, followed by arsenic, uranium and molybdenum [11,37]. This study also suggested regular co-occurrence of these contaminants. For example, selenium was found to be elevated with uranium, and these contaminants were accumulating in the soils and affecting long-term suitability for crop production. This was one of the first reports to correlate trace element contamination in water sources used for irrigation to soil quality. These findings led to the appreciation of the intricate complexities of irrigation water quality and its role in ensuring proper crop growth and long-term food quality. These studies mainly focused on the impact of water quality on crop productivity and soil quality, while effects to food quality and safety were just beginning to be recognized.

3. Impact of Contemporary Irrigation Water Quality Issues on Soil and Crop Quality

Irrigation water quality has mainly been characterized with respect to effects on crop growth and yield, though an emerging and pressing issue relates to plant uptake and soil enrichment with inorganic and organic contaminants (Figure 2). These "new" issues with respect to irrigation water quality can lead to food quality and safety concerns, as well as affect crop growth and yield [38–41]. Wastewater reuse for irrigation contributes to increasing incidence of organic microcontaminants [42], such as pharmaceuticals and other synthetic organics in soils and crops. Increasing reliance on groundwater also contributes to the probability for elevated concentrations of natural geogenic contaminants such as arsenic and selenium in irrigation water and soils. Understanding the occurrence and fate of these new contaminants in irrigation water sources is paramount in limiting the effects to modern agricultural products [43]. Long term impacts to soil and crop quality (see Figure 2) need to be understood.

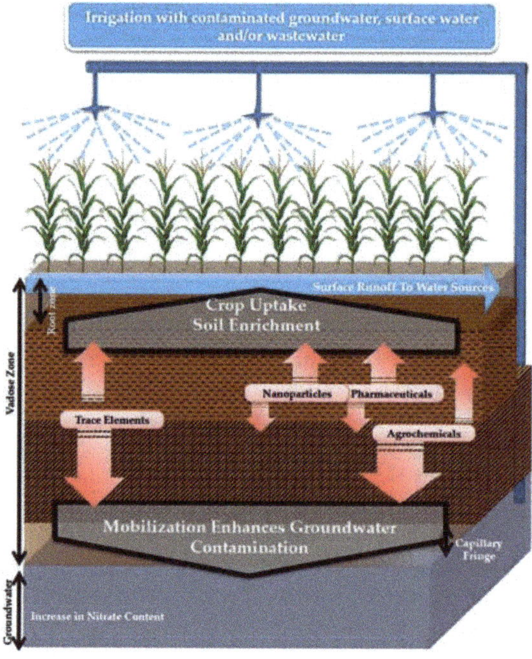

Figure 2. A conceptual model of the impact of inadequate quality of irrigation water sources on soil and crop quality.

3.1. Emerging Contaminants: Organic Pollutants

3.1.1. Pharmaceuticals

Traces of pharmaceuticals and personal care products have been identified in a variety of freshwater sources, including drinking water [44], groundwater [45], and surface water [45]. Pharmaceuticals can enter the water system from various sources, including direct disposal and human excretion into sewers leading to elevated concentrations of pharmaceuticals in wastewater [46]. Pharmaceuticals often detected in sewage sludge include non-steroidal anti-inflammatory drugs (NSAIDs), blood thinners, psychiatric drugs, antidiuretics and β-blockers [47–49]. Plant uptake of a wide variety of pharmaceutical groups like NSAIDs, antihistamine, β-blockers, calcium channel blockers, antiepileptics, steroid hormones, antidepressants, antineoplastic agents, anti-itch compounds, x-ray contrast agents, lipid-lowering agents, benzodiazepines, tranquilizers and veterinary drugs from soil and contaminated water has been observed and studied [50–52]. Wu et al. reported that a primary pathway for contamination by pharmaceuticals in food crops is through irrigation water [53,54]. For example, a recent study found traces of carbamazepine, caffeine, lamotrigine, gabapentin and acesulfame in a variety of vegetables grown with treated wastewater in Jordan [55]. Treated wastewater is well known to contain a large variety of pharmaceuticals and personal care products, many of which are known to accumulate in food crops [56,57]. The occurrence of these and other synthetic organic chemicals is likely to increase in water supplies, especially in areas with water scarcity, and irrigation with contaminated water will lead to soil contamination and plant uptake.

3.1.2. Antibiotics

Environmental contamination by antibiotic residues in food production systems is a growing problem worldwide, and the potential implications to proliferation of antibiotic resistance have been the subject of multiple reviews and opinion articles [58–60]. The occurrence of persistent antibiotic residues in various water sources [61,62] is well documented which not only includes municipal [63,64], agricultural [65,66] and hospital sewage [67,68] but also groundwater [69] and surface water [70–72]. The concentration range of antibiotics is generally measured at ng L^{-1} to a few µg L^{-1} in many water sources [73], though concentrated wastewater can have much higher levels [65]. Recent studies have demonstrated that plants can take up antibiotics (like amoxicillin, ketoconazole, lincomycin, oxytetracycline, sulfamethoxazole, sulfonamides, and tetracyclines) [74–76] and antibiotic contaminated irrigation water can play a significant role in the uptake [77]. The environmental fate and transport of antibiotics depend on various physical properties such as water solubility, lipophilicity, volatility and sorption potential [77].

The implications to human health due to the presence of antibiotics in food crops is not clear, but other potential adverse impacts include allergic reactions, disruption of digestive function and chronic toxic effects as a result of prolonged low-level exposure [78–81]. One of the major concerns for the increasing prevalence of antibiotics in the environment is the development and spread of antibiotic-resistant gene and bacteria [82], which is discussed in Section 3.2.2. Clearly, the absence of antibiotic residues in irrigation water cannot be assumed.

3.1.3. Steroids

Land application of livestock manure can contribute to accumulation of steroid hormones [83] and veterinary pharmaceuticals. Very low concentrations of natural steroid hormones, such as estrone, 17α-estradiol, 17β-estradiol, estriol, testosterone, androstenedione and progesterone that occur in animal waste and wastewater have been documented as accumulating in soil [84]. Traces of steroid hormones have also been reported in groundwater [85] and surface water [86], and often in treated municipal wastewater [87–91]. Wastewater treatment plants are known to discharge these hormones into river water and also other recipients [92]. Laboratory experiments have suggested that traces of steroid hormones and pharmaceuticals can be taken up in crops [93] and recent studies of food crops irrigated with treated municipal water have confirmed this can occur in the field. Further work is needed to understand the significance and impact of these chemicals in the environment and to human health [94], though at present, the reported concentrations are relatively low in comparison to other contaminants.

3.1.4. Agrochemicals

Regular use of pesticides in irrigated crops is also likely to lead to the occurrence of these residues in irrigation water and food crops, especially in regions where regulation and training in proper application of these substances is lacking. Application of large quantities of agrichemicals and improper management can create a substantial effect on the environment. The leaching and runoff of agrochemicals is a potential source of groundwater and surface water contamination [95,96]. The occurrence of agrichemical residues in vegetables has been documented [97] and their uptake by crops is well studied [98,99] and regulated. Leaching of nitrate from fertilizer over application to groundwater below is well reported, and accumulation of reactive nitrogen is also thought to initiate mobilization of other geogenic contaminants [15].

3.1.5. Cyanotoxins and Mycotoxins

Cyanotoxins, which comprise a large range of naturally produced organic compounds, are produced and released by cyanobacteria when they are present in large quantities (blooms) and especially when these organisms die off and decay in surface water. Cyanobacteria also referred to as blue-green algae, naturally occur in all freshwater ecosystems [100]. Warmer temperatures coupled with high nutrient concentrations are thought to favor conditions for algae blooms to form in surface water. Of many different groups of cyanotoxins, hepatoxic cyclic peptides collectively known as microcystins are the most commonly studied cyanotoxins, which cause a wide range of symptoms in humans [101]. Other studies have also shown that cyanotoxins, which include hepatotoxic microcystins and neurotoxic compounds such as anatoxin-a and beta-*N*-methylamino-alanine, can make their way to human and animal food chain from contaminated reservoirs [102–105]. A recent review has summarized the extent of the literature investigating the fate in soils, and agricultural crops [103]. It seems quite clear that toxins can accumulate in plants, including food crops and under some conditions can also inhibit plant growth [102,103,106–108]. Though there are many gaps, and only a handful of studies have investigated this route for exposure. There is evidence for human health effects through consumption of plants contaminated with cyanotoxins by irrigation using surface water sources impacted by cyanotoxins.

Mycotoxins are naturally occurring fungal toxins (chemicals), which can cause a variety of adverse health effects to both humans and livestock. A few mycotoxins are known or suspected carcinogens. Fungi do pose potential hazards to human health. However, there were relatively few studies of mycotoxins in water sources until recently. Fungal contamination has been observed in drinking water [109] and recently it was reported that untreated surface water can be breeding place for these fungi, generating mycotoxins [110]. Kolpin et al. led a broad scale study on the occurrence of mycotoxins across streams in the United States (US) [111]. Their study concluded that the ecotoxicological effects from long-term, low-level exposures to mycotoxins are poorly understood and would require further investigation. Mycotoxin uptake in rice has been studied [112] and these chemicals have been reported to be present in various food grains [113,114]. The prevalence of mycotoxins in surface water makes it an important consideration regarding modern water quality of irrigation as mycotoxin health hazards are widely reported.

3.2. Biological Contaminants: Bacteria, Virus and Antibiotic Resistance

3.2.1. Pathogens

Often because of its nutrient content and accessibility, untreated wastewater from municipal and domestic sources containing excessive levels of pathogens is often directly used for irrigation in developing countries [6]. Untreated wastewater generally carries a high pathogen load compared to other irrigation water sources. Risks from pathogen (bacteria, viruses, or protozoan or larger organisms) contamination to irrigation water quality will continue to be a topic of primary concern [115–117], and it is impossible to adequately address this topic in a few paragraphs. Pathogenic microorganisms in irrigation water likely pose the greatest acute risk to human health and will continue to be a concern especially in freshly-eaten produce. Pathogens are biological organisms that may influence modern-day irrigation water quality. Pathogen contamination is generally related to surface water sources, but groundwater may also be under threat if it is recharged with wastewater sources [116]. The complexity of reproducibly measuring microbiological contamination of irrigation water has made monitoring difficult. Several different types of pathogens have been detected in diverse irrigation water sources including bacteria (e.g., *Salmonella* and *Escherichia coli*), protozoa (e.g., *Cryptosporidium* and *Giardia*), as well as viruses (e.g., noroviruses)) [118,119]. Irrigation of food crops with surface water clearly has the highest potential for contaminating freshly eaten produce, and this topic has had the greatest research and regulatory effort in recent years.

There have been quite a few comprehensive reviews emphasizing irrigation water as a source of pathogenic microorganisms in fresh produce [120–123]. Between 1973 and 2012, the Centers for Disease Control and Prevention reported 606 leafy-vegetable associated pathogenic outbreaks (norovirus (55% of outbreaks), Shiga toxin-producing *Escherichia coli* (STEC) (18%), and *Salmonella* (11%)), with 20,003 associated illness and 19 deaths [124]. From 2013 to 2017, the number of outbreaks (mainly from norovirus (32%) STEC (23%), and *Salmonella* (32%)) associated with leafy greens and vegetables decreased to 21, with 699 illness and five deaths [125]. In 2018, 272 infections were reported from two outbreaks (*E. Coli*) associated with romaine lettuce resulting in five deaths [126,127], and another multi-state outbreak was linked to parasite *Cyclospora*, which reported 511 cases of infection [128]. However, a 2014 risk-based review conducted in California suggests that recycled water quality criteria, along with proper agricultural management practices do not lead to increased public health risk [129]. In the US, the Center for Produce Safety has published information on the factors that affect the microbiological safety of agricultural water [130]. The Foodborne Disease Outbreak Surveillance System (FDOSS) has an online tool, the National Outbreak Report System (NORS), which keeps track of outbreaks in the United States. Reports of pathogen contamination from inadequately treated wastewater have also been documented in developing countries [8]. The occurrence of pathogens in water used to irrigate food crops is considered a severe problem affecting human health both in both developing and even in developed countries [131]. Groundwater sources are generally considered less vulnerable to contamination by pathogenic microorganisms, while surface water and wastewater have a much higher potential for contamination. Farmers utilizing surface water for food crops, which are consumed raw should follow proper mitigation strategies to control contamination [132]. The method utilized for irrigation has a substantive role in pathogenic contamination of crops. For example, subsurface drip may have the lowest risk as the water is generally applied at the root zone, unlike other methods (e.g., sprinkler irrigation) where the edible portions of crops can come in contact with contaminated water [133]. New and more intensive monitoring approaches and potential disinfection and treatment techniques for surface water used to irrigate food crops are needed to improve food safety [8].

3.2.2. Antibiotic Resistance

The World Health Organization (WHO) has listed antibiotic resistance among today's biggest threats for global health, food safety, and development, as this threatens the ability to treat common infectious diseases [134]. Antibiotic resistome is defined as the sum of all genes directly or indirectly contributing to antibiotic resistance both in the clinics and the environment [135]. Aquatic ecosystems are regarded as a primary reservoir of antibiotic-resistant bacteria (ARB) [136]. The presence of ARB and their resistance determinants in surface water sources have been well documented and is generally linked to nearby wastewater treatment plant effluent [137–141]. Wastewater treatment plants enrich ARB and their resistance determinants as it favors exchange of antibiotic resistance genes (ARG) among bacteria and selection of resistant strains [142]. In a recent study, it was found that multidrug-resistant (MDR) bacteria were found to be more prevalent in surface waters than in treated wastewater [143].

Irrigation water is one of the major sources for contamination of fresh produce with antibiotic resistance bacteria [58,144,145]. Similar to pathogens, the incidence of ARB contamination is higher when using overhead sprinklers as water can directly come in contact with fresh produce. When fresh produce is consumed raw it can act as an ideal vector for exposure. The diversity of ARB present in fresh produce is significant and can have a severe impact on human health.

3.3. Inorganic Contaminants: Geogenic Source and Nanomaterials

3.3.1. Geogenic Contaminants in Irrigation Water

Selected naturally occurring geogenic contaminants, such as boron, arsenic and selenium, have been the subject of much previous work focused on irrigation water quality [146,147], especially in areas with extensive use of groundwater. With the exception of boron, trace element contaminants were not studied with respect to soil quality, crop productivity and phytotoxicity [9]. Boron is an essential trace element for plant growth, but elevated boron concentrations (>1 mg/L) in irrigation water can cause stunted growth and reduced productivity in sensitive crops such as wheat. The occurrence of geogenic contaminants is a growing contemporary issue because of the potential impact on food quality and human health. Concentrations of geogenic contaminants have likely been increasing over time in a variety of irrigation water sources, often due to increasing agricultural intensification [148,149]. Researchers have reported that groundwater in China has elevated levels of arsenic; this water is used for irrigating feed crops [150]. Similarly, studies in the United States have reported higher levels of uranium [15] and arsenic [151] in its aquifers. India, Bangladesh and Vietnam have widespread arsenic contamination in groundwater used for drinking and irrigation, especially in areas where the use of contaminated water has led to contaminated soils and crops [152–154]. Presently there are no federal guidelines regulating geogenic contaminant levels in irrigation water except in the case of direct wastewater reuse [155,156]. The increasing levels of contaminants in soils and irrigation water are a growing issue across the globe, and there is little work to date regarding strategies to mitigate accumulation in plants and food crops. Contaminant uptake by crops has been well studied and plant uptake has even been used as a remediation method, viz. phytoremediation. However, mitigation strategies for prevention of geogenic contaminant uptake by plants have received scant attention in the literature. Plant uptake and accumulation of specific trace elements may not affect plant growth, but accumulation and consumption may pose hazards to animal or humans.

Table 1 summarizes the estimated ranges of aqueous concentrations of many geogenic elements with respect to guidelines and recommendations of water use. Irrigation water guidelines compiled by the Food and Agriculture Organization (FAO) in 1976 are generally based on the toxic effects to crops and plant growth [157]. FAO has recommended using these values as guidelines for irrigation utilizing groundwater and surface water sources, but not for irrigation with wastewater containing measurable levels of trace elements [157]. Wastewater guidelines set by FAO (Table 1) are equivalent to irrigation water quality recommendations in 1976, though may contradict newer guidelines for special constituents of wastewater [157]. Early recommendations rarely consider uptake of trace elements by crops and the impact on food quality, which may affect human health. Recommended concentrations to ensure consistent food quality are absent, as even low concentrations of geogenic contaminants can impact food quality (Table 1) [158–172]. Continuous use of irrigation water with low concentrations of geogenic contaminants concentrations can result in soil enrichment and affect food crop quality [158–172].

Table 1. Common geogenic trace element contaminants in irrigation water with current regulatory and recommended levels (NA: Not Available, AL: Action level, CCC: Criterion continuous concentration, CMC: Criterion maximum concentration, SDWR: Secondary drinking water regulations). (* Few examples of plant uptake, ** in µg, *** greenhouse experiment concentration was equivalent to natural condition).

Trace Element	EPA Drinking Water Guideline ($\mu g\ L^{-1}$) [173]	Regulatory Limits for Wastewater ($\mu g\ L^{-1}$) [156,174,175]	Freshwater CCC, CMC Limits for Aquatic Life ($\mu g\ L^{-1}$) [176]	Recommended Maximum Concentrations of Trace Elements in Irrigation Waters ($\mu g\ L^{-1}$) [157]	Reported Ranges for Trace Elements in Glacier Aquifer System of USA (Includes Wells Used for Agriculture) ($\mu g\ L^{-1}$) [177]	Reported Ranges of Water Known to Impact Food and Soil Quality * ($\mu g\ L^{-1}$)
Arsenic	10	100	340, 150	100	0.09–340	58.7 [158], 0.2–164 [159]
Boron	NA	NA	NA	1000	NA	3351–16,000 [165–167]
Cadmium	5	10	1.8, 0.72	10	0.018–1	<1–3200 [168]
Cobalt	NA	50	NA	50	0.007–95	0.21–0.81 [169]
Copper	1300 (AL)	200	NA	200	0.126–127	10–133 ** [170]
Chromium (III) or (VI)	100 (Cr)	100 (Cr)	570, 74 (III) 16, 11 (VI)	100 (Cr)	0.4–22 (Cr)	1–46 (VI) [171] ≤250 *** [172]
Iron	300 (SDWR)	5000	NA, 1000	5000	3–38,100	-
Lead	15 (AL)	5000	65, 2.5	5000	0.04–9.0	≤140 [160]
Lithium	NA	2500	NA	2500	0.040–126	-
Manganese	500 (SDWR)	200	NA	200	0.056–28,200	≤100 [160]
Nickel	-	200	470, 52	200	0.035–56	≤50 [160]
Selenium	50	20	NA	20	0.173–223	0.12–341 [161]
Silver	100 (SDWR)	NA	3.2, NA	NA	NA	-
Uranium	30	NA	NA	NA	0.009–162	1–1200 [162]
Zinc	5000 (SDWR)	2000	120, 120	2000	0.536–1000	~130 [168]

It is likely that irrigation with elevated levels of geogenic contaminants leads to contaminant enrichment in crops. For example, Bundschuh et al. [178] compiled data from different regions of South America known for high occurrences of arsenic in groundwater and surface water. Their study indicated that arsenic concentrations in edible plants and crops were associated with the elevated arsenic concentrations in soil and irrigation waters. The study showed that regions with high arsenic concentrations in surface water and groundwater relate directly to accumulation in plants, fish, livestock meat, milk, and milk products. Their study indicates that there is a need for more rigorous studies in evaluating pathways of arsenic exposure through the food chain in Latin America and other regions.

Factors such as arsenic speciation, type and composition of soil, and plant species also plays a significant role in crop uptake [179]. An interesting aspect of arsenic transformation is where arsenic present in soil may occur as oxidized arsenate (As(V)) but may become reduced to arsenite (As(III)) after uptake in crops [180]. Arsenite is regarded as 25–60 times more toxic to humans than arsenate [181]. Arsenic, by far, is the most studied geogenic contaminant in crops, especially rice, and is included in many review articles detailing its impact of food quality and human health [182–186]. Ongoing research is focused on managing arsenic uptake by crops [154,187,188]. Other geogenic contaminants in groundwater, such as uranium, are less well studied with respect to food contamination.

Elevated levels of selenium may be toxic, though selenium is an essential micronutrient for crops. Selenium is known to accumulate in crops grown on soils with high selenium content [161,189], and selenium enrichment has been reported in soils throughout the United States [190]. In their recent study, Wang et al. [191] found that volatile organic compounds released by plant growth-promoting rhizobacteria increases both selenium and iron uptake. Uranium is another geogenic, potentially toxic contaminant [192] and studies in nutrient culture show its uptake by crops [193,194]. Several studies confirmed that irrigation water contaminated with uranium has an impact on crop quality though less to soil contamination by uranium [195–197]. Lead, mercury [38], chromium [198] and cadmium [199] are also known to accumulate in crops grown on soils with high levels of these contaminants. While there is extensive information in the literature on uptake by plants, most studies have focused on the general type of contaminant, or its accumulation in crops, use in phytoremediation of soils and the pathway of uptake within the crop. Contaminated irrigation water is capable of enriching surface soil with these contaminants [200] and likely enhances availability for uptake by crops [197]. There are few federal guidelines for trace element limits in foods in the United States, other than in arsenic, lead, cadmium, and mercury in wastewater [201], and the evidence suggests there is an urgent need for irrigation water quality standards that include geogenic contaminants.

3.3.2. Engineered Nanomaterials

Earth is rich with natural nanomaterials and it is estimated that thousands of megatons move through the hydrosphere annually [14]. Natural nanoparticles in water can easily pass through conventional membrane filter pore sizes of 0.2-μm and may not be accounted for as nanoparticles [202], which adds to the complexity in understanding their impact on crop health. In the past decade, production and use of engineered nanoparticles have also risen significantly and continues to trend upwards. Nanoparticles are used in a wide variety of contemporary products, ranging from electronics and cosmetics to processed food. This proliferation in use of nanoparticles has paved their way to increasing occurrence in water sources [203,204]. Nanotechnology is also used widely for water treatment for both groundwater and surface water sources [205,206], but its repercussions are still not well understood [207]. In 2018, the occurrence of nanoparticle size plastics (nanoplastics) were critically reviewed with respect to human health and growing global occurrence in freshwater [208,209]. Analytical methods capable of detecting and quantifying nanoparticles in complex aqueous matrices are lacking, increasing the challenge in tracking fate and transport of these particles [207]. In 2005, Oberdörster et al. [210] reviewed the interaction of nanomaterials in regards to human health. Still, the full-scale toxicity of natural or engineered nanomaterials is not well understood in the context of complex biosystems [14]. The size of nanomaterials (100 nm or less in size) is the main factor [14] which makes studying nanoparticle contamination challenging. Therefore, the paradigm of risk assessment of nanoparticles needs to be reevaluated with respect to the unique challenges involved in monitoring environmental pathways and assessing impacts on human health.

Broadly, engineered nanoparticles can be classified into carbon and metal-based nanoparticles. Nanoparticles derived from carbon (e.g., carbon nanotubes) simultaneously act like particles and high molecular weight organic compounds. Metal oxides or metal nanoparticles must be evaluated for their metal-related chemotoxicity, and also toxicity arising due to their particulate form. These dual behaviors again add to the complexity of studying and assessing risk. Nanotoxicity can elicit significant effects to human health. Nanoparticles can be carcinogenic and may produce reactive species inside the body [211]. Ganguly et al. have recently reviewed the toxicity of nanomaterials, emphasizing major exposure pathways [211].

The nanoparticle life cycle is poorly understood. Therefore, the Environmental Protection Agency (EPA) has allowed limited manufacturing of new materials by administrative orders or new use rules under the Toxic Substances Control Act. These new rules have significantly expanded since their inception in 2010, which is on par with our increase in understanding. However, the use of and occurrence of nanomaterials in irrigation water is not monitored or regulated [206,212].

Nanotechnology usage is vital in modern civilization and will have a substantial impact on the world economy [213]. It is projected that the nanomaterials market will reach 55 billion US dollars by year 2022 [213]. Production of engineered nanoparticles is expected to rise in the coming years, making these contaminants more likely to occur in different water sources, both conventional (surface and/or groundwater) and in treated wastewater. Both carbon-based and metal-based nanoparticles are of concern and it has been reported these particles are persistent in water [214,215]. Nanomaterial transformation in fresh water systems is an active field of research [216]. Nanoparticle assessments have concluded that fine particulate matter occurs in a variety of water sources [217–221], so its impact on irrigation water, accumulation in soil, and potential for crop uptake is of paramount significance.

The growing availability of engineered nanomaterials/particles is relevant to understand how these nanosized particles may impact water and food quality in agriculture. The use of nanomaterials in agriculture is also increasing, and little data is available to understand how occurrence in irrigation water may influence crops [222,223]. There is recent work on uptake of both carbon and metal-based nanoparticles by plants. For example, multi-walled carbon nanotubes have been observed in broccoli, resulting in a positive impact on plant growth [224] but potentially creating health concerns. Carbon nanotubes are also known to act as a carrier for other contaminants (like organochlorine pesticide, etc.) in plants and enhance contaminant translocation [225]. These contrasting effects of carbon nanotubes have been well summarized in a recent review article by Vithanage et al. [226].

Metal oxide nanoparticles, which form a bulk of engineered nanomaterials, have been well studied under the purview of plant uptake [227]. Metal oxides like titanium dioxide, silver oxide, iron oxide, copper oxide and metal nanoparticles have been shown to accumulate in a variety of food crops and have even been detected in commercial produce [228–232]. A recent review by Ma et al. describes studies of nanoparticle uptake by crops and their occurrence in the final produce [40]. Similar to carbon-based nanoparticles, metal or metal oxide-based nanoparticles are also known to be beneficial to plant growth and have been marketed as nanofertilizers [233]. For example, iron oxide nanoparticles have shown to be a potential iron source for peanut crops [234]. Nanoparticles are also known to induce oxidative stress in crops [235]. A recent study by Liu et al. suggested that combinations of nanomaterials might have different impacts on the soil microcosm compared to a single nanomaterial [236]. In actual conditions, mixed nanomaterials will be more prevalent in soil and irrigation water. In addition to engineered nanomaterial, natural nanomaterials of silver are found in groundwater [237]. Life cycle assessment of engineered titanium dioxide nanoparticles showed that their impact is not just limited to crop uptake, but exhibit marine aquatic ecotoxicity and human toxicity [238,239]. Nanoscale is an important factor in ensuring uptake of fertilizer [240,241] but the long-term effects of nanoagrochemicals have yet to be studied [242]. While nanoagrochemicals may be beneficial to crops [243], they may have undesired effects on the environment and on human health [242].

There are many questions and few answers for understanding the effects of nanosized contaminants in irrigated agriculture. Future studies should be focused on understanding retention times and fate in water, plants and soils, including degradation and transformation rates, and biological effects of different forms. Naturally-formed nanoparticles can occur in irrigation water and in food crops and it is clear this route for exposure needs to be better understood and monitored. Are nanoparticles easily broken down in the environment or are they stable, do they form aggregates, and what is the accumulation in both soil and water sources? If they accumulate, how do they impact soil health and modern irrigation techniques? Nanoparticle occurrence and behavior in irrigation water sources, soils and plants is clearly an emerging area of research.

4. Changing Quality of Water Sources Used for Irrigation

This discussion on the changing composition and quality of different water sources for irrigation—be it conventional sources like surface- or groundwater, or nonconventional source like reclaimed water—signifies an urgent need for increased efforts to monitor the quality of agricultural water [244]. Recently, there have been various efforts to address changing water quality [9] by incorporating measurement of new contaminant types (e.g., arsenic, fluoride) [10,245,246] in water quality guidelines. Still, it is challenging to address the complexity of irrigation water quality covering different forms of geogenic and emerging contaminants. In the coming decades, the concentrations of specific contaminants will likely increase and continue to affect the quality of water sources for irrigation. Impending climate change may make the situation more extreme as drought and water scari]city may concentrate contaminants in water. The increasing prevalence of cyanotoxins may increase due to more intensive agriculture, fertilization and water use. Although there are strict guidelines for contaminant concentrations in drinking water, few guidelines exist for the use of water for any irrigation, including irrigation of food crops. The substantial expense of monitoring, including sampling and laboratory analysis, for "new" recently identified contaminants is a reason to implement a more practical approach for monitoring irrigation water quality, incomparison to the complex and expensive framework currently used for drinking water in developed countries.

Establishing regulations and clear guidelines for irrigation water quality in different countries or individual states of the United States for conventional sources of water is necessary, but not sufficient, to ensure healthy produce for human consumption. Proper food quality can only be insured if water sources are regulated and regularly tested. Testing can also be used to monitor accumulation of contaminants in soil. Moreover, plant tissue should also be checked periodically for contaminant uptake to ensure appropriate produce quality. Presently in the United States, reclaimed wastewater is regulated for irrigation usage [247,248] but almost any other source may be used without restriction. Guidelines are provided for pathogen levels in reclaimed wastewater [174]. There are recommendations for limits of geogenic contaminants in irrigation water in traditional sources [157,249] but there is a complete lack of recommendations or information on new and emerging organic contaminants or nanomaterials. There is a whole range of second and third generation nanomaterials proposed for commercial uses (e.g., nanocomposites and multi-element materials). These new nanomaterials will occur in waste streams and we have little understanding of their fate or toxicity. Moreover, the basis for existing recommendations is questionable, as concentrations far lower than some recommended values have been shown to be biomagnified in crops (see Table 1).

The present review focuses on the changing quality of water used for irrigation, and clearly there are many gaps to be addressed in future research. High priority should be given to research focused on improving our understanding and address the increasing occurrence of geogenic contaminants, pathogens and other biological contaminants in irrigation water, especially as they relate to food crops. Increased wastewater reuse for agricultural purposes will likely increase the occurrence of biologically active organic contaminants such as pharmaceuticals and antibiotics, and the effect on food crops and human health is also a major research gap. The dynamic nature of the chemical composition of different irrigation water makes it very important that relevant and periodic data is available for both conventional and nonconventional water sources. A more comprehensive set of water quality guidelines needs to be created incorporating our present understanding of the occurrence and effects of emerging contaminants. There should be specific recommendations for monitoring and tests at to check irrigation water quality used in agriculture. These recommendations can be soil specific, crop specific and water specific. The availability of new guidelines would help ensure better food quality, as the next generation will not only need larger quantities of irrigation water to feed the growing population, but health concerns may rise as we resort to the use of low-quality water to irrigate food and feed crops.

Author Contributions: A.M. led writing of this article, including extensive literature investigation on irrigation water quality, orginal draft preparation and sources on geogenic and nanomaterial contaminants. D.D.S. provided conceptualization of the need for a review on this topic, editing and input on literature sources for emerging contaminants. C.R. provided additional input on wastewater contaminants, nanomaterials and additional editing.

Funding: This work is based on research that was partially supported by the Nebraska Agricultural Experiment Station with funding from the Hatch Multistate Research (Accession Number 1011588) through the USDA National Institute of Food and Agriculture.

Acknowledgments: Malakar thanks NET for salary support. The author thank feedback on drafts by Jason White and Sushil Kanel, as well as editorial comments from Erin Haackker and Lacey Bodnar.

Conflicts of Interest: The authors declare no conflict of interest.

References

1. Dieter, C.A.; Maupin, M.A.; Caldwell, R.R.; Harris, M.A.; Ivahnenko, T.I.; Lovelace, J.K.; Barber, N.L.; Linsey, K.S. *Estimated Use of Water in the United States in 2015*; U.S. Geological Survey: Reston, VA, USA, 2018.

2. *World Agriculture: Towards 2015/2030 an FAO Perspective*; Bruinsma, J. (Ed.) Earthscan Publications Ltd.: London, UK, 2003; ISBN 9251048355.

3. FAO. How to Feed the World in 2050. *Insights Expert Meet. FAO* **2009**, *2050*, 1–35. [CrossRef]

4. Sauvé, S.; Desrosiers, M. A review of what is an emerging contaminant. *Chem. Cent. J.* **2014**, *8*, 15. [CrossRef] [PubMed]

5. National Research Council. *Identifying Future Drinking Water Contaminants*; National Academies Press: Washington, DC, USA, 1999; ISBN 978-0-309-06432-3.

6. WWAP (United Nations World Water Assessment Programme). *The United Nations World Water Development Report 2017. Wastewater: The Untapped Resource*; United Nations Educational, Scientific and Cultural Organization: Paris, France, 22 March 2017.

7. Hass, A.; Mingelgrin, U.; Fine, P. Heavy metals in soils irrigated with wastewater. In *Treated Wastewater in Agriculture: Use and Impacts on the Soil Environment and Crops*; Wiley-Blackwell: Hoboken, NJ, USA, 2010; ISBN 9781405148627.

8. Allende, A.; Monaghan, J. Irrigation water quality for leafy crops: A perspective of risks and potential solutions. *Int. J. Environ. Res. Public Health* **2015**, *12*, 7457–7477. [CrossRef] [PubMed]

9. Singh, S.; Ghosh, N.C.; Gurjar, S.; Krishan, G.; Kumar, S.; Berwal, P. Index-based assessment of suitability of water quality for irrigation purpose under Indian conditions. *Environ. Monit. Assess.* **2018**, *190*, 29. [CrossRef] [PubMed]

10. Islam, M.A.; Romić, D.; Akber, M.A.; Romić, M. Trace metals accumulation in soil irrigated with polluted water and assessment of human health risk from vegetable consumption in Bangladesh. *Environ. Geochem. Health* **2018**, *40*, 59–85. [CrossRef] [PubMed]

11. Seiler, R.L.; Skorupa, J.P.; Naftz, D.L.; Nolan, B.T. *Irrigation-Induced Contamination of Water, Sediment, and Biota in the Western United States—Synthesis of Data from the National Irrigation Water Quality Program*; U.S. Geological Survey Professional Paper1655; U.S. Geological Survey: Reston, VA, USA, 2003.

12. Servin, A.D.; De la Torre-Roche, R.; Castillo-Michel, H.; Pagano, L.; Hawthorne, J.; Musante, C.; Pignatello, J.; Uchimiya, M.; White, J.C. Exposure of agricultural crops to nanoparticle CeO_2 in biochar-amended soil. *Plant Physiol. Biochem.* **2017**, *110*, 147–157. [CrossRef] [PubMed]

13. Zhao, Q.; Ma, C.; White, J.C.; Dhankher, O.P.; Zhang, X.; Zhang, S.; Xing, B. Quantitative evaluation of multi-wall carbon nanotube uptake by terrestrial plants. *Carbon* **2017**, *114*, 661–670. [CrossRef]

14. Hochella, M.F.; Mogk, D.W.; Ranville, J.; Allen, I.C.; Luther, G.W.; Marr, L.C.; McGrail, B.P.; Murayama, M.; Qafoku, N.P.; Rosso, K.M.; et al. Natural, incidental, and engineered nanomaterials and their impacts on the Earth system. *Science* **2019**, *363*, eaau8299. [CrossRef] [PubMed]

15. Nolan, J.; Weber, K.A. Natural Uranium Contamination in Major U.S. Aquifers Linked to Nitrate. *Environ. Sci. Technol. Lett.* **2015**, *2*, 215–220. [CrossRef]

16. Mateo-Sagasta, J.; Marjani, S.; Turral, H.; Burke, J. *Water Pollution from Agriculture: A Global Review Executive Summary*; The Food and Agriculture Organization of the United Nations: Rome, Italy; The International Water Management Institute: Colombo, Sri Lanka, 2017.

17. Dexssen, A.; Dole, R.B. *Ground Water in LaSalle and McMullen Counties*; Texas. U.S. Geol. Surv. Water-Supply Paper 375-G; U.S. GPO: Washington, DC, USA, 1916.

18. Schwennesen, A.T.; Forbes, R.H. *Ground Water in San Simon Valley, Arizona and New Mexico*; Water-Supply Paper 425-A; U.S. GPO: Washington, DC, USA, 1917.

19. Clark, W.O. *Ground Water for Irrigation in the Morgan Hill Area, California*; Water-Supply Paper 400-E; U.S. GPO: Washington, DC, USA, 1917.

20. Scofield, C.S.; Headley, F.B. Quality of irrigation water in relation to land reclamation. *J. Agric. Res.* **1921**, *21*, 265–278. [CrossRef]

21. Sturdy, D.; Calton, W.E.; Milne, G. A chemical survey of the waters of Mount Meru, Tanganyika Territory, especially with regard to their qualities for irrigation. *J. East Afr. Uganda Nat. Hist. Soc.* **1932**, *45–46*, 1–38.

22. Taylor, E.M.; Puri, A.N.; Asghar, A.G. Soil deterioration in the canal-irrigated areas of the Punjab. I. Equilibrium between calcium and sodium ions in base-exchange reactions. *Res. Publ.* **1934**, *4*, 7.

23. Mados, L. The qualifications of irrigation waters. *Mezogazdasagi Kut* **1940**, *12*, 121–131.

24. Eaton, F.M.; McCallum, R.D.; Mayhugh, M.S. *Quality of Irrigation Waters of the Hollister area of California with Special Reference to Boron Content and Its Effect on Apricots and Prunes*; Technical Bulletin; United States Department Agriculture: Washington, DC, USA, 1941; Volume 746, p. 59.

25. Wilcox, L.V. The quality of water for irrigation use. *U.S. Dept. Agr. Tech. Bull.* **1948**, *962*, 40.

26. Pacheco, J.d.l.R.; Lopez-Rubio, F.B. Analysis of waters for agricultural uses. *Inf. Quim. Anal.* **1949**, *3*, 90–96.

27. Thorne, D.W.; Thorne, J.P. Changes in composition of irrigated soils as related to the quality of irrigation waters. *Soil Sci. Soc. Am. Proc.* **1949**, *18*, 92–97. [CrossRef]

28. Lewis, G.C.; Juve, R.L. Some effects of irrigation-water quality on soil characteristics. *Soil Sci.* **1956**, *81*, 125–137. [CrossRef]

29. Pearson, H.E.; Huberty, M.R. Response of citrus to irrigation with waters of different chemical characteristics. *Proc. Am. Soc. Hortic. Sci.* **1959**, *73*, 248–256.

30. Babcock, K.L.; Carlson, R.M.; Schulz, R.K.; Overstreet, R. A study of the Effect of irrigation water composition on soil properties. *Hilgardia* **1959**, *29*, 155–170. [CrossRef]

31. Longenecker, D.E.; Lyerly, P.J. Chemical characteristics of soils of west Texas as affected by irrigation water quality. *Soil Sci.* **1959**, *87*, 207–216. [CrossRef]

32. Bernstein, L. Quantitative assessment of irrigation water quality. In *Water Quality Criteria*; Bramer, H., Ed.; ASTM International: West Conshohocken, PA, USA, 1967; pp. 51–65.

33. Park, D.M.; White, S.A.; McCarty, L.B.; Menchyk, N.A. *Interpreting Irrigation Water Quality Reports*; CU-14-700; Clemson University Cooperative Extension: Clemson, SC, USA, 2014.

34. Frenkel, H. Reassessment of Water Quality Criteria for Irrigation. *Ecol. Stud. Anal. Synth.* **1984**, *51*, 142–172.

35. Bauder, T.; Waskom, R.; Davis, J.; Sutherland, P. Irrigation water quality criteria. *Crop Ser. Irrig. Fact Sheet* **2007**, *506*, 10–13.

36. Feltz, H.; Engberg, R.; Sylvester, M. Investigations of water quality, bottom sediment, and biota associated with irrigation drainage in the western United States. In Proceedings of the International Symposium on the Hydrologic Basis for Water Resources Management, Beijing, China, 23–26 October 1990. Publ. no. 197.

37. Seiler, R.L. Synthesis of data from studies by the national irrigation water-quality program. *J. Am. Water Resour. Assoc.* **1996**, *32*, 1233–1245. [CrossRef]

38. Tangahu, B.V.; Sheikh Abdullah, S.R.; Basri, H.; Idris, M.; Anuar, N.; Mukhlisin, M. A review on heavy metals (As, Pb, and Hg) uptake by plants through phytoremediation. *Int. J. Chem. Eng.* **2011**, *2011*, 31. [CrossRef]

39. Intawongse, M.; Dean, J.R. Uptake of heavy metals by vegetable plants grown on contaminated soil and their bioavailability in the human gastrointestinal tract. *Food Addit. Contam.* **2006**, *23*, 36–48. [CrossRef] [PubMed]

40. Ma, C.; White, J.C.; Zhao, J.; Zhao, Q.; Xing, B. Uptake of engineered nanoparticles by food crops: Characterization, mechanisms, and implications. *Annu. Rev. Food Sci. Technol.* **2018**, *9*, 129–153. [CrossRef]

41. Calderón-Preciado, D.; Matamoros, V.; Bayona, J.M. Occurrence and potential crop uptake of emerging contaminants and related compounds in an agricultural irrigation network. *Sci. Total Environ.* **2011**, *412–413*, 14–19. [CrossRef]

42. Sedlak, D.L.; Gray, J.L.; Pinkston, K.E. Peer reviewed: Understanding microcontaminants in recycled water. *Environ. Sci. Technol.* **2000**, *34*, 508A–515A. [CrossRef]

43. IOM (Institute of Medicine) and NRC (National Research Council). *A Framework for Assessing Effects of the Food System*; The National Academic Press: Washington, DC, USA, 2015; ISBN 978-0-309-30780-2.

44. World Health Organization. *Pharmaceuticals in Drinking Water*; WHO: Geneva, Switerland, 2012; ISBN 9789241502085.

45. Balakrishna, K.; Rath, A.; Praveenkumarreddy, Y.; Guruge, K.S.; Subedi, B. A review of the occurrence of pharmaceuticals and personal care products in Indian water bodies. *Ecotoxicol. Environ. Saf.* **2017**, *137*, 113–120. [CrossRef]

46. Yang, Y.; Ok, Y.S.; Kim, K.H.; Kwon, E.E.; Tsang, Y.F. Occurrences and removal of pharmaceuticals and personal care products (PPCPs) in drinking water and water/sewage treatment plants: A review. *Sci. Total Environ.* **2017**, *596–597*, 303–320. [CrossRef]

47. Fijalkowski, K.; Rorat, A.; Grobelak, A.; Kacprzak, M.J. The presence of contaminations in sewage sludge—The current situation. *J. Environ. Manag.* **2017**, *203*, 1126–1136. [CrossRef] [PubMed]

48. Subedi, B.; Balakrishna, K.; Joshua, D.I.; Kannan, K. Mass loading and removal of pharmaceuticals and personal care products including psychoactives, antihypertensives, and antibiotics in two sewage treatment plants in southern India. *Chemosphere* **2017**, *167*, 429–437. [CrossRef] [PubMed]

49. Subedi, B.; Lee, S.; Moon, H.B.; Kannan, K. Emission of artificial sweeteners, select pharmaceuticals, and personal care products through sewage sludge from wastewater treatment plants in Korea. *Environ. Int.* **2014**, *68*, 33–40. [CrossRef] [PubMed]

50. Madikizela, L.M.; Ncube, S.; Chimuka, L. Uptake of pharmaceuticals by plants grown under hydroponic conditions and natural occurring plant species: A review. *Sci. Total Environ.* **2018**, *636*, 477–486. [CrossRef] [PubMed]

51. Wu, X.; Dodgen, L.K.; Conkle, J.L.; Gan, J. Plant uptake of pharmaceutical and personal care products from recycled water and biosolids: A review. *Sci. Total Environ.* **2015**, *536*, 655–666. [CrossRef]

52. Tasho, R.P.; Cho, J.Y. Veterinary antibiotics in animal waste, its distribution in soil and uptake by plants: A review. *Sci. Total Environ.* **2016**, *563–564*, 366–376. [CrossRef] [PubMed]

53. Wu, X.; Conkle, J.L.; Ernst, F.; Gan, J. Treated wastewater irrigation: Uptake of pharmaceutical and personal care products by common vegetables under field conditions. *Environ. Sci. Technol.* **2014**, *48*, 11286–11293. [CrossRef]

54. Santiago, S.; Roll, D.M.; Ray, C.; Williams, C.; Moravcik, P.; Knopf, A. Effects of soil moisture depletion on vegetable crop uptake of pharmaceuticals and personal care products (PPCPs). *Environ. Sci. Pollut. Res.* **2016**, *23*, 20257–20268. [CrossRef]

55. Riemenschneider, C.; Al-Raggad, M.; Moeder, M.; Seiwert, B.; Salameh, E.; Reemtsma, T. Pharmaceuticals, their metabolites, and other polar pollutants in field-grown vegetables irrigated with treated municipal wastewater. *J. Agric. Food Chem.* **2016**, *64*, 5784–5792. [CrossRef]

56. Colon, B.; Toor, G.S. A review of uptake and translocation of pharmaceuticals and personal care products by food crops irrigated with treated wastewater. *Adv. Agron.* **2016**, *140*, 75–100.

57. Calderón-Preciado, D.; Jiménez-Cartagena, C.; Matamoros, V.; Bayona, J.M. Screening of 47 organic microcontaminants in agricultural irrigation waters and their soil loading. *Water Res.* **2011**, *45*, 221–231. [CrossRef] [PubMed]

58. Christou, A.; Agüera, A.; Bayona, J.M.; Cytryn, E.; Fotopoulos, V.; Lambropoulou, D.; Manaia, C.M.; Michael, C.; Revitt, M.; Schröder, P.; et al. The potential implications of reclaimed wastewater reuse for irrigation on the agricultural environment: The knowns and unknowns of the fate of antibiotics and antibiotic resistant bacteria and resistance genes—A review. *Water Res.* **2017**, *123*, 448–467. [CrossRef] [PubMed]

59. Williams-Nguyen, J.; Sallach, J.B.; Bartelt-Hunt, S.; Boxall, A.B.; Durso, L.M.; McLain, J.E.; Singer, R.S.; Snow, D.D.; Zilles, J.L. Antibiotics and antibiotic resistance in agroecosystems: State of the science. *J. Environ. Qual.* **2016**, *45*, 394–406. [CrossRef] [PubMed]

60. Durso, L.M.; Cook, K.L. Impacts of antibiotic use in agriculture: What are the benefits and risks? *Curr. Opin. Microbiol.* **2014**, *19*, 37–44. [CrossRef] [PubMed]

61. Kümmerer, K. Antibiotics in the aquatic environment—A review—Part I. *Chemosphere* **2009**, *75*, 417–434. [CrossRef] [PubMed]

62. Kümmerer, K. Antibiotics in the aquatic environment—A review—Part II. *Chemosphere* **2009**, *75*, 435–441. [CrossRef]

63. Behera, S.K.; Kim, H.W.; Oh, J.E.; Park, H.S. Occurrence and removal of antibiotics, hormones and several other pharmaceuticals in wastewater treatment plants of the largest industrial city of Korea. *Sci. Total Environ.* **2011**, *409*, 4351–4360. [CrossRef] [PubMed]

64. Zhou, L.J.; Ying, G.G.; Liu, S.; Zhao, J.L.; Yang, B.; Chen, Z.F.; Lai, H.J. Occurrence and fate of eleven classes of antibiotics in two typical wastewater treatment plants in South China. *Sci. Total Environ.* **2013**, *452–453*, 365–376. [CrossRef]

65. Aga, D.S.; Lenczewski, M.; Snow, D.; Muurinen, J.; Sallach, J.B.; Wallace, J.S. Challenges in the measurement of antibiotics and in evaluating their impacts in agroecosystems: A critical review. *J. Environ. Qual.* **2016**, *45*, 407–419. [CrossRef]

66. Burkholder, J.A.; Libra, B.; Weyer, P.; Heathcote, S.; Kolpin, D.; Thorne, P.S.; Wichman, M. Impacts of waste from concentrated animal feeding operations on water quality. *Environ. Health Perspect.* **2007**, *115*, 308–312. [CrossRef]

67. Chang, X.; Meyer, M.T.; Liu, X.; Zhao, Q.; Chen, H.; Chen, J.A.; Qiu, Z.; Yang, L.; Cao, J.; Shu, W. Determination of antibiotics in sewage from hospitals, nursery and slaughter house, wastewater treatment plant and source water in Chongqing region of Three Gorge Reservoir in China. *Environ. Pollut.* **2010**, *158*, 1444–1450. [CrossRef] [PubMed]

68. Duong, H.A.; Pham, N.H.; Nguyen, H.T.; Hoang, T.T.; Pham, H.V.; Pham, V.C.; Berg, M.; Giger, W.; Alder, A.C. Occurrence, fate and antibiotic resistance of fluoroquinolone antibacterials in hospital wastewaters in Hanoi, Vietnam. *Chemosphere* **2008**, *72*, 968–973. [CrossRef] [PubMed]

69. Hirsch, R.; Ternes, T.; Haberer, K.; Kratz, K.L. Occurrence of antibiotics in the aquatic environment. *Sci. Total Environ.* **1999**, *225*, 109–118. [CrossRef]

70. Kolpin, D.W.; Furlong, E.T.; Meyer, M.T.; Thurman, E.M.; Zaugg, S.D.; Barber, L.B.; Buxton, H.T. Pharmaceuticals, hormones, and other organic wastewater contaminants in U.S. streams, 1999–2000: A national reconnaissance. *Environ. Sci. Technol.* **2002**, *36*, 1202–1211. [CrossRef] [PubMed]

71. Yan, C.; Yang, Y.; Zhou, J.; Liu, M.; Nie, M.; Shi, H.; Gu, L. Antibiotics in the surface water of the Yangtze Estuary: Occurrence, distribution and risk assessment. *Environ. Pollut.* **2013**, *175*, 22–29. [CrossRef] [PubMed]

72. Deng, W.; Li, N.; Zheng, H.; Lin, H. Occurrence and risk assessment of antibiotics in river water in Hong Kong. *Ecotoxicol. Environ. Saf.* **2016**, *125*, 121–127. [CrossRef] [PubMed]

73. Zuccato, E.; Castiglioni, S.; Bagnati, R.; Melis, M.; Fanelli, R. Source, occurrence and fate of antibiotics in the Italian aquatic environment. *J. Hazard. Mater.* **2010**, *179*, 1042–1048. [CrossRef]

74. Ahmed, M.B.M.; Rajapaksha, A.U.; Lim, J.E.; Vu, N.T.; Kim, I.S.; Kang, H.M.; Lee, S.S.; Ok, Y.S. Distribution and accumulative pattern of tetracyclines and sulfonamides in edible vegetables of cucumber, tomato, and lettuce. *J. Agric. Food Chem.* **2015**, *63*, 398–405. [CrossRef]

75. Chitescu, C.L.; Nicolau, A.I.; Stolker, A.A.M. Uptake of oxytetracycline, sulfamethoxazole and ketoconazole from fertilised soils by plants. *Food Addit. Contam. Part A* **2013**, *30*, 1138–1146. [CrossRef]

76. Sallach, J.B.; Bartelt-Hunt, S.L.; Snow, D.D.; Li, X.; Hodges, L. Uptake of antibiotics and their toxicity to lettuce following routine irrigation with contaminated water in different soil types. *Environ. Eng. Sci.* **2018**, *35*. [CrossRef]

77. Azanu, D.; Mortey, C.; Darko, G.; Weisser, J.J.; Styrishave, B.; Abaidoo, R.C. Uptake of antibiotics from irrigation water by plants. *Chemosphere* **2016**, *157*, 107–114. [CrossRef] [PubMed]

78. Phillips, I.; Casewell, M.; Cox, T.; De Groot, B.; Friis, C.; Jones, R.; Nightingale, C.; Preston, R.; Waddell, J. Does the use of antibiotics in food animals pose a risk to human health? A critical review of published data. *J. Antimicrob. Chemother.* **2004**, *53*, 28–52. [CrossRef] [PubMed]

79. Kuppusamy, S.; Kakarla, D.; Venkateswarlu, K.; Megharaj, M.; Yoon, Y.E.; Lee, Y.B. Veterinary antibiotics (VAs) contamination as a global agro-ecological issue: A critical view. *Agric. Ecosyst. Environ.* **2018**, *257*, 47–59. [CrossRef]

80. Bedford, M. Removal of antibiotic growth promoters from poultry diets: Implications and strategies to minimise subsequent problems. *Worlds Poult. Sci. J.* **2000**, *56*, 347–365. [CrossRef]

81. Schuijt, T.J.; van der Poll, T.; de Vos, W.M.; Wiersinga, W.J. The intestinal microbiota and host immune interactions in the critically ill. *Trends Microbiol.* **2013**, *21*, 221–229. [CrossRef] [PubMed]

82. Davies, J.; Davies, D. Origins and evolution of antibiotic resistance. *Microbiol. Mol. Biol. Rev.* **2010**, *74*, 417–433. [CrossRef]

83. Zhang, F.S.; Xie, Y.F.; Li, X.W.; Wang, D.Y.; Yang, L.S.; Nie, Z.Q. Accumulation of steroid hormones in soil and its adjacent aquatic environment from a typical intensive vegetable cultivation of North China. *Sci. Total Environ.* **2015**, *538*, 423–430. [CrossRef] [PubMed]

84. Donk, S.v.; Biswas, S.; Kranz, W.; Snow, D.; Bartelt-Hunt, S.; Mader, T.; Shapiro, C.; Shelton, D.; Tarkalson, D.; Zhang, T.; et al. Transport of steroid hormones in the vadose zoneafter land application of beef cattle manure. *Biol. Syst. Eng. Pap. Publ.* **2013**, *56*, 1327–1338.

85. Bartelt-Hunt, S.; Snow, D.D.; Damon-Powell, T.; Miesbach, D. Occurrence of steroid hormones and antibiotics in shallow groundwater impacted by livestock waste control facilities. *J. Contam. Hydrol.* **2011**, *123*, 94–103. [CrossRef]

86. Torres, N.H.; Aguiar, M.M.; Ferreira, L.F.R.; Américo, J.H.P.; Machado, Â.M.; Cavalcanti, E.B.; Tornisielo, V.L. Detection of hormones in surface and drinking water in Brazil by LC-ESI-MS/MS and ecotoxicological assessment with Daphnia magna. *Environ. Monit. Assess.* **2015**, *187*, 379. [CrossRef]

87. Pauwels, B.; Noppe, H.; De Brabander, H.; Verstraete, W. Comparison of steroid hormone concentrations in domestic and hospital wastewater treatment plants. *J. Environ. Eng.* **2008**, *134*, 933–936. [CrossRef]

88. Servos, M.R.; Bennie, D.T.; Burnison, B.K.; Jurkovic, A.; McInnis, R.; Neheli, T.; Schnell, A.; Seto, P.; Smyth, S.A.; Ternes, T.A. Distribution of estrogens, 17β-estradiol and estrone, in Canadian municipal wastewater treatment plants. *Sci. Total Environ.* **2005**, *336*, 155–170. [CrossRef] [PubMed]

89. Andersen, H.; Siegrist, H.; Halling-Sørensen, B.; Ternes, T.A. Fate of estrogens in a municipal sewage treatment plant. *Environ. Sci. Technol.* **2003**, *37*, 4021–4026. [CrossRef] [PubMed]

90. Baronti, C.; Curini, R.; D'Ascenzo, G.; Di Corcia, A.; Gentili, A.; Samperi, R. Monitoring natural and synthetic estrogens at activated sludge sewage treatment plants and in a receiving river water. *Environ. Sci. Technol.* **2000**, *34*, 5059–5066. [CrossRef]

91. Sellin, M.K.; Snow, D.D.; Akerly, D.L.; Kolok, A.S. Estrogenic compounds downstream from three small cities in Eastern Nebraska: Occurrence and biological effect. *J. Am. Water Resour. Assoc.* **2009**, *45*, 14–21. [CrossRef]

92. Yarahmadi, H.; Duy, S.V.; Hachad, M.; Dorner, S.; Sauvé, S.; Prévost, M. Seasonal variations of steroid hormones released by wastewater treatment plants to river water and sediments: Distribution between particulate and dissolved phases. *Sci. Total Environ.* **2018**, *635*, 144–155. [CrossRef] [PubMed]

93. Zheng, W.; Wiles, K.N.; Holm, N.; Deppe, N.A.; Shipley, C.R. Uptake, Translocation, and Accumulation of Pharmaceutical and Hormone Contaminants in Vegetables. In *ACS Symposium Series*; American Chemical Society: Washington, DC, USA, 2014; Volume 1171, pp. 167–181.

94. Adeel, M.; Song, X.; Wang, Y.; Francis, D.; Yang, Y. Environmental impact of estrogens on human, animal and plant life: A critical review. *Environ. Int.* **2017**, *99*, 107–119. [CrossRef]

95. Rose, S.C.; Carter, A.D. Agrochemical leaching and water contamination. In *Conservation Agriculture*; Springer: Dordrecht, The Netherlands, 2003; pp. 417–424.

96. Jimoh, O.D.; Ayodeji, M.A.; Mohammed, B. Effects of agrochemicals on surface waters and groundwaters in the Tunga-Kawo (Nigeria) irrigation scheme. *Hydrol. Sci. J.* **2003**, *48*, 1013–1023. [CrossRef]

97. Yu, Y.; Hu, S.; Yang, Y.; Zhao, X.; Xue, J.; Zhang, J.; Gao, S.; Yang, A. Successive monitoring surveys of selected banned and restricted pesticide residues in vegetables from the northwest region of China from 2011 to 2013. *BMC Public Health* **2017**, *18*, 91. [CrossRef]

98. Retention, uptake, and translocation of agrochemicals in plants. In *ACS Symposium Series*; Myung, K.; Satchivi, N.M.; Kingston, C.K. (Eds.) American Chemical Society: Washington, DC, USA, 2014; Volume 1171, ISBN 0-8412-2972-4.

99. Juraske, R.; Castells, F.; Vijay, A.; Muñoz, P.; Antón, A. Uptake and persistence of pesticides in plants: Measurements and model estimates for imidacloprid after foliar and soil application. *J. Hazard. Mater.* **2009**, *165*, 683–689. [CrossRef]

100. Elmore, S.A.; Boorman, G.A. Environmental toxicologic pathology and human health. *Haschek Rousseaux's Handb. Toxicol. Pathol.* **2013**, 1029–1049. [CrossRef]

101. US Protection Agency. *Cyanobacteria and Cyanotoxins: Information for Drinking Water Systems*; EPA-810F11001; USEPA: Washington, DC, USA, 2014.

102. Crush, J.R.; Briggs, L.R.; Sprosen, J.M.; Nichols, S.N. Effect of irrigation with lake water containing microcystins on microcystin content and growth of ryegrass, clover, rape, and lettuce. *Environ. Toxicol.* **2008**, *23*, 246–252. [CrossRef] [PubMed]

103. Corbel, S.; Mougin, C.; Bouaïcha, N. Cyanobacterial toxins: Modes of actions, fate in aquatic and soil ecosystems, phytotoxicity and bioaccumulation in agricultural crops. *Chemosphere* **2014**, *96*, 1–15. [CrossRef] [PubMed]

104. Loftin, K.A.; Graham, J.L.; Hilborn, E.D.; Lehmann, S.C.; Meyer, M.T.; Dietze, J.E.; Griffith, C.B. Cyanotoxins in inland lakes of the United States: Occurrence and potential recreational health risks in the EPA National Lakes Assessment 2007. *Harmful Algae* **2016**, *56*, 77–90. [CrossRef] [PubMed]

105. Al-Sammak, M.A.; Hoagland, K.D.; Cassada, D.; Snow, D.D. Co-occurrence of the cyanotoxins BMAA, DABA and anatoxin-a in Nebraska reservoirs, fish, and aquatic plants. *Toxins* **2014**, *6*, 488–508. [CrossRef] [PubMed]

106. Miller, A.; Russell, C. Food crops irrigated with cyanobacteria-contaminated water: An emerging public health issue in Canada. *Environ. Heal. Rev.* **2017**, *60*, 58–63. [CrossRef]

107. Saqrane, S.; Oudra, B. CyanoHAB occurrence and water irrigation cyanotoxin contamination: Ecological impacts and potential health risks. *Toxins* **2009**, *1*, 113–122. [CrossRef] [PubMed]

108. Abeysiriwardena, N.M.; Gascoigne, S.J.L.; Anandappa, A. Algal bloom expansion increases cyanotoxin risk in food. *Yale J. Biol. Med.* **2018**, *91*, 129–142. [PubMed]

109. Al-Gabr, H.M.; Zheng, T.; Yu, X. Fungi contamination of drinking water. *Rev. Environ. Contam. Toxicol.* **2014**, *228*, 121–139. [PubMed]

110. Oliveira, B.R.; Mata, A.T.; Ferreira, J.P.; Barreto Crespo, M.T.; Pereira, V.J.; Bronze, M.R. Production of mycotoxins by filamentous fungi in untreated surface water. *Environ. Sci. Pollut. Res.* **2018**, *25*, 17519–17528. [CrossRef]

111. Kolpin, D.W.; Hoerger, C.C.; Meyer, M.T.; Wettstein, F.E.; Hubbard, L.E.; Bucheli, T.D. Phytoestrogens and mycotoxins in Iowa streams: An examination of underinvestigated compounds in agricultural basins. *J. Environ. Qual.* **2010**, *39*, 2089–2099. [CrossRef] [PubMed]

112. Rao, G.J.; Govindaraju, G.; Sivasithamparam, N.; Shanmugasundaram, E.R.B. Uptake, translocation and persistence of mycotoxins in rice seedlings. *Plant Soil* **1982**, *66*, 121–123. [CrossRef]

113. Mohammad-Hasani, F.; Mirlohi, M.; Mosharraf, L.; Hasanzade, A. Occurrence of aflatoxins in wheat flour specified for sangak bread and its reduction through fermentation and baking. *Qual. Assur. Saf. Crop. Foods* **2016**, *8*, 1–8. [CrossRef]

114. Tola, M.; Kebede, B. Occurrence, importance and control of mycotoxins: A review. *Cogent Food Agric.* **2016**. [CrossRef]

115. Tanaka, H.; Asano, T.; Schroeder, E.D.; Tchobanoglous, G. Estimating the safety of wastewater reclamation and reuse using enteric virus monitoring data. *Water Environ. Res.* **1998**, *70*, 39–51. [CrossRef]

116. Asano, T.; Cotruvo, J.A. Groundwater recharge with reclaimed municipal wastewater: Health and regulatory considerations. *Water Res.* **2004**, *38*, 1941–1951. [CrossRef] [PubMed]

117. Lothrop, N.; Bright, K.R.; Sexton, J.; Pearce-Walker, J.; Reynolds, K.A.; Verhougstraete, M.P. Optimal strategies for monitoring irrigation water quality. *Agric. Water Manag.* **2018**, *199*, 86–92. [CrossRef]

118. Jongman, M.; Chidamba, L.; Korsten, L. Bacterial biomes and potential human pathogens in irrigation water and leafy greens from different production systems described using pyrosequencing. *J. Appl. Microbiol.* **2017**, *123*, 1043–1053. [CrossRef]

119. Truchado, P.; Hernandez, N.; Gil, M.I.; Ivanek, R.; Allende, A. Correlation between *E. coli* levels and the presence of foodborne pathogens in surface irrigation water: Establishment of a sampling program. *Water Res.* **2018**, *128*, 226–233. [CrossRef]

120. Steele, M.; Odumeru, J. Irrigation water as source of foodborne pathogens on fruit and vegetables. *J. Food Prot.* **2004**, *67*, 2839–2849. [CrossRef]

121. Pachepsky, Y.; Shelton, D.R.; McLain, J.E.T.; Patel, J.; Mandrell, R.E. Irrigation waters as a source of pathogenic microorganisms in produce: A review. *Adv. Agron.* **2011**, *113*, 75–141. [CrossRef]

122. Park, S.; Szonyi, B.; Gautam, R.; Nightingale, K.; Anciso, J.; Ivanek, R. Risk factors for microbial contamination in fruits and vegetables at the preharvest level: A systematic review. *J. Food Prot.* **2012**, *75*, 2055–2081. [CrossRef] [PubMed]

123. Jongman, M.; Korsten, L. Irrigation water quality and microbial safety of leafy greens in different vegetable production systems: A review. *Food Rev. Int.* **2018**, *34*, 308–328. [CrossRef]

124. Herman, K.M.; Hall, A.J.; Gould, L.H. Outbreaks attributed to fresh leafy vegetables, United States, 1973–2012. *Epidemiol. Infect.* **2015**, *143*, 3011–3021. [CrossRef] [PubMed]

125. National Outbreak Reporting System (NORS) Dashboard | CDC. Available online: https://wwwn.cdc.gov/norsdashboard/ (accessed on 25 March 2019).

126. Multistate Outbreak of *E. coli* O157:H7 Infections Linked to Romaine Lettuce (Final Update) | Investigation Notice: Multistate Outbreak of *E. coli* O157:H7 Infections April 2018 | *E. coli* | CDC. Available online: https://www.cdc.gov/ecoli/2018/o157h7-04-18/index.html (accessed on 25 March 2019).

127. Outbreak of *E. coli* Infections Linked to Romaine Lettuce | *E. coli* Infections Linked to Romaine Lettuce | November 2018 | *E. coli* | CDC. Available online: https://www.cdc.gov/ecoli/2018/o157h7-11-18/index.html (accessed on 25 March 2019).

128. FDA Investigation of a Multistate Outbreak of Cyclospora Illnesses Linked to Fresh Express Salad Mix Served at McDonald's Ends | FDA. Available online: https://www.fda.gov/food/outbreaks-foodborne-illness/fda-investigation-multistate-outbreak-cyclospora-illnesses-linked-fresh-express-salad-mix-served#Cyclospora (accessed on 3 July 2019).

129. Olivieri, A.W.; Seto, E.; Cooper, R.C.; Cahn, M.D.; Colford, J.; Crook, J.; Debroux, J.-F.; Mandrell, R.; Suslow, T.; Tchobanoglous, G.; et al. Risk-based review of California's water-recycling criteria for agricultural irrigation. *J. Environ. Eng.* **2014**, *140*, 04014015. [CrossRef]

130. Leaman, S.; Gorny, J.; Wetherington, D.; Belkris, H. *Agricultural Water: Five Year Research Review*; Center for Produce Safety: Davis, CA, USA, 2014.

131. U.S. EPA. *Regulations Governing Agricultural Use of Municipal Wastewater and Sludge*; National Academy Press: Washington, DC, USA, 1996; ISBN 0309054796.

132. Jones, L.A.; Worobo, R.W.; Smart, C.D. UV light inactivation of human and plant pathogens in unfiltered surface irrigation water. *Appl. Environ. Microbiol.* **2014**, *80*, 849–854. [CrossRef] [PubMed]

133. Uyttendaele, M.; Jaykus, L.A.; Amoah, P.; Chiodini, A.; Cunliffe, D.; Jacxsens, L.; Holvoet, K.; Korsten, L.; Lau, M.; McClure, P.; et al. Microbial hazards in irrigation water: Standards, norms, and testing to manage use of water in fresh produce primary production. *Compr. Rev. Food Sci. Food Saf.* **2015**, *14*, 336–356. [CrossRef]

134. World Health Organization World Health Organization (WHO): Antibiotic Resistance—Fact Sheet. Available online: https://www.who.int/en/news-room/fact-sheets/detail/antibiotic-resistance (accessed on 26 June 2019).

135. Perry, J.A.; Wright, G.D. The antibiotic resistance "mobilome": Searching for the link between environment and clinic. *Front. Microbiol.* **2013**, *4*, 138. [CrossRef]

136. Marti, E.; Variatza, E.; Balcazar, J.L. The role of aquatic ecosystems as reservoirs of antibiotic resistance. *Trends Microbiol.* **2014**, *22*, 36–41. [CrossRef]

137. Bergeron, S.; Brown, R.; Homer, J.; Rehage, S.; Boopathy, R. Presence of antibiotic resistance genes in different salinity gradients of freshwater to saltwater marshes in southeast Louisiana, USA. *Int. Biodeterior. Biodegrad.* **2016**, *113*, 80–87. [CrossRef]

138. Pepper, I.; Brooks, J.P.; Gerba, C.P. Antibiotic resistant bacteria in municipal wastes: Is there reason for concern? *Environ. Sci. Technol.* **2018**, *52*, 3949–3959. [CrossRef] [PubMed]

139. Zhang, X.X.; Zhang, T.; Fang, H.H.P. Antibiotic resistance genes in water environment. *Appl. Microbiol. Biotechnol.* **2009**, *82*, 397–414. [CrossRef] [PubMed]

140. Fahrenfeld, N.; Ma, Y.; O'Brien, M.; Pruden, A. Reclaimed water as a reservoir of antibiotic resistance genes: Distribution system and irrigation implications. *Front. Microbiol.* **2013**, *4*, 130. [CrossRef] [PubMed]

141. Aslan, A.; Cole, Z.; Bhattacharya, A.; Oyibo, O.; Aslan, A.; Cole, Z.; Bhattacharya, A.; Oyibo, O. Presence of antibiotic-resistant *Escherichia coli* in wastewater treatment plant effluents utilized as water reuse for irrigation. *Water* **2018**, *10*, 805. [CrossRef]

142. Gekenidis, M.-T.; Qi, W.; Hummerjohann, J.; Zbinden, R.; Walsh, F.; Drissner, D. Antibiotic-resistant indicator bacteria in irrigation water: High prevalence of extended-spectrum beta-lactamase (ESBL)-producing Escherichia coli. *PLoS ONE* **2018**, *13*, e0207857. [CrossRef] [PubMed]

143. Farkas, A.; Bocoş, B.; Butiuc-Keul, A. Antibiotic resistance and intI1 carriage in waterborne enterobacteriaceae. *Water Air Soil Pollut.* **2016**, *227*, 251. [CrossRef]

144. Olaimat, A.N.; Holley, R.A. Factors influencing the microbial safety of fresh produce: A review. *Food Microbiol.* **2012**, *32*, 1–19. [CrossRef]

145. Vital, P.G.; Zara, E.S.; Paraoan, C.E.M.; Dimasupil, M.A.Z.; Abello, J.J.M.; Santos, I.T.G.; Rivera, W.L. Antibiotic resistance and extended-spectrum beta-lactamase production of escherichia coli isolated from irrigationwaters in selected urban farms in Metro Manila, Philippines. *Water* **2018**, *10*, 548. [CrossRef]

146. Fipps, G. *Irrigation Water Quality Standards and Salinity Management Strategies*; Texas Agriculture Extension Service 7-96 Rev edition; Texas A&M University System: College Station, TX, USA, 1996.

147. Stoner, J.D. *Water-Quality Indices for Specific Water Uses*; Circulur 770; United States Department of the Interior, Geological Survey: Arlington, VA, USA, 1978. [CrossRef]

148. Ayotte, J.D.; Gronberg, J.A.M.; Apodaca, L.E. *Trace Elements and Radon in Groundwater across the United States, 1992–2003*; Scientific Investigations Report 2011–5059; U.S. Geological Survey: Reston, VA, USA, 2011; Volume i–xi, pp. 1–115.

149. Welch, A.H.; Westjohn, D.B.; Helsel, D.R.; Wanty, R.B. Arsenic in ground water of the United States: Occurrence and geochemistry. *Ground Water* **2000**, *38*, 589–604. [CrossRef]

150. Rodríguez-Lado, L.; Sun, G.; Berg, M.; Zhang, Q.; Xue, H.; Zheng, Q.; Johnson, C.A. Groundwater arsenic contamination throughout China. *Science* **2013**, *341*, 866–868. [CrossRef]

151. Selck, B.J.; Carling, G.T.; Kirby, S.M.; Hansen, N.C.; Bickmore, B.R.; Tingey, D.G.; Rey, K.; Wallace, J.; Jordan, J.L. Investigating anthropogenic and geogenic sources of groundwater contamination in a semi-arid alluvial basin, Goshen Valley, UT, USA. *Water Air Soil Pollut.* **2018**, *229*, 186. [CrossRef]

152. Signes-Pastor, A.J.; Mitra, K.; Sarkhel, S.; Hobbes, M.; Burló, F.; De Groot, W.T.; Carbonell-Barrachina, A.A. Arsenic speciation in food and estimation of the dietary intake of inorganic arsenic in a rural village of West Bengal, India. *J. Agric. Food Chem.* **2008**, *56*, 9469–9474. [CrossRef] [PubMed]

153. Erban, L.E.; Gorelick, S.M.; Fendorf, S. Arsenic in the multi-aquifer system of the Mekong Delta, Vietnam: Analysis of large-scale spatial trends and controlling factors. *Environ. Sci. Technol.* **2014**, *48*, 6081–6088. [CrossRef] [PubMed]

154. Huhmann, B.; Harvey, C.F.; Uddin, A.; Choudhury, I.; Ahmed, K.M.; Duxbury, J.M.; Ellis, T.; van Geen, A. Inversion of high-arsenic soil for improved rice yield in Bangladesh. *Environ. Sci. Technol.* **2019**, *53*, 3410–3418. [CrossRef] [PubMed]

155. Pick, T. *Assessing Water Quality for Human Consumption, Agriculture, and Aquatic Life Uses*; United States Department of Agriculture: Washington, DC, USA, 2011.

156. U.S. EPA. *Guidelines for Water Reuse 2012*; US Agency for International Development: Washington, DC, USA, 2012; p. 643.

157. Ayers, R.S.; Westcot, D.W. *Water Quality for Agriculture*; Food and Agricultural Organization, United Nations: Rome, Italy, 1976.

158. Kandakji, T.; Udeigwe, T.K.; Dixon, R.; Li, L. Groundwater-induced alterations in elemental concentration and interactions in semi-arid soils of the Southern High Plains, USA. *Environ. Monit. Assess.* **2015**, *187*, 665. [CrossRef] [PubMed]

159. Scanlon, B.R.; Nicot, J.P.; Reedy, R.C.; Kurtzman, D.; Mukherjee, A.; Nordstrom, D.K. Elevated naturally occurring arsenic in a semiarid oxidizing system, Southern High Plains aquifer, Texas, USA. *Appl. Geochem.* **2009**, *24*, 2061–2071. [CrossRef]

160. Malan, M.; Müller, F.; Cyster, L.; Raitt, L.; Aalbers, J. Heavy metals in the irrigation water, soils and vegetables in the Philippi horticultural area in the Western Cape Province of South Africa. *Environ. Monit. Assess.* **2015**, *187*, 1–8. [CrossRef] [PubMed]

161. Gupta, M.; Gupta, S. An overview of selenium uptake, metabolism, and toxicity in Plants. *Front. Plant Sci.* **2017**, *7*, 1–14. [CrossRef]

162. Hakonson-Hayes, A.C.; Fresquez, P.R.; Whicker, F.W. Assessing potential risks from exposure to natural uranium in well water. *J. Environ. Radioact.* **2002**, *59*, 29–40. [CrossRef]

163. Islam, S.M.A.; Fukushi, K.; Yamamoto, K. Contamination of agricultural soil by arsenic containing irrigation water in Bangladesh: Overview of status and a proposal for novel biological remediation. *WIT Trans. Biomed. Heal.* **2006**, *6*, 295–316. [CrossRef]

164. Jeambrun, M.; Pourcelot, L.; Mercat, C.; Boulet, B.; Pelt, E.; Chabaux, F.; Cagnat, X.; Gauthier-Lafaye, F. Potential sources affecting the activity concentrations of 238U, 235U, 232Th and some decay products in lettuce and wheat samples. *J. Environ. Monit.* **2012**, *14*, 2902–2912. [CrossRef] [PubMed]

165. Bañuelos, G.S.; Ajwa, H.A.; Caceres, L.; Dyer, D. Germination responses and boron accumulation in germplasm from Chile and the United States grown with boron-enriched water. *Ecotoxicol. Environ. Saf.* **1999**, *43*, 62–67. [CrossRef] [PubMed]

166. Rhoades, J.D.; Bingham, F.T.; Letey, J.; Hoffman, G.J.; Dedrick, A.R.; Pinter, P.J.; Replogle, J.A. Use of saline drainage water for irrigation: Imperial Valley study. *Agric. Water Manag.* **1989**, *16*, 25–36. [CrossRef]

167. Hopkins, B.G.; Horneck, D.A.; Stevens, R.G.; Ellsworth, J.W.; Sullivan, D.M. *Managing Irrigation Water Quality for Crop Production in the Pacific Northwest*; PNW597-E; USDA: Washington, DC, USA, 2007. Available online: https://catalog.extension.oregonstate.edu/sites/catalog/files/project/pdf/pnw597.pdf (accessed on 19 March 2019).

168. Hem, J.D. Chemistry and occurrence of cadmium and zinc in surface water and groundwater cadmium is reported of compounds in rice. *Water Resour. Res.* **1972**, *8*, 661–679. [CrossRef]

169. Alexakis, D. Assessment of water quality in the Messolonghi-Etoliko and Neochorio region (West Greece) using hydrochemical and statistical analysis methods. *Environ. Monit. Assess.* **2011**, *182*, 397–413. [CrossRef] [PubMed]

170. Irmak, S. Copper correlation of irrigation water, soils and plants in the Cukurova Region of Turkey. *Int. J. Soil Sci.* **2009**, *4*, 46–56. [CrossRef]

171. Manning, A.H.; Mills, C.T.; Morrison, J.M.; Ball, L.B. Insights into controls on hexavalent chromium in groundwater provided by environmental tracers, Sacramento Valley, California, USA. *Appl. Geochem.* **2015**, *62*, 186–199. [CrossRef]

172. Stasinos, S.; Zabetakis, I. The uptake of nickel and chromium from irrigation water by potatoes, carrots and onions. *Ecotoxicol. Environ. Saf.* **2013**, *91*, 122–128. [CrossRef]

173. USEPA. *2018 Edition of the Drinking Water Standards and Health Advisories*; USEPA: Washington, DC, USA, 2018.

174. Gurel, M.; Iskender, G.; Ovez, S.; Arslan-Alaton, I.; Tanik, A.; Orhon, D. A global overview of treated wastewater guidelines and standards for agricultural reuse. *Fresenius Environ. Bull.* **2007**, *16*, 590–595.

175. Pescod, M.B. *Wastewater Treatment and Use in Agriculture*; Food and Agricultural Organization, United Nations: Rome, Italy, 1992; ISBN 9251031355.

176. US EPA. National Recommended Water Quality Criteria—Aquatic Life Criteria Table. Available online: https://www.epa.gov/wqc/national-recommended-water-quality-criteria-aquatic-life-criteria-table (accessed on 27 December 2018).

177. Groschen, G.E.; Arnold, T.L.; Morrow, W.S.; Warner, K.L. *Occurrence and Distribution of Iron, Manganese, and Selected Trace elements in Ground Water in the Glacial Aquifer System of the Northern United States*; U.S. Geological Survey Scientific Investigations Report 2009–5006; USGS: Reston, VA, USA, 2008.

178. Bundschuh, J.; Nath, B.; Bhattacharya, P.; Liu, C.W.; Armienta, M.A.; Moreno López, M.V.; Lopez, D.L.; Jean, J.S.; Cornejo, L.; Lauer Macedo, L.F.; et al. Arsenic in the human food chain: The Latin American perspective. *Sci. Total Environ.* **2012**, *429*, 92–106. [CrossRef]

179. Meharg, A.A.; Rahman, M. Arsenic contamination of Bangladesh paddy field soils: Implications for rice contribution to arsenic consumption. *Environ. Sci. Technol.* **2003**, *37*, 229–234. [CrossRef] [PubMed]

180. Finnegan, P.M.; Chen, W. Arsenic toxicity: The effects on plant metabolism. *Front. Physiol.* **2012**, *3*, 182. [CrossRef] [PubMed]

181. Kim, J.Y.; Davis, A.P.; Kim, K.W. Stabilization of available arsenic in highly contaminated mine tailings using iron. *Environ. Sci. Technol.* **2003**, *37*, 189–195. [CrossRef] [PubMed]

182. Bakhat, H.F.; Zia, Z.; Fahad, S.; Abbas, S.; Hammad, H.M.; Shahzad, A.N.; Abbas, F.; Alharby, H.; Shahid, M. Arsenic uptake, accumulation and toxicity in rice plants: Possible remedies for its detoxification: A review. *Environ. Sci. Pollut. Res.* **2017**, *24*, 9142–9158. [CrossRef] [PubMed]

183. Arslan, B.; Djamgoz, M.B.A.; Akün, E. ARSENIC: A review on exposure pathways, accumulation, mobility and transmission into the human food chain. *Rev. Environ. Contam. Toxicol.* **2017**, *243*, 27–51. [PubMed]

184. Zhao, F.J.; Ma, J.F.; Meharg, A.A.; McGrath, S.P. Arsenic uptake and metabolism in plants. *New Phytol.* **2009**, *181*, 777–794. [CrossRef] [PubMed]

185. Zhao, F.-J.; McGrath, S.P.; Meharg, A.A. Arsenic as a food chain contaminant: Mechanisms of plant uptake and metabolism and mitigation strategies. *Annu. Rev. Plant Biol.* **2010**, *61*, 535–559. [CrossRef] [PubMed]

186. Chakraborty, S.; Alam, M.O.; Bhattacharya, T.; Singh, Y.N. Arsenic accumulation in food crops: A potential threat in Bengal Delta Plain. *Water Qual. Expo. Heal.* **2014**, *6*, 233–246. [CrossRef]

187. Brammer, H. Mitigation of arsenic contamination in irrigated paddy soils in South and South-East Asia. *Environ. Int.* **2009**, *35*, 856–863. [CrossRef]

188. Polizzotto, M.L.; Birgand, F.; Badruzzaman, A.B.M.; Ali, M.A. Amending irrigation channels with jute-mesh structures to decrease arsenic loading to rice fields in Bangladesh. *Ecol. Eng.* **2015**, *74*, 101–106. [CrossRef]

189. Winkel, L.H.E.; Johnson, C.A.; Lenz, M.; Grundl, T.; Leupin, O.X.; Amini, M.; Charlet, L. Environmental selenium research: From microscopic processes to global understanding. *Environ. Sci. Technol.* **2012**, *46*, 571–579. [CrossRef] [PubMed]

190. USGS. *Geochemical and Mineralogical Data for Soils of the Conterminous United States*; USGS: Reston, VA, USA, 2014.

191. Wang, J.; Zhou, C.; Xiao, X.; Xie, Y.; Zhu, L.; Ma, Z. Enhanced iron and selenium uptake in plants by volatile emissions of *Bacillus amyloliquefaciens* (BF06). *Appl. Sci.* **2017**, *7*, 85. [CrossRef]

192. Mitchell, N.; Pérez-Sánchez, D.; Thorne, M.C. A review of the behaviour of U-238 series radionuclides in soils and plants. *J. Radiol. Prot.* **2013**, *33*, R17–R48. [CrossRef] [PubMed]

193. Soudek, P.; Petrova, T.; Benesova, D.; Dvorakova, M.; Vanek, T. Uranium uptake by hydroponically cultivated crop plants. *J. Environ. Radioact.* **2011**, *102*, 598–604. [CrossRef] [PubMed]

194. Boghi, A.; Roose, T.; Kirk, G.J.D. A model of uranium uptake by plant roots allowing for root-induced changes in the soil. *Environ. Sci. Technol.* **2018**, *52*, 3536–3545. [CrossRef] [PubMed]

195. Hayes, A.C.; Fresquez, P.R.; Whicker, W.F. *Uranium Uptake Study, Nambe, New Mexico: Source Document*; Los Alamos National Laboratory: Los Alamos, NM, USA, 2000.

196. Neves, O.; Abreu, M.M. Are uranium-contaminated soil and irrigation water a risk for human vegetables consumers? A study case with *Solanum tuberosum* L., *Phaseolus vulgaris* L. and *Lactuca sativa* L. *Ecotoxicology* **2009**, *18*, 1130–1136. [CrossRef] [PubMed]

197. Neves, M.O.; Abreu, M.M.; Figueiredo, V. Uranium in vegetable foodstuffs: Should residents near the Cunha Baixa uranium mine site (Central Northern Portugal) be concerned? *Environ. Geochem. Health* **2012**, *34*, 181–189. [CrossRef]

198. Gomes, M.A.d.C.; Hauser-Davis, R.A.; Suzuki, M.S.; Vitória, A.P. Plant chromium uptake and transport, physiological effects and recent advances in molecular investigations. *Ecotoxicol. Environ. Saf.* **2017**, *140*, 55–64. [CrossRef]

199. Song, Y.; Jin, L.; Wang, X. Cadmium absorption and transportation pathways in plants. *Int. J. Phytoremediation* **2017**, *19*, 133–141. [CrossRef]

200. Amrhein, C.; Mosher, P.A.; Brown, A.D. The effects of redox on Mo, U, B, V, and As solubility in evaporation pond soils. *Soil Sci.* **1993**, *155*, 249–255. [CrossRef]

201. Metals-U.S. Food & Drug Administration. Available online: https://www.fda.gov/Food/FoodborneIllnessContaminants/Metals/default.htm (accessed on 18 February 2019).

202. Lapworth, D.J.; Stolpe, B.; Williams, P.J.; Gooddy, D.C.; Lead, J.R. Characterization of suboxic groundwater colloids using a multi-method approach. *Environ. Sci. Technol.* **2013**, *47*, 2554–2561. [CrossRef] [PubMed]

203. Praetorius, A.; Scheringer, M.; Hungerbühler, K. Development of environmental fate models for engineered nanoparticles—A case study of TiO_2 nanoparticles in the Rhine River. *Environ. Sci. Technol.* **2012**, *46*, 6705–6713. [CrossRef] [PubMed]

204. González-Gálvez, D.; Janer, G.; Vilar, G.; Vílchez, A.; Vázquez-Campos, S. *The Life Cycle of Engineered Nanoparticles*; Springer: Cham, Switzerland, 2017; pp. 41–69.

205. Thomé, A.; Reddy, K.R.; Reginatto, C.; Cecchin, I. Review of nanotechnology for soil and groundwater remediation: Brazilian perspectives. *Water Air Soil Pollut.* **2015**, *226*, 121. [CrossRef]

206. Gehrke, I.; Geiser, A.; Somborn-Schulz, A. Innovations in nanotechnology for water treatment. *Nanotechnol. Sci. Appl.* **2015**, *8*, 1–17. [CrossRef] [PubMed]

207. Troester, M.; Brauch, H.-J.; Hofmann, T. Vulnerability of drinking water supplies to engineered nanoparticles. *Water Res.* **2016**, *96*, 255–279. [CrossRef] [PubMed]

208. Alimi, O.S.; Farner Budarz, J.; Hernandez, L.M.; Tufenkji, N. Microplastics and nanoplastics in aquatic environments: Aggregation, deposition, and enhanced contaminant transport. *Environ. Sci. Technol.* **2018**, *52*, 1704–1724. [CrossRef] [PubMed]

209. Lehner, R.; Weder, C.; Petri-Fink, A.; Rothen-Rutishauser, B. Emergence of nanoplastic in the environment and possible impact on human health. *Environ. Sci. Technol.* **2019**, *53*, 1748–1765. [CrossRef] [PubMed]

210. Oberdörster, G.; Oberdörster, E.; Oberdörster, J. Nanotoxicology: An emerging discipline evolving from studies of ultrafine particles. *Environ. Health Perspect.* **2005**, *113*, 823–839. [CrossRef]

211. Ganguly, P.; Breen, A.; Pillai, S.C. Toxicity of nanomaterials: Exposure, pathways, assessment, and recent advances. *ACS Biomater. Sci. Eng.* **2018**, *4*, 2237–2275. [CrossRef]

212. Environmental Protection Agency. *Technical Fact Sheet—Nanomaterials*; USEPA: Washington, DC, USA, 2017. Available online: https://www.epa.gov/sites/production/files/2014-03/documents/ffrrofactsheet_emergingcontaminant_nanomaterials_jan2014_final.pdf (accessed on 2 February 2019).

213. Inshakova, E.; Inshakov, O. World market for nanomaterials: Structure and trends. *MATEC Web Conf.* **2017**. [CrossRef]

214. Hyung, H.; Kim, J.-H. Dispersion of C60 in natural water and removal by conventional drinking water treatment processes. *Water Res.* **2009**, *43*, 2463–2470. [CrossRef] [PubMed]

215. Zhang, Y.; Chen, Y.; Westerhoff, P.; Hristovski, K.; Crittenden, J.C. Stability of commercial metal oxide nanoparticles in water. *Water Res.* **2008**, *42*, 2204–2212. [CrossRef] [PubMed]

216. Hedberg, J.; Blomberg, E.; Odnevall Wallinder, I. In the search for nanospecific effects of dissolution of metallic nanoparticles at freshwater-like conditions: A critical review. *Environ. Sci. Technol.* **2019**, *53*, 4030–4044. [CrossRef] [PubMed]

217. Gottschalk, F.; Sonderer, T.; Scholz, R.W.; Nowack, B. Modeled environmental concentrations of engineered nanomaterials (TiO_2, ZnO, Ag, CNT, Fullerenes) for different regions. *Environ. Sci. Technol.* **2009**, *43*, 9216–9222. [CrossRef] [PubMed]

218. Blaser, S.A.; Scheringer, M.; MacLeod, M.; Hungerbühler, K. Estimation of cumulative aquatic exposure and risk due to silver: Contribution of nano-functionalized plastics and textiles. *Sci. Total Environ.* **2008**, *390*, 396–409. [CrossRef] [PubMed]

219. Brar, S.K.; Verma, M.; Tyagi, R.D.; Surampalli, R.Y. Engineered nanoparticles in wastewater and wastewater sludge—Evidence and impacts. *Waste Manag.* **2010**, *30*, 504–520. [CrossRef] [PubMed]

220. Baalousha, M.; Yang, Y.; Vance, M.E.; Colman, B.P.; McNeal, S.; Xu, J.; Blaszczak, J.; Steele, M.; Bernhardt, E.; Hochella, M.F. Outdoor urban nanomaterials: The emergence of a new, integrated, and critical field of study. *Sci. Total Environ.* **2016**, *557–558*, 740–753. [CrossRef] [PubMed]

221. Min Park, C.; Hoon Chu, K.; Her, N.; Jang, M.; Baalousha, M.; Heo, J.; Yoon, Y. Occurrence and removal of engineered nanoparticles in drinking water treatment and wastewater treatment processes. *Sep. Purif. Rev.* **2017**, *46*, 255–272. [CrossRef]

222. Kaphle, A.; Navya, P.N.; Umapathi, A.; Daima, H.K. Nanomaterials for agriculture, food and environment: Applications, toxicity and regulation. *Environ. Chem. Lett.* **2018**, *16*, 43–58. [CrossRef]

223. Pourzahedi, L.; Pandorf, M.; Ravikumar, D.; Zimmerman, J.B.; Seager, T.P.; Theis, T.L.; Westerhoff, P.; Gilbertson, L.M.; Lowry, G.V. Life cycle considerations of nano-enabled agrochemicals: Are today's tools up to the task? *Environ. Sci. Nano* **2018**, *5*, 1057–1069. [CrossRef]

224. Martínez-Ballesta, M.C.; Zapata, L.; Chalbi, N.; Carvajal, M. Multiwalled carbon nanotubes enter broccoli cells enhancing growth and water uptake of plants exposed to salinity. *J. Nanobiotechnol.* **2016**, *14*, 42. [CrossRef] [PubMed]

225. Chen, G.; Qiu, J.; Liu, Y.; Jiang, R.; Cai, S.; Liu, Y.; Zhu, F.; Zeng, F.; Luan, T.; Ouyang, G. Carbon nanotubes act as contaminant carriers and translocate within plants. *Sci. Rep.* **2015**, *5*, 15682. [CrossRef] [PubMed]

226. Vithanage, M.; Seneviratne, M.; Ahmad, M.; Sarkar, B.; Ok, Y.S. Contrasting effects of engineered carbon nanotubes on plants: A review. *Environ. Geochem. Health* **2017**, *39*, 1421–1439. [CrossRef] [PubMed]

227. Lv, J.; Christie, P.; Zhang, S. Uptake, translocation, and transformation of metal-based nanoparticles in plants: Recent advances and methodological challenges. *Environ. Sci. Nano* **2019**, *6*, 41–59. [CrossRef]

228. Larue, C.; Khodja, H.; Herlin-Boime, N.; Brisset, F.; Flank, A.M.; Fayard, B.; Chaillou, S.; Carrière, M. Investigation of titanium dioxide nanoparticles toxicity and uptake by plants. *J. Phys. Conf. Ser.* **2011**, *304*, 012057. [CrossRef]

229. Siddiqi, K.S.; Husen, A. Plant response to engineered metal oxide nanoparticles. *Nanoscale Res. Lett.* **2017**, *12*, 92. [CrossRef] [PubMed]

230. Rastogi, A.; Zivcak, M.; Sytar, O.; Kalaji, H.M.; He, X.; Mbarki, S.; Brestic, M. Impact of metal and metal oxide nanoparticles on plant: A critical review. *Front. Chem.* **2017**, *5*, 78. [CrossRef]

231. Zhu, H.; Han, J.; Xiao, J.Q.; Jin, Y. Uptake, translocation, and accumulation of manufactured iron oxide nanoparticles by pumpkin plants. *J. Environ. Monit.* **2008**, *10*, 713–717. [CrossRef]

232. Tripathi, D.K.; Tripathi, A.; Shweta; Singh, S.; Singh, Y.; Vishwakarma, K.; Yadav, G.; Sharma, S.; Singh, V.K.; Mishra, R.K.; et al. Uptake, Accumulation and toxicity of silver nanoparticle in autotrophic plants, and heterotrophic microbes: A concentric review. *Front. Microbiol.* **2017**, *8*, 7. [CrossRef]

233. Dimkpa, C.O.; Bindraban, P.S. Nanofertilizers: New products for the industry? *J. Agric. Food Chem.* **2018**, *66*, 6462–6473. [CrossRef]

234. Rui, M.; Ma, C.; Hao, Y.; Guo, J.; Rui, Y.; Tang, X.; Zhao, Q.; Fan, X.; Zhang, Z.; Hou, T.; et al. Iron oxide nanoparticles as a potential iron fertilizer for peanut (*Arachis hypogaea*). *Front. Plant Sci.* **2016**, *7*, 815. [CrossRef] [PubMed]

235. Motyka, O.; Štrbová, K.; Olšovská, E.; Seidlerová, J. Influence of Nano-ZnO Exposure to Plants on L-Ascorbic Acid Levels: Indication of Nanoparticle-Induced Oxidative Stress. *J. Nanosci. Nanotechnol.* **2019**, *19*, 3019–3023. [CrossRef] [PubMed]

236. Liu, J.; Williams, P.C.; Goodson, B.M.; Geisler-Lee, J.; Fakharifar, M.; Gemeinhardt, M.E. TiO$_2$ nanoparticles in irrigation water mitigate impacts of aged Ag nanoparticles on soil microorganisms, Arabidopsis thaliana plants, and Eisenia fetida earthworms. *Environ. Res.* **2019**, *172*, 202–215. [CrossRef] [PubMed]

237. Hu, G.; Cao, J. Occurrence and significance of natural ore-related Ag nanoparticles in groundwater systems. *Chem. Geol.* **2019**, *515*, 9–21. [CrossRef]

238. Céspedes, C.; Yeo, M.-K. Life cycle assessment of a celery paddy macrocosm exposed to manufactured Nano-TiO$_2$. *Toxicol. Environ. Health Sci.* **2018**, *10*, 288–296. [CrossRef]

239. Wu, F.; Zhou, Z.; Hicks, A.L. Life cycle impact of titanium dioxide nanoparticle synthesis through physical, chemical, and biological routes. *Environ. Sci. Technol.* **2019**, *53*, 4078–4087. [CrossRef] [PubMed]

240. Solanki, P.; Bhargava, A.; Chhipa, H.; Jain, N.; Panwar, J. Nano-fertilizers and their smart delivery system. In *Nanotechnologies in Food and Agriculture*; Springer International Publishing: Cham, Switzerland, 2015; pp. 81–101.

241. Rai, M.; Ribeiro, C.; Mattoso, L.; Duran, N. Nanotechnologies in food and agriculture. *Nanotechnol. Food Agric.* **2015**, 1–347. [CrossRef]

242. Kah, M.; Kookana, R.S.; Gogos, A.; Bucheli, T.D. A critical evaluation of nanopesticides and nanofertilizers against their conventional analogues. *Nat. Nanotechnol.* **2018**, *13*, 677–684. [CrossRef]

243. White, J.C.; Gardea-Torresdey, J. Achieving food security through the very small. *Nat. Nanotechnol.* **2018**, *13*, 627–629. [CrossRef]

244. Jackson, R.B.; Carpenter, S.R.; Dahm, C.N.; McKnight, D.M.; Naiman, R.J.; Postel, S.L.; Running, S.W. Water in a changing world. *Ecol. Appl.* **2001**, *11*, 1027–1045. [CrossRef]

245. RamyaPriya, R.; Elango, L. Evaluation of geogenic and anthropogenic impacts on spatio-temporal variation in quality of surface water and groundwater along Cauvery River, India. *Environ. Earth Sci.* **2018**, *77*, 1–17. [CrossRef]

246. Etheridge, A.B.; MacCoy, D.E.; Weakland, R.J. *Water-Quality and Biological Conditions in Selected Tributaries of the Lower Boise River, Southwestern Idaho, Water Years 2009—12 Scientific Investigations Report 2014—5132*; U.S. Geological Survey: Reston, VA, USA, 2014; pp. 1–70.

247. *Use of Reclaimed Water and Sludge in Food Crop Production*; National Academies Press: Washington, DC, USA, 1996; ISBN 978-0-309-05479-9.

248. Crook, J.; Surampalli, R.Y. Water reclamation and reuse criteria in the U.S. *Water Sci. Technol.* **1996**, *33*, 451–462. [CrossRef]

249. Bortolini, L.; Maucieri, C.; Borin, M. A tool for the evaluation of irrigation water quality in the arid and semi-arid regions. *Agronomy* **2018**, *8*, 23. [CrossRef]

Review

Sustainable Irrigation in Agriculture: An Analysis of Global Research

Juan F. Velasco-Muñoz [1] , José A. Aznar-Sánchez [1,*] , Ana Batlles-delaFuente [1] and Maria Dolores Fidelibus [2]

1   Department of Economy and Business, Research Centre CAESCG and CIAIMBITAL, University of Almería, 04120 Almería, Spain
2   Department of Civil, Environmental, Land, Building Engineering and Chemistry, Polytechnic University of Bari, 70126 Bari, Italy
*   Correspondence: jaznar@ual.es; Tel.: +34-950-015-192

Received: 25 June 2019; Accepted: 20 August 2019; Published: 23 August 2019

Abstract: Irrigated agriculture plays a fundamental role as a supplier of food and raw materials. However, it is also the world's largest water user. In recent years, there has been an increase in the number of studies analyzing agricultural irrigation from the perspective of sustainability with a focus on its environmental, economic, and social impacts. This study seeks to analyze the dynamics of global research in sustainable irrigation in agriculture between 1999 and 2018, including the main agents promoting it and the topics that have received the most attention. To do this, a review and a bibliometric analysis were carried out on a sample of 713 articles. The results show that sustainability is a line of study that is becoming increasingly more prominent within research in irrigation. The study also reveals the existence of substantial differences and preferred topics in the research undertaken by different countries. The priority issues addressed in the research were climatic change, environmental impact, and natural resources conservation; unconventional water resources; irrigation technology and innovation; and water use efficiency. Finally, the findings indicate a series of areas related to sustainable irrigation in agriculture in which research should be promoted.

Keywords: sustainable irrigation; bibliometric analysis; climate change; innovation and technology; water use efficiency; unconventional water resources

---

1. Introduction

The current global context is conditioned by the growth of the world's population and the progressive and continuous deterioration of the environment. This creates the challenge of ensuring the supply of basic resources, such as food and water, and sustainable development [1], where water plays an essential role in the survival of human society [2] and contributes to the provision of a wide range of services on which the wellbeing of society is based [3–5]. However, water resources are subject to severe degradation due to many factors, such as the consequences of global climate change, alterations in the use of land, agricultural and urban expansion, and overexploitation due to economic development [6–8]. In parallel with this degradation and overexploitation of ecosystems and water resources, the demand for the services supplied by these resources is expected to increase.

Agricultural ecosystems are the principal suppliers of food, but they are also the main users of water resources on a global level [9,10]. These ecosystems use between 60% and 90% of the available water, depending on the climate and economic development of the region [11,12]. The global area dedicated to irrigated crops is estimated to be 275 million hectares, with an upward growth trend of 1.3% per year [13]. This accounts for just 23% of farmed land; however, 45% of total food production is obtained through these types of crops [14,15]. It has been estimated that in order to satisfy the food demand in 2050, world production must increase by 70% [16]. In a scenario of low production, in order to fulfil this objective, it will be necessary to increase the use of water resources on a global level by 53% [17]—around 50% in developing countries and 16% in developed countries [18]—keeping the current values of variables like productivity and technology.

Currently, different approaches are being used to address the challenges of food provision and the supply of water for different uses and to maintain an environmental balance. Some works point to the development of measures to control demand so that irrigation water sustainability can be reached. The development of efficient water markets can be an optimal measure in underdeveloped areas and with a high level of water scarcity, like in South Africa [19,20]. The implementation of joint restrictions based on the establishment of quotas and the payment of fees can be an effective control system for the use of agriculture water in developed regions specialized in the production of high-quality crops and where overexploitation of water resources is currently taking place [21]. Regarding water supply, many authors recommend the joint use of different water resources and the development of infrastructures as nonconventional water sources [22,23]. Another line of research is focused on the improvement of the efficiency of water use and the development of clean production models that guarantee sustainability from social and economic perspectives [24,25]. In order to achieve this objective, the whole irrigation process must be analyzed. This process covers different phases beginning with the water source and ending with its use for agriculture. Zhang et al. [26] identified three phases in irrigation: The first includes the extraction of water from the source and its transfer through channels to the point of use; the second consists of the distribution of the water to the root system to facilitate its absorption by crops (this includes both traditional irrigation using floods and furrows and modern irrigation through drip systems and microsprinklers); and the third covers the whole crop-growing process, whereby the water is transported from the roots to the rest of the plant. The goal is to save resources through minimizing water losses during these three phases and to improve the efficiency in the use of water resources.

The so-called "Science of Sustainability" also studies how to address these challenges. It is defined as "a discipline that points the way towards a sustainable society" and is "aimed at understanding the fundamental character of interactions between natural, human, and social systems, covers a wide range of academic disciplines", for the development of agricultural systems and the sustainable use of water [27–29]. At the end of the 1990s, sustainability was used as a characteristic to describe ecosystems, referring to the capacity to maintain the flow of services in different environmental, economic, and social contexts [30]. When it is applied to the management of water resources in agriculture, sustainability is considered to be a series of practices that increase crop yield and minimize water losses [31]. The objectives of the sustainable management of water resources in agriculture consider the continuity of the agricultural system from physical and biological perspectives, as well as the economic efficiency of the use of the resources and social participation in the decision-making processes [32]. An evaluation of a change in water use requires, therefore, a multidisciplinary approach that includes an analysis of the body of water under study in order to understand the possible impacts on the quantity and quality of the water and the timetable of the different uses. A comprehensive evaluation of the marginal productivity of water is also required, together with an analysis of its nonmarketable value, such as that derived from ecosystem services [33].

In recent years, there has been an increase in the number of studies analyzing agricultural irrigation from the perspective of sustainability with a focus on its environmental, economic, and social impacts. The objective of this study is to analyze the dynamics of the research on sustainable irrigation in agriculture over the last twenty years. In order to fulfil this objective, a two-fold analysis was undertaken: quantitatively through a bibliometric analysis; and qualitatively through a systemic review based on keyword analysis. The study analyzes the evolution of the number of articles published, the main authors, institutions and countries that promote this research field, the disciplines involved in the research, the main lines of research, the differences in academic approach and the countries considered, and the main issues that affect the research in this field.

Bibliometric analysis was introduced by Garfield in the 1950s [34], and its objective is to identify, classify, and evaluate the principal components within a specific research field [35]. Bibliometry combines tools of quantitative analysis to study the trends of a research topic and identify the main driving agents and the relevance of their publications [36,37]. In bibliometric analyses, three types of indicators can be distinguished, which were defined by Durieux and Gevenois [38]: productivity indicators, relevance indicators, and structural indicators. In addition to these indicators, different approaches exist in bibliometric analysis. Co-occurrence, co-citation, and bibliographic coupling analysis are among the traditional approaches. This extended methodology can be considered as a new one in some research areas. This has also continuously been developing. In this sense, this work introduces some new methodological aspects which provide a contribution regarding previous works—in fact, the sample search process, a mixed quantitative and qualitative review, and the production of keyword networks to identify main trends per country. The results of this study provide a basis on which to establish priorities and to develop new projects in future research on this topic.

2. Methodology

In order to conduct this study, a traditional approach based on co-occurrence was selected, which included the assessment of productivity, quality, and structural indicators. In this approach, first, the agents with the highest number of publications were identified, and second, the impact of the publications of these authors was analyzed. This type of analysis, particularly with respect to journals, is highly interesting for researchers, given that it constitutes a way to assess the relevance of the journals in which authors publish their studies [39]. Finally, we used mapping techniques to analyze the structure of the network between different agents. The Scopus database was used to select the sample of studies to analyze. This database has proven to be the most suitable for our area of study, enabling us to ensure the selection of a representative sample of the studies carried out on sustainable irrigation (SI). Furthermore, it is easy to access, allows the visualization and analysis of data, and allows data to be downloaded in different formats for subsequent processing using software applications [40]. Nevertheless, if some works on SI are not indexed in the Scopus database, they have not been considered in our sample.

The term used to carry out the search was "sustainable irrigation", and this selection was based on previous studies on the same topic [41–43]. This term was searched for under authors' keywords and titles. The study period selected was 1999 to 2018. Research activity in this topic peaked during these years. Furthermore, this period immediately followed the 1st World Water Forum held in Marrakesh in 1997, which is considered to be one of the main landmarks in this field. Only documents until 2018 were included so that complete annual periods could be compared. In order to avoid duplication, the sample only included original articles [44]. It is worth pointing out that a different search query could give rise to different results. The search was carried out in January 2019. The sample of this study was composed of 713 articles. In addition, a search of articles on "irrigation" was also carried out with the same restrictions in order to analyze the relative importance of sustainability within this general theme. Figure 1 shows an outline of the methodology on which this study was based.

The analyzed variables were the number of articles, their years of publication, all of the authors of the articles, the institutions and countries of all of the authors, the subject areas in which Scopus classifies the studies, the name of the journals in which they were published, and the keywords. After downloading this information, the first task was to eliminate duplications. The names of authors and institutions can be found in different formats. This can lead to errors when counting these records. Therefore, these two variables were analyzed, and the different records were regrouped so that the same author and institution were not counted more than once. Once the information had been refined, different tables and figures were drawn up, and the analysis of the data was conducted. The programs used were Excel (version 2016) and SciMAT (v1.1.04) (University of Granada, Granada, Spain). The tool used to create the network maps was VOSviewer, which is widely used in this type of study [42]. Finally, keyword analysis was used to extract the principal research trends [45]. The terms were regrouped in order to eliminate duplications due to plurals, hyphens, words in upper case letters, etc. For the grouping of keywords by topics, standardized grouping algorithms were used with the following tools: Vosviewer (Association strength) and SciMAT (network analysis).

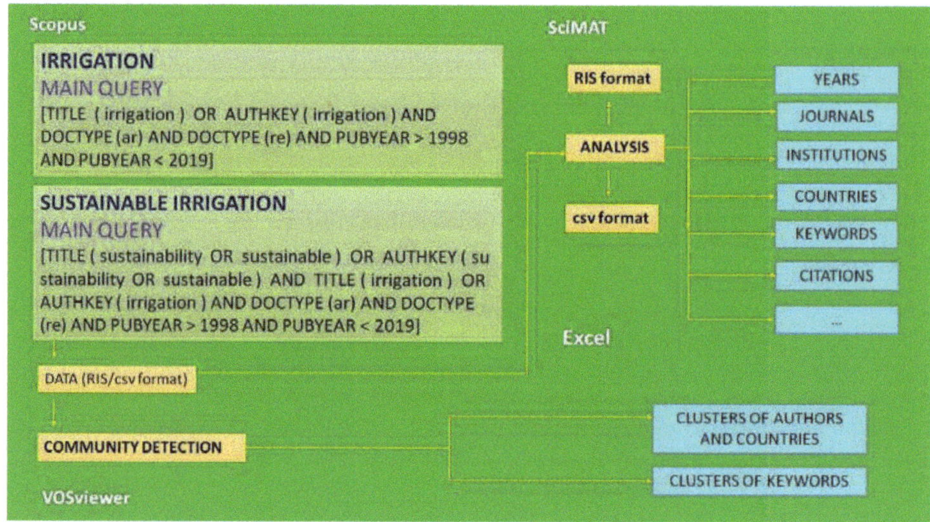

Figure 1. Outline of the methodological development of this study.

As for the methodology, this work includes some novel aspects compared to previous studies dealing with a similar topic. Firstly, regarding the sample selection of articles to be analyzed, some previous studies made a search based on titles, abstracts, and keywords [41–43,46,47]. In this work, the search was conducted in the fields of title and authors' keywords. Furthermore, before getting the final sample, it was checked that all included articles were related to the actual SI research. Secondly, most works on bibliometric reviews include the analysis of keywords. Nevertheless, this is the first search analysis which aims at detecting SI research trends based on the disciplinary approach of the study and the country where the research was conducted. Finally, the work includes a quantitative review based on a bibliometric analysis, as well as a qualitative one based on the traditional review.

3. Results

3.1. Evolution of the General Characteristics of Research on Sustainable Irrigation (SI)

Table 1 shows the evolution of the main variables related to research on SI during the period 1999–2018. During the studied period, relevant events like international declarations and congresses decisively influenced on the sustainability research. The Kyoto Protocol (UNFCCC, 2008), which commits world countries to reduce greenhouse gas emissions, should be highlighted, as well as the Economics of Ecosystems and Biodiversity of 2010; the Rio +20 of 2012; the Millennium Development Goals of the United Nations (UN, 2015), which provides guidelines for improving livelihoods and the environment globally; or the Paris Agreement on Climate Change of 2016; among others. These happenings additionally stimulate research on this topic [48]. This could also explain the existence of peaks regarding the publication of articles on SI research, like in 2017. A further reason explaining the higher number of published articles in 2017 compared to 2018 is that the sample selection was conducted in January 2019. The Scopus database updates itself continuously and, at the time of the sample search, not all published articles in 2018 had been registered. If the sample selection were to be performed at the end of 2019, the number of published and indexed articles on SI in Scopus in 2018 would increase.

Table 1. Main characteristics of sustainable irrigation (SI) research.

Year	Articles	Authors	Journals	Countries	Citation	Average Citation [1]
1999	6	13	6	5	0	0.0
2000	11	22	9	7	14	0.8
2001	15	30	11	10	5	0.6
2002	15	43	11	14	14	0.7
2003	12	24	12	11	35	1.2
2004	15	33	15	13	47	1.6
2005	20	57	16	15	76	2.0
2006	35	112	23	25	130	2.5
2007	22	68	19	15	181	3.3
2008	31	89	24	20	211	3.9
2009	25	56	18	21	289	4.8
2010	47	135	39	29	331	5.2
2011	37	118	31	29	442	6.1
2012	37	107	26	22	517	7.0
2013	45	149	38	27	650	7.9
2014	63	214	47	38	743	8.5
2015	53	202	46	30	901	9.4
2016	68	244	55	34	1268	10.5
2017	88	325	58	37	1515	11.4
2018	68	292	45	42	1707	12.7

[1] Total number of citations accumulated to date divided by the total number of articles published to date.

In general terms, we observed a growth trend in all of the variables analyzed, which indicates the development of this line of research. More than 45% of the total number of studies in the sample are concentrated in the last five years of the period analyzed. In order to confirm the growth of this field of study, the evolution of the number of articles on SI during the period of analysis was compared with all of the articles published on irrigation and all of the articles published on sustainability. Figure 2 shows the percentage of annual variation in the number of articles published in these lines of research. The average annual growth of the articles on irrigation was 1.6%, the one of articles on sustainability 3.8%, while that of articles on SI was 5.2%. This enabled us to confirm that SI is a line of study that is becoming increasingly more prominent within research in irrigation and in sustainability in general. These results agree with other works on water and sustainability [1,39,49].

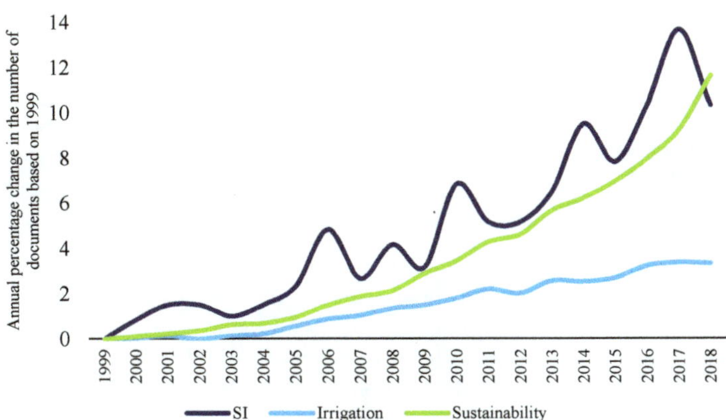

Figure 2. Comparative trends in irrigation, sustainability, and SI research.

With respect to the rest of the variables included in Table 1, the average number of authors per article doubled from two at the beginning of the period to four at the end. The number of journals in which articles on SI were published increased from six in 1999 to 45 in 2018. The number of countries also grew during the period analyzed (from five in 1999 to 42 in 2018). The annual number of references increased from 0.8 in 2000 to 12.7 in 2018.

3.2. Evolution of Research in SI by Subject Area

Figure 3 shows the evolution of the main subject areas into which the articles on SI included in the Scopus database were classified. It should be noted that an article may belong to more than one category. From the beginning of the period, the category in which the highest number of studies were classified was Environmental Sciences, which accounted for almost 65% of the total sample. The second largest block of studies was classified in the Agricultural and Biological Sciences category, with 44.3% of the total sample. In third place was the Social Sciences category with 21.1% of the articles. These three categories have dominated research on SI since the beginning of the studied period. However, in contrast to some previous works [37,39,48], our results revealed that over the last five years, the Earth and Planetary Sciences, Engineering, Energy, and Economics categories have begun to gain relevance, although none of them include more than 15% of the total articles in the sample. The Scopus classification distinguishes between the following categories: Business, Management, and Accounting; and Economics, Econometrics, and Finance, which also differ from Social Sciences. For the purpose of simplification, we grouped these two categories into only one and termed it "Economics".

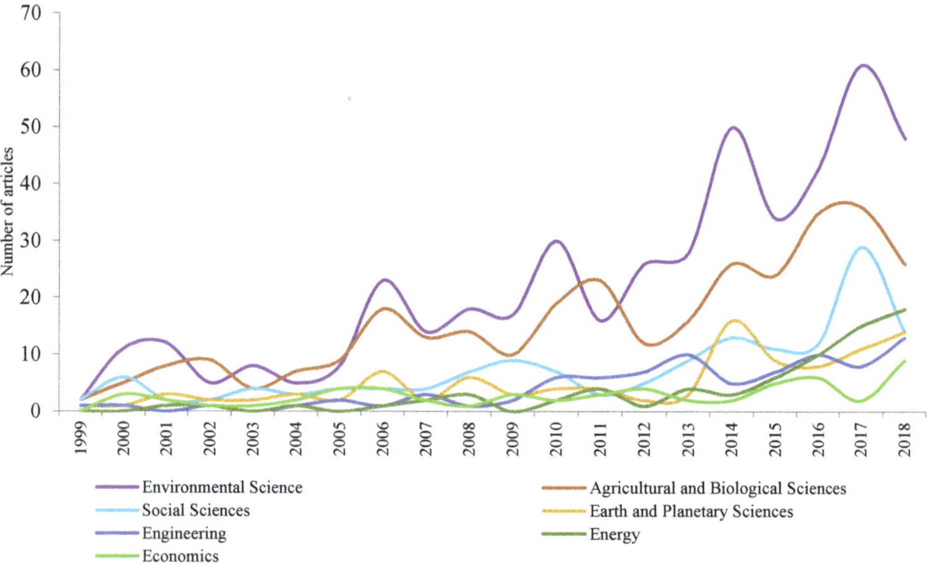

Figure 3. Comparative trends of subject categories related to SI research.

The keyword analysis revealed that there is a series of commonly used terms in research on SI, irrespective of the approach of the study. These terms include, among others, agriculture, alternative agriculture, climate change, crops, groundwater, irrigation system, salinity, sustainable development, water conservation, water management, water resources, water supply, water use, water use efficiency, and water quality. When we took the analysis beyond these terms, we identified a group of keywords used specifically by each discipline.

In the studies classified in the category of Environmental Sciences, there was an emphasis on the state of the soil (soils, soil moisture), aquifers, and surface water. With respect to processes, from the environmental approach, recycling and wastewater reclamation were prominent. In terms of methodology, the keywords that stood out were numerical model and decision making, particularly related to the management of the available water resources (water budget, water availability). With regard to the geographical dimension, China, the United States, and India were particularly prominent, as were the regions of Eurasia and Asia.

In the studies classified within the category of Agricultural and Biological Sciences, technical terms were predominant. Studies in this category mostly focused on soils and groundwater. There was particular emphasis on different types of crops (Triticum aestivum, Zea mays, fruit, Gossypium hirsutum, and rice) and irrigation processes (deficit irrigation, drainage, drip irrigation, leaching, waterlogging, agricultural irrigation). Furthermore, from the agronomic perspective, the environmental dimension was also considered (Environmental Impact). In these studies, China and the United States stood out, together with the regions of Africa and Asia.

The studies carried out from a Social Sciences approach had a more multidisciplinary perspective. They focused primarily on the stakeholders, water demands, and food security. However, technical concepts were also prominent, particularly those related to irrigation and water management (drip irrigation, sustainable water management), crops (Triticum aestivum), and economics (water economics). Unlike the over categories, land use was found to be one of the prominent subjects of this group of studies. A focus on management and decision making at different levels was also characteristic of these studies (governance approach, water planning, policy making, resource management, decision support system). With respect to geographical distribution, the United States, China, India, Spain, and Australia were among the most cited countries, and Asia, Europe and Africa stood out on a regional level.

Finally, the studies carried out based on an economic approach (Economic Sciences) were the most multidisciplinary, including technical, social, and environmental aspects. Among the main themes analyzed in these studies, we found food supply, food security, the development and innovation of irrigation systems (agricultural technology, irrigation performance) and management processes (integrated resource management, managed change, project management, strategic change, strategic management, strategic planning, decision making), and issues related to economic and social management (efficiency, investment, performance, economic and social effects) and the environment (environmental impact, environmental sustainability).

3.3. Most Relevant Journals in the Research on SI

Table 2 shows the main characteristics of the most prolific journals in the field of SI. The group of journals with the highest number of articles published on SI accounted for 25.7% of the total articles in the sample. This indicates that there is a high level of dispersion in terms of the journals that publish articles on this subject area. The leading journal in terms of the total number of articles published during the whole period analyzed was *Agricultural Water Management*, with a total of 52 articles on SI. This journal has the highest H index and the most citations of the journals with articles published in this area, and a Scimago journal rank (SJR) factor of 1.272. It published its first issue on this subject in 2001. Since then, it has remained among the top positions in terms of the number of articles published on SI, becoming the leader in 2014. The journal in second place was *Irrigation and Drainage*, with a total of 30 articles on SI. This journal published its first article on SI in 2001 and was the most prolific journal until 2005. It has the second highest H index, an average of 10.4 citations per article and an SJR index of 0.342. The third journal was *Sustainability*, with 17 articles on SI. This journal is among the most recently incorporated journals, as its first article on SI was published in 2013. However, in only five years, it rose to third position in terms of the number of articles for the whole of the period of study. It has an average of 3.8 citations per article, an H index of 6 and an SJR of 0.537. The journal with the highest average number of citations per article was *Science of the Total Environment* with 33.8; followed by the *Journal of Hydrology* with 31.5 and *Water Resources Management* with 29.4.

Table 2. Main characteristics of the most active journals related to SI research.

Journal	Articles	SJR [1]	H index [2]	Country	Citation	Average Citation [3]	1st Article	Last Article
Agricultural Water Management	52	1.272 (Q1)	21	Netherlands	1128	21.7	2001	2018
Irrigation and Drainage	30	0.342 (Q2)	10	USA	313	10.4	2001	2018
Sustainability	17	0.537 (Q2)	6	Switzerland	65	3.8	2013	2018
Water	15	0.634 (Q1)	8	Switzerland	129	8.6	2009	2018
Water Policy	11	0.461 (Q2)	6	UK	74	6.7	2005	2018
Water Resources Management	11	1.185 (Q1)	9	Netherlands	323	29.4	2000	2015
Acta Horticulturae	10	0.198 (Q3)	2	Belgium	17	1.7	2011	2018
Journal of Hydrology	10	1.832 (Q1)	8	Netherlands	315	31.5	2010	2017
Journal of Cleaner Production	9	1.467 (Q1)	5	Netherlands	49	5.4	2015	2018
Journal of Irrigation and Drainage Engineering	9	0.521 (Q2)	4	USA	47	5.2	2007	2017
Science of The Total Environment	9	1.546 (Q1)	6	Netherlands	304	33.8	2004	2018

[1] Scimago Journal Rank 2017; [2] only sample documents; [3] total number of citations divided by the total number of articles.

3.4. Most Relevant Countries in Research on SI

Table 3 shows the principle characteristics of the articles on SI from the most prolific countries. During the analyzed period, the United States was the leading country in research on SI in terms of the number of articles, with a total of 143. The country with the second highest number of articles was India, with a total of 74. This was followed by Australia with 67, Spain with 61, and Italy with 55. Due to the differences in terms of the size and economic development of the different countries, these data were analyzed to determine the number of articles per capita, measured as the number of articles per million inhabitants. Based on this variable, Australia was shown to be the most productive country with 2.7 articles per million inhabitants. This was followed by the Netherlands with 1.5, Spain with 1.3, Italy with 0.9, and the United Kingdom with 0.8. France was shown to be the country with the most citations per article, with 23.9, followed by the United Kingdom with 22.5, Iran with 21.9, the Netherlands with 18.6, and the United States with 18.1.

Table 3. Main characteristics of the most active countries related to SI research.

Country	Articles	Average per Capita Articles [1]	Citation	Average Citation [2]	H Index [3]	1st Article	Last Article	% of Cultivated Area Equipped for Irrigation (Ranking Countries) [4]
USA	143	0.439	2585	18.1	25	1999	2018	16.94 (72)
India	74	0.055	688	9.3	14	1999	2018	41.54 (38)
Australia	67	2.724	941	14.0	17	2000	2018	5.72 (110)
Spain	61	1.310	815	13.4	14	2004	2018	21.61 (64)
Italy	55	0.908	359	6.5	9	2002	2018	44.22 (35)
China	52	0.038	791	15.2	15	2004	2018	51.48 (28)
UK	51	0.772	1147	22.5	16	1999	2018	3.41 (126)
Germany	36	0.435	509	14.1	12	2004	2018	5.65 (111)
France	29	0.432	692	23.9	10	2000	2018	14.53 (82)
Japan	26	0.205	347	13.3	8	2001	2017	54.96 (25)
Netherlands	26	1.518	483	18.6	11	2001	2018	46.85 (31)
Brazil	22	0.105	163	7.4	6	2006	2018	5.79 (108)
Canada	22	0.599	179	8.1	8	2005	2018	2.44 (134)
Iran	19	0.234	416	21.9	8	2009	2018	51.88 (27)
South Africa	19	0.335	109	5.7	7	2002	2018	12.93 (83)

[1] Total number of articles per million inhabitants; [2] total number of citations divided by the total number of articles; [3] only sample documents; [4] FAO Aquastat (2019), last available data.

The percentage of cultivated area ready for irrigation per country has been included in the last column of the table, as well as the position they have in the world ranking regarding this variable. It can be stated that these countries do not occupy leading positions as far as irrigation-equipped cultivated land is concerned. From the available information about 177 countries, Japan is the country with the highest percentage of irrigation-equipped land surface—54.96%—, reaching the 25th place—followed by Iran with 51.88% (27th place), China with 51.48% (28th place), and Netherlands with 46.85% (31st place). However, some countries leading research on SI place themselves on lower positions within the irrigation-equipped cultivated surface ranking. This is the case for the USA with 16.94% (72nd place), Australia with 5.72% (110th place), or the UK with 3.41% (126th place).

Table 4 shows the principal variables related to the international collaboration of countries with the highest numbers of articles. The average percentage of articles carried out through international collaboration was 50.3%. The countries with the highest percentage of studies carried out in collaboration were Canada with 81.8%, France with 79.3%, Germany with 75.1%, the Netherlands with 65.4%, and China with 55.8%. The United States was found to have the largest collaboration network, with 33 different collaborators. In addition, similarly to Australia, this country forms part of the group of the main collaborators of 10 of the 15 countries in the table. These data reveal the global nature of research in this subject area, with very high percentages and extensive collaboration networks on a global level. The majority of the countries obtained a higher average number of citations per article when they worked in collaboration with other countries. The articles produced through collaboration obtained an average of 14.6 citations as opposed to 13.5 citations of noncollaborative articles. When comparing these results to those of related works on irrigation and water [37,39,48], it can be observed that studies on sustainability trigger a higher level of international cooperation.

Table 4. International collaboration of the most active countries related to SI research.

Country	Percentage of Collaboration [1]	Number of Collaborators	Main Collaborators	Average Citation	
				Collaboration [2]	Noncollaboration [3]
USA	45.5	33	China, Italy, Mexico, Canada, Sweden	23.1	13.9
India	21.6	15	USA, Ethiopia, France, UK, Australia	11.6	8.7
Australia	52.2	22	China, Canada, Spain, USA, Bangladesh	18.6	9.1
Spain	40.9	18	Portugal, Australia, Germany, Italy, Belgium	22.1	7.3
Italy	40.0	19	USA, Netherlands, Spain, France, South Africa	5.0	7.5
China	55.8	16	USA, Australia, Canada, Bangladesh, Germany	15.7	14.6
UK	47.1	28	India, Italy, Netherlands, Philippines, Spain	22.3	22.6
Germany	75.1	30	Uzbekistan, USA, Spain, Switzerland, Australia	15.6	9.9
France	79.3	21	India, USA, Australia, Belgium, Italy	19.1	42.2
Japan	38.5	9	Thailand, USA, Vietnam, Australia, Egypt	9.9	15.5
Netherlands	65.4	16	Italy, Australia, China, Germany, Pakistan	23.6	9.1
Brazil	31.8	7	Spain, USA, Argentina, Germany, Italy	14.7	4.0
Canada	81.8	13	Australia, China, USA, Denmark, India	9.2	3.3
Iran	26.3	4	Australia, USA, Germany, Netherlands	2.2	28.9
South Africa	52.6	13	Australia, Italy, Belgium, Bolivia, Denmark	6.2	5.2

[1] Number of articles made through international collaboration divided by the total number of articles; [2] number of citations obtained by articles made through international collaboration divided by the number of articles; [3] number of citations obtained for articles not made through international collaboration divided by the number of articles.

Figure 4 shows a network map of the collaborations carried out between countries, where the size of the circle represents the number of documents per country and the color corresponds to the cluster formed by the different groups of countries. Three clusters can be distinguished, led by the United States, Australia, and Spain in terms of the number of articles. The first (shown in blue) includes some of the most prolific countries, such as India, Italy, China, France, Japan, and the Netherlands, and others, such as Mexico, Egypt, and Bangladesh. Together with Australia, the second cluster (shown in red) includes Canada, Iran, South Africa, Sweden, Switzerland, Pakistan, Sri Lanka, and Uzbekistan. The group led by Spain (shown in green) includes some European countries, such as the United Kingdom, Germany, Belgium, Portugal, and Greece, as well as countries in the Mediterranean basin, such as Israel, Jordan, Morocco, and Turkey, and others, such as Brazil, Thailand, and New Zealand.

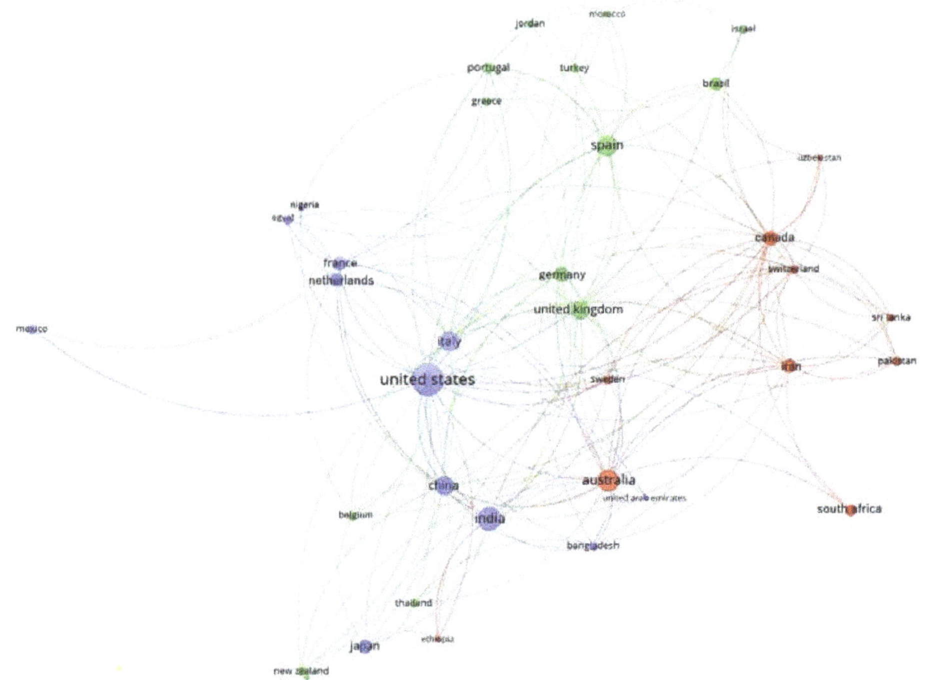

Figure 4. Main relationships between countries in SI research.

A keyword analysis was used to detect the preferences in the research conducted by the countries included in Table 4 (Table 5). We established that there is a group of terms that make up a general line from which the different specific topics are derived.

In the studies conducted in the United States, the central themes included food supply and food security (food-supply, food-security), the conservation of natural resources and environmental impacts (conservation-of-natural-resources, environmental-impact), and land use (land-use). Agronomic issues, such as crop evapotranspiration and productivity, were prominent. Both surface water and groundwater were studied with special emphasis on the availability of the resource, the water budget, the water table, and water stress (surface-water, water-budget, water-availability, water-table, water-stress). The term most used in relation to methodology was numerical model (numerical-model), and the term most used with regard to crops was Zea mays (Zea-mays). No noteworthy geographical terms other than the United States were identified.

Table 5. Main keywords of the most active countries related to SI research.

Country	Keywords
USA	crop-yield, surface-water, alternative-agriculture, numerical-model, food-security, water-budget, wastewater-reclamation, water-availability, food-supply, evapotranspiration, conservation-of-natural-resources, environmental-impact, water-table, water-stress, Zea-mays, land-use
India	optimization, irrigation-planning, water-table, waterlogging, drainage, Maharashtra, irrigation-projects, rice, arid-regions, ecosystems, fertilizers, nutrient, vegetables, waste-water, South-Asia, surface-water
Australia	Australasia, wastewater, irrigation-efficiency, Murray-Darling-basin, environmental-protection, evapotranspiration, food-supply, hydrogeology, runoff, water-availability, controlled-study, hydrology, water-treatment, rice, recycling, Canada
Spain	energy-efficiency, fruit, irrigation-networks, semiarid-region, profitability, southern-Europe, soil-moisture, deficit-irrigation, carbon-dioxide, drip-irrigation, water-economics, economic-analysis, water-productivity, decision-support-systems, energy-resources, stakeholder
Italy	deficit-irrigation, decision-support-system, southern-Europe, farms, fruit, orchard, Mediterranean-environment, forestry, stem-water-potential, crop-yield, dicotyledon, stomatal-conductance, water-stress, drainage, decision-making, environmental-impact
China	alternative-agriculture, Zea-mays, Triticum-aestivum, Xinjiang-Uygur, North-China-plain, decision-making, irrigation-district, evapotranspiration, environmental-protection, hydrological-modelling, ecology, landforms, soil-moisture, integrated-approach, water-availability, uncertainty
UK	alternative-agriculture, environmental-impact, water-treatment, drainage, Africa, environment, wastewater, cost-benefit-analysis, rice, wetland, recycling, runoff, drainage-and-irrigation, arid-regions, crop-yield, hydrocarbon
Germany	food-supply, Triticum-aestivum, Uzbekistan, alternative-agriculture, arid-region, fertilizer, oasis, climate-models, economic-and-social-effects, drip-irrigation, agricultural-intensification, food-security, leaching, common-pool-resource, cropping-system, greenhouse-gas
France	farming-system, stakeholder, groundwater-overexploitation, evapotranspiration, deficit-irrigation, decision-making, environmental-policy, governance-approach, surface-water, public-private-partnership, pricing-policy, water-economics, chemical-composition, dynamic-model, linear-programming, ecophysiological-responses
Japan	irrigation-development, water-users'-organization, rainfall, sustainable-rice-production, institutional-development, Triticum-aestivum, Zea-mays, water-policy, participatory-irrigation-management, saline-water-irrigation, soil-water-salinity, sorghum, electrical-conductivity, semiarid-region, sorghum-bicolour, drought
Netherlands	drainage, Triticum-aestivum, food-production, recirculations, well, groundwater-abstraction, agricultural-extension, smallholder, agricultural-development, food-supply, agricultural-management, decision-making, rain, alternative-agriculture, catchments, networking-system
Brazil	biofuel, expansion, water-availability, environmental-impact, bioenergy, biomass-power, Cerrado, sugar-cane, glycine-max, sustainable-production, Saccharum-officinarum, carbon-dioxide, chemistry, evapotranspiration, metabolism, rainwater
Canada	alternative-agriculture, water-policies, sensitivity-analysis, stochastic-programming, decision-making, environmental-protection, uncertainty-analysis, water-availability, water-stress, Gossypium-hirsutum, food-security, global-perspective, food-production, Triticum-aestivum, water-sharing, yield-response
Iran	crop-yield, cropping-pattern, economic-and-social-effects, rainfall, food-supply, Zea-mays, FAO, optimum-decision, surface-water-resources, recycling, genetic-algorithm, GHG-emission, irrigation-district, farmers-motivation, untreated-wastewater-irrigation, Bayesian-networks
South Africa	Sub-Saharan-Africa, wastewater, evapotranspiration, water-economics, drought, simulation, controlled-study, electric-conductivity, economic-and-social-effects, GIS, water-balance, computer-simulation, food-supply, project-management, mine-water, environmental-impact

The articles from India were found to place more emphasis on new alternative agriculture systems (alternative-agriculture) and the agricultural activity in arid areas (arid-regions) on an ecosystem level (ecosystems). Agronomic aspects such as the nutrition and fertilization of crops (fertilizers, nutrient) and their productivity (crop-yield, productivity) were also prominent. Both surface and groundwater were studied (surface-water), as was the planning and development of irrigation projects for optimizing the resource (irrigation-planning, irrigation-project, optimization), with particular emphasis on irrigation processes (drainage, waterlogging, water-table) and alternative water sources (wastewater). The term most used in relation to methodology was numerical model (numerical-model), and with regard to crops, the most used terms were rice and vegetables (rice, vegetables). Prominent geographical terms, such as the State of Maharashtra or the region of South Asia, were identified.

The studies conducted in Australia covered food supply (food-supply), runoff (runoff), and environmental protection (environmental-protection). Agronomic issues, such as crop evapotranspiration (evapotranspiration) and irrigation efficiency (irrigation-efficiency), were prominent. With respect to water resources, particular emphasis was placed on the availability of the resources, the water budget (water-availability, water-budget), and the use of alternative sources through recycling and wastewater treatment (wastewater, water-treatment, recycling). From a methodological point of view, the hydrological and geological approaches were prominent (hydrology, hydrogeology), as were controlled studies (controlled-study), and in terms of crops, the prominent term was rice (rice). Noteworthy geographical terms, such as the basin of the river Murray–Darling (Murray-Darling-basin), the region of Australasia and the country of Canada, were identified.

In the case of Spain, relevant topics were the management of the use of energy resources (energy-efficiency, energy-resources), the productivity of water (water-productivity), efficiency (efficiency), semiarid regions (semiarid-regions), and environmental protection (environmental-protection). Agronomic issues such as soil moisture (soil-moisture) were prominent, as were innovations in agricultural and irrigation systems (alternative-agriculture, irrigation-networks, drip-irrigation, deficit-irrigation) and irrigation efficiency (irrigation-efficiency). In terms of methodology, the economic approach was prominent (water-economics, economic-analysis, profitability), as was the social approach (stakeholders, decision-support-systems). With respect to crops, fruit was the most relevant term (fruit). Noteworthy geographical terms, such as the regions of Southern Europe and Eurasia (Southern-Europe, Europe, Eurasia), were identified.

The studies conducted in Italy were similar to those conducted in Spain. The main differences were found in certain agronomic terms related to crops (crop-yield, dicotyledon, stomatal-conductance, stem-water-potential) or irrigation (drainage, soil-moisture). The studies were carried out on a farm level (farms, orchard) and focused on the Mediterranean environment. In terms of methodology, the numerical models focused on decision making (decision-making) were noteworthy. With respect to crops, as well as fruit (fruit), Zea mays was also prominent. The central themes included water stress (water-stress), the environmental impact (environmental-impact), and forestry (forestry).

The articles conducted in China placed greater emphasis on environmental protection (environmental-protection), agricultural activity in arid regions (arid-regions), decision-making processes (decision-making), and issues at the district level (irrigation-district). Agronomic issues such as crop evapotranspiration (evapotranspiration) and crop productivity (crop-yield) were also prominent. With respect to water, the central theme was the level of water resources (water-level), while in the methodological area, hydrogeological models were prominent (hydrogeological-modelling). With respect to crops, rice, corn maize and wheat were found to be noteworthy (rice, Zea-mays, Triticum-aestivum, maize). The most prominent geographical terms were the regions of the North China Plain and Xinjiang Uygur.

The central themes in the studies conducted in the United Kingdom were crop yield (crop-yield); the environment and the assessment of the environmental impact (environment, environmental-impact), particularly in arid regions (arid-regions); and hydrocarbon (hydrocarbon). Processes related to water and irrigation, such as drainage (drainage), runoff (runoff), and the recycling of water in wetlands and wastewater treatment (water-treatment, wastewater, wetlands-recycling), stood out. The most prominent term related to methodology was cost benefit analysis (cost-benefit-analysis), and in terms of crops, the study of rice was predominant (rice). This may be due to the country's connection with Asian countries (Asia, Eurasia, and South Asia).

The articles conducted in Germany, as in the case of other countries, were conditioned by its collaborative ties with other nations. The main themes included food supply and food security (food-supply, food security); the new alternative agricultural systems and the intensification of agricultural activity (alternative-agriculture, agricultural-intensification); agricultural activity in arid regions (arid-regions, oasis); and the effects of agricultural activity on economic and social levels and environmental pollution (economic-and-social-effects, greenhouse-gas, particle-size). The prominent agronomic aspects were fertilizers (fertilizers), leaching (leaching), and drip irrigation (drip-irrigation). The most commonly used term with respect to methodology was related to the study of the effects of climate change: climate models (climate-models). Further fields of interest are common pool resources; and, with respect to crops, studies on wheat (Triticum-aestivum). One of Germany's principal collaborators was China, which is why this country appeared prominently among the keywords of German studies. Similarly, Uzbekistan is one of Germany's main trading partners and also appears among the keywords.

The studies carried out in France particularly focused on the policy and institutional dimension (environmental-policy, decision-making, governance-approach, public-private-partnership, pricing-policy). French studies analyzed the level of exploitation (farming-system) and contemplated the different agents involved (stakeholders). The priority issues included the overexploitation of groundwater and surface water (groundwater-overexploitation, surface-water); the chemical composition (chemical-composition); crop evapotranspiration (evapotranspiration) and the response to possible alterations (ecophysiological-responses); and deficit irrigation (deficit-irrigation). In terms of methodology, dynamic models and linear programming were prominent (dynamic-model, linear-programming), together with economic issues related to water (water-economics).

Studies conducted in Japan considered issues such as the development of irrigation, particularly through participative processes on both institutional and irrigation water user levels (irrigation-development, water-users'-organization, participatory-irrigation-management, institutional-development, water-policy). The Japanese studies analyzed aspects related to the salinity of irrigation water and soil (saline-water-irrigation, soil-water-salinity) and the electrical conductivity of water (electrical-conductivity). Water shortages due to drought, particularly in semiarid regions, were also found to be relevant issues (drought, semi-arid-region). The studies from this country analyzed the use of rainwater as a source for irrigation (rainfall). It is the country with the highest number of crops, and the study of the sustainable production of rice was found to be particularly relevant (sustainable-rice-production, Triticum-aestivum, Zea-mays, sorghum, sorghum-bicolour).

In the Netherlands, the primary topics identified were the management and development of agriculture, particularly towards the use of new alternative systems (agricultural-development, agricultural-management, alternative-agriculture), and there was special emphasis on the extension of agricultural practices (agricultural-extension). Other subject areas of many of this country's studies were the security and production of food (food-supply, food-production). The crop that is most studied was wheat (Triticum-aestivum). The articles focused on groundwater and rainwater with respect to water resources (well, groundwater-abstraction, rain), recirculation processes and network development (drainage, recirculations, networking-system) and the studies on the level of the basins were predominant (catchments). On a social level, the point of view of the small farmers (smallholders) was given special attention, particularly in relation to the decision-making processes (decision-making).

The studies conducted in Brazil contemplated the environmental impacts (environmental-impact), water availability (water-availability), and the expansion of the agricultural activity (agricultural-expansion). With respect to crop processes, evapotranspiration, metabolism, sustainable production, and carbon dioxide were prominent (carbon-dioxide, sustainable-production, evapotranspiration, metabolism). This country has published a large number of studies on the use of rainwater for irrigation (rainwater). Particularly noteworthy is the research on crops related to the use of biomass for different purposes (biofuel, bioenergy, biomass-power, sugar-cane, glycine-max, saccharum-officinarum). On a geographic level, studies in the region of Cerrado were predominant. Additionally, the most prominent methodological approach was found to be chemistry (chemistry).

In Canada, the most relevant themes were the new forms of agriculture (alternative-agriculture), environmental protection (environmental-protection), food security (food-security, food-production), and decision making (decision-making, water-policies). The global perspective of this line of research (global-perspective) was found to be noteworthy. From a methodological point of view, the sensitivity models, stochastic programming, and uncertainty analysis were prominent (sensitivity-analysis, stochastic-programming, uncertainty-analysis). The predominant terms with respect to water were water availability, water stress, and water sharing. With regard to crops, wheat and cotton stood out (Gossypium-hirsutum, Triticum-aestivum).

In the case of Iran, concern for the food supply was found to be prominent, despite its relationship with the FAO—Food and Agriculture Organization of the United Nations (food-supply, FAO, crop-yield). On a methodological level, genetic algorithms stood out (genetic-algorithm), together with processes for optimizing decisions (optimum-decision), and Bayesian networks (Bayesian-networks). The economic and social levels were represented through the motivation of farmers and the assessment of the economic and social effects (economic-and-social-effects, farmers-motivation). In the agronomic field, cropping patterns and the use of untreated wastewater were priority areas (cropping-pattern, untreated-wastewater-irrigation). In addition to wastewater, the combined integral use of surface water, recycled water, and rainwater for irrigation was prominent (rainfall, surface-water-resources, recycling).

Finally, the articles from South Africa were based on controlled studies, geographical information systems, and computer simulations (computer-simulation, simulation, controlled-study, GIS). The environmental impacts (environmental-impact), particularly those related to the use of mine water and wastewater (mine-water, wastewater); food supply (food-supply); project management (project-management); and the economic and social effects related to irrigation (economic-and-social-effects) were priority themes. With respect to the agronomic dimension, the evapotranspiration processes, drought, electric conductivity, and water balance stood out (evapotranspiration, drought, electric-conductivity, water-balance).

3.5. Most Relevant Institutions in Research on SI

Table 6 shows the main characteristics of the institutions with the highest number of articles on SI. The Chinese Academy of Sciences holds the first position with 14 articles. This institution has accumulated a total of 376 citations in these articles, with an average of 26.9 citations per article and an H index of 7. The institution with the second largest number of articles is the Commonwealth Scientific and Industrial Research Organisation—Land and Water (CSIRO Land and Water), with a total of 12 studies published. It has 212 citations, an average of 17.7 citations per article, and an H index of 8. In third place is the University of South Australia, with 10 articles. This institution has accumulated a total of 24 citations, an average of 2.4 citations per article, and an H index of 3. The institution with the largest number of citations and the highest average citations per article in its studies on SI is the University of Texas, with a total of 562 citations and 93.7 citations per article.

With respect to the international collaboration of institutions, the average percentage of articles carried out jointly was 46.6%. The institutions with the highest percentage of articles carried out in collaboration were the IHE Delft Institute for Water Education and the Universidade de Lisboa (University of Lisbon) with 83.3%. These two institutions were followed by the University of South Australia with 80.1%, Columbia University with 75.1%, China Agricultural University and Texas A and M University with 66.7%, and the University of California with 62.5%. The average number of citations of the jointly-written articles of the group of 22 institutions was 18.3 as opposed to 16.7 citations for the rest. The institutions with the highest number of citations in articles written in collaboration were the University of Texas, the University of California, and China Agricultural University.

Table 6. Main characteristics of the most active institutions related to SI research.

Institution	Country	Articles	Citation	Average Citation [1]	H Index [2]	Percentage of Collaboration [3]	Average Citation	
							Collaboration [4]	Noncollaboration [5]
Chinese Academy of Sciences	China	14	376	26.9	7	35.7	22.2	29.4
CSIRO Land and Water	Australia	12	212	17.7	8	41.7	21.2	15.1
University of South Australia	Australia	10	24	2.4	3	80.1	2.5	2.0
USDA Agricultural Research Service, Washington DC	USA	9	86	9.6	5	22.2	8.5	9.9
Wageningen University and Research Centre	Netherlands	9	117	13.0	6	55.6	16.2	9.0
Indian Institute of Technology Kharagpur	India	9	163	18.1	7	33.3	6.0	24.2
University of California, Riverside	USA	8	406	50.8	5	62.5	49.6	52.7
Columbia University in the City of New York	USA	8	97	12.1	5	75.1	13.2	9.0
Universidad de Cordoba	Spain	7	73	10.4	7	28.6	14.5	8.8
University of California, Davis	USA	7	87	12.4	3	42.9	26.0	2.3
Università degli Studi della Basilicata	Italy	7	16	2.3	3	14.3	0.0	2.7
Harran Üniversitesi	Turkey	6	77	12.8	3	0.0	0.0	12.8
University of Texas at Austin	USA	6	562	93.7	5	50.0	110.7	76.7
China Agricultural University	China	6	104	17.3	3	66.7	26.0	0.0
Universidad de Almeria	Spain	6	40	6.7	4	50.0	2.7	10.7
Texas A and M University	USA	6	42	7.0	3	66.7	5.5	10.0
IHE Delft Institute for Water Education	Netherlands	6	142	23.7	5	83.3	22.2	31.0
Ben-Gurion University of the Negev	Israel	6	66	11.0	6	16.7	6.0	12.0
Alma Mater Studiorum Università di Bologna	Italy	6	155	25.8	3	16.7	13.0	28.4
Commonwealth Scientific and Industrial Research Organization	Australia	6	92	15.3	3	50.0	16.7	14.0
Northwest A & F University	China	6	42	7.0	5	50.0	7.0	7.0
Universidade de Lisboa	Portugal	6	65	10.8	5	83.3	13.0	0.0

[1] The total number of citations divided by the total number of articles; [2] only sample documents; [3] the number of articles produced through international collaboration divided by the total number of articles; [4] the number of citations obtained by articles produced through international collaboration divided by the number of articles; [5] the number of citations obtained for articles not made through international collaboration divided by the number of articles.

3.6. Most Relevant Authors in Research on SI

Table 7 shows the main characteristics of the authors who have produced the highest numbers of articles on SI. A large number of authors published articles on SI within the study period, but the number of publications per author was small. The three authors with the most articles were Henning Bjornlund from the University of South Australia, Bartolomeo Dichio from the Università degli Studi della Basilicata, and Ajay Kumar R. Singh from the Indian Institute of Technology Kharagpur. Although they published the same number of articles, there were large differences in terms of the relevance of the publications of the different authors. Bjornlund has a total of 20 citations for his articles, Dichio 11, and Singh 145. James D. Oster of the University of California was the author with the largest number of citations with a total of 369; and the highest average number of citations per article with 73.8. He was followed by Mohammad Valipour from the Islamic Azad University, with a total of 366 citations and 73.2 per article. In third place was Dennis Wichelns from the Stockholm Environment Institute with 189 total citations and 37.8 citations per article. The author with the oldest publication was Bart Schultz of the IHE Delft Institute for Water Education, who published his first article in 2001. This author, who has written four articles on SI, has accumulated a total of 122 citations and published his last article on this subject in 2005. The authors who have made more recent contributions to this line of research are P. Amparo López-Jiménez and Modesto Pérez-Sánchez from the Universitat Politècnica de València.

Table 7. Major characteristics of the most active authors related to SI research.

Author	Articles	Citation	Average Citations [1]	H Index [2]	Country	Affiliation [3]	1st Article	Last Article
Bjornlund, Henning	6	20	3.3	2	Australia	University of South Australia	2010	2017
Dichio, Bartolomeo	6	11	1.8	2	Italy	Università degli Studi della Basilicata	2010	2018
Singh, Ajay Kumar R.	6	145	24.2	5	India	Indian Institute of Technology Kharagpur	2010	2017
Oster, James D.	5	369	73.8	3	USA	University of California	2003	2013
Scholz, Miklas	5	49	9.8	4	UK	University of Salford	2006	2018
Valipour, Mohammad	5	366	73.2	5	Iran	Islamic Azad University	2015	2017
Wichelns, Dennis	5	189	37.8	4	Sweden	Stockholm Environment Institute	2002	2014
Xiloyannis, Cristos	5	11	2.2	2	Italy	Università degli Studi della Basilicata	2010	2018
Annandale, John George	4	28	7.0	3	South Africa	Universiteit van Pretoria	2002	2017
Aydogdu, Mustafa Hakki	4	15	3.8	3	Turkey	Harran Üniversitesi	2015	2017
López-Jiménez, P. Amparo	4	50	12.5	3	Spain	Universitat Politècnica de València	2016	2018
Montanaro, Giuseppe	4	9	2.3	1	Italy	Università degli Studi della Basilicata	2010	2018
Pérez-Sánchez, Modesto	4	50	12.5	3	Spain	Universitat Politècnica de València	2016	2018
Schultz, Bart	4	122	30.5	4	Netherlands	IHE Delft Institute for Water Education	2001	2005
Vico, Giulia	4	46	11.5	3	Sweden	Swedish University of Agricultural Sciences	2011	2018

[1] Total number of citations divided by the total number of articles; [2] only sample documents; [3] last verified affiliation.

3.7. Main Issues in SI Research

The most relevant issues in SI research were determined based on the analysis of keywords. These issues included the concern for the state of natural resources, including water and soil and their conservation; the impact of agriculture on the environment and the consequences of climate change; the use of nonconventional water resources as an alternative for irrigation, including desalinated seawater, reused water, and harvested rainwater; the developments in innovation and technology for irrigation systems, particularly drip irrigation and deficit irrigation; and, finally, the improvements in the efficiency of the use of irrigation water. Below is an overview of the research carried out on these four priority issues. The weight of each priority research line has been established through the repetition number of the main keywords within each research line as the article average of the sample. It has to be taken into account that an article can be classified under different topics. For example, an article which analyzes water efficiency regarding nonconventional water resources shows as the most weighted research line climatic change, environmental impact, and natural resource conservation, with a 56.8% out of the total sample works; followed by water use efficiency with 42.5%, irrigation technology and innovation with 37.6%, and unconventional water resources with 31.2%.

3.7.1. Climatic Change, Environmental Impact, and Natural Resource Conservation

It is predicted that the consequences derived from global climate change will be alterations in precipitation cycles, triggering long-term droughts, more frequent and more intense extreme phenomena, and water supply imbalances [39,50]. Furthermore, these consequences will be reflected in agriculture by way of variations in soil humidity and in the evapotranspiration and runoff flows [14,51]. The United Nations report on the development of global water resources of 2015 estimates that there will be a drinking water shortage of 40% on a global level by the year 2030 [52].

Bad practice in agriculture produces a series of impacts that can have consequences on environmental, economic, and social levels. As well as water use, current irrigation agriculture requires the addition of fertilizers and other chemical products [53]. When the use of chemical products is incomplete or inefficient or when excessive water is applied, the resulting filtration ends up in drainage systems or in the groundwater recharge areas under the cultivated land [54]. The most deteriorated ecosystems currently include the majority of the groundwater bodies on a global scale. These water resources have enabled the development of agricultural activity in arid and semiarid regions and also in more humid regions where there are mismatches between precipitation and the needs of the crops [55]. In recent decades, the intensification of agriculture has given rise to a fall in piezometric levels, the development of salinization processes, seawater intrusion, and pollution by agricultural nitrates, among other effects [23].

Due to the estimated increase in the amount of fresh water required to meet the future irrigation demands, a drastic reduction in biodiversity is expected to take place, together with an increase in the salinity or flooding of soil, a loss in the flow of complementary services provided by the ecosystems, and the degradation of water sources and ecosystems in general [56,57]. On a social level, an increase in the vulnerability and inequality between users is expected [58,59].

In order to mitigate these adverse effects and to contribute to the conservation of the ecosystems, important legislation is currently being developed on a global level. Among the objectives established by the United Nations for the Horizon 2030 on Sustainable Development is one specifically related to water and sanitation (ODS 6), which addresses aspects ranging from water shortage to water use efficiency [60]. The Horizon 2020 Plan of the European Parliament includes the requirement for sustainable production in agricultural systems [61]. Many countries, including the United States, China, India, and Costa Rica, have consolidated payment systems for environmental services provided by agricultural ecosystems with the objective of conserving water resources in good condition.

3.7.2. Unconventional Water Resources

The current scenario is one in which so-called conventional water sources are being exhausted and degraded in large parts of the world. These water sources include both surface water (rivers, lakes, reservoirs) and groundwater (aquifers). The principal option for increasing the water supply for irrigation consists of using alternative sources, also called nonconventional water sources [23]. These other sources include the reuse of urban and industrial water, the desalination of seawater, and rainwater harvesting. In recent years, water from nontraditional sources has become a competitive option in the supply of quality water for irrigation, particularly in arid and semiarid regions [62] The use of these types of resources has a series of advantages—two in particular. First, the contribution of these new water sources represents an increase in the supply of the resource, which is capable of satisfying the growing demand of the different sectors (urban supply, agricultural activity, tourism sector, industrial sector, and environmental requirements) [63]. Second, the use of water from alternative sources should serve to diminish the use of traditional water sources so that the state of deterioration of the wetlands, rivers, and aquifers can be restored or at least alleviated [64]. If these two advantages are to be efficient, they have to be accompanied by demand control. In addition to these two principal functions, the use of nonconventional water resources gives rise to other advantages. They provide a greater reliability in the supply, they supply higher quality water, they can generate increases in crop yields, they contribute to ensuring the stability of agricultural incomes, and they can have positive effects on seawater intrusion processes in the aquifers [65–67].

The use of each of these alternatives also gives rise to a series of limitations and disadvantages. Evidently, the construction of seawater desalination plants is only feasible in coastal areas. A wastewater treatment plant requires a volume of a large enough size for the facility to be viable [64]. Therefore, the use of this resource is not possible in areas where the activities generating wastewater (population nuclei, industrial facilities, livestock farms, etc.) do not have sufficient water use [63]. Thus, despite water reuse being the ideal way of maintaining continuous use of the resource, this type of facility is not appropriate for many rural areas where the population is dispersed. The main problem of rainwater harvesting systems is the low volume of water that can be supplied in comparison with the demand [67]. Furthermore, the seasonality of rain in many regions means that the water must be stored for long periods of time for use when needed. To these limitations we must also add the high cost of water derived through these systems. The installation costs are usually very high, and we must also take into account the cost of production. In the case of desalination and reuse, these costs usually establish a price for water that is much higher than the price of conventional resources [68]. This means that many farmers throughout the world are not willing to pay the price of the water unless there is no alternative available. Studies have been carried out on experience with the use of desalinated seawater, reused water or harvested rainwater all over the world [69–75].

3.7.3. Irrigation Technology and Innovation

Irrigation technology has evolved continuously over the last few decades. Flood irrigation, sprinkler irrigation, furrow irrigation, and drip irrigation are some of the methods that have emerged, and their advantages and disadvantages have been studied with respect to different types of crops, soils, and climatic conditions. New technologies have given rise to the development of comprehensive automated systems that combine the use of tensiometers, lysimeters, software applications, and even geographical information systems. However, drip irrigation and deficit irrigation are the terms that appear among the most used keywords.

De Wrachienb et al. [76] date the beginning of drip irrigation systems to the 1940s in Australia. The development of this system came about after the emergence of polypropylene tubes. It was not until two decades later that this system was improved in Israel, from where it was exported all over the world. Currently, thanks to automation and the use of microcontrollers, sensors, and integrated systems, this method has been perfected, and the drip irrigation system is now considerably more advantageous than traditional systems such as flood irrigation or sprinklers [77–79]. The main contribution of this system is that it enables a substantial saving in the use of water for irrigation, which enables the development or expansion of agricultural activity in arid and semiarid regions, where it would not be possible otherwise [80,81]. Another advantage is that it can prevent evaporation, as it supplies water directly to the roots of plants [82]. Different studies show that the use of drip irrigation increases the marketable yield and quality of crops and stabilizes production when deficit irrigation is used and that fertigation through drip irrigation helps to reduce the use of fertilizers and, therefore, the risk of pollution due to leachate [80,83]. Salvador and Aragüés [84] analyzed the advantages and disadvantages of the use of underground drip irrigation systems. They demonstrated their usefulness, profitability, and sustainability and indicated that the design, handling, and maintenance of this system, together with the quality of the irrigation water and type of soil, are fundamental aspects that determine their sustainability. On the other hand, Puy et al. [85] indicated that this type of system can have harmful consequences in terms of the degradation of the soil or the production of greenhouse gas emissions.

Deficit irrigation was introduced as a measure to limit the vegetative growth of crops [86]. This irrigation technique has been fully developed, and it is used extensively [87]. This method has been used with both drip irrigation and microsprinkling on different crops and can be combined with remote sensing technology or infrared techniques to produce significant water savings while crop yields remain unaffected. Du et al. [88] analyzed the use of deficit irrigation as a sustainable strategy for managing water resources in agriculture for food security in China. These authors concluded that the current understanding of physiological processes enables the deficit irrigation methods to be adjusted to different crops and environments in order to increase water use efficiency and the yield and quality of crops. Many studies have been carried out on this subject area [89–93].

Though many authors support drip irrigation as a sustainability measure, some recent studies question it. Perry et al. [94] confirm the "zero-sum game" hypothesis which argues that the impact of high-technology watering in a farm increases the demand of local water and land production at the expense of water availability and production in other places. Furthermore, due to the advantageous effects of drip irrigation, it makes water more affordable and, at the time, it allows irrigating larger areas, obtaining greater profits, and shift to more valuable crops. The most foreseeable impact of water efficiency improvement will be the increase of current water demands. In this sense, water scarcity would remain difficult to manage. Paul et al. [95], in their review of the rebound effects on the management of land and cultivation soils, found evidence for the presence of rebound effects and the Jevons paradox, together with productivity increases and efficiency of irrigation water due to technological innovations. Further studies agree with these results [96,97].

3.7.4. Water Use Efficiency

All of these innovations have the objective of improving water use efficiency for irrigation. In the year 2000, Kofi Annan, the Secretary General of the United Nations, proposed a "Blue Revolution in Agriculture" that was proposed to be capable of increasing productivity per unit of water. This strategy became known by the slogan "more crop per drop" [24]. According to Yang [98], obtaining the ideal water efficiency for irrigating crops involves the reduction of losses caused by evaporation, runoff, and underground draining while increasing production. Zhang et al. [26] indicated that the use of technology to save irrigation water not only saves water and increases production but also improves the nutritional value of agricultural products and guarantees food safety by improving the environmental conditions. Water use efficiency in agriculture generally implies a reduction in water use to meet a specific production objective or to increase the production of a specific water supply [99]. The aim of improving water use efficiency is to increase food production, boost financial gains, and guarantee the supply of ecosystem services at lower social and environmental costs per unit of water used [100,101]. The practices used to achieve this objective include rainwater harvesting, complementary irrigation, deficit irrigation, and the use of precision irrigation techniques and practices to conserve groundwater [24,102]. The priority areas where it is possible to significantly increase the productivity of water include areas with a high level of poverty and a low level of water productivity; areas with physical water shortage, where competition for water is high; areas with limited development of water resources, where the high yields of additional water have a considerable impact; and areas with degraded ecosystems driven by water, such as depleting water tables and dried-up rivers [103,104].

Among the different improvements developed over the last few decades, the use of drip irrigation has been fundamental in the improvement of water use efficiency and saving. Different studies have shown that drip irrigation has a water-saving potential of between 18% and 75%. According to Narayanamoorthy [82], drip irrigation saves an average of 25 to 75% of water compared to flood irrigation. Similar results were found, although with different percentages, in studies by Ibragimov et al. [105], Maisiri et al. [106], Yazar et al. [107], or Peterson and Ding [108], Abdulai et al. [109], Cremades et al. [110], and Jalota et al. [111].

4. Conclusions

This study presented the dynamics of global research in sustainable irrigation in agriculture over the last two decades, the main agents promoting it, and the topics that have received the most attention. The main concerns stated in the Introduction section related to the improvement of irrigation water use in order to increase food production, the world overexploitation of water resources, and the effects of global climate change. Our analysis verified how these questions are addressed by countries taking into account interdisciplinary approaches, and it also proved how these questions are mirrored in the main research lines on SI.

The results of the analysis of the principal variables revealed that the study of sustainable irrigation has grown in recent years in all of the variables considered: articles, authors, journals, institutions, and countries. Despite the fact that the growth trend in this topic is higher than that of general research in irrigation, an even greater research effort using a sustainability-based approach is required to further knowledge in this area. Traditionally, studies on sustainability have focused on one of the areas of which it is composed, namely, the environmental, social or economic dimensions. In the study of irrigation, the dominant area has been the environmental dimension, far more than the social or economic perspectives. The studies that analyzed just one of these dimensions provide highly useful information, but this information is only partial. It is necessary to integrate the three aspects of sustainability in order to gain full knowledge of the feasibility of certain practices, not only in terms of their environmental impacts but also with respect to income generation for farmers and the wellbeing of the community.

The keyword study revealed the existence of diversity between studies carried out using specific approaches and in different countries. In general, the study of environmental impacts and climate change, water availability, the improvement in efficiency, sustainable development, food supply, and the conservation of water bodies, particularly aquifers that have deteriorated, are common themes. However, certain practices, such as deficit irrigation or drip irrigation and aspects related to energy consumption and certain crops, are priority issues for particular countries. The methodological approaches used and the tools applied are other points of differentiation of the research carried out by each country. The keyword analysis showed four main research lines on SI: climatic change, environmental impact, and natural resource conservation; unconventional water resources; irrigation technology and innovation; and water use efficiency. Due to the large number of analyzed documents and the scope of this work, an in-depth content analysis per topic has not been undertaken. It will be highly interesting for future studies in order to provide more detailed information of these four specific topics.

As a final conclusion, we believe that certain aspects of the research on sustainable irrigation in agriculture in each of the dimensions of sustainability should be promoted. From a technical point of view, innovation and technology have furthered the development of irrigation systems and new available water sources that can contribute to improving the efficiency of water use and the sustainability of rural areas, particularly agricultural activity in arid regions. However, effort should be made to make this technology accessible, as its cost is economically unfeasible for small-scale agriculture in many countries. New water sources, such as those derived from desalination, reuse and rainwater harvesting systems, are very expensive for farmers compared to traditional sources. The production processes for desalination and reuse should be improved, particularly with respect to energy consumption in order to bring down the final price of the water. Furthermore, although the use of these nonconventional water resources has proved to have a series of advantages for the crops and the soil, this knowledge has not been transmitted to the farmers, and therefore, they are still reluctant to use it for irrigation. Greater effort should be made to communicate the results of the research to society. Finally, greater knowledge of the environmental impacts of irrigation-related practices in different areas on plot, district, basin and regional levels is needed. Water bodies are connected to each other, so certain practices that generate a small impact on river source areas can have a multiplying effect and be experienced in the underground bodies of coastal areas.

Author Contributions: The four authors have equally contributed to this paper. All authors have revised and approved the final manuscript.

Funding: This research received no external funding.

Acknowledgments: This work was partially supported by the Spanish Ministry of Economy and Competitiveness and the European Regional Development Fund by means of the research project ECO2017-82347-P, and by the Research Plan of the University of Almería through a Postdoctoral Contract to Juan F. Velasco Muñoz.

Conflicts of Interest: The authors declare no conflict of interest.

References

1. Hossain, M.S.; Pogue, S.J.; Trenchard, L.; Van Oudenhoven, A.P.E.; Washbourne, C.L.; Muiruri, E.W.; Tomczyk, A.M.; García-Llorente, M.; Hale, R.; Hevia, V.; et al. Identifying future research directions for biodiversity, ecosystem services and sustainability: Perspectives from early-career researchers. *Int. J. Sustain. Dev. World Ecol.* **2018**, *25*, 249–261. [CrossRef]
2. Manju, S.; Sagar, N. Renewable energy integrated desalination: A sustainable solution to overcome future fresh-water scarcity in India. *Sustain. Energy. Rev.* **2017**, *73*, 594–609. [CrossRef]
3. Millennnium Ecosystem Assessment (MA). *Ecosystems and Human Well-Being: Biodiversity Synthesis*; World Resources Institute: Washington, DC, USA, 2005.
4. Wang, M.H.; Li, J.; Ho, Y.S. Research articles published in water resources journals: A bibliometric analysis. *Desalin. Water Treat.* **2011**, *28*, 353–365. [CrossRef]
5. Flávio, H.M.; Ferreira, P.; Formigo, N.; Svendsen, J.C. Reconciling agriculture and stream restoration in Europe: A review relating to the EU Water Framework Directive. *Sci. Total Environ.* **2017**, *596–597*, 378–395. [CrossRef] [PubMed]
6. Zhang, Y.; Chen, H.; Lu, J.; Zhang, G. Detecting and predicting the topic change of Knowledge-based Systems: A topic-based bibliometric analysis from 1991 to 2016. *Knowl. Based Syst.* **2017**, *133*, 255–268. [CrossRef]
7. Damkjaer, S.; Taylor, R. The measurement of water scarcity: Defining a meaningful indicator. *Ambio* **2017**, *46*, 513–531. [CrossRef] [PubMed]
8. Liu, J.; Wang, Y.; Yu, Z.; Cao, X.; Tian, L.; Sun, S.; Wu, P. A comprehensive analysis of blue water scarcity from the production, consumption, and water transfer perspectives. *Ecol. Indic.* **2017**, *72*, 870–880. [CrossRef]
9. Forouzani, M.; Karami, E. Agricultural water poverty index and sustainability. *Agron. Sustain. Dev.* **2011**, *31*, 415–432. [CrossRef]
10. Fu, H.Z.; Wang, M.H.; Ho, Y.S. Mapping of drinking water research: A bibliometric analysis of research output during 1992–2011. *Sci. Total Environ.* **2013**, *443*, 757–765. [CrossRef]
11. Pedro-Monzonís, M.; Solera, A.; Ferrer, J.; Estrela, T.; Paredes-Arquiola, J. A review of water scarcity and drought indexes in water resources planning and management. *J. Hydrol.* **2015**, *527*, 482–493. [CrossRef]
12. Adeyemi, O.; Grove, I.; Peets, S.; Norton, T. Advanced monitoring and management systems for improving sustainability in precision irrigation. *Sustainability* **2017**, *9*, 353. [CrossRef]
13. Hedley, C.B.; Knox, J.W.; Raine, S.R.; Smith, R. Water: Advanced irrigation technologies. In *Encyclopedia of Agriculture and Food Systems*, 2nd ed.; Elsevier Academic Press: San Diego, CA, USA, 2014; pp. 378–406. ISBN 978-0-444-52512-3.

14. Zhang, Y.; Zhang, Y.; Shi, K.; Yao, X. Research development, current hotspots, and future directions of water research based on MODIS images: A critical review with a bibliometric analysis. *Environ. Sci. Pollut. Res. Int.* **2017**, *24*, 15226–15239. [CrossRef] [PubMed]

15. Gago, J.; Douthe, C.; Coopman, R.E.; Gallego, P.P.; Ribas-Carbo, M.; Flexas, J.; Escalona, J.; Medrano, H. UAVs challenge to assess water stress for sustainable agriculture. *Agric. Water Manag.* **2015**, *153*, 9–19. [CrossRef]

16. Wu, W.; Ma, B. Integrated nutrient management (INM) for sustaining crop productivity and reducing environmental impact: A review. *Sci. Total Environ.* **2015**, *512–513*, 415–427. [CrossRef] [PubMed]

17. De Fraiture, C.; Wichelns, D. Satisfying future water demands for agriculture. *Agric. Water Manag.* **2010**, *97*, 502–511. [CrossRef]

18. Fischer, G.; Tubiello, F.N.; van Velthuizen, H.; Wiberg, D.A. Climate change impacts on irrigation water requirements: Effects of mitigation, 1990–2080. *Technol. Forecast. Soc.* **2007**, *74*, 1083–1107. [CrossRef]

19. Matchaya, G.; Nhamo, L.; Nhlengethwa, S.; Nhemachena, C. An Overview of Water Markets in Southern Africa: An Option for Water Management in Times of Scarcity. *Water* **2019**, *11*, 1006. [CrossRef]

20. Graveline, N. Combining flexible regulatory and economic instruments for agriculture water demand control under climate change in Beauce. *Water Resour. Econ.* **2019**, 100143. [CrossRef]

21. Jothibasu, A.; Anbazhagan, S. Hydrogeological assessment of the groundwater aquifers for sustainability state and development planning. *Environ. Earth Sci.* **2018**, *77*, 88. [CrossRef]

22. Singh, A. Conjunctive use of water resources for sustainable irrigated agriculture. *J. Hydrol.* **2014**, *519*, 1688–1697. [CrossRef]

23. Aznar-Sánchez, J.A.; Belmonte-Ureña, L.J.; Velasco-Muñoz, J.V.; Valera, D.L. Aquifer Sustainability and the Use of Desalinated Seawater for Greenhouse Irrigation in the Campo de Níjar, Southeast Spain. *Int. J. Environ. Res. Public Health* **2019**, *16*, 898. [CrossRef] [PubMed]

24. Morison, J.I.L.; Baker, N.R.; Mullineaux, P.M.; Davies, W.J. Improving water use in crop production. *Philos. Trans. R. Soc. Lond. B Biol. Sci.* **2017**, *363*, 639–658. [CrossRef] [PubMed]

25. Melo-Zurita, M.L.; Thomsen, D.C.; Holbrook, N.J.; Smith, T.F.; Lyth, A.; Munro, P.G.; de Bruin, A.; Seddaiu, G.; Roggero, P.P.; Baird, J.; et al. Global water governance and climate change: Identifying innovative arrangements for adaptive transformation. *Water* **2018**, *10*, 29. [CrossRef]

26. Zhang, B.; Fu, Z.; Wang, J.; Zhang, L. Farmers' adoption of water-saving irrigation technology alleviates water scarcity in metropolis suburbs: A case study of Beijing, China. *Agric. Water Manag.* **2019**, *212*, 349–357. [CrossRef]

27. Komiyama, H.; Takeuchi, K. Sustainability science: Building a new discipline. *Sustain. Sci.* **2006**, *1*, 1–6. [CrossRef]

28. Yarime, M.; Takeda, Y.; Kajikawa, Y. Towards institutional analysis of sustainability science: A quantitative examination of the patterns of research collaboration. *Sustain. Sci.* **2010**, *5*, 115–125. [CrossRef]

29. Juwana, I.; Muttil, N.; Perera, B.J.C. Indicator-based water sustainability assessment—A review. *Sci. Total Environ.* **2012**, *438*, 357–371. [CrossRef]

30. Becker, B. *Sustainability Assessment: A Review of Values, Concepts and Methodological Approaches*; Issues in Agriculture 10; World Bank-Consultative Group on International Agriculture Research (CGIAR): Washington, DC, USA, 1997.

31. Mancosu, N.; Snyder, R.L.; Kyriakakis, G.; Spano, D. Water scarcity and future challenges for food production. *Water* **2015**, *7*, 975–992. [CrossRef]

32. Ioris, A.A.R.; Hunter, C.; Walker, S. The development and application of water management sustainability indicators in Brazil and Scotland. *J. Environ. Manag.* **2008**, *88*, 1190–1201. [CrossRef]

33. Ward, F.A.; Michelsen, A. The economic value of water in agriculture: Concepts and policy applications. *Water Policy* **2002**, *4*, 423–446. [CrossRef]

34. Huang, L.; Zhang, Y.; Guo, Y.; Zhu, D.; Porter, A.L. Four dimensional science and technology planning: A new approach based on bibliometrics and technology roadmapping. *Technol. Forecast. Soc. Chang.* **2014**, *81*, 39–48. [CrossRef]

35. Aznar-Sánchez, J.A.; Belmonte-Ureña, L.J.; López-Serrano, M.J.; Velasco-Muñoz, J.F. Forest ecosystem services: An analysis of worldwide research. *Forests* **2018**, *9*, 453. [CrossRef]

36. Li, W.; Zhao, Y. Bibliometric analysis of global environmental assessment research in a 20-year period. *Environ. Impact Assess. Rev.* **2015**, *50*, 158–166. [CrossRef]

37. Aznar-Sánchez, J.A.; Belmonte-Ureña, L.J.; Velasco-Muñoz, J.F.; Manzano-Agugliaro, F. Economic analysis of sustainable water use: A review of worldwide research. *J. Clean Prod.* **2018**, *198*, 1120–1132. [CrossRef]

38. Durieux, V.; Gevenois, P.A. Bibliometric Indicators: Quality Measurements of Scientific Publication. *Radiology* **2010**, *255*, 342. [CrossRef] [PubMed]

39. Velasco-Muñoz, J.F.; Aznar-Sánchez, J.A.; Belmonte-Ureña, L.J.; López-Serrano, M.J. Advances in water use efficiency in agriculture: A bibliometric analysis. *Water* **2018**, *10*, 377. [CrossRef]

40. Aznar-Sánchez, J.A.; Velasco-Muñoz, J.F.; Belmonte-Ureña, L.J.; Manzano-Agugliaro, F. The worldwide research trends on water ecosystem services. *Ecol. Indic.* **2019**, *99*, 310–323. [CrossRef]

41. Tancoigne, E.; Barbier, M.; Cointet, J.P.; Richard, G. The place of agricultural sciences in the literature on ecosystem services. *Ecosyst. Serv.* **2014**, *10*, 35–48. [CrossRef]

42. Velasco-Muñoz, J.V.; Aznar-Sánchez, J.A.; Belmonte-Ureña, L.J.; Román-Sánchez, I.M. Sustainable water use in agriculture: A review of worldwide research. *Sustainability* **2018**, *10*, 1084. [CrossRef]

43. Hassan, S.U.; Haddawy, P.; Zhu, J. A bibliometric study of the world's research activity in sustainable development and its sub-areas using scientific literature. *Scientometrics* **2014**, *99*, 549–579. [CrossRef]

44. Cossarini, D.M.; MacDonald, B.H.; Wells, P.G. Communicating marine environmental information to decision makers: Enablers and barriers to use of publications (grey literature) of the Gulf of Maine Council on the Marine Environment. *Ocean Coastal Manag.* **2014**, *96*, 163–172. [CrossRef]

45. Aznar-Sánchez, J.A.; Velasco-Muñoz, J.F.; Belmonte-Ureña, L.J.; Manzano-Agugliaro, F. Innovation and technology for sustainable mining activity: A worldwide research assessment. *J. Clean Prod.* **2019**, *221*, 38–54. [CrossRef]

46. Cogato, A.; Meggio, F.; Migliorati, M.; Marinello, F. Extreme Weather Events in Agriculture: A Systematic Review. *Sustainability* **2019**, *11*, 2547. [CrossRef]

47. Cui, X.; Guo, X.; Wang, Y.; Wang, X.; Zhu, W.; Shi, J.; Lin, C.; Gao, X. Application of remote sensing to water environmental processes under a changing climate. *J. Hydrol.* **2019**, *574*, 892–902. [CrossRef]

48. Aznar-Sánchez, J.A.; Piquer-Rodríguez, M.; Velasco-Muñoz, J.F.; Manzano-Agugliaro, F. Worldwide research trends on sustainable land use in agriculture. *Land Use Pol.* **2019**, *87*, 104069. [CrossRef]

49. Zhou, X.Y. Spatial explicit management for the water sustainability of coupled human and natural systems. *Environ. Pollut.* **2019**, *251*, 292–301. [CrossRef] [PubMed]

50. Sillmann, J.; Roeckner, E. Indices for extreme events in projections of anthropogenic climate change. *Clim. Chang.* **2008**, *86*, 83–104. [CrossRef]

51. Mitrică, B.; Mitrică, E.; Enciu, P.; Mocanu, I. An approach for forecasting of public water scarcity at the end of the 21st century, in the Timiş Plain of Romania. *Technol. Forecast. Soc. Chang.* **2017**, *118*, 258–269. [CrossRef]

52. United Nations World Water Assessment Programme (WWAP). *Water for a Sustainable World*; The United Nations World Water Development Report; UNESCO: Paris, France, 2015.

53. Wichelns, D.; Oster, J.D. Sustainable irrigation is necessary and achievable, but direct costs and environmental impacts can be substantial. *Agric. Water Manag.* **2006**, *86*, 114–127. [CrossRef]

54. Hadas, A.; Hadas, A.; Sagiv, B.; Haruvy, N. Agricultural practices, soil fertility management modes and resultant nitrogen leaching rates under semi-arid conditions. *Agric. Water Manag.* **1999**, *42*, 81–95. [CrossRef]

55. Sears, L.; Caparelli, J.; Lee, C.; Pan, D.; Strandberg, G.; Vuu, L.; Lawell, C.Y. Jevons' Paradox and Efficient Irrigation Technology. *Sustainability* **2018**, *10*, 1590. [CrossRef]

56. Singh, A. Decision support for on-farm water management and long-term agricultural sustainability in a semi-arid region of India. *J. Hydrol.* **2010**, *391*, 63–76. [CrossRef]

57. Kögler, F.; Söffker, D. Water (stress) models and deficit irrigation: System-theoretical description and causality mapping. *Ecol. Model.* **2017**, *361*, 135–156. [CrossRef]

58. Richard-Ferroudji, A.; Faysse, N.; Bouzidi, Z.; Menon, R.T.P.; Rinaudo, J.D. The DIALAQ project on sustainable groundwater management: A transdisciplinary and transcultural approach to participatory foresight. *Curr. Opin. Environ. Sustain.* **2016**, *20*, 56–60. [CrossRef]

59. García-Caparrós, P.; Contreras, J.I.; Baeza, R.; Segura, M.L.; Lao, M.T. Integral management of irrigation water in intensive horticultural systems of Almería. *Sustainability* **2017**, *9*, 2271. [CrossRef]

60. Vanham, D.; Hoekstra, A.Y.; Wada, Y.; Bouraoui, F.; de Roo, A.; Mekonnen, M.M.; van de Bund, W.J.; Batelaan, O.; Pavelic, P.; Bastiaanssen, W.G.M.; et al. Physical water scarcity metrics for monitoring progress towards SDG target 6.4: An evaluation of indicator 6.4.2. "Level of water stress". *Sci. Total Environ.* **2018**, *613–614*, 218–232. [CrossRef] [PubMed]

61. Geoghegan-Quin, M. Role of Research & Innovation in Agriculture. 2013. European Commission-SPEECH/13/505. Available online: http://europa.eu/rapid/press-release_SPEECH-13-505_en.htm (accessed on 20 January 2019).

62. Aznar-Sánchez, J.A.; Belmonte-Ureña, L.J.; Valera, D.L. Perceptions and Acceptance of Desalinated Seawater for Irrigation: A Case Study in the Níjar District (Southeast Spain). *Water* **2017**, *9*, 408. [CrossRef]

63. Ghaffour, N.; Missimer, T.M.; Amy, G.L. Technical review and evaluation of the economics of water desalination: Current and future challenges for better water supply sustainability. *Desalination* **2013**, *309*, 197–207. [CrossRef]

64. Zepeda-Quintana, D.S.; Loeza-Rentería, C.M.; Munguía-Vega, N.E.; Esquer-Peralta, J.; Velazquez-Contreras, L.E. Sustainability strategies for coastal aquifers: A case study of the Hermosillo Coast aquifer. *J. Clean Prod.* **2018**, *195*, 1170–1182. [CrossRef]

65. Duarte, T.K.; Minciardi, R.; Robba, M.; Sacile, R. Optimal control of coastal aquifer pumping towards the sustainability of water supply and salinity. *Sustain. Water Qual. Ecol.* **2015**, *6*, 88–100. [CrossRef]

66. Martínez-Granados, D.; Calatrava, J. Combining economic policy instruments with desalinisation to reduce overdraft in the Spanish Alto Guadalentín aquifer. *Water Policy* **2017**, *19*, 341–357. [CrossRef]

67. Jorreto, S.; Sola, F.; Vallejos, A.; Sánchez-Martos, F.; Gisbert, J.; Molina, L.; Rigol, J.P.; Pulido-Bosch, A. Evolution of the geometry of the freshwater-seawater interface in a coastal aquifer affected by an intense pumping of seawater. *Geogaceta* **2017**, *62*, 87–90.

68. Quintana, J.; Tovar, J. Evaluación del acuífero de Lima (Perú) y medidas correctoras para contrarrestar la sobreexplotación. *Bol. Geológico Y Min.* **2002**, *113*, 303–312.

69. Assouline, S.; Shavit, U. Effects of management policies, including artificial recharge, on salinization in a sloping aquifer: The Israeli Coastal Aquifer case. *Water Resour. Res.* **2004**, *40*, W04101. [CrossRef]

70. Boisson, A.; Villesseche, D.; Baisset, M.; Perrin, J.; Viossanges, M.; Kloppmann, W.; Chandra, S.; Dewandel, B.; Picot-Colbeaux, G.; Rangarajan, R.; et al. Questioning the impact and sustainability of percolation tanks as aquifer recharge structures in semi-arid crystalline context. *Environ. Earth. Sci.* **2015**, *73*, 7711–7721. [CrossRef]

71. Khezzani, B.; Bouchemal, S. Variations in groundwater levels and quality due to agricultural over-exploitation in an arid environment: The phreatic aquifer of the Souf oasis (Algerian Sahara). *Environ. Earth Sci.* **2018**, *77*, 142. [CrossRef]

72. Rupérez-Moreno, C.; Senent-Aparicio, J.; Martinez-Vicente, D.; García-Aróstegui, J.L.; Cabezas-Calvo-Rubio, F.; Pérez-Sánchez, J. Sustainability of irrigated agriculture with overexploited aquifers: The case of Segura basin (SE, Spain). *Agric. Water Manag.* **2017**, *182*, 67–76. [CrossRef]

73. Salcedo-Sánchez, E.R.; Esteller, M.V.; Garrido-Hoyos, S.E.; Martínez-Morales, M. Groundwater optimization model for sustainable management of the Valley of Puebla aquifer, Mexico. *Environ. Earth Sci.* **2013**, *70*, 337–351. [CrossRef]

74. Palacios-Vélez, O.L.; Escobar-Villagrán, B.S. La sustentabilidad de la agricultura de riego ante la sobreexplotación de acuíferos. *Tecnol. Y Cienc. del Agua* **2016**, *7*, 5–16.

75. Reca, J.; Trillo, C.; Sánchez, J.A.; Martínez, J.; Valera, D. Optimization model for on-farm irrigation management of Meditarranean greenhouse crops using desalinated and saline water from different sources. *Agric. Syst.* **2018**, *166*, 173–183. [CrossRef]

76. De Wrachien, W.; Medicia, M.; Lorenzini, G. The Great Potential of Micro-Irrigation Technology for Poor-Rural Communities. *Irrigat. Drain. Syst. Eng.* **2014**, *3*, e124. [CrossRef]

77. Guerbaoui, M.; Afou, Y.; Ed-Dahhak, A.; Lachhab, A.; Bouchikhi, B. Pc-based automated drip irrigation system. *Int. J. Eng. Sci. Technol.* **2013**, *5*, 221–225.

78. De Wrachien, W.; Lorenzini, G.; Medici, M. Sprinkler irrigation systems: State-of-the-art of kinematic analysis and quantum mechanics applied to water jets. *Irrig. Drain.* **2013**, *62*, 407–413. [CrossRef]

79. Lorenzini, G.; Saro, O. Thermal fluid dynamic modelling of a water droplet evaporating in air. *Int. J. Heat Mass Transf.* **2013**, *62*, 323–335. [CrossRef]

80. Surendran, U.; Jayakumar, M.; Marimuthu, S. Low cost drip irrigation: Impact on sugarcane yield, water and energy saving in semiarid tropical agro ecosystem in India. *Sci. Total Environ.* **2016**, *573*, 1430–1440. [CrossRef] [PubMed]

81. Kalpakian, J.; Legrouri, A.; Ejekki, F.; Doudou, K.; Berrada, F.; Ouardaoui, A.; Kettani, D. Obstacles facing the diffusion of drip irrigation technology in the Middle Atlas region of Morocco. *Int. J. Environ. Stud.* **2014**, *71*, 63–75. [CrossRef]

82. Narayanamoorthy, A. Economic Viability of Drip Irrigation: An Empirical Analysis from Maharashtra. *Indian J. Agric. Econ.* **1997**, *52*, 728–739.

83. Lamm, F.R. Cotton, tomato, corn, and onion production with subsurface drip irrigation: A review. *Trans. ASABE* **2016**, *59*, 263–278. [CrossRef]

84. Salvador, R.; Aragüés, R. Estado de la cuestión del riego por goteo enterrado: Diseño, manejo, mantenimiento y control de la salinidad del suelo. *ITEA-Inf. Tec. Econ. Agrar.* **2013**, *109*, 395–407. [CrossRef]

85. Puy, A.; García-Avilés, J.M.; Balbo, A.L.; Keller, M.; Riedesel, S.; Blum, D.; Bubenzer, O. Drip irrigation uptake in traditional irrigated fields: The edaphological impact. *J. Environ. Manag.* **2017**, *202*, 550–561. [CrossRef]

86. Holzapfel, E.A.; Pannunzio, A.; Lorite, I.; Silva de Oliveira, A.; Farkas, I. Design and Management of Irrigation Systems. *Chil. J. Agric. Res.* **2009**, *69*, 17–25. [CrossRef]

87. Fereres, E.; Soriano, M.A. Deficit irrigation for reducing agricultural water use. *J. Exp. Bot.* **2007**, *58*, 147–159. [CrossRef] [PubMed]

88. Du, T.; Kang, S.; Zhang, J.; Davies, W.J. Deficit irrigation and sustainable water-resource strategies in agriculture for China's food security. *J. Exp. Bot.* **2015**, *66*, 2253–2269. [CrossRef] [PubMed]

89. Girona, J.; Mata, M.; Marsal, J. Regulated deficit irrigation during the kernel-filling period and optimal irrigation rates in almond. *Agric. Water Manag.* **2005**, *75*, 152–167. [CrossRef]

90. Hutmacher, R.B.; Nightingale, H.I.; Rolston, D.E.; Biggar, J.W.; Dale, F.; Vail, S.S.; Peters, D. Growth and yield responses of almond (Prunus amygdalus) to trickle irrigation. *Irrig. Sci.* **1994**, *14*, 117–126. [CrossRef]

91. Bassoi, L.H.; Hopmans, J.W.; de C. Jorge, L.A.; de Alencar, C.M.; Silva, J. Grapevine root distribution in drip and microsprinkler irrigation. *Sci. Agric.* **2003**, *60*, 377–387. [CrossRef]

92. Sepulcre-Cantó, G.; Zarco-Tejada, P.J.; Jiménez-Muñoz, J.C.; Sobrino, J.A.; de Miguel, E.; Villalobos, F.J. Detection of water stress in an olive orchard with thermal remote sensing imagery. *Agric. For. Meteorol.* **2006**, *136*, 31–44. [CrossRef]

93. Falkenberg, N.; Piccinni, G.; Cothren, J.T.; Leskovar, D.I.; Rush, C.M. Remote sensing of biotic and abiotic stress for irrigation management of cotton. *Agric. Water Manag.* **2007**, *87*, 23–31. [CrossRef]

94. Perry, C.; Steduto, P.; Karajeh, F. *Does Improved Irrigation Technology Save Water? A Review of the Evidence*; Food and Agriculture Organization of the United Nations: Cairo, Egypt, 2017.

95. Paul, C.; Techen, A.; Robinson, J.; Helming, K. Rebound effects in agricultural land and soil management: Review and analytical framework. *J. Clean. Prod.* **2019**, *227*, 1054–1067. [CrossRef]

96. Grafton, R.Q.; Williams, J.; Perry, C.J.; Molle, F.; Ringler, C.; Steduto, P.; Udall, B.; Wheeler, S.A.; Wang, Y.; Garrick, D.; et al. The paradox of irrigation efficiency. *Science* **2018**, *361*, 748–750. [CrossRef]

97. Berbel, J.; Pedraza, V.; Giannoccaro, G. The trajectory towards basin closure of a European river: Guadalquivir. *Int. J. River Basin Manag.* **2013**, *11*, 111–119. [CrossRef]

98. Yang, C. Technologies to improve water management for rice cultivation to cope with climate change. *Crop Environ. Bioinform.* **2012**, *9*, 193–207.

99. Ma, H.; Shi, C.; Chou, N. China's water utilization efficiency: An analysis with environmental considerations. *Sustainability* **2016**, *8*, 516. [CrossRef]

100. Boutraa, T. Improvement of water use efficiency in irrigated agriculture: A review. *J. Agron.* **2010**, *9*, 1–8. [CrossRef]

101. Xue, J.; Guan, H.; Huo, Z.; Wang, F.; Huang, G.; Boll, J. Water saving practices enhance regional efficiency of water consumption and water productivity in an arid agricultural area with shallow groundwater. *Agric. Water Manag.* **2017**, *194*, 78–89. [CrossRef]

102. Attwater, R.; Derry, C. Achieving resilience through water recycling in peri-urban agriculture. *Water* **2017**, *9*, 223. [CrossRef]

103. Molden, D.; Oweis, T.; Steduto, P.; Bindraban, P.; Hanjra, M.A.; Kijne, J. Improving agricultural water productivity: Between optimism and caution. *Agric. Water Manag.* **2010**, *97*, 528–535. [CrossRef]

104. Fang, S.; Jia, R.; Tu, W.; Sun, Z. Assessing factors driving the change of irrigation water-use efficiency in China based on geographical features. *Water* **2017**, *9*, 759. [CrossRef]

105. Ibragimov, N.; Evett, S.R.; Esanbekov, Y.; Kamilov, B.S.; Mirzaev, L.; Lamers, J.P.A. Water use efficiency of irrigated cotton in Uzbekistan under drip and furrow irrigation. *Agric. Water Manag.* **2007**, *90*, 112–120. [CrossRef]

106. Maisiri, N.; Sanzanje, A.; Rockstrom, J.; Twomlow, S.J. On farm evaluation of the effect of low cost drip irrigation on water and crop productivity compared to conventional surface irrigation system. *Phys. Chem. Earth* **2005**, *30*, 783–791. [CrossRef]

107. Yazar, A.; Sezen, S.M.; Sesveren, S. LEPA and trickle irrigation of cotton in the Southeast Anatolia Project (GAP) area in Turkey. *Agric. Water Manag.* **2002**, *54*, 189–203. [CrossRef]

108. Peterson, J.M.; Ding, Y. Economic adjustments to groundwater depletion in the high plains: Do water-saving irrigation systems save water? *Am. J. Agric. Econ.* **2005**, *87*, 147–159. [CrossRef]

109. Abdulai, A.; Owusu, V.; Bakang, J.E.A. Adoption of safer irrigation technologies and cropping patterns: Evidence from Southern Ghana. *Ecol. Econ.* **2011**, *70*, 1415–1423. [CrossRef]

110. Cremades, R.; Wang, J.; Morris, J. Policies, economic incentives and the adoption of modern irrigation technology in China. *Earth Syst. Dyn.* **2015**, *6*, 399–410. [CrossRef]

111. Jalota, S.K.; Singh, K.B.; Chahal, G.B.S.; Gupta, R.K.; Chakraborty, S.; Sood, A.; Ray, S.S.; Panigrahy, S. Integrated effect of transplanting date, cultivar and irrigation on yield, water saving and water productivity of rice (Oryza sativa L.) in Indian Punjab: Field and simulation study. *Agric. Water Manag.* **2009**, *96*, 1096–1104. [CrossRef]

 *water*

Article

Water Resources Allocation Systems under Irrigation Expansion and Climate Change Scenario in Awash River Basin of Ethiopia

Mohammed Gedefaw [1,2,*], Hao Wang [2], Denghua Yan [2,*], Tianling Qin [2], Kun Wang [2], Abel Girma [1,2], Dorjsuren Batsuren [1,3] and Asaminew Abiyu [1]

[1] College of Environmental Science & Engineering, Donghua University, Shanghai 200336, China; abelethiop@yahoo.com (A.G.); batsuren@seas.num.edu.mn (D.B.); asaminew@yahoo.com (A.A.)

[2] State Key Laboratory of Simulation and Regulation of Water Cycle in River Basin, China Institute of Water Resource and Hydropower Research, Beijing 100038, China; wanghao@iwhr.com (H.W.); tianling406@163.com (T.Q.); pingguo88wangkun@163.com (K.W.)

[3] Department of Environment and Forest Engineering, School of Engineering and Applied Sciences, National University of Mongolia, Ulaanbaatar 210646, Mongolia

* Correspondence: mohammedgedefaw@gmail.com (M.G.); yandh@iwhr.com (D.Y.)

Received: 11 July 2019; Accepted: 17 September 2019; Published: 20 September 2019

Abstract: Rational allocation of water resources is very essential to cope with water scarcity. The optimal allocation of limited water resources is required for various purposes to achieve sustainable development. The Awash River Basin is currently faced with a scarcity of water due to increasing demands, urbanization, irrigation expansion, and variability of climates. The excessive abstraction of water resources in the basin without proper assessing of the available water resources contributed to water scarcity. This paper aimed to develop a water evaluation and planning (WEAP) model to allocate the water supplies to demanding sectors based on an economic parameter to maximize the economic benefits. The water demands, water shortages, and supply alternatives were analyzed under different scenarios. Three scenarios were developed, namely reference (1981–2016), medium-term development (2017–2030), and long-term development (2031–2050) future scenarios with the baseline period (1980). The results of this study showed that the total quantity of water needed to meet the irrigation demands of all the stations was 306.96 MCM from 1980 to 2016. Seasonally, March, April, May, and June require the maximum irrigation water demand. However, July, August, and September require minimum demand for water because of the rainy season. The seasonal unmet demand is observed in all months, which ranged from 6×10^6 m^3 to 35.9×10^6 m^3 in August and May respectively. The trend of streamflow in Melka Kuntre was a statistically significant increasing trend after 2008 (Z = 5.33) whereas the trends in other gauge stations showed a relatively decreasing trend. The results also showed that future water consumption would greatly increase in the Awash River Basin. The prevention of future water shortages requires the implementation of water-saving measures and the use of new water supply technologies. The findings of this study will serve as a reference for water resources managers and policy and decision makers.

Keywords: water allocation; WEAP model; scenario; climate change; Awash River Basin

---

1. Introduction

The increasing demand of water resources in the world is the main problem for the sustainable utilization of water resources [1]. Water scarcity is mainly caused by over-exploitation of water resources, population growth, pollution, and increasing demand for economic development [2,3]. The ever-increasing population and economic development put more stress on the hydrological cycle and water resources, particularly in the river basin [1]. These pressures also cause the decline of available water resources in the basin. Climate change has also affected the hydrological cycles and water resources [4,5]. Recently, water resources allocation got much attention across the globe since climate change and population growth pushed to their natural limits [6]. Still, the scope of the problem becomes a research area in many regions of the world [7–16].

Efficient and optimal allocation of water resources plays a great role in balancing the demand and supply of water resources based on economic development [2]. However, the allocation fails to meet the acquired demand when the water demand exceeds the available water. The problems become worse with increasing water demand and economic growth.

The Awash River Basin is the most utilized basin in Ethiopia. It covers a total area of 114,123 km^2 that encompasses five regional states; Amhara, Oromia, Afar, SNNP, and Somali including two administrative councils, Addis Ababa and Dire Dawa [17]. The mean annual rainfall of the basin ranges from 100 to 1700 mm with great spatiotemporal variation. The basin also has a potential of 8.2 BCM and 10.3 BCM surface and groundwater, respectively, with 300 m exploration [18]. The temporal variation shares 71% and 29% of the rainy season (June to October) and dry season (November to May) respectively. The Awash River is the biggest contributor of water for the Awash River Basin. The river basin water is becoming scarce due to increasing demands and poor water resources management [19]. Hence, there is a need to allocate the water among the demanding sectors and to build water storage systems such as water harvesting, reservoirs, and dams. Water resources management models play a great role in addressing the water shortage of river basins by formulating prior allocation of water rights [20–22].

Nowadays, a water resource is modeled with various models such as MODSIM [23], WAS [20], CWAM [24], WEAP [25], and MOEA [6]. In this study, the water evaluation and planning (WEAP) model was chosen as it incorporates different hydrological components in data scarce areas, and because of its flexibility, simplicity, inclusiveness, and possibility of modeling the impact of climate change scenario on reservoirs and evaluating water resources using a scenario-based system. The model also simulated domestic, irrigation, and ecological water consumption in time and space as compared to other allocation models. The WEAP model resolves problems faced by water resources managers and planners using a scenario-based system by providing a set of objects and procedures which can be applied to reservoirs, river basins, and watersheds [19,26].

Previously, several distributed hydrological models have been applied to the Awash River Basin to model the hydrological characteristics. For example, Berhe et al. (2012), modeled the Awash River Basin using the MODSIM based allocation model under three different scenarios and the findings suggested that the model could be effective for water allocation [23]. Adeba et al. (2015), also tried to assess the water scarcity of the basin using the SWAT model [25]. Similarly, Mesfin et al. (2018), modeled the upper Awash River Basin using the SWAT model to evaluate the climate forecast system reanalysis weather data for watershed modeling [27]. Karimi, P., et al. (2015) [28], investigated the spatial evapotranspiration, rainfall, and land use data in water accounting that focused on the impact of the error in remote sensing measurements on water accounting and information provided to policymakers.

The studies done so far do not accurately model the water resources of the basin to allocate and address the water scarcity. Therefore, this study assessed the water scarcity of the basin under irrigation expansion and climate change scenarios for sustainable availability of water in the basin in the future. Furthermore, the study strengthened the concept that improving integrated water resources management through optimal and efficient water resources allocation is the key to overcoming the water shortage during dry periods.

Therefore, this paper aimed to develop a water allocation model to allocate the water supplies to demanding sectors based on an economic parameter to maximize the economic benefits.

2. Materials and Methods

2.1. Description of the Study Area

The Awash River Basin is one of the 12 river basins of Ethiopia which is found between latitudes of 7°53′ N and 12° N and longitudes of 37°57′ E and 43°25′ E [29]. The basin constitutes the central and northern part of the Rift Valley and is bounded to the west, southeast, and south by the Blue Nile, the Rift Valley lakes, and the Wabeshebele basins, respectively [17]. It covers a total area of 110,000 km^2, with a length of 1200 km [23]. The basin is a home of about 15 million inhabitants [17]. This basin has been the most highly utilized basin in Ethiopia since modern agriculture was introduced, as early as the 1950s [30]. The basin is divided into upper, middle, and lower valleys. The mean annual rainfall of the basin varies from 1600 mm northeast of Addis Ababa to 160 mm in the northern part of the basin (Figure 1). The distribution of rainfall is bimodal in the middle and lower parts of the basin and unimodal in the upper part [29]. The distribution of rainfall in the basin is influenced by the Inter Tropical Convergence Zone (ITCZ).

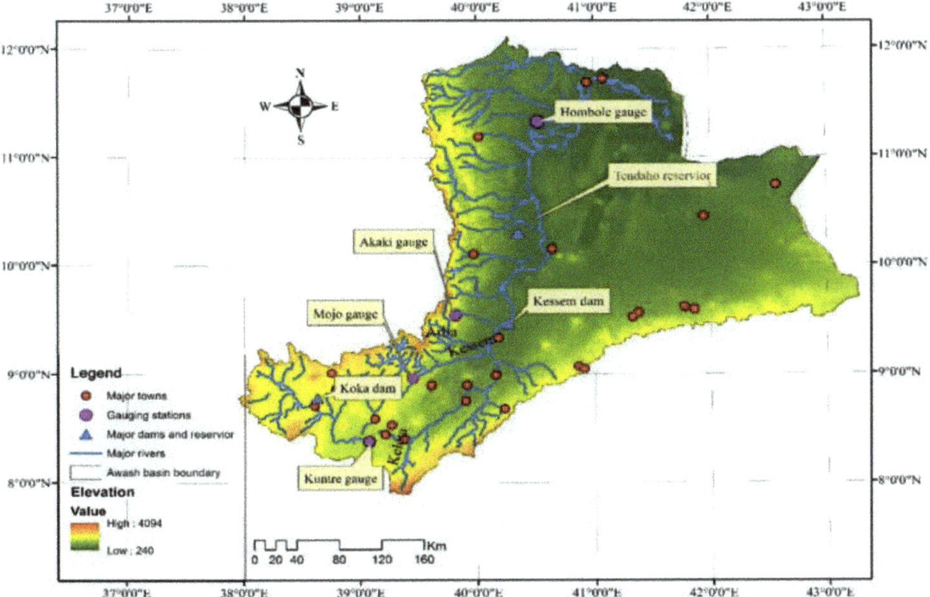

Figure 1. Location map of the Awash River Basin.

The mean surface water resource of the Awash River Basin is approximately 4.9 × 10^8 m^3 [31]. Irrigation used 44% from the surface water resources. More than 70% of large-scale irrigated agriculture in Ethiopia is found in this basin. The irrigation potential of the basin is estimated to be 206,000 ha as reported in the ministry of water and energy office. The total mean annual evaporation is 1810 mm and 2348 mm in the upper and lower parts of the basin [31]. The estimated mean annual runoff within the basin is about 4.6 km^3 [17]. Many rivers are functional only in rainy seasons (July to September) especially in lowland parts of the basin. However, the Mojo, Akaki, Kessem, Kebena, and Mile rivers are functional throughout the year. Since the population are highly dependent on rainfed agriculture, this has made the population and the economy vulnerable to impacts of climate change and droughts [32].

2.2. Sources of Data

Different datasets were used to establish the WEAP hydrological model for the study basin (Table 1). These data include the Digital Elevation Model (DEM) of the Awash River Basin, hydrological, climate, remote sensing, and water consumptions. All the necessary data for this manuscript were provided after quality control. The stations were also selected based on completeness of the data during the study periods. The land use data were obtained from the Awash River Basin master plan office. Previous studies also obtained the raw data from the same place for their research works.

Table 1. Input datasets of the water evaluation and planning (WEAP) modeling in the Awash River Basin.

Data Item	Description	Sources
Meteorological data (1980–2016)	Precipitation, temperature	National Meteorological Agency of Ethiopia
Hydrological data (1980–2014)	Reservoirs, data of Gauging stations	Department of water resources and hydrology of Ethiopia
Remote sensing data	Digital Elevation Model (DEM) of Awash River Basin	Department of GIS and Remote sensing
Water demand data	✓ Water use rate ✓ Population number ✓ Water consumption ✓ Agricultural sector ✓ Urban sector ✓ Land use data	Ministry of water resources and energy of Ethiopia

The climatic data such as precipitation, temperature, wind speed, and humidity were collected from the Ethiopian National Meteorological Agency (NMA). The streamflow/discharge data was acquired from the department of water resources and hydrology, which was used for calibration and validation of the basin. Water use, population, and other data were collected from various socioeconomic surveys and the statistical agency of Ethiopia, which are essential to analyze the water demand, water coverage, and unmet demand of water in the basin. The irrigation water demand of irrigated sites was also obtained from the basin authority office, as well as literature to compute the water requirement and water scarcity.

2.3. Methods

This study aims to develop the WEAP model for optimal allocation of water resources among competing sites. The WEAP model is one of the powerful tools used to evaluate the existing and planned water resources development in a given watershed. The model is used to identify water scarcity areas that cause conflict and can simulate a water allocation policy. Water allocation priority rules are set within the WEAP model based on either first come first served, or specific use or user, and/or making allocation proportional to demand. As it is generic, the model is not capable of capturing every fine distinction or detail of a water resource system and as such is best applied to scenario screening and pre and feasibility levels of analysis rather than to detailed design and permitting tasks. To allow simulation of water allocation, the elements that comprise the water demand–supply system and their spatial relationship are characterized for the catchment under consideration. It also helps to understand the available water resources demand for current and future development scenarios. The WEAP model also helps to provide a system for maintaining water demand and supply information in addition to the simulation of demand and flows. The hydrological systems in the WEAP model of Awash River Basin are depicted as nodes and links. The main river is drawn as a series of nodes, showing points of inflows from each catchment and river confluence linked to each other by river reaches.

It also calibrates and validates the streamflow in four gauging stations (Akaki, Hombole, Melka Kuntre, and Modjo) in the Awash River Basin. Three scenarios were established after developing the WEAP model to predict the water demands until 2050. The model was run on a monthly basis. The baseline or current scenario 1980–2016 was set to estimate the irrigation water demand. The WEAP model for the Awash River Basin was set up to simulate the current/base year (1980) condition and three subsequent scenarios: the reference scenario (1981–2016), the medium-term development (2017–2030), and the long-term development (2031–2050). For each development scenario, the outputs analyzed include the water demand satisfaction of the irrigation demanding sector and spatiotemporal variations in water shortage. The methodology of the study is shown in (Figure 2).

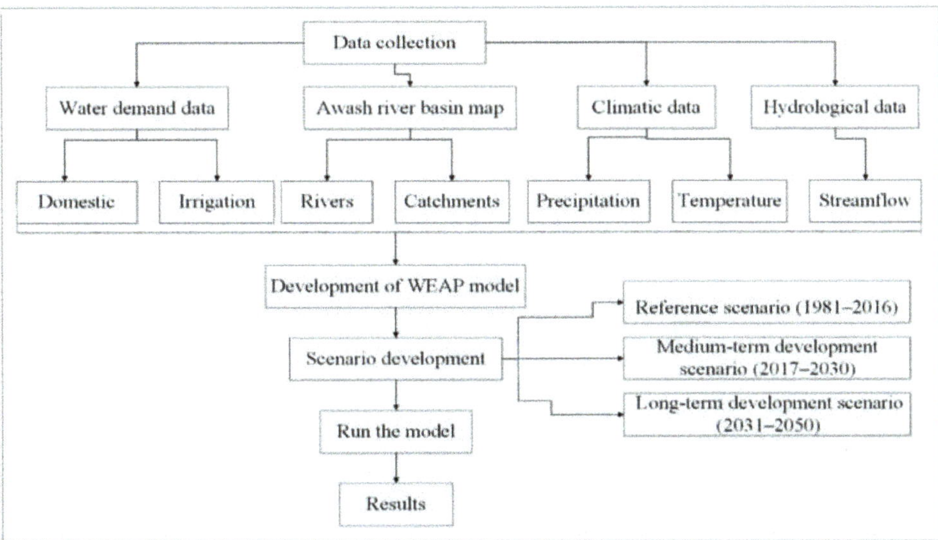

Figure 2. Methods of estimating water demands through the WEAP model.

The WEAP model provides an integrated assessment of climate, hydrology, water resources allocation, and watershed management [33]. It also addresses various issues such as water resources, water demands analysis in different sectors, provides priorities in water allocation, reservoir operation, and management. It solves the water allocation challenges at user-defined periods, either monthly or yearly based on linear programming structures [34].

2.4. Modeling Set up and Key Assumptions

1. All the demanding sites are given equal priority in the provision of water regardless of differences in financial returns expected from each scheme.
2. The model also contains four streamflow gauging stations, six transmission links, and six runoff/infiltration lines.
3. Six irrigation demand sites are also included in the model (Figure 3).
4. The model includes one main river (Awash River) and small tributary rivers (Figure 4).

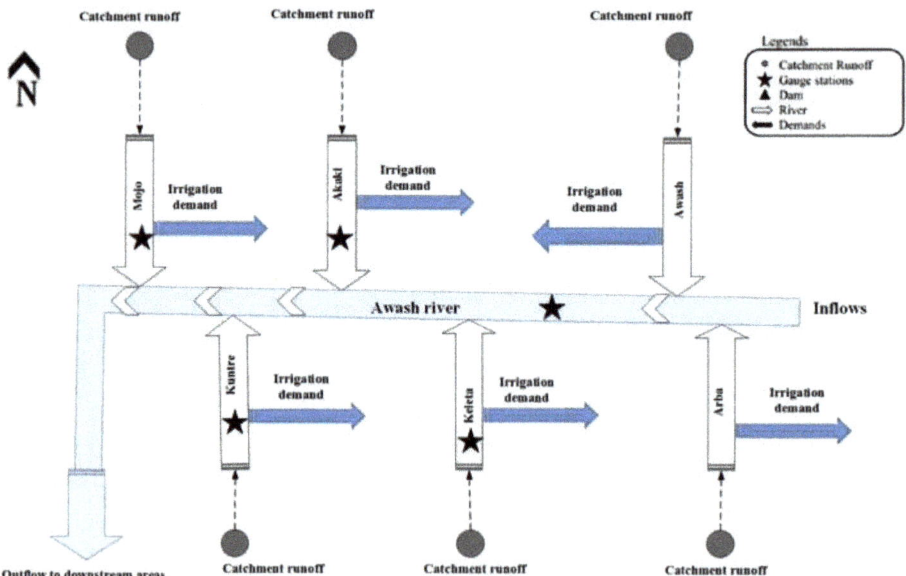

Figure 3. Water supply service structure and irrigation demands in the catchment.

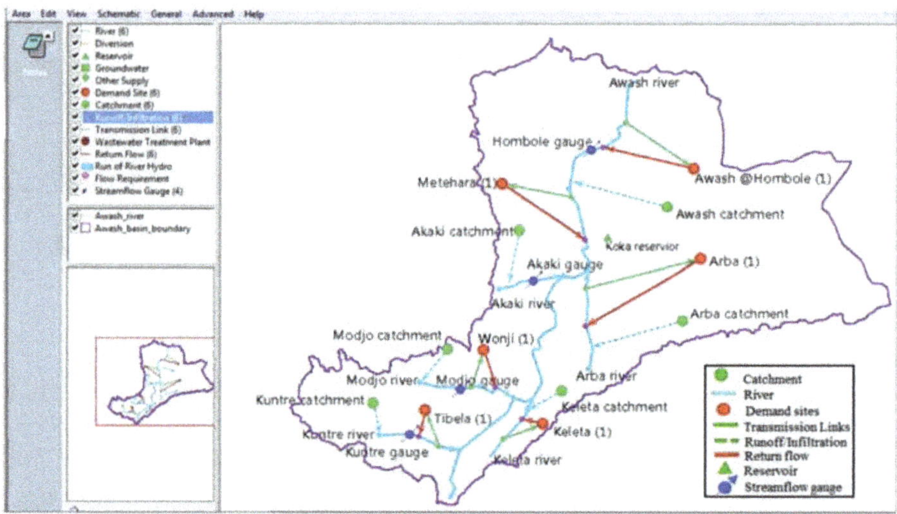

Figure 4. A snapshot of the WEAP model configuration showing the demand sites in the Awash River Basin.

3. Results

3.1. Analysis of Irrigation Water Demand under Irrigation Expansion Scenario

The simulation of the WEAP model provides the mean monthly and annual demands for downstream irrigation schemes. The scenario showed the full utilization of the irrigation potential from the Koka reservoir for 58,660 ha of land including the recent expansion coverage of 36,266 ha of irrigated land.

The total quantity of water needed to meet the irrigation demands of all the selected stations was 306.96 MCM from 1980 to 2016 under the current scenario. Seasonally, March, April, May, and June require the maximum irrigation water demand. However, July, August, and September require the minimum demand for water because of the Kiremt (wet) season. The extremely high unmet demand for water in May for the Tibela site was observed. This is probably due to a shortage of rainfall in Tibela since May is characterized as a dry month in the area. The average monthly supply of water delivered for each station are illustrated in Figure 5 without including the losses.

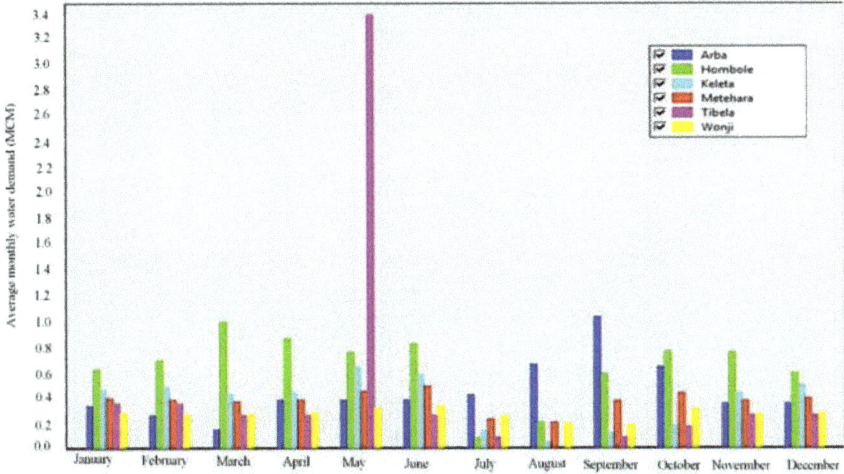

Figure 5. Mean monthly current demands without including losses.

3.2. Spatiotemporal Occurrence of Unmet Demand under Each Irrigation Site

The results of this study showed that the seasonal unmet demand was observed in all months of the year. This is ranged from 6×10^6 m^3 to 35.9×10^6 m^3 in August and May, respectively. The maximum water shortage occurred in May and declined relatively to an optimal level in July, September, and October. This indicates that, except during the rainy season (July to September), the water demand is not adequately met for each station (Table 2).

Table 2. Unmet demand across stations (MCM).

Irrigation Schemes	Arba	Hombole	Keleta	Metehara	Tibela	Wonji	Total
January	0.44	0.82	0.62	0.52	0.47	0.37	3.23
February	0.34	0.92	0.64	0.50	0.47	0.35	3.23
March	0.20	1.31	0.57	0.49	0.34	0.35	3.26
April	0.50	1.14	0.58	0.50	0.34	0.37	3.44
May	0.50	1.00	0.84	0.59	4.51	0.43	7.88
June	0.50	1.08	0.77	0.65	0.34	0.44	3.79
July	0.56	0.11	0.18	0.31	0.11	0.33	1.60
August	0.87	0.28	0.08	0.27	0.00	0.26	1.76
September	1.37	0.77	0.16	0.50	0.11	0.24	3.16
October	0.84	1.01	0.24	0.58	0.23	0.40	3.30
November	0.47	1.00	0.58	0.50	0.34	0.35	3.24
December	0.47	0.78	0.65	0.65	0.34	0.36	3.13
Total	7.06	10.23	5.90	5.90	7.62	4.27	41.0

3.3. Analysis of Unmet Demands under Each Scenario

All scenarios showed an increase of unmet demand throughout the simulation years. The gap between the current reference scenario and the long-term future development scenario was very high during simulation periods (Table 3). The demand was fully satisfied during wet/rainy seasons and high-water shortage happened in dry seasons. A significant increase of unmet demand was observed in the Wonji and Tibela demanding sites.

Table 3. Temporal occurrence of unmet demand across the basin (MCM).

Scenario	Reference (1980–2016)	Medium Term Development (2017–2030)	Long Term Development (2031–2050)
January	0.09	6.30	6.30
February	0.09	6.39	6.39
March	0.10	7.03	7.03
April	0.11	7.63	7.63
May	0.12	8.27	8.27
June	0.13	9.57	9.57
July	9.40	12.57	12.57
August	16.50	24.40	24.40
September	0.23	16.40	16.40
October	0.11	7.75	7.75
November	0.10	6.91	6.91
December	0.09	6.43	6.43
Summary	**27.07**	**119.65**	**119.65**

3.4. Analysis of Water Demand under Climate Scenario

The results showed that the average temperature was increased in the long-term climate scenario from the baseline scenario. It was increased from 22.08 °C (1980–2016) to 24.04 °C (2017–2030) and 26.49 °C for the long-term future development (2030–2050) scenario. The increased average temperature was mainly observed in the month of May.

The average temperature was increased by 1.96 °C and 4.41 °C in 2017–2030 and 2030–2050 from the reference scenario, respectively. Thus, the change in average temperature would cause a change in the reservoir surface temperature and significant evaporation loss.

The average precipitation in the basin also showed a fluctuation from the reference scenario. The average precipitation was 1733.33, 2433.33, and 2525.00 mm under the reference, medium-term development, and medium-term development future scenario, respectively. The precipitation would be increasing in the long-term development future scenario (2030–2050) as compared to the other two scenarios.

As far as the trends of evaporation in the reservoir are concerned, the modeled results showed decreasing and increasing trends in some months. The maximum and minimum evaporation was observed as 53.8 and 16.5 Mm^3 in October and August, respectively. The evaporation under each scenario was also 33.73, 35.21, and 35.71 Mm^3 in the 1980–2016, 2017–2030, and 2030–2050 scenarios, respectively. The simulation results of annual evaporation in each scenario was 404.80, 422.50, and 428.50 Mm^3, respectively. The increase of evaporation in the reservoir was due to the increase of temperature under the climate change scenario. The climate change scenario with respect to temperature, precipitation, and evaporation is shown in (Figure 6).

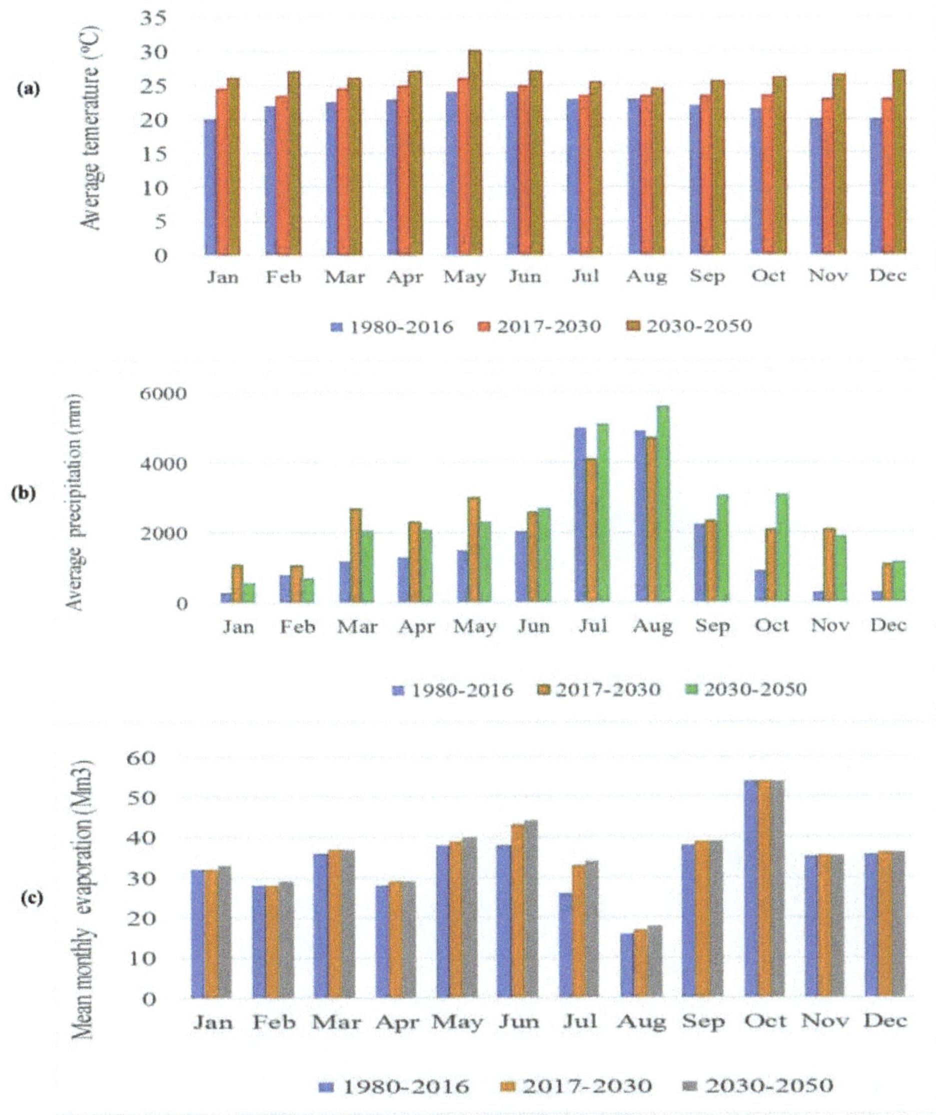

Figure 6. Average monthly temperature, precipitation, and evaporation under the three scenarios. (**a**) Average temperature, (**b**) average precipitation, and (**c**) mean monthly evaporation.

3.5. Model Calibration and Validation

The water balance of the basin was estimated based on climatic data analysis from 1980–2016. The base year is considered from January to December 2016, which is taken as a normal hydrological year. The results showed that the total water balance of the basin during the study period was found to be 427 MCM. The values of parameters for the calibration results in the WEAP model is shown (Table 4).

Table 4. The calibration parameters' possible ranges used in the WEAP model.

Parameters	Model Ranges	Optimal Range	Unit
Soil water capacity	0–higher	0–1200	mm
Root zone conductivity	Default = 20	10–50	Mm/month
Deep water conductivity	0.1–higher	20	Mm/month
Runoff resistance factor	0–1000 (default = 20)	0–100	
Preferred flow direction	0–1 (default = 0.15)	0.5–1	
Initial Z_1	0–100		%
Initial Z_2	0–100		%

The WEAP model was calibrated and validated with monthly observed streamflow from 1980–1999 and 2000–2014, respectively (Figure 7). In order to simulate the streamflow values in the model, the crop coefficient of (kc) of different land covers, soil water capacity, root zone conductivity, and preferred flow direction were manually calibrated using the default values of the WEAP model (Table 4). Hence, during the calibration period of this study, the values of the coefficient of determination (R^2) ranged from 0.73 to 0.91, with an average value of 0.82. On the other hand, the value of the Nash–Sutcliffe efficiency (NSE) was 0.65 and 0.83, the minimum and maximum values, respectively, with an average value of 0.74. However, during validation periods, 0.54 and 0.91 are the minimum and maximum values of the coefficient of determination (R^2), with an average value of 0.73. As far as the values of the Nash–Sutcliffe efficiency (NSE), 0.50 and 0.84 were the minimum and maximum values respectively, with an average value of 0.67. From these values, we have observed that in both calibration and validation periods, the values correspond to a perfect match between the observed and the modeled streamflow values. Thus, the performance of the model is acceptable between the trends of observed and simulated streamflow in both calibration and validation of the study basin. This also helps to accurately project the prediction of future discharges based on the future scenario set. The statistical values of Akaki (R^2 = 0.73; NSE = 0.65) were less than other gauging stations; Kuntre (R^2 = 0.85; NSE = 0.73), Hombole (R^2 = 0.89; NSE = 0.83), and Modjo (R^2 = 0.91; NSE = 0.85) (Table 5). This may be due to urbanization.

Table 5. Statistical values of stations for calibration and validation.

Gauging Stations	Calibration		Validation	
	R2	NSE	R^2	NSE
Hombole	0.89	0.83	0.91	0.84
Melka Kuntre	0.85	0.73	0.86	0.73
Akaki	0.73	0.65	0.63	0.5
Modjo	0.91	0.85	0.54	0.68

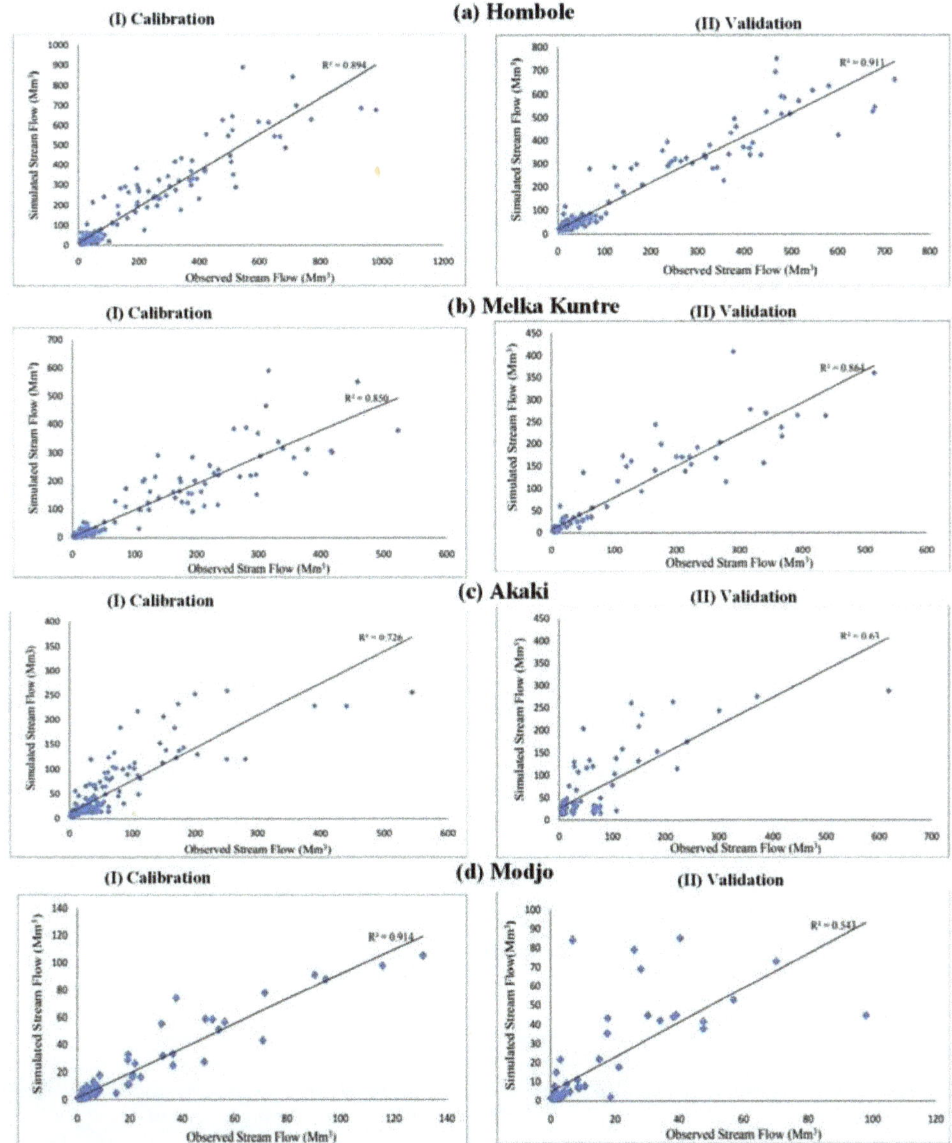

Figure 7. Calibration and validation of the four gauging stations and the entire Awash River Basin.

3.6. Shortage of Water in the Basin

There was a spatiotemporal variation of streamflow across the catchments during the study periods. Catchments which faced water shortage were identified in this analysis. The water shortage is very critical to meet the water demand of the sites in the Awash River Basin. All the demanding sites in the Awash River Basin were unmet at the end of the long-term development scenario (2030–2050). The annual water deficiency of Arba, Hombole, Keleta, Metehara, Tibela, and Wonji was 261.07, 378.43, 218.14, 219.74, 281.86, and 157.86 MCM, respectively. As far as temporal water deficiency is concerned, the biggest water shortage was observed in May.

4. Discussion

The findings of this study showed that the overall unmet demand under the long-term development future scenario by 2050 was from 6×10^6 m^3 to 35.9×10^6 m^3 in August and May, respectively. Water deficiency was observed in the dry season, especially in May. All scenarios showed an increase of unmet demand throughout the simulation years. The water demand was fully satisfied under each scenario because of the priority given for water allocation for each demanding site. Water shortage is aggravated by the expansion of irrigation lands and this, in turn, resulted in the failure of production. The spatiotemporal unmet demand of each demand site under the current scenario is shown (Figure 8). The population in the Awash River Basin is projected to increase and put more pressure on the limited water resources of the basin. With this trend of water consumption, the basin could face more water deficiency in the future. Sustainable and integrated water resources management approaches in the basin may overcome the observed problems that could affect the water resources. This may be done by developing reservoirs to store more water for the dry, low-flow season and create awareness among the public to use the water efficiently and sustainably. Strategic exploitation of additional water supplies and a paradigm reformation of policy of the basin management, which encourages sustainable use of water resources of the basin, should be done to improve the water shortage.

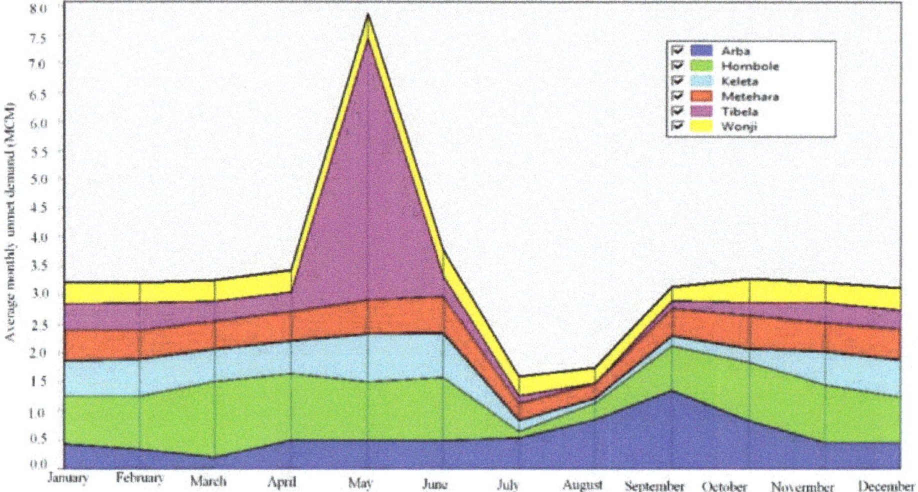

Figure 8. Mean monthly unmet demands of the current scenario for each irrigation sites.

Decrease of streamflow at the main outlet of the Awash River Basin could cause high pressure on available water resources. Thus, all the basin sections are impacted by the risks and cause conflicts between upstream and downstream dwellers [31]. No measurement is taken by the concerned entity of the basin to enhance the streamflow of the basin. The Awash River Basin has the most suitable space for future development of irrigation and will likely face huge population increases and industrial expansion. This will cause increased water consumption and large waterwork capacity deficits for the basin. As a result, prioritizing building and expansion of more water supply lines or reservoirs should be considered. Normally, warmer and wetter scenarios of the Awash River Basin are expected to increase the river discharge substantially and could serve to alleviate current local water shortages. The results of this study are consistent with other findings, such as [27,30,34,35]. The WEAP model in this study could not model reservoir water quality and cannot account for stream attenuation. Therefore, this can be a future research work for researchers to model the reservoir water quality by developing appropriate hydrological models in the study basin.

5. Conclusions

This study formulated the water allocation networks under an irrigation expansion and climate change scenario using the WEAP model for the Awash River Basin. The study also calibrated and validated the streamflow successfully with a reasonable range of R^2 and NSE. The annual water balance of the basin was also determined by considering the main parameters that could affect the water availability of the basin. Population growth, urbanization, industrialization, and agricultural intensification further magnify the conflicts of water among users. Equitable allocation of water among competing sites and finding alternatives to improve the water availability of the basin must be given attention. Water harvesting, soil, and water conservation to minimize the rate of runoff and digging boreholes are some options to enhance the water capacity of the basin.

The present study showed that the basin is faced with water shortage when meeting the requirements of the competing users. Previous studies also showed similar results, which was a water shortage of 1.27 BCM/year in 2011 and 2.82 BCM/year in 2012 [25]. The results of the present study also provide insights into the vulnerability of the available water resources of the Awash River Basin. Therefore, this study gives a direction for decision and policy makers to maintain the existing water resources and reduce further ecological threats that prevail in the Awash River Basin due to the scarcity of water resources.

Author Contributions: M.G. made substantial contributions to the design, idea generating, analysis, interpretation, and drafting of the original manuscript. D.Y. commented on the draft manuscript and supervised the whole work. H.W. was a resource person of the project. T.Q. and K.W. participated in the methodology and software, and A.G., D.B., and A.A. participated in designing this study. The final manuscript before submission was checked and approved by all the authors.

Funding: This research was funded by the National Key Research and Development Project (Grant No. 2016YFA0601503).

Acknowledgments: The authors would like to thank the National Meteorological Service Agency of Ethiopia for providing the raw meteorological data.

References

1. Cai, Y.; Xu, M.; Wang, X.; Yue, W.; Li, C. Optimal water utilization and allocation in industrial sectors based on water footprint accounting in Dalian City, China. *J. Clean. Prod.* **2017**, *176*, 1283–1291.
2. Divakar, L.; Babel, M.S.; Perret, S.R.; Gupta, A.D. Optimal allocation of bulk water supplies to competing use sectors based on economic criterion—An application to the Chao Phraya River Basin, Thailand. *J. Hydrol.* **2011**, *401*, 22–35. [CrossRef]
3. Roozbahani, R.; Schreider, S.; Abbasi, B. Optimal water allocation through a multi-objective compromise between environmental, social, and economic preferences. *Environ. Model. Softw.* **2015**, *64*, 18–30. [CrossRef]
4. Wang, X.; Luo, Y.; Sun, L.; Zhang, Y. Assessing the effects of precipitation and temperature changes on hydrological processes in a glacier-dominated catchment. *Hydrol. Process.* **2015**, *29*, 4830–4845. [CrossRef]
5. Pettinotti, L.; de Ayala, A.; Ojea, E. Benefits From Water Related Ecosystem Services in Africa and Climate Change. *Ecol. Econ.* **2018**, *149*, 294–305. [CrossRef]
6. Yan, D.; Ludwig, F.; Huang, H.Q.; Werners, S.E. Many-objective robust decision making for water allocation under climate change. *Sci. Total Environ.* **2017**, *607*, 294–303. [CrossRef] [PubMed]
7. Bangash, R.F.; Passuello, A.; Hammond, M.; Schuhmacher, M. Water allocation assessment in low flow river under data scarce conditions: A study of hydrological simulation in Mediterranean basin. *Sci. Total Environ.* **2012**, *440*, 60–71. [CrossRef]
8. Ramsar Convention Secretariat. Water allocation and management planning: Guidelines for the allocation and management of water for maintaining the ecological functions of wetlands. In *Ramsar Handbooks for Wise Use of Wetland*; Ramsar Convention Secretariat: Gland, Switzerland, 2010; Volume 10.
9. Hu, Z.; Wei, C.; Yao, L.; Li, L.; Li, C. A multi-objective optimization model with conditional value-at-risk constraints for water allocation equality. *J. Hydrol.* **2016**, *542*, 330–342. [CrossRef]

10. Alauddin, M. Optimization of Water-Allocation Networks with Multiple Contaminants using Genetic Algorithm. *Int. J. Biol. Chem. Sci.* **2014**, *1*, 7–14.

11. Huang, Q.; Huo, Z.; Xu, X.; Huang, G.; Jiang, Y. Assessment of irrigation performance and water productivity in irrigated areas of the middle Heihe River basin using a distributed agro-hydrological model. *Agric. Water Manag.* **2014**, *147*, 67–81.

12. Jiang, Y.; Xu, X.; Huang, Q.; Huo, Z.; Huang, G. Optimizing regional irrigation water use by integrating a two-level optimization model and an agro-hydrological model. *Agric. Water Manag.* **2016**, *178*, 76–88. [CrossRef]

13. Leonard, L.; Duffy, C.J. Essential terrestrial variable data workflows for distributed water resources modeling. *Environ. Model. Softw.* **2013**, *50*, 85–96. [CrossRef]

14. Li, M.; Fu, Q.; Singh, V.P.; Liu, D. An interval multi-objective programming model for irrigation water allocation under uncertainty. *Agric. Water Manag.* **2018**, *196*, 24–36. [CrossRef]

15. Yandamuri, S.R.; Srinivasan, K.; Murty Bhallamudi, S. Multiobjective Optimal Waste Load Allocation Models for Rivers Using Nondominated Sorting Genetic Algorithm-II. *J. Water Resour. Plan. Manag.* **2006**, *132*, 133–143. [CrossRef]

16. Ren, C.; Guo, P.; Tan, Q.; Zhang, L. A multi-objective fuzzy programming model for optimal use of irrigation water and land resources under uncertainty in Gansu Province, China. *J. Clean. Prod.* **2017**, *164*, 85–94. [CrossRef]

17. Gedefaw, M.; Wang, H.; Yan, D.; Song, X.; Yan, D.; Dong, G.; Wang, J.; Girma, A.; Ali, B.; Batsuren, D.; et al. Trend Analysis of Climatic and Hydrological Variables in the Awash River Basin. *Ethiopia* **2018**, *10*, 1554. [CrossRef]

18. Mekonen, A.; Gebremeskel, T.; Mengistu, A.; Fasil, E.; Melkamu, M. Irrigation water pricing in Awash River Basin of Ethiopia: Evaluation of its impact on scheme-level irrigation performances and willingness to pay. *Afr. J. Agric. Res.* **2015**, *10*, 554–565. [CrossRef]

19. Davijani, M.H.; Banihabib, M.E. Multi-Objective Optimization Model for the Allocation of Water Resources in Arid Regions Based on the Maximization of Socioeconomic Efficiency. *Water Resour. Manag.* **2016**, *30*, 927–946. [CrossRef]

20. Yan, Z.; Zhou, Z.; Sang, X.; Wang, H. Water replenishment for ecological flow with an improved water resources allocation model. *Sci. Total Environ.* **2018**, *643*, 1152–1165. [CrossRef]

21. Masih, I.; Uhlenbrook, S.; Turral, H.; Karimi, P. Analysing streamflow variability and water allocation for sustainable management of water resources in the semi-arid Karkheh river basin, Iran. *Phys. Chem. Earth* **2009**, *34*, 329–340. [CrossRef]

22. Yang, Z.F.; Sun, T.; Cui, B.S.; Chen, B.; Chen, G.Q. Environmental flow requirements for integrated water resources allocation in the Yellow River Basin, China. *Commun. Nonlinear Sci. Numer. Simul.* **2009**, *14*, 2469–2481. [CrossRef]

23. Berhe, F.T.; Melesse, A.M.; Hailu, D.; Sileshi, Y. Catena MODSIM-based water allocation modeling of Awash River Basin, Ethiopia. *Catena* **2013**, *109*, 118–128. [CrossRef]

24. Wang, L.; Fang, L.; Hipel, K.W. Basin-wide cooperative water resources allocation. *Eur. J. Oper. Res.* **2008**, *190*, 798–817. [CrossRef]

25. Adeba, D. Assessment of water scarcity and its impacts on sustainable development in Awash basin, Ethiopia. *Sustain. Water Resour. Manag.* **2015**, *1*, 71–87. [CrossRef]

26. Ki, B.; Mengistu, G.; Hendrik, G. Climate Risk Management Climate change and population growth impacts on surface water supply and demand of Addis Ababa, Ethiopia. *Clim. Risk Manag.* **2017**, *18*, 21–33. [CrossRef]

27. Basin, A. Evaluation of the Climate Forecast System Reanalysis Weather Data for Watershed Modeling in Upper Awash Basin, Ethiopia. *Water* **2018**, *10*, 725. [CrossRef]

28. Karimi, P.; Bastiaanssen, W.G.M.; Sood, A.; Hoogeveen, J.; Peiser, L.; Bastidas-Obando, E.; Dost, R.J. Spatial evapotranspiration, rainfall and land use data in water accounting—Part 2: Reliability of water acounting results for policy decisions in the Awash Basin. *Hydrol. Earth Syst. Sci.* **2015**, *19*, 533–550. [CrossRef]

29. Hailu, R.; Tolossa, D.; Alemu, G. Water institutions in the Awash basin of Ethiopia: The discrepancies between rhetoric and realities. *Int. J. River Basin Manag.* **2018**, *16*, 107–121. [CrossRef]

30. Edossa, D.C.; Babel, M.S.; Gupta, A.D. Drought analysis in the Awash River Basin, Ethiopia. *Water Resour. Manag.* **2010**, *24*, 1441–1460. [CrossRef]

31. Mersha, A.N. Evaluating the Impacts of IWRM Policy Actions on Demand Satisfaction and Downstream Water Availability in the Upper Awash Basin, Ethiopia. *Water* **2018**, *10*, 892. [CrossRef]
32. Gedefaw, M.; Yan, D.; Wang, H.; Qin, T.; Wang, K. Analysis of the Recent Trends of Two Climate Parameters over Two Eco-Regions of Ethiopia. *Water* **2019**, *11*, 161. [CrossRef]
33. Xiao-jun, W.; Jian-yun, Z.; Elmahdi, A.; Rui-min, H.E. Water demand forecasting under changing environment: A System Dynamics approach. In *Risk in Water Resources Management, Proceedings of Symposium H03 held during IUGG2011, Melbourne, Australia, July 2011*; IAHS: Edinburgh, UK, 2011.
34. Adgolign, T.B.; Rao GV, R.S.; Abbulu, Y. WEAP modeling of surface water resources allocation in Didessa. *Sustain. Water Resour. Manag.* **2016**, *2*, 55–70. [CrossRef]
35. Hussen, B.; Mekonnen, A.; Murlidhar, S. Integrated water resources management under climate change scenarios in the sub-basin of Abaya-Chamo, Ethiopia. *Model. Earth Syst. Environ.* **2018**, *4*, 221–240. [CrossRef]

Article

Crop Performance and Water Productivity of Transplanted Rice as Affected by Seedling Age and Seedling Density under Alternate Wetting and Drying Conditions in Lao PDR

Rubenito Lampayan [1,2,*], Phetmanyseng Xangsayasane [3] and Crisanta Bueno [2,4]

[1] College of Engineering and Agro-Industrial Technology, University of the Philippines Los Baños, Laguna 4031, Philippines
[2] International Rice Research Institute (IRRI), Los Baños, Laguna 4031, Philippines
[3] Agricultural Research Center, National Agriculture and Forestry Research Institute, Napok, Vientiane 0101, Laos
[4] Institute of Crop Science (ICrops), College of Agriculture and Food Science, University of the Philippines Los Baños, Laguna 4031, Philippines
* Correspondence: rmlampayan@up.edu.ph; Tel.: +63-(049)-536-2387

Received: 9 July 2019; Accepted: 24 August 2019; Published: 31 August 2019

Abstract: Drought is common under rainfed lowlands in Lao People's Democratic Republic, and with the uncertain onset of rains during the wet season, delay in transplanting results in yield reduction. This study aims to explore ways to ameliorate the negative influence of delayed transplanting on rice crop. A field experiment was conducted for two wet seasons to investigate the effect of seedling age and seedling density on crop performance in terms of grain yield and water productivity. The experiment was laid out in a split–split plot design in four replicates, with seedling age as the main plot, seedling density as the subplot, and varieties as the sub-sub plot. In both years, there were significant seedling age and variety interactions on grain yield. Higher grain yields were observed with older seedlings having stronger tillering propensity. Seedling density did not affect grain yields in both years, but on grain yield components. Shorter duration variety received less supplemental irrigation than longer duration varieties. Late transplanting improved total water productivity but decreased irrigation water productivity due to harvesting delay. The total crop growth duration (from sowing to maturity) was prolonged with transplanting delay. However, the total stay of plants in the main field (from transplanting to maturity) was reduced by 3–5 d for every 10 d delay in transplanting. The results indicated that a good selection of varieties and increasing seedling density improve crop performance and water productivity with delayed transplanting.

Keywords: delayed transplanting; seedling age; seedling density; wet season

---

1. Introduction

Efficient water use in rice cultivation is a prerequisite to sustain food security for the rice-consuming population of the world. In recent years, increasing water scarcity has been a major threat to rice production in Asia, where by 2025, about 15–20 million of irrigated rice is estimated to suffer [1]. If today's food production and environmental trends continue, crises in many parts of the world will arise. Action should now be taken to improve water use in agriculture to address severe water challenges for the next 50 years [2].

Rice is a key staple in the Lao People's Democratic Republic (PDR) and is an important component of food security efforts in the country. Lao PDR has one of the world's highest per capita consumption of rice, with around 179 kg per capita per year recorded in 2007 [3]. Rice production in the country is the primary source of livelihood for 724,000 producers. The rainfed lowland rice system dominates with only 13% of the total area being irrigated, considering the total paddy harvested area of 830,000 ha in 2011 [4]. Availability and access to water have been identified as the major constraints to the improvement of rice-based farming systems. Lao PDR has seen a high incidence of significant floods and droughts, which severely affected agricultural production in the country [5].

Rice is the biggest user of water in agriculture and in fact, one of the biggest users of the world's fresh water resources [1]. Most rice fields are under conventional continuous flooded conditions [6], which leads to high amounts of surface runoff, seepage, and percolation losses that account for 50%–80% of total water input [7]. With decreasing water availability for agriculture and with increasing demand for rice, water input in rice production should be reduced and water productivity must be increased. Many water-saving practices, including alternate wetting and drying (AWD), have been identified and promoted for widescale dissemination in Asia to reduce water input and increase water productivity [8]. AWD has been successfully evaluated and introduced at the farmers' demonstration fields in the drought-prone southern provinces of Lao PDR in the 2011 and 2012 dry seasons, respectively [9]. Comparison between AWD and the farmers' water management practice of continuous flooding resulted to similar yields, but a 19%–25% water input reduction with AWD was observed. These results were consistent with what had been reported in other countries that tried the technology [8]. Seedling age at transplanting is an important factor to consider in attaining the uniform crop stand [10] and for regulating growth and yield [11]. When rice seedlings are transplanted at the right age, optimum tillering and growth are achieved. However, if transplanting is delayed, fewer tillers are produced during the vegetative stage resulting to poor yield [12]. "Delayed transplanted rice" or "rice with old seedling age" is the term usually used when transplanted seedling age is more than 25 d [13]. Delayed transplanted rice is common in rainfed lowland fields or in irrigated areas in Lao PDR [5]. The annual cropping cycle in Vientiane province begins either in May or June, depending on the onset of rains, with the preparation of the nursery seedbed and the sowing of seeds for the nursery. In Lao PDR, seedlings are usually transplanted about 30 d or more after sowing. However, the untimely release of irrigation in both dry and wet seasons in irrigated areas or the late onset of rainfall in rainfed areas in the wet season for land preparation and crop establishment activities result in a delay in the transplanting of seedlings. Delayed transplanting may result in yield reduction [13,14].

In this paper, we hypothesize that increasing seedling density at transplanting and use of varieties with stronger tillering propensity ameliorates the effects of delayed transplanting on crop performance. Moreover, delayed transplanting reduces irrigation requirement and increases irrigation water productivity during the rainy season. To test this hypothesis, a field experiment was conducted to evaluate the interactive effects of seedling age, seedling density, and variety on post-transplanted rice crop development, grain yield, and water productivity.

2. Materials and Methods

2.1. Site Description

A field experiment was conducted at the Agricultural Research Center (ARC) of the National Agriculture and Forestry Research Institute (NAFRI) in Vientiane, Lao PDR to investigate the effect of seedling age (SA) and seedling density (SD) on post-transplanting performance of selected waxy Lao rice varieties in terms of crop growth, grain yield, water input, and water productivity. The experimental area (16°29′ N, 104°49′ E) in the research center was characterized by loam soil (21% clay and 39% silt), which is typical for lowland rice. The soil properties are shown in Table 1. The area was previously used as a production plot and was cropped with lowland rice in both wet and dry seasons. Vientiane province has a monsoonal climate, with the southwest monsoons being associated with distinct wet (May–October) and dry (November–April) seasons. Mean annual rainfall at the experimental area (taken at a nearby agrometeorological station 200 m away) is about 1790 mm, of which 1500 mm (89% of total rainfall) falls during the wet season, and only about 197 mm (or 11%) falls during the dry season. Annual mean solar radiation in the area is about 20 MJ m^{-2} d^{-1}, with a range of 18–24 MJ m^{-2} d^{-1}. Temperatures increase gradually from around 28–30 °C in January to about 35 °C in April. Peak monthly maximum temperature is recorded in April, immediately before the start of the wet-season rains. The temperature remains above 30 °C between April and October and starts to decline from late October. Minimum temperatures also follow a similar pattern, with 18–20 °C in January to about 26–27 °C in April, then gradually declining from late October [5].

Table 1. Soil properties of the top soil layer (0–20 cm) in the experimental site.

Soil Property	Mean
% clay	21.0
% silt	39.0
% sand	40.0
pH	6.0
Organic C (%)	2.3
Total N (%)	6.9
Available P (mg kg^{-1})	28.4
Available K (mg kg^{-1})	29.1
CEC (meq 100^{-1}g)	4.5

2.2. Experimental Setup and Treatment Details

The field experiment was implemented for two wet seasons (2014 and 2015) and was laid out in a split–split plot design with three replicates. In this experiment, seedling age at the time of transplanting (SA) was assigned as the main plot, seedling density (SD) as the subplot, and variety (V) as the sub-subplot. Four levels of SA were used: 15 (SA_{15}), 25 (SA_{25}), 35 (SA_{35}), and 45 (SA_{45}) d-old seedlings; three levels for SD: One (SD_1), three (SD_3), and five (SD_5) seedlings per hill; and three different waxy or glutinous varieties: IRUBN0300–63–5–4 (V_1), TDK10239–SSD4–303–1 (V_2), and TDK-8 (V_3). V_1 was considered a high-tillering variety, while V_2 and V_3 were relatively low and medium tillering varieties, respectively. V_3 (TDK-8) is a variety widely grown in southern Lao PDR because of its good eating quality (used as check variety in this experiment), while V_1 and V_2 were two of the most promising waxy rice varieties tested at ARC. A total of 108 plots (4 SA × 3 SD × 3 V × 3 replications) were established in each year of the field experimentation, with a sub-sub plot size of 12 m^2 (2 m × 6 m). Only the main plot (SA) was separated by bunds, but a buffer space of about 0.5 m was established in each sub-subplot.

The experimental field was the same for both years and was prepared with one dry plowing (after all crop residues were removed from the field), followed by land soaking and two wet harrowing operations to achieve thorough puddling in each year. Bunds and canals were constructed and plastic linings were installed to 40 cm depth in the sides of the bunds to reduce seepage losses around the main plot. Transplanting schedules were based on seedling age treatments. At a 10 d transplanting interval, SA_{15} was transplanted first, then followed by SA_{25}, SA_{35}, and SA_{45} treatments, respectively. Transplanting dates during 2014 were the following: 1 July (SA_{15}), 10 July (SA_{25}), 20 July (SA_{35}), and 30 July (SA_{45}); whereas during 2015, transplanting dates were 2 July (SA_{15}), 12 July (SA_{25}), 23 July (SA_{35}), and 2 August (SA_{45}). Transplanting of seedlings was carefully done at a regular spacing of 20 cm × 20 cm in all treatments to minimize seedling transplanting shock and to allow the plants to recover quickly in the experimental field. The number of seedlings planted per hill varied according to the seedling density treatments (one, three, and five seedlings per hill) of the experiment.

2.3. Seedbed Management

Seedlings used in the experiment were grown in the seedbed adjacent to the experimental field. Three well-puddled seedbed plots were prepared with one seedbed plot for each variety used. Each plot was equally divided into four divisions for the different seedling ages. To raise healthy and vigorous seedlings in the seedbed, seeding rate and fertilizer management recommendations from [14] were used. Basal fertilizer was applied (2 g N + 3 g P + 2 g K m^{-2}) in the seedbed using complete fertilizer (14–14–14). After testing for germination rate, the pregerminated seeds of the three varieties were carefully sown in the seedbeds separately by variety at a uniform seeding rate of 25 g m^{-2}. After sowing, the seedbeds were kept wet all the time to avoid hardening of the soil, which may damage seedling roots when pulled out. The seedbeds were flooded as seedling height increased. Seedlings were carefully hand-pulled 1 d before transplanting, and the pulled seedlings with intact roots were placed in plastic trays with water. Only needed seedlings were hand-pulled for specific seedling age treatments. Seedbeds were continuously managed until the transplanting of the last seedling age treatment (SA$_{45}$) in the main field was completed.

2.4. Water Management in the Main Field

During the first 3 wk after transplanting, soil was kept saturated to promote better seedling establishment. Thereafter, water management based on safe AWD practice was used at the vegetative stage and after the flowering stage, where timing of irrigation was based on water depth in the field water tubes installed in each plot. This was done by irrigating the plots only when field water depths in the field were about 15 cm below the ground surface. During the flowering stage, plots were under continuous flooding (1–5 cm depth) to avoid possible yield losses. Water stress at the flowering stage may induce spikelet sterility and will thus result in yield loss. Near the end of crop maturity (10–15 d before harvesting), terminal drainage was implemented.

2.5. Fertilizer and Other Cultural Management

In each season, a total of 60 kg ha^{-1} of nitrogen (N) was applied in three splits: (1) Basal or before transplanting (30 kg ha^{-1}), (2) 21–23 d after transplanting (DAT) or mid-tillering of crop growth stage (15 kg ha^{-1}), and (3) panicle initiation (15 kg ha^{-1}). Basal application of phosphorus (P) and potassium (K) were also done at 30 kg ha^{-1} each. Complete (14–14–14) and urea (46–0–0) fertilizers were used as source of N,P and K for basal application, whereas urea was used as a source of N during the crop growth period. All fertilizers were applied during flooded soil condition in all treatments. To keep the experiment field weed-free, preemergence herbicide (butachlor) was applied a few days after transplanting, and then spot hand weeding was done during the vegetative growth stage until a full canopy cover was achieved. Plants were also protected from pests and diseases using agrochemicals.

2.6. Data Collection and Calculation

Field water tubes were used to guide the implementation of AWD [6], and to estimate the amount of irrigation input in the field experiment. Field water depths in the main plots of the experiment were regularly monitored using field water tubes that were installed at a depth of 15 cm below the ground surface. Monitoring of field water depths was done every other day, between 8:00 and 9:00 AM for all treatments from transplanting until 15 d before harvesting. Water level inside the tube was measured from the top to the level of the water inside the tube. To get the actual depth of water inside the tube, the reading will be subtracted from the height of the AWD tube that protruded above the soil surface.

Irrigation dates were all noted and during irrigation events, field water tubes were read before and after each irrigation. The irrigation water input under AWD conditions was then computed using a procedure outlined by [15] as follows:

$$I = d_f - ((\theta_s - \theta_i) \times D), \tag{1}$$

where I = irrigation (mm); d_f = final water depth above soil surface (mm); θ_s = soil water content at saturation (cc/cc); θ_i = soil water content when field water falls below the ground surface (cc cc^{-1}), which was assumed as the field capacity, especially when the perched water table is 15 cm or more from the soil surface; and D = depth of the perched water table.

Since flooded conditions were essential during flowering stage, irrigation water input was computed as:

$$I = d_f - d_s, \tag{2}$$

where d_s = initial field water depth (mm) above the soil surface.

Total water input was the sum of total irrigation and rainfall from transplanting to 15 d before harvesting. Rainfall was taken from the automatic weather station (Vantage Pro2 Weather Station, Davis Instruments Corp, USA) installed adjacent to the experimental area. Groundwater depths were also monitored using three observation wells (2-inch diameter PVC pipe, 2 m long, driven down to 1.5 m) installed along the bunds in the upper, middle, and lower portions of the experiment field. Reading of the water levels in the observation wells was also done every 2 d, from transplanting to 15 d before harvest between 8:00 AM and 9:00 AM. The water level in the observation wells was measured from the top to the level of the water inside the tube. To get the actual depth of the groundwater, the reading was subtracted from the height of the tube that protruded above the soil surface.

Plant height and number of tillers per hill were monitored every 2 wk from transplanting to flowering. Phenology dates (mid-tillering, panicle initiation, flowering, grain filling, and physiological maturity) were monitored for each sub-subplot. At harvest, crop cut samples were taken from 110 h at the center of each plot to determine grain yield. Moisture content of the grains was measured with a digital grain moisture meter (OGA Electric Co., Ltd, Japan), and grain yield was calculated at 14% moisture content. Grain yield components, number of panicles m^{-2}, % spikelet sterility, number of spikelet per panicle, 1000-grain weight, and total biomass of grains and vegetative matter were determined in five-hill samples near the harvest area. Harvest index (in percent) was also determined by dividing grain yield by total biomass and then multiplying it by 100. Irrigation water productivity (WP_I) and total water productivity (WP_{I+R}) were calculated as kg grain m^{-3} water input. For WP_I, water input was from the sum of all irrigations, while for WP_{I+R}, water input was from total rainfall and the sum of all irrigations received by the plants from transplanting up to 15 d after harvesting.

Selected weather data (rainfall, minimum, and maximum temperatures) were taken from the agrometeorological station in the ARC. Seasonal means and sums were reported based on the actual growth duration per treatment.

2.7. Statistical Analysis

The data were subjected to an analysis of variance [16], using International Rice Research Institute's (IRRI) open-access software for statistical analysis, which was implemented in the R statistical package [17]. Treatment means were separated using the least significant difference (LSD) tests and compared at the $p \leq 0.05$ level of significance. Significant interactions of factors used in this study were reported.

3. Results

3.1. Weather

The total wet season rainfall was higher in 2015 (1993.0 mm) than in 2014 (1478.0 mm). Most of the rains occurred during the wet season (May–October), with about 89.3% and 73.7% of the total rainfall in 2014 and 2015, respectively. During the crop growth period (July–October), however, the total accumulated rainfall was higher in 2014 (1078.4 mm) than in 2015 (905.7 mm) as shown in Figure 1, where the highest accumulated monthly rainfall was observed in July (451.9 mm) during 2014 and in September (322.7 mm) during 2015. Mean daily maximum and minimum temperatures during the crop growth period in 2014 and 2015 were relatively similar during the same period (July–September), ranging from 27.5 to 35.3 °C (Figure 1). However, the mean daily minimum temperature was lower in 2014 (23.4 °C) than in 2015 (24.9 °C).

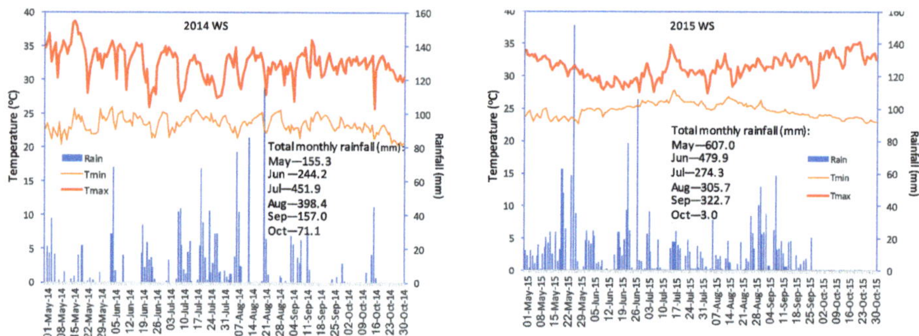

Figure 1. Daily maximum and minimum temperature, and daily rainfall in the study site, 2014 and 2015.

3.2. Hydrological Conditions and Water Level Changes

Groundwater table depths were shallow during wet seasons, especially from July to September. Similar trends of groundwater level fluctuations were observed for both years. As shown in Figure 2, groundwater depth below the ground surface fluctuated between 21 and 33 cm in July, between 18 and 57 cm in August, and between 20 and 63 cm in September, gradually declining after the second week of September.

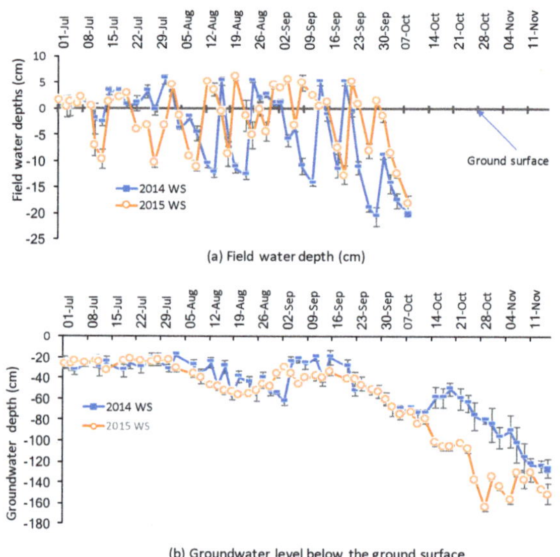

Figure 2. Average field water (**a**) and groundwater (**b**) level fluctuations in the study site during 2014 and 2015. Field water level fluctuations presented were the average of $SA_{15} \times SD_1 \times V_1$ plots, while average groundwater level was from the three groundwater observation wells installed in the site.

In October, groundwater table depths ranged from 51 to 84 cm during 2014 and 60 to 163 cm in 2015. In terms of field water depths, (using average water level data for $SA_{15} \times SD_1 \times V_1$ plots as example), it fluctuated between 5 cm (above the surface) and −15 cm (below the ground surface) during the crop growth stage in both years. Water levels were slightly above the ground surface (0–2 cm) during the first week after transplanting; after the plants fully recovered from transplanting shock, water levels fluctuated between 6 cm and −15 cm. While rice plants received ample rains in 2014 and 2015, the rains were not evenly distributed during the crop growth period. As a result, the paddy soil surface experienced five drying cycles in 2014 and about six drying cycles in 2015. This means that there was some form of AWD condition of the soil surface in the field experiment. Field water level fluctuations in September and October (Figure 2) were due to the application of supplemental irrigation as the rains became limited.

3.3. Crop Growth and Development

On average, total crop growth duration (sowing to harvest) was longer in 2015 than in 2014 by about 4 d (Table 2). SA and V significantly affected crop growth duration among the treatments. No significant interactions between treatments were revealed in the ANOVA. The number of days from sowing to panicle initiation and from sowing to harvest was highest in SA_{45} and lowest in SA_{15} between SA treatments. Mean differences in total crop growth duration between SA_{45} and SA_{15} were 17 d in 2014 and 20 d in 2015. The mean difference between SA_{35} and SA_{15} (10–13 d) was observed in both years, while a difference of 6–7 d was observed in both years between SA_{25} and SA_{15}. In terms of plant growth duration in the main field (from transplanting to harvest), the older the seedlings at transplanting, the shorter the crop duration. This means that SA_{45} had the shortest duration and SA_{15} had the longest duration in the main field in both years. A difference of 4–5 d was observed between SA_{15} and SA_{25}, 7–10 d between SA_{15} and SA_{35}, and 10–13 d between SA_{15} and SA_{45}. In terms of varieties, V_3 had the longest crop growth duration, followed by V_2 and V_1, respectively.

Table 2. Crop duration from sowing to panicle initiation and harvest, 2014 and 2015 wet seasons.

Season/ Treatment	2014			2015		
	Sowing to PI (d)	Sowing to Harvest (d)	Transplanting to Harvest (d)	Sowing to PI (d)	Sowing to Harvest (d)	Transplanting to Harvest (d)
Seedling age						
SA_{15}	58 d	117 d	102 a	61 d	119 d	104 a
SA_{25}	68 c	122 c	97 b	68 c	125 c	100 b
SA_{35}	73 b	127 b	92 c	79 b	132 b	97 c
SA_{45}	84 a	134 a	89 c	89 a	139 a	94 d
Seedling density						
SD_1	70 a	124 a	94 a	74 a	127 a	97 a
SD_3	72 a	126 a	96 a	74 a	129 a	99 a
SD_5	71 a	126 a	96 a	76 a	130 a	100 a
Variety						
V_1	68 b	123 b	93 b	71 c	125 c	95 c
V_2	69 b	125 b	95 b	73 b	129 b	99 b
V_3	75 a	127 a	97 a	79 a	132 a	102 a
ANOVA results						
SA	*	*	*	*	*	*
SD	ns	ns	ns	ns	ns	ns
V	*	*	*	*	*	*
SA × SD	ns	ns	ns	ns	ns	ns
SA × V	ns	ns	ns	ns	ns	ns
SD × V	ns	ns	ns	ns	ns	ns
SA × SD × V	ns	ns	ns	ns	ns	ns

[1] Within each column, season and treatment, means followed by the same letter are not significantly different at $p < 0.05$. In the ANOVA results, single (*) asterisk means that the F-value was significant at 5% level, while ns means not-significant.

In general, the average number of tillers (per hill) was higher in 2015 than in 2014 (Table 3). The tillering ability of the rice plants was significantly influenced by seedling age, seedling density, and variety. There was, however, no significant interactions among treatments on tiller count (number of tillers per hill) in both years. In 2014, across seedling age treatments, average tiller count at around panicle initiation stage (42–44 DAT) was highest in SA_{15} (11.7) and lowest in SA_{45} (6.6), while SA_{25} and

SA_{35} had the same number of tillers (10.6; Table 3). In 2015, a similar trend was observed: Highest average tiller count was seen in SA_{15} (15), and the lowest was observed in SA_{45} (9.8). SA_{25} and SA_{35} had similar tiller counts (12.7). During 2014 and at the panicle initiation stage, SD_1 provided the lowest tiller count across SD treatments, while SD_5 had the highest count, although not significantly it was different from SD_3. In 2015, differences in tiller count among seedling age treatments were significant, with average tiller counts being highest in SD_5 (14.8), followed by SD_3 (12.9). The lowest was noted in SD_1 (10.5). Across varieties, as expected, V_1 provided the significantly highest average tiller count in both years (10.9 in 2014 and 14.5 in 2015). In 2014, tiller counts of V_2 and V_3 were similar, while in 2015, V_2 had higher tiller count than V_3.

Table 3. Mean comparison of plant height and tiller count at the panicle initiation stage, 2014 and 2015 wet seasons [1].

Treatment	Plant Height (cm)		Tiller Per Hill (no.)	
	2014	2015	2014	2015
Seedling age				
SA_{15}	55.4 a	63.1 a	11.7 a	15.0 a
SA_{25}	53.3 a	63.0 a	10.5 b	12.9 b
SA_{35}	47.0 b	55.5 b	10.7 b	12.7 b
SA_{45}	38.3 c	54.2 b	6.6 c	9.8 c
Seedling density				
SD_1	44.7 a	55.8 a	6.5 b	10.5 c
SD_3	49.3 a	59.3 a	10.6 a	12.9 b
SD_5	51.4 a	62.0 a	11.9 a	14.8 a
Variety				
V_1	41.9 b	53.2 c	10.9 a	14.5 a
V_2	53.3 a	60.9 b	9.5 a b	13.1 b
V_3	50.2 a	63.0 a	9.2 a b	11.7 c
Interaction significance				
$SA \times SD$	ns	ns	ns	ns
$SA \times V$	ns	ns	ns	ns
$SD \times V$	ns	ns	ns	ns
$SA \times SD \times V$	ns	ns	ns	ns

[1] Within each column and treatment, means followed by the same letter are not significantly different at $p < 0.05$. In the interaction significance, ns means that the F-value is not significant.

On the average, plants were taller in 2015 than in 2014 (Table 3). At the panicle initiation stage, only seedling age and variety significantly influenced plant height. No significant interactions among treatments on plant height were observed. In both years, early transplanted seedlings (SA_{15} and SA_{25}) were significantly taller than late transplanted ones (SA_{35} and SA_{45}). In 2014, the mean difference in height between SA_{15} and SA_{45} was about 15 cm, while in 2015, the difference was about 9 cm. No significant difference in plant height was found between SA_{15} and SA_{25} in both years; but between SA_{35} and SA_{45}, a difference of about 9 cm was observed in 2014. Across varieties, plant heights of V_2 and V_3 were similar in both years and were, on average, significantly taller than V_1 by 10 cm.

3.4. Grain Yield and Grain Yield Components

Comparing between years, grain yields were significantly lower in 2014 (3.9 t ha^{-1}) than in 2015 (4.5 t ha^{-1}). There was also a significant interaction between year and SA in the experiment (ANOVA results not shown). As shown in Figure 3, the highest value was observed in SA_{15} in 2014 (albeit not significantly different from the SA_{25}, SA_{35} and SA_{45} during 2014, and SA_{45} in 2015), while in 2015, the highest value was obtained in SA_{15} (although not significantly different from SA_{25} and SA_{35} in 2015).

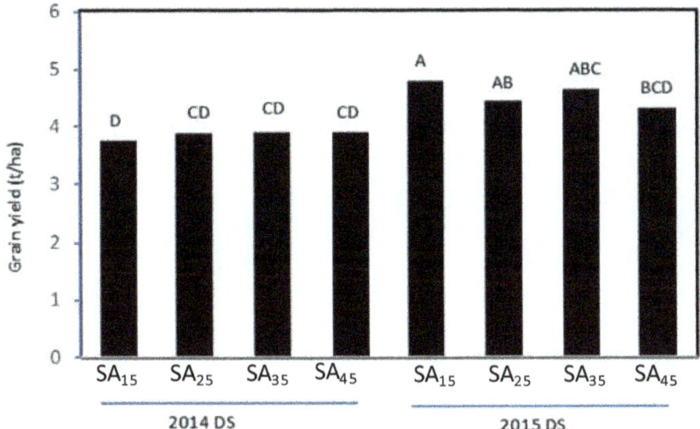

Figure 3. Comparison of yields across treatments and years (2014 and 2015). Columns with common letter are not significantly different at $p < 0.05$.

Among treatment factors, ANOVA results indicated that V significantly affected grain yield in each year (data not shown), while SA and SD did not. Higher (although not statistically different) yields were observed in V_2 and V_3, while lower grain yields were observed in V_1 in both years (Table 4). A significant SA × V interaction on grain yield was also observed in both years. Comparing SA at each level of V during 2015, a lower grain yield of SA_{45} (albeit not significantly different from SA_{25}) was observed in the V_3 level, while in the V_1 and V_2 levels, grain yields in all SA treatments were similar (Table 5). In 2014, however, there were mixed yield trends when comparing SA at each level of V: Significantly lower yields for younger seedlings (SA_{15}) in V_1 and V_2, while lower yields in late transplanted seedlings (SA_{45}) in V_3. In 2015, comparing V at each level of SA showed significantly lower yields in V_1 compared with the other two varieties (particularly under SA_{15}, SA_{25}, and SA_{35}), while grain yields in V_2 and V_3 were the same under seedling age levels. Similarly, in 2014, grain yields were lower in V_1, although not significantly different from V_3 under SA_{25} and SA_{45}.

In terms of grain yield components, no significant effect of year (2014 vs. 2015) was found in any of the yield components. However, in terms of experiment treatment effects in each year, the number of panicles (per m^2) was significantly influenced by all three factors (SA, SD, and V) during 2014 and by two factors (SD and V) in 2015. However, no significant interaction effects of treatments were found in both years. As shown in Table 4, across seedling age the significantly highest number of panicles was observed in SA_{25} particularly in 2014, and the lowest in SA_{45} (although not significantly different from SA_{15} and SA_{35} in 2014 and from SA_{15}, SA_{25}, and SA_{35} in 2015). Across seedling densities, a greater number of panicles per m^2 were observed with higher seedling densities than with lower seedling densities. In both years, the highest number of panicles was observed in SD_5 (although not significantly different from SD_3 in 2015); it was significantly lowest in SD_1. Across varieties, V_1 consistently produced the significantly highest number of panicles compared with the other varieties; no significant difference in the number of panicles was found between V_2 and V_3 varieties in both years.

Table 4. Effects of seedling age, seedling density and variety on grain yield, grain yield components, and harvest index of rice, 2014 and 2015 [1].

Season	Treatment	Panicle (no. m^{-2})	Spikelet (no. panicle^{-1})	1000-Grain Weight (g)	Filled Grain (%)	Grain Yield (t ha^{-1})	Harvest Index (%)
		Seedling Age					
	SA$_{15}$	180.6 b	106.2 c	29.1 a	79.3 c	3.7 a	39 a
	SA$_{25}$	192.2 a	109.0 b	29.7 a	82.2 b	3.9 a	39 a
	SA$_{35}$	180.8 b	107.6 b c	29.3 a	83.9 a	3.9 a	36 a
	SA$_{45}$	176.2 b	114.1 a	28.8 a	80.5 c	3.9 a	42 a
		Seedling Density					
2014	SD$_1$	167.3 c	120.1 a	28.9 a	80.4 a	3.9 a	40 a
	SD$_3$	177.0 b	107.0 b	29.2 a	82.3 a	3.8 a	40 a
	SD$_5$	203.0 a	100.5 c	29.5 a	81.8 a	3.9 a	36 a
		Variety					
	V$_1$	236.3 a	97.5 c	21.2 c	82.6 a	3.5 b	42 a
	V$_2$	150.1 b	108.5 b	36.9 a	79.4 a	4.1 a	38 a
	V$_3$	161.0 b	121.6 a	29.5 b	82.5 a	3.9 a	37 a
		Interaction Significance					
	SA × SD	ns	ns	ns	ns	ns	ns
	SA × V	ns	ns	ns	*	**	ns
	SD × V	ns	ns	ns	ns	ns	ns
	SA × SD × V	ns	*	ns	ns	ns	ns
		Seedling age					
	SA$_{15}$	182.0 a	108.8 b	33.1 a	78.9 a	4.8 a	31 a
	SA$_{25}$	196.0 a	114.3 a	28.9 b	79.3 a	4.4 a	32 a
	SA$_{35}$	179.2 a	104.24 b	28.8 b	79.0a	4.7 a	32 a
	SA$_{45}$	179.0 a	94.2 c	28.7 b	79.4 a	4.3 a	32 a
		Seedling density					
2015	SD$_1$	168.3 b	111.2 a	31.1 a	80.0 a	4.6 a	35 a
	SD$_3$	192.8 a	103.2 b	29.6 a	79.9 a	4.5 a	31 b
	SD$_5$	193.0 a	101.8 b	29.0 a	77.6 a	4.5 a	30 b
		Variety					
	V$_1$	227.2 a	100.1 a	22.8 c	78.4 b	4.3 b	35 a
	V$_2$	161.9 b	108.7 a	36.6 a	75.5 b	4.7 a	31 a
	V$_3$	165.0 b	107.6 a	30.3 b	83.7 a	4.7 a	30 a
		Interaction significance					
	SA × SD	ns	ns	**	ns	ns	ns
	SA × V	ns	ns	ns	ns	*	ns
	SD × V	ns	ns	ns	ns	ns	ns
	SA × SD × V	ns	ns	ns	ns	ns	ns

[1] In a column and within the same treatment factor, means with the same lower case letter are not significantly different at $p < 0.05$. In the interaction significance, double (**) and single (*) asterisks mean that the F-value are significant at 1% and 5% level, respectively, and ns means not significant.

Table 5. Effects of seedling age × variety (SA × V) interaction on grain yield (2014 and 2015) and percent filled spikelet (2014) and of seedling age × seedling density (SA × SD) interaction on 1000–grain weight (2015) [1].

Seedling Age	Grain Yield (t ha^{-1})			1000-Grain Weight (g)			Filled Spikelet (%)		
	V_1	V_2	V_3	SD_1	SD_3	SD_5	V_1	V_2	V_3
2014									
SA_{15}	3.4b C	3.8b B	4.1ab A				78.6b B	77.0b B	82.4a A
SA_{25}	3.6ab B	4.2a A	3.8bc B				84.4a A	78.6b B	83.7a A
SA_{35}	3.4b B	4.1a A	4.2a A				84.0a A	83.0a A	84.9a A
SA_{45}	3.7a B	4.4a A	3.7c B				83.4a A	78.9b B	79.0b B
2015									
SA_{15}	4.6a B	4.9a A	5.0a A	38.6a A	31.2a B	29.4a C			
SA_{25}	4.2a B	4.5a A	4.6ab A	28.2b A	29.2b A	29.4a A			
SA_{35}	4.2a B	4.9a A	4.9a A	28.7b A	29.0b A	28.8a A			
SA_{45}	4.3a A	4.4a A	4.3b A	28.7b A	28.9b A	28.5a A			

[1] In a column and within the same year, means with the same lowercase letter are not significantly different (comparison of SA at each level of V for grain yield and percent filled spikelet, and comparison of SA at each level of SD for 1000-grain weight). In a row and within the same year, means with the same uppercase letter are not significantly different (comparison of V at each level of SA for grain yield and filled spikelet, and comparison of SD at each level of SA for 1000-grain weight).

Significant effects of SA, SD, and V were observed on the number of spikelets per panicle (spikelet count) in 2014, while in 2015 only significant effects of SA and SD on spikelet count were noted. No significant interactions were found among factors on the spikelet count in both years. Across SA treatments in 2014, the spikelet count was highest in SA_{45}, followed by SA_{25}, then SA_{35} and SA_{15}, although there was no significant difference between SA_{15} and SA_{35} and between SA_{25} and SA_{35} (Table 4). In 2015, there was an opposite trend: SA_{45} provided the lowest spikelet count among the four SA treatments. Highest spikelet count was recorded in SA_{25}, followed by SA_{15} and SA_{35}, but as in 2014, the difference between SA_{15} and SA_{35} was not significant. Across SD, highest spikelet count was observed in SD_1 and the lowest was seen in SD_5 in both years. The difference in spikelet count between SD_3 and SD_5 was only significant in 2014. Across varieties, the highest spikelet count was found in V_3 and the lowest in V_1 in 2014. Spikelet counts were similar in all varieties used in the 2015 experiment.

In 2014, the 1000-grain weight (1000 GW) was significantly influenced by V, while by all three factors (SA, SD, and V) in 2015. There was a significant SA × SD interaction on 1000 GW in 2015. In both years, consistently highest 1000 GW was produced in V_2, with the lowest seen in V_1 (Table 4). In 2015, comparison of SA at each SD level (Table 5) indicated that SA_{15} was significantly highest among the four SA treatments under SD_1 and SD_3, whereas similar values were observed among SA_{25}, SA_{35}, and SA_{45}. Under SD_5, no difference in 1000 GW was found among the SA treatments. Comparing SD at each level of SA, a significant difference in 1000 GW was only found under SA_{15} level, where $SD_5 < SD_3 < SD_1$; the differences in means across SD were not significant at the SA_{25}, SA_{35}, and SA_{45} levels.

The percentage of filled spikelet was significantly influenced by SA and V in 2014 and in 2015 by V. There was also a significant SA × V interaction on percent filled spikelet during 2014. Comparing SA at each level of V (Table 5), the lowest percent filled spikelet was observed in SA_{15} under V_1 and V_2 levels, although SA_{15} was not significantly different from SA_{25} and SA_{45}, especially under V_2. However, under V_3, SA_{15}, SA_{25}, and SA_{35} were similar but were significantly higher than SA_{45}. Comparing V at each level of SA (Table 5), significant differences in percent filled spikelet between varieties were observed in SA_{15}, SA_{25}, and SA_{45}. Percent filled spikelet of V_3 was significantly higher than that of V_1 and V_2 under SA_{15}, and significantly higher than V_2 under SA_{25}. However, V_3 was statistically similar to V_2 under SA_{45}, but significantly lower than V_1. Between V_1 and V_2, values were similar under SA_{15}, but V_1 was significantly higher than V_2 under SA_{25} and SA_{45}. Mean harvest index (HI), was relatively higher in 2014 (38.7%) than in 2015 (32%), since both grain yield and dry biomass were higher in 2015 (data not shown). All treatments and their interactions did not influence HI in 2014, but, in 2015, HI was significantly influenced by seedling density. SD_I had higher HI compared with other SD treatments (Table 4), whereas SD_1, SD_3, and SD_5 were not significantly different.

3.5. Water Input and Water Productivity

Rice plants that were planted earlier received more rainfall than those planted later because more rainfall events occurred from July to early September (Table 6 and Figure 1). In 2014, total rainfall received by SA_{15} was 65 mm, 123 mm, and 363 mm more than SA_{25}, SA_{35}, and SA_{45}, respectively. Seedling age and variety significantly affected seasonal irrigation, and total water (irrigation plus rainfall) inputs in both years, while seedling density and interactions among the treatments were not significant. In 2015, the differences in total rainfall between SA_{15} and the other seedling age treatments were 96 mm (SA_{25}), 192 mm (SA_{35}), and 279 mm (SA_{45}). In general, more rainfall was received by plants in 2014 than in 2015 in all SA treatments, ranging from 4% to 20%, with the highest percent difference observed in SA_{35} and the lowest in SA_{45}.

Table 6. Irrigation and rainfall water input during 2014 and 2015 under different seedling age by a variety treatments, ARC, Vientiane, Lao PDR [1].

Treatment	2014			2015		
	Total Irrigation (mm)	Total Rainfall (mm)	Total Water Input (mm)	Total Irrigation (mm)	Total Rainfall (mm)	Total Water Input (mm)
Seedling age						
SA_{15}	439.3 d	1009.2 a	1448.5 a	317.0 d	902.8 a	1219.8 a
SA_{25}	469.5 c	944.4 b	1413.9 b	456.3 c	807.0 b	1263.3 a
SA_{35}	506.4 b	886.2 c	1392.6 b	522.4 b	711.1 c	1233.5 a
SA_{45}	596.9 a	647.2 d	1244.1 c	577.3 a	623.7 d	1201.0 a
Seedling density						
SD_1	500.2 a	876.9 a	1377.1 a	501.3 a	761.2 a	1262.5 a
SD_3	537.7 a	875.6 a	1413.3 a	488.9 a	761.2 a	1250.1 a
SD_5	505.2	862.7 a	1367.9 a	500.3 a	761.2 a	1261.5 a
Variety						
V_1	493.3 b	863.7 b	1357.0 a	465.3 b	759.4 a	1224.7 a
V_2	505.7 a	866.5 b	1372.2 a	470.3 b	759.4 a	1229.7 a
V_3	510.1 a	884.8 a	1394.9 a	524.3 a	764.7 a	1289.0 a
Interaction significance						
SA × SD	ns	ns	ns	ns	ns	ns
SA × V	ns	ns	ns	ns	ns	ns
SD × V	ns	ns	ns	ns	ns	ns
SA × SD × V	ns	ns	ns	ns	ns	ns

[1] In a column and within the same treatment, means with the same letter (lower case) are not significantly different at $p < 0.05$. In the interaction significance, ns means that the F-value is not significant.

With the abundance of rain in the wet season, total irrigation inputs received by the plants were 30%–45% of the total water input in 2014 and 26%–48% in 2015, with delayed transplanted rice receiving a higher fraction of the water input from irrigation than early transplanted seedlings. The lowest amount of irrigation input was received by SA_{15}, whereas older seedlings (SA_{35} and SA_{45}) received more irrigation input in both years (Table 6). The total irrigation input was higher by 36% in SA_{45}, 15% in SA_{35}, and 7% in SA_{25} in 2014 in comparison with SA_{15}. Higher percent differences between irrigation inputs were observed in 2015, wherein against SA_{15}, irrigation input was higher by 82%, 65%, and 44% in SA_{45}, SA_{35}, and SA_{25}, respectively. On a per season basis, the total water input was 1244–1449 mm in 2014 (while it was 1201–1333 mm in 2015). Among SA treatments, SA_{45} received the lowest and SA_{15} received the highest total water input in 2014; in 2015, total water input was similar among SA treatments. V_3 received the highest water input among the varieties tested and V_1 received the lowest, particularly from irrigation, in both years. However, we did not find any significant difference in total rainfall and total water input between varieties during the field experiment.

Irrigation water productivity (WP_I) and total water productivity (WP_{I+R}) were only significantly influenced by V and by SA × V interactions in both years. In 2014, WP_I and WP_{I+R} values ranged from 0.78 to 1.01 kg m^{-3} and from 0.23 to 0.40 kg m^{-3}, respectively; in 2015, the corresponding range were 0.80–1.01 kg m^{-3} and 0.32–0.40 kg m^{-3}. Comparing SA at each level V in 2014, the WP_I of SA_{45} was significantly higher than those of SA_{35} and SA_{15} in V_1 and significantly higher than that of SA_{15} in V_2. However, SA_{45}'s WP_I under V_3 became significantly lower than those of SA_{35} and SA_{15} (Table 7). Comparing V at each level of SA, V_1 had higher WP_I than V_2 at all levels of SA; it was also higher than V_3 at SA_{15} and SA_{35} levels, respectively, in 2014. In 2015, there were no significant differences in WP_I among SA treatments in each variety level. However, comparing WP_I of the different varieties V at each level of SA, V_1 had the lowest and V_3 (although not significantly different from V_2) had the highest WP_I values. In terms of WP_{I+R}, the significantly highest WP_{I+R} value was found in SA_{45} in all levels of varieties in 2014 (Table 7). The lowest values were observed in SA_{15} and SA_{35} under the V_1 level and in SA_{15} and SA_{25} under the V_3 levels; SA_{15}, SA_{25}, and SA_{35} were similar under the V_2 level. Comparing WP_{I+R} of the different varieties V at each level of SA in 2014, V_1 had the lowest values, particularly under SA_{15} and SA_{35} levels, respectively, while, V_2 and V_3 were similar (except in SA_{45}). In 2015, there were no significant differences in WP_{I+R} among SA treatments under V_1 and V_3 levels. Significant differences were only observed under the V_2 level, wherein the most delayed transplanted seedlings (SA_{45}) produced higher WP_{I+R} than did the earliest transplanted seedlings (SA_{15}) in 2015 (Table 7). Comparing varieties at each level of SA, lower values of WP_{I+R} were observed with V_1 under the SA_{25} and SA_{35} levels, whereas no significant differences in WP_{I+R} were found among varieties under SA_{15} and SA_{45}.

Table 7. Seedling age × variety (SA × V) interaction on irrigation water productivity (WP_I) and total water productivity (WP_{I+R}) during 2014 and 2015 wet seasons, respectively [1].

Season/Seedling Age	Irrigation Water Productivity (WP_I)			Total Water Productivity (WP_{I+R})		
	V_1	V_2	V_3	V_1	V_2	V_3
2014						
SA_{15}	0.78 b B	0.89 b A	0.94 a b A	0.23 c B	0.27 b A	0.29 c A
SA_{25}	0.84 ab B	0.96 ab A	0.89 bc AB	0.28 b A	0.30 b A	0.28 c A
SA_{35}	0.79 b B	0.96 ab A	0.97 a A	0.24 bc B	0.30 b A	0.32 ab A
SA_{45}	0.87 a B	1.01 a A	0.86 c A	0.33 a B	0.40 a A	0.33 a B
2015						
SA_{15}	0.89 a B	0.99 a A	1.01 a A	0.32 a A	0.33 b A	0.36 a A
SA_{25}	0.97 a A	1.01 a A	0.98 a A	0.32 a B	0.36 ab AB	0.37 a A
SA_{35}	0.80 a B	0.92 a A	0.97 a A	0.33 a B	0.40 a A	0.40 a A
SA_{45}	0.82 a A	0.84 a A	0.82 a A	0.37 a A	0.37 a b A	0.36 a A

[1] In a column and within the same year, means with the same lowercase letter are not significantly different (comparison of SA at each level of V); in a row and within the same year, means with the same uppercase letter are not significantly different (comparison of V at each level of SA).

4. Discussion

In our experiment, seedling age significantly affected the tillering dynamics of the rice plants. Younger seedlings (SA_{15}) had the highest tiller count, while older seedlings (SA_{45}) had the least in both years, regardless of planting density and variety used. This indicates that rice seedling age is vital in determining tiller occurrence and confirmed the findings of [18] that, with increasing seedling age, tillering was depressed, thereby resulting in reduced tiller number. Increasing seedling density ameliorated the effects of delayed transplanting on tillering performance in our experiment. Transplanting with three (SD_3) to five (SD_5) seedlings per hill produced more tillers per hill than with one seedling per hill (SD_1), and therefore the tiller number of older seedlings could be induced by increasing seedling density at transplanting. The result is in line with the findings of [13,19], but the effect may vary with seasons [20].

Seedling age also affected crop growth duration. In the present study, the phenological development of the rice plants was delayed with the transplanting of older seedlings, resulting in longer crop duration. The panicle initiation and physiological maturity were delayed by 5–11 d and 5–7 d, respectively, for every 10-d delay of transplanting in both years. Similar findings were reported by [14]: An increase of 6–9 d of vegetative phase (sowing to panicle initiation) with 10-d-old seedlings vs. 30-d-old seedlings in Central Luzon Philippines during two dry seasons, and by 8 d during one wet season of field experimentation. With increased total crop duration (sowing to physiological maturity) using older seedlings, the expected shortening of the stay of the crop in the main field did not correspond to the number of days of delay in transplanting. In our study, crop duration in the main field was only shortened by 3–5 d for every 10-d delay in transplanting. Seedling density did not show any effect on the total rice growth duration, regardless of the variety used. Seedling age could be an important factor in determining phenological stages and crop duration. Indisputably, varietal characteristics dictated the tiller number, plant height, and growth duration.

Many research studies have shown that the use of younger seedlings (not older than 25 d old) produced positive impacts on grain yield [21–23], although many authors have also contradicted this [24–26]. In our study, a significant SA × V interaction on grain yield was found in both years. This means that the influence of seedling age on grain yield depends on the variety used. The use of varieties with a stronger tillering propensity ameliorated the effects of delayed transplanting on crop yield.

Availability of water is crucial in deciding whether to transplant rice seedlings at an early or later stage of the season. When the water supply is uncertain, delay in transplanting becomes inevitable. Rainfall is abundant during the wet season in Lao PDR. However, much of these rains occur from May to September, when crops are in their vegetative growth stage. In our study site, the rains were not evenly distributed during the rainy months and the number of dry spell periods interrupted the regular monsoon rainfalls, which occurred particularly during critical crop growth stages. Therefore, supplemental irrigation is vital to avoid water stress during periods with no rains. In our study, we hypothesized that the shortening of stay in the main field had implications on total water input of the rice crop. However, our results indicated that the shortened stay of older seedlings in the main field did not translate in irrigation reduction. Higher rainfall occurred from July to August in 2014, and from July to September in 2015, which favored early transplanting. Although with relatively shorter stay in the main field, older seedlings were harvested later than younger seedlings; an additional one or two supplemental irrigations were applied (data not shown) before terminal irrigation. Compared with older seedlings (SA_{45}), younger seedling age treatment (SA_{15}) received about 56% more rainfall and about 26% less irrigation in 2014 and about 45% more rainfall and 45% less irrigation in 2015. At the end of the season, total water input (rainfall plus irrigation) was similar among the SA treatments, considering that older seedlings received less rainfall but more irrigation, while younger seedlings received more rainfall but less irrigation. A similar study conducted in the Philippines during the dry season (with comparable seasonal rainfall in Lao PDR) has indicated that the total irrigation input decreased with older seedlings but not in the wet season when rain could not be easily controlled in the field [14]. A significant difference in irrigation input was also found among the varieties used in the experiment. As expected, irrigation received by the shorter duration variety was lower than that of the longer duration variety (i.e., $V_1 > V_2 > V_3$) in both years because of one additional irrigation needed for V_3 before terminal drainage was implemented (data not shown). In terms of water productivity, a significant SA × V interaction was found in both years.

There is a common understanding among farmers that planting old seedlings (especially with TDK 8) could result in yield losses. The experiment results further confirmed their suspicion. The inference from the results can be given only to the selected varieties used in the experiment. However, as the seeding age at transplantation is more important for better crop productivity, breeders should take note on testing this effect more systematically before releasing the varieties for rainfed lowland rice cultivation.

The overall water management practice in the experimental field was safe AWD and although rainfall was abundant, the field experiments attained an average of five to six wetting and drying cycles in the main field. Safe AWD was developed to reduce irrigation water input without sacrificing rice grain yields particularly in water-short areas and seasons. In our experiment, we have demonstrated that safe AWD is possible during the wet season, although the wetting and drying cycles were fewer compared to what we expected during the dry season with better control of irrigation.

5. Conclusions

Grain yield was significantly influenced by the interaction of seedling age and variety and their interactions subjected to AWD conditions during the study. Grain yields of V_1 (IRUBN0300-63-5-4) and V_2 (TDK10239-SSD4-303-1) increased with increasing seedling age, while grain yield of V_3 (TDK-8) decreased with increasing seedling age, especially in 2014. The effect of seedling age on grain yields largely depended on the variety used particularly during the wet season in Lao PDR. Regardless of the variety, the seedling density treatments (one, three, and five seedlings per hill) had significant effects on tillering dynamics but not on grain yield. Moreover, the use of varieties with a stronger tillering propensity ameliorated the effects of delayed transplanting on crop performance. Late transplanting of seedlings improved total water productivity, but it might decrease irrigation water productivity due to the delay in harvesting. Total crop growth duration was extended with late transplanting, which required one or two supplemental irrigations until harvest when rainfall is not available. However, the total duration of stay of the rice plants in the main field was only reduced by 3–5 d for every 10 d delay in transplanting. Proper selection of varieties to use under delayed transplanting coupled with appropriate seedling density can improve crop performance, yield, and water productivity. More so, safe AWD can be implemented in rainfed areas, thus a high potential for methane emissions reduction during wet seasons.

Author Contributions: R.L. designed and supervised the research, analyzed the data and wrote the paper. P.X. co-supervised the research and advised on the varietal treatments. C.B. advised on the methodologies and commented to improve the manuscript.

Funding: The research was supported by funding provided to the International Rice Research Institute by the Swiss Agency for Development and Cooperation for the CORIGAP project (Grant no. 81016734).

Acknowledgments: We thank Duangsavang Lovanxay (ARC, NAFRI) and Sonephom Xayachack (ERRDC, NAFRI) for their assistance in facilitating the field experiment establishment and management, data collection, soil analysis and initial data processing. We also sincerely thank the anonymous reviewers for their helpful comments that improved the manuscript.

References

1. Tuong, T.P.; Bouman, B.A.M. Rice production in water scarce environments. In *Water Productivity in Agriculture: Limits and Opportunities for Improvement*; IKijne, J.W., Barker, R., Molden, D., Eds.; CABI: Wallingford, UK, 2003; pp. 53–56.
2. Comprehensive Assessment of Water Management in Agriculture. *Water for Food, Water for Life: A Comprehensive Assessment of Water Management in Agriculture*; Earthscan: London, UK; International Water Management Institute: Colombo, Sri Lanka, 2007; Volume 645, pp. 7–10.

3. Ramasawamy, S.; Armstrong, J. *Lao PDR Trend Food Insecurity Assessment—Lao PDR Expenditure and Consumption Survey (2002/03 and 2007/08 LECS (draft))*; European Commission; Lao Statistics Bureau (Lao PDR); FAO: Rome, Italy, 2012.

4. Eliste, P.; Santos, N. *Lao People's Democratic Republic Rice Policy Study*; World Bank: Washington, DC, USA, 2012. (In English)

5. Basnayake, J.; Inthavong, T.; Kam, S.P.; Fukai, S.; Schiller, J.M.; Chanphengxay, M. Climatic diversity within rice environments in Laos. In *Rice in Laos*; Schiller, J.M., Chanphenxay, M.B., Linquist, B., Appa Rao, S., Eds.; International Rice Research Institute: Los Banos, Philippines, 2006; p. 457.

6. Bouman, B.A.M.; Lampayan, R.M.; Tuong, T.P. *Water Management in Irrigated Rice. Coping with Water Scarcity*; International Rice Research Institute: Los Banos, Philippines, 2007; p. 53.

7. Sharma, P.K. Effect of periodic moisture stress on water-sue efficiency in wetland rice. *Oryza* **1989**, *26*, 252–257.

8. Lampayan, R.M.; Rejesus, R.M.; Singleton, G.R.; Bouman, B.A.M. Adoption and economics of alternate wetting and drying water management for irrigated lowland rice. *Field Crop. Res.* **2015**, *170*, 95–108. [CrossRef]

9. Lampayan, R.M.; Inthavong, T.; Somsamay, V.; Quicho, E.; Eberbach, P. Farmer-participatory evaluation of alternate wetting and drying (AWD) water-saving technology in Southern Laos. *Lao J. Agric. For.* **2014**, *30*, 121–135.

10. Paddalia, C.R. Effect of age of seedling on the growth and yield of transplanted rice. *Oryza* **1980**, *18*, 165–167.

11. Bassi, G.; Rang, A.; Joshi, S.P. Effect of seedling age on flowering of cytoplasmic male sterile and restore lines of rice. *Int. Rice Res. Notes* **1994**, *19*, 4–8.

12. Mobasser, H.R.; Tari, D.B.; Muhammad, V.; Ali, E. Effect of seedling age and planting space on yield and yield components of rice (Neda variety). *Asian J. Plant Sci.* **2007**, *6*, 438–440.

13. Liu, W.H.; Zhou, X.; Li, Z.; Xin, C. Effects of seedling age and cultivation density on agronomic characteristics and grain yield of mechanically transplanted rice. *Sci. Rep.* **2017**, *7*, 14072. [CrossRef] [PubMed]

14. Lampayan, R.M.; Faronilo, J.E.; Tuong, T.P.; Espiritu, A.J.; de Dios, J.L.; Bayot, R.S. Effects of seedbed management and delayed transplanting of rice seedlings on crop performance, grain yield, and water productivity. *Field Crop. Res.* **2015**, *183*, 303–314. [CrossRef]

15. Stuart, A.M.; Pame, A.P.; Vithoonjit, D.; Viriyangkura, L.; Pithuncharurnlap, J.; Meesang, N.; Suksiri, P.; Singleton, G.R.; Lampayan, R.M. The application of best management practices increases the profitability and sustainability of rice farming in the central plains of Thailand. *Field Crop. Res.* **2018**, *220*, 78–87. [CrossRef]

16. Gomez, K.A.; Gomez, A.A. *Statistical Procedures for Agricultural Research*, 2nd ed.; John Wiley and Sons: New York, NY, USA, 1984.

17. *STAR Version 2.0.1. Statistical Tools for Agricultural Research (STAR)*; Biometrics and Breeding Informatics, PBGB Division, International Rice Research Institute: Los Baños, Philippines, 2014.

18. Mishra, A.; Salokhey, V.M. Seedling characteristics and the early growth of transplanted rice under different water regimes. *Exp. Agric.* **2008**, *44*, 1–19. [CrossRef]

19. Li, G.H.; Zhang, Z.; Yang, C.; Liu, Z.; Wang, S.; Ding, Y. Population characteristics of high–yielding rice under different densities. *Agron. J.* **2016**, *108*, 1415–1423. [CrossRef]

20. Huang, M.; Yang, C.; Ji, Q.; Jiang, L.; Tan, J.; Li, Y. Tillering responses of rice to plant density and nitrogen rate in a subtropical environment of southern China. *Field Crop. Res.* **2013**, *149*, 187–192. [CrossRef]

21. Thanunathan, K.; Sivasubramanian, V. Age of seedling and crop management practices for high density (HD) grain rice. *Crop. Res.* **2002**, *24*, 421–424.

22. Mandal, B.K.; Sainik, T.R.; Ray, P.K. Effect of age of seedling and level of nitrogen on the productivity of rice. *Oryza* **1984**, *21*, 225–232.

23. Singh, R.S.; Singh, S.B. Effect of age of seedlings, N–levels and time of application on growth and yield of rice under irrigated condition. *Oryza* **1999**, *36*, 351–354.

24. Kewat, M.L.; Agrawal, S.B.; Agrawal, K.K.; Sharma, R.S. Effect of divergent plant spacings and age of seedlings on yield and economics of hybrid rice (*Oryza sativa*). *Indian J. Agron.* **2002**, *47*, 367–371.

25. Chandra, D.; Manna, G.B. Effect of planting date, seedling age, and planting density in late planted wet season rice. *Int. Rice Res. Notes* **1988**, *6*, 30.
26. Sarangi, S.K.; Maji, B.; Singh, S.; Burman, D.; Mandal, S.; Sharma, D.K.; Singh, U.S.; Ismail, A.M.; Haefele, S.M. Improved nursery management further enhances the productivity of stress–tolerant rice varieties in coastal rainfed lowlands. *Field Crop. Res.* **2015**, *174*, 61–70. [CrossRef]

Article

Water Use and Rice Productivity for Irrigation Management Alternatives in Tanzania

Stanslaus Terengia Materu [1], Sanjay Shukla [2,*], Rajendra P. Sishodia [2], Andrew Tarimo [1] and Siza D. Tumbo [1]

[1] Department of Engineering Sciences and Technology, Sokoine University of Agriculture, P. O. Box 3003, Chuo Kikuu, Morogoro, Tanzania; stanslaus_materu@yahoo.com (S.T.M.); andrewtarimo2@yahoo.co.uk (A.T.); siza.tumbo@gmail.com (S.D.T.)

[2] Agricultural and Biological Engineering Department, University of Florida, 2685 State Road 29 N, Immokalee, FL 34142, USA; rpsishodia@ufl.edu

* Correspondence: sshukla@ufl.edu; Tel.: +1-239-658-3425

Received: 28 April 2018; Accepted: 27 June 2018; Published: 1 August 2018

Abstract: Rice production is important for global food security but given its large water footprint, efficient irrigation management strategies need to be developed. Expansion of rice growing area is larger than any other crop in Africa due to increasing demand for rice. Three rice irrigation management alternatives with the system of rice intensification (SRI) were field-evaluated against the conventional continuously flooded system (CF) in Tanzania. Production systems included: (1) CF (50 mm ponding depth for the entire season); (2) SRI (40 mm ponding for 3 days and no irrigation for next 5 days); (3) 80% SRI (80% of the SRI ponding); and (4) 50% SRI (50% of the SRI ponding). Experimental evaluation of the four systems was conducted for both wet and dry seasons. For the dry season, the SRI and 80% SRI produced higher yields of 9.68 tons/ha and 11.45 tons/ha and saved 26% and 35% of water, respectively compared to the CF (8.69 tons/ha). The yield advantage of the 80% SRI and SRI over the CF was less during the wet season with 6.01 tons/ha and 5.99 tons/ha of production, and water savings of 30% and 14%, respectively compared to the CF (5.64 tons/ha). The 50% SRI had lowest yield of all for both seasons, 7.48 tons/ha and 4.99 tons/ha for the dry and wet seasons, respectively. Statistically, the 80% SRI treatment outperformed all other treatments over the two seasons with an additional yield of 1.57 tons/ha and 33% (345 mm) water savings compared to the CF. Economic productivity of water (US$/ha-cm) over two seasons was highest for the 80% SRI ($20.27/ha-cm), while it was lowest for the CF ($12.89/ha-cm). Water saved by converting from the CF to the 80% SRI (1.98 million ha-cm) can support a 50% expansion in the current rice irrigated area in Tanzania. Even without irrigation expansion, the 80% SRI can increase rice production by 1.5 million tons annually while enhancing water availability for industrial and environmental uses (e.g., ecological preserves) and help achieve food security in Tanzania and the greater sub-Saharan Africa.

Keywords: Africa; deficit irrigation; food security; system of rice intensification; water conservation; water productivity

---

1. Introduction

Water is a valuable resource that is becoming increasingly scarce due to growing population and intensifying agriculture [1]. Water scarcity is challenging the ability of countries to meet the increasing food demand [2]. Globally, agriculture is the largest consumer ($\approx$70%) of freshwater accounting for 90% of consumptive water use [3,4]. Of the three main food crops (maize, wheat, and rice), rice is the most important crop especially in developing countries [5]. Given its large water footprint, practices that can reduce water inputs for rice production such as deficit irrigation need to be explored. Deficit irrigation is a technique used to minimize water losses and increase water efficiency, especially in areas where there is insufficient water supply for irrigation. Deficit irrigation management involves inducing marginal stress, except in critical growth stages where crop yield might be negatively affected [6].

Expansion of rice growing area is larger than any other crop in Africa due to its increasing demand. [7]. Tanzania is the largest (947,303 km^2) country in East Africa and accounts for 9% (2.6 million ton) of African rice production (30.8 million ton) [8]. However, due to a rising gap between production and consumption, many African countries, including Tanzania, are becoming increasingly dependent on rice imports [9,10]. At the same time, increasing irrigation withdrawals and spatial and temporal variability in rainfall and surface flows are causing water scarcity in many parts of Tanzania such as the Pangani and Rufiji River basin [11,12]. The Pangani and Rufiji rivers support majority of irrigated agriculture in Tanzania and support almost entire hydroelectric generation in Tanzania (Mtera, Kidatu and Kihansi plants) [12]. Growing population, increasing food demands, and increased rainfall variability due to changed climate is likely to exacerbate water availability in the future. There is a need to develop alternative farming systems that can increase or sustain rice yields with reduced water footprint to ensure the food security in Tanzania.

Field water use for rice typically ranges from 1000–2000 mm [13], which is 2–3 times of other cereal crops. In rice production systems, a large quantity of water is lost through evapotranspiration, surface runoff, seepage, and deep percolation [14]. Several water-saving irrigation techniques have been developed for rice [15]. For instance, in Asia, the most widely adopted water-saving practice is aerobic rice production system. Although the aerobic rice system reduces water use it also results in lower yields compared to lowland flooded rice [16]. This practice also has some limitations related to soil type, rice variety, and socio-economic constraints [17]. Other strategies being pursued to reduce rice water requirements include alternate wetting and drying (AWD) and saturated soil culture [18]. Studies show that AWD can reduce crop water requirements while maintaining or even increasing the yield as compared to the conventional flooded system [19,20]. The AWD is an irrigation practice where water is applied to attain certain depth of ponding after which the field is left unirrigated for some time (e.g., 5 to 7 days) to dry out or drain. In the traditional continuously flooded (CF) system, water is applied at a frequency that will maintain a certain depth (e.g., 5 cm) of ponding throughout the season. Under the CF, more than 50% of irrigation water is lost through seepage, deep percolation, and excessive unproductive evaporation [21].

The system of rice intensification (SRI) is a relatively new production practice that has been adopted by many farmers in Asia and Sub-Saharan Africa [22–25]. The SRI is a combination of agronomic practices comprising of land preparation, seed selection, nursery establishment, transplanting of young age seedlings (8 to 12 days), wider plant spacing, AWD, and frequent weeding [22,26]. The SRI practice has been reported to substantially increase the yields as compared to the CF system [27]. The AWD irrigation technique used in the SRI production system can reduce water use by minimizing evaporation and deep percolation losses. The SRI combined with AWD has a potential to reduce water application and yet increase or sustain current yields in Tanzania. A range of AWD regimes are possible with SRI. However, limited research has been conducted on the evaluation of SRI in combination with different AWD regimes (e.g., ponding depth) in Tanzania. The current SRI practice lacks specific information regarding the irrigation management needed to achieve optimum yield. The goal of this study was to find water sustainable rice production systems in Tanzania by comparing water and yield metrics of conventional continuously flooded rice with SRI production system under different AWD regimes during wet (February-June) and dry (September-January) seasons. Specifically, the study attempts to answer the following questions: (a) Can SRI-AWD combination with reduced ponding depth significantly increase crop yield and reduce irrigation requirement compared to CF? (b) Which irrigation management strategy (scheduling and ponding depth) provides highest water productivity?

2. Methods and Materials

2.1. Study Area

The experiment was conducted in Morogoro, located 200 km south west of the city of Dar es Salaam. The experimental fields were located at the research farm of Sokoine University of Agriculture (SUA) (latitude = 37°39′26″ E and longitude = 6°51′5″ S) at an altitude of 510 m above mean sea level. The average annual temperature at the site is 23 °C with a minimum of 15 °C in July and a maximum of 32 °C in November and December. The mean relative humidity (1971–2000) for the area is 73%. Rice is grown during two seasons, dry (September–January) and wet (February–June) seasons. These two growing seasons have different irrigation needs due to differences in rainfall (1971–2000 average wet season rainfall = 53 cm and dry season = 38 cm) and evaporative demands. The seasonal mean (1971–2000) relative humidity is 66% and 78% for the dry and wet seasons, respectively. Humidity, wind speed, solar radiation, and temperature data, measured at the SUA meteorological station, were used to calculate reference evapotranspiration (ETo) using FAO Penman-Monteith method [28] (Figure 1). The average ETo for the wet and dry seasons are 52 and 64 cm, respectively. Given this weather variability, the optimum irrigation strategy for the dry season is likely to be different from the wet season. Although water availability limits large-scale production, better yields promote rice production during the dry season.

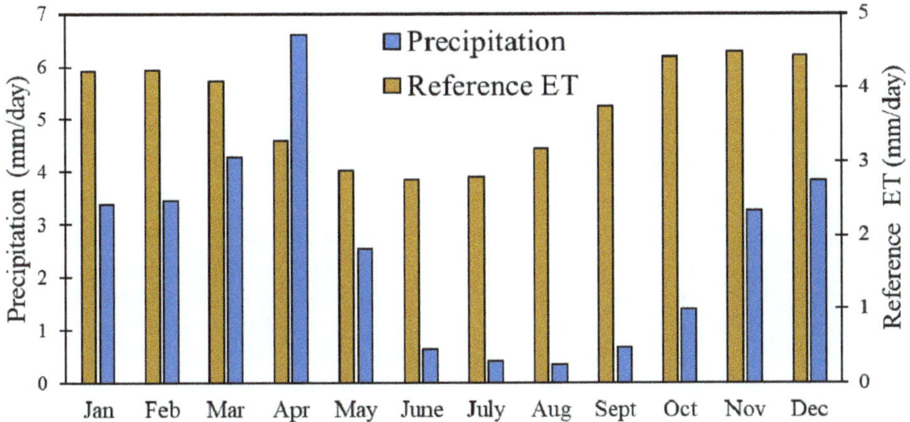

Figure 1. Average daily rainfall and reference ET (1971–2000) for Morogoro, Tanzania.

The experiment was conducted during the dry and wet seasons of 2012–2013. For 2013 wet season, the experimental site was moved to another plot within the same research farm due to land availability issues. Annual rainfall was measured at the site using a standard rain gauge. Another set of rainfall measurements were taken at the meteorological station at SUA. The rainfall system is bimodal, characterized by two rainfall peaks with short rains from October to December and long rains from March to May (Figure 1). Measured rainfall at the site was 489 mm and 1379 mm during the dry and wet season, respectively. High rainfall in the wet season disrupted the AWD cycle for all the SRI treatments. The top soil (0–30 cm) at the research site is a dark brown, clay loam (47% clay, 7% silt and 46% sand). The soil is acidic with a pH of 5.6. The volumetric water content at field capacity and permanent wilting point is 40.1% and 28.7%, respectively for the top soil (0–30 cm). The saturated hydraulic conductivity of the top 30 cm soil is 0.35 mm/h. Soil texture (sand, silt and clay percent) and bulk density data obtained from collected soil samples were used to estimate soil water retention and saturated hydraulic conductivity parameters [29].

2.2. Experimental Design

The layout of the experiment was a complete randomized block design (CRBD) with factorial arrangement of four treatments with three replications. A total of 12 experimental fields (area each field = 40 m^2) were identified within a rice field. Four production systems (treatments) with differential irrigation volumes evaluated in this study included: (1) continuous flooding (CF, Control) (maintain 50 mm ponding depth for the entire season); (2) SRI (maintain 40 mm ponding depth for three consecutive days followed by no irrigation for five days); (3) 80% SRI (maintain 80% of the SRI ponding depth i.e., 32 mm for three consecutive days followed by no irrigation for five days); and (4) 50% SRI (maintain 50% of the SRI ponding depth i.e., 20 mm for three consecutive days followed by no irrigation for five days). All the SRI systems involved alternate wetting and drying (AWD) during the initial stages. After the panicle initiation stage, continuous flooding was practiced in all the SRI treatments to maintain 20 mm ponding depth until plant senescence to ensure better grain filling. Important dates for the two seasons are shown in Table 1.

Table 1. Crop management and development stages for dry and wet seasons.

Event	Period (DAT *)	Dry Season	Wet Season
Field Preparations	-	15 September 2012	8 February 2013
Nursery	-	24 September 2012	18 February 2013
Transplanting	-	6 October 2012	1 March 2013
Tillering	0–46	21 November 2012	16 April 2013
Panicle initiation	47–59	3 December 2012	29 April 2013
Flowering	60–72	15 December 2012	13 May 2013
Grain filling	73–90	1 January 2013	27 May 2013
Harvesting	111 (Wet season) 113 (Dry season)	26 January 2013	19 June 2013

*: DAT-days after transplanting.

The SRI treatments involved transplanting 12-day seedlings, with two leaves. The seedlings were transplanted carefully and quickly to minimize seedling damage. The number of seedling per hill was one for all the SRI treatments; this allowed optimum growth without competition for nutrients. For the CF treatment, based on the conventional practice in Tanzania, three seedlings per hill were planted. For both the CF and SRI treatments, fertilizer was applied at a rate of 50 kg/ha (N), 50 kg/ha (K_2O) and 50 kg/ha (P_2O_5) before the last puddling event. Weeding was done manually at 12-day intervals using a spike-toothed harrow. To encourage greater root and canopy growth, plant-to-plant and row-to-row spacing was maintained at 25 cm for all the SRI and CF treatments.

2.3. Sri Management

Each experimental plot was leveled to allow uniform water ponding. The soil was kept saturated for five days, and then rotavated. The field was harrowed twice at an interval of three days to ensure proper soil-water mixture. Twelve-days-old seedlings were transplanted before the emergence of a third leaf. Care was taken to separate the seedling from the seedbed to avoid damage to the young root. One seedling was planted per hill at a depth of two cm on the 25 cm square grid. Between the transplanting and appearance of panicles, three to five days irrigation cycle was followed, i.e., the field was irrigated for three consecutive days and then left to dry for five days. The goal was to keep the soil moist but not saturated to allow air to get into the soil for improved soil health and root growth. After panicle initiation, irrigation was applied to maintain 20 mm ponding for all three SRI treatments.

2.4. Irrigation and Soil Moisture Measurements

Irrigation water volume for each treatment was measured using a propeller type flow meter. Soil moisture was measured every 15-min using a capacitance probe (EnviroScan, Sentek Technologies,

Stepney, Australia; sensors at 10, 20, 30, 60, 80, and 90 cm below the soil surface) in one of the plots for each of the four treatments. Manufacturer provided calibration equation for clay loam soil was used to measure the soil moisture content at multiple depths. Capacitance-type soil moisture probes provide reasonable soil moisture measurements even without site-specific calibration (3–4% accuracy) and therefore, could be used for irrigation scheduling [30,31]. Measured soil moisture data from capacitance probes have been successfully used to determine irrigation water requirements and scheduling for agricultural crops [32,33]. Furthermore, measured average maximum soil moisture (saturation moisture) within and below the root zone at the study site was 48–60% which is close to the estimated saturation moisture content of 45–50%. Therefore, measured soil moisture was assumed to adequately represent the actual soil moisture content and its variation during the growing seasons. The capacitance probes were connected to a CR206 datalogger (Campbell Scientific Inc., Logan, UT, USA) to store the data. For the dry season, soil moisture was measured from 3 November 2012 to 30 January 2013. During the wet season, soil moisture data could only be measured from 4 March–6 May 2013 due to theft of the datalogger.

2.5. Plant and Yield Observations

Five plants were randomly selected from each plot to measure plant height and number of tillers during each development stage. At grain maturity, the field was drained to allow the soil to dry before harvesting. Three one m quadrants were selected in each plot for yield measurements. Dry biomass (oven dried for 24 h or more until no change in weight) of different plant organs (stem, leaves and panicles) was weighed. Length and width of leaves were measured manually to estimate leaf area index (LAI) using the method by Yin et al. (2000) [34].

2.6. Data Analysis

All statistical analyses were conducted using SAS [35]. The data were analyzed using Tukey-Kramer test for comparing pair wise differences of means [36]. Variables compared included water applied (mm), crop yield (kg/ha), soil moisture (%vol.), LAI, above ground biomass (kg/ha) and economic productivity of water (US$/ha-cm). The economic productivity of water was calculated using the income (I, $) from crop yield and volume of water applied (ha-cm, irrigation plus rainfall) [37] as:

$$Economic\ productivity\ of\ water = \frac{I(\$)}{Water\ applied\ (\text{Ha} - \text{cm})}$$

Average farm gate paddy price of 550 Tanzanian Shillings TZS/kg [38] and prevailing exchange rate (1 US$ = 1600 TZS) during 2013 was used to calculate the income (I).

3. Results and Discussion

3.1. Plant Growth

Plant height for the CF and SRI treatments were significantly ($p < 0.05$) higher than both the 80% SRI and 50% SRI for both seasons (Table 2). Plant height for the CF was 23% and 63% higher than the 80% SRI and 50% SRI, respectively during the dry season. Plant height in the wet season for the CF was 11% and 30% greater than the 80% SRI and 50% SRI, respectively. More ponding depth or higher water availability is the likely reason for higher plant height in the CF as it can increase plant nutrient uptake and plant height [39,40]. Similar plant heights for the CF and SRI are likely due to similar soil moisture or water availability in the root zone.

The number of tillers for the 80% SRI was significantly higher ($p < 0.05$) than rest of the three treatments in both seasons (Table 2). Based on the statistical analyses results, the number of tillers can be arranged as 80% SRI > SRI = CF > 50% SRI (Table 2). For the dry season, the 80% SRI had 40% more tillers than the CF and 93% more tillers than the 50% SRI. Shortening of the vegetative stage duration has been shown to result in increased tillers [41]. High number of tillers per hill for the 80% SRI indicates higher potential yield than the rest of the treatments. Because panicles are attached to tillers, the number of tillers are usually an indicator of yield; the higher the number of tillers, the higher the potential for increased yield. The advantage of the SRI method in enhancing tiller numbers was observed by many researchers [42,43]. Transplanting of younger seedlings, higher plant spacing and soil aeration due to wetting and drying cycle promotes root growth and tillers under the SRI system [44]. Results for the wet season were similar to the dry season however the numerical differences between the treatments were much lower. Part of this difference was due to rainfall that masked the effect of differential irrigation input. Overall, the 80% SRI outperformed all other treatments in the number of tillers indicating higher yield potential than other treatments.

Biomass (Figure 2) and LAI (Figure 3) followed similar trends. Based on the results from statistical analyses, the order for the dry and wet seasons biomass were 80% SRI > CF > SRI > 50% SRI and 80% SRI > SRI > CF > 50% SRI (Figure 2), respectively. Higher biomass leads to higher accumulation of non-structural carbohydrate in the culms and leaf cover which can rapidly be trans-located to the panicle during the initial stage of grain filling and can increase the potential for higher crop yield [45]. Overall, key plant growth parameters such as biomass, LAI and number of tillers indicate the best plant performance for the 80% SRI treatment followed by the SRI for both seasons.

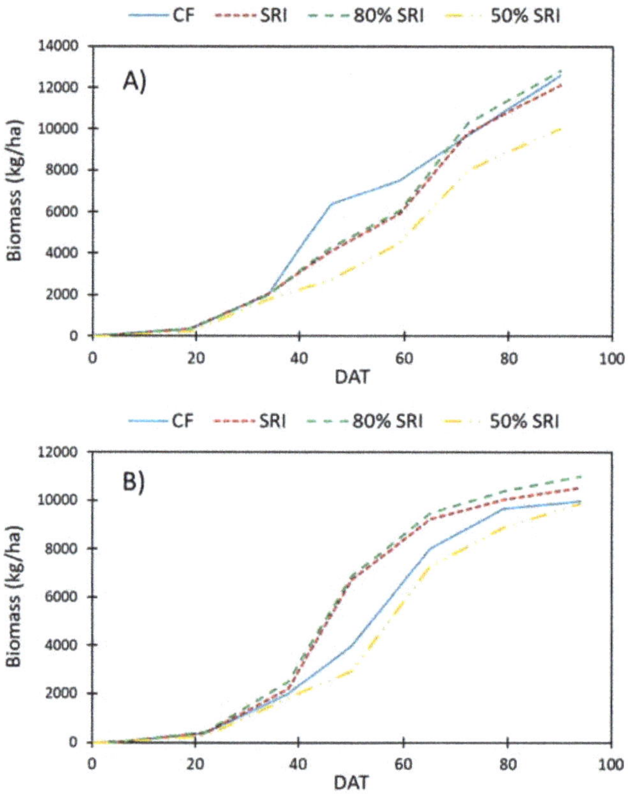

Figure 2. (**A**) Dry season (October 2012 to January 2013) and (**B**) wet season (February 2013 to June 2013) total biomass for continuous flooding (CF), system of rice intensification (SRI), 80% SRI, and 50% SRI treatments. The 80% SRI and 50% SRI refers to 80% and 50% of the SRI ponding depth, respectively. 0 to 46, 47 to 60, 61 to 75, and 77 to 94 days after transplanting (DAT) corresponds to vegetative, flowering, panicle initiation, and senescence stages, respectively.

Table 2. Plant height and number of tillers for the dry and wet seasons.

Treatments *	Plant Height (m)		Number of Tillers	
	Dry Season	Wet Season	Dry Season	Wet Season
CF	0.49 [a]	0.52 [a]	40 [a]	28 [a]
SRI	0.44 [a]	0.48 [a]	42 [a]	31 [a]
80% SRI	0.40 [b]	0.47 [b]	56[b]	38 [b]
50% SRI	0.30 [c]	0.40 [c]	29 [c]	25 [c]

* CF—Continuously flooded, SRI—System of Rice Intensification, 80% SRI—80% of SRI ponding depth, and 50% SRI—50% of SRI ponding depth. Note: Treatments with different letters (superscripts) were significantly different at 0.05 significance level.

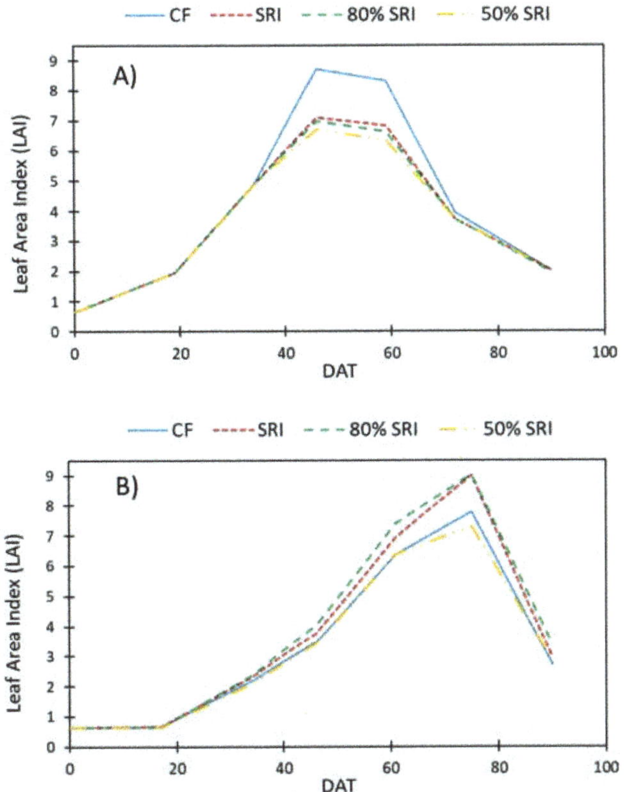

Figure 3. (**A**) Dry season (October 2012 to January 2013) and (**B**) wet season (February 2013 to June 2013) leaf area index (LAI) for continuous flooding (CF), system of rice intensification (SRI), the 80% SRI, and 50% SRI treatments. The 80% SRI and 50% SRI refers to 80% and 50% of the SRI ponding depth, respectively. Days after transplanting (DAT) 0 to 46, 47 to 60, 61 to 75, and 77 to 90 corresponds to vegetative, panicle initiation, flowering, and senescence stages, respectively.

3.2. Yield

For both seasons, the 80% SRI had statistically higher yield than other treatments while the 50% SRI had the lowest yield ($p < 0.05$; Figure 4). Following comparisons were statistically significant ($p < 0.05$) for both seasons: 80% SRI > CF, CF > 50% SRI, SRI > 50% SRI. For the dry season, yield for the 80% SRI was significantly higher than the SRI ($p = 0.01$). However, frequent rainfall events between transplanting and panicle initiation during the wet season (March–May 2013) resulted in similar soil moisture (Figures 5 and 6) in the root zone for all the SRI treatments which is the likely reason for small yield differences between the SRI, 50% SRI, and 80% SRI.

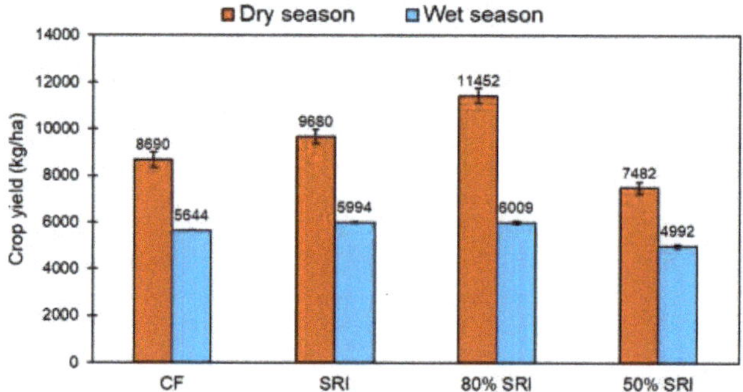

Figure 4. Dry and wet seasons rice crop yields for continuous flooding (CF), system of rice intensification (SRI), 80% SRI, and 50% SRI. 80% SRI and 50% SRI refers to 80% and 50% of the SRI ponding depth, respectively.

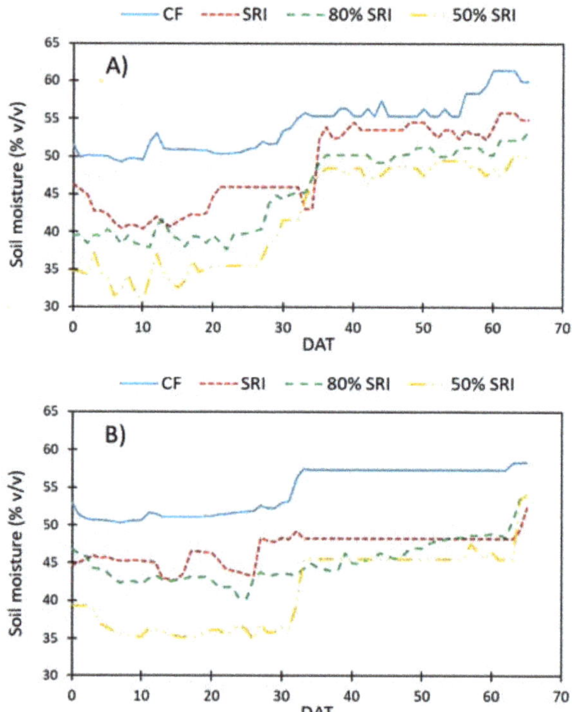

Figure 5. Daily soil moisture at 30 cm depth during the (**A**) dry season (October 2012 to January 2013) and (**B**) wet season (February 2013 to June 2013). CF is continuous flooding, SRI is system of rice intensification, 80% SRI and 50% SRI refers to 80% and 50% of the SRI ponding depth, respectively DAT is days after transplanting.

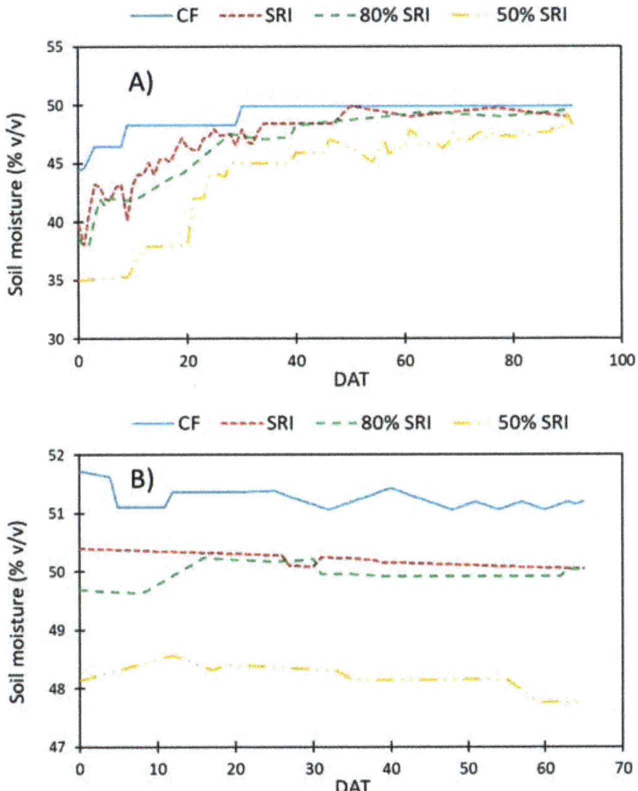

Figure 6. (**A**) Dry season (October 2012 to January 2013) and (**B**) wet season (February 2013 to June 2013) daily average soil moisture below the root zone (60 cm) against days after transplanting (DAT). The CF is continuous flooding, SRI is system of rice intensification, and 80% SRI and 50% SRI refers to 80% and 50% of the SRI ponding depth, respectively.

The wet season yields were lower than dry season mainly due to frequent rainfall between March and May 2013 resulting in saturated to near saturated soil moisture conditions during wet season for all treatments (Figures 5 and 6). The lowest yield observed for the 50% SRI in the dry season was higher than the highest yield for the 80% SRI during the wet season. Very little rainfall during November–December 2012 helped maintain the target soil moisture for all three SRI treatments (Figures 5 and 6) resulting in higher yields during the dry season as compared to the wet season.

The 80% SRI treatment produced 32% and 6% more yield than the CF for the dry and wet seasons, respectively. Although the 80% SRI treatment had almost the same yield as the SRI for the wet season, it had 18% more yield than the SRI treatment for the dry season. Higher yield for SRI is in agreement with observations from other studies [42,46,47] that noted higher grain yield when younger seedlings (8 to 12 days old) are transplanted at spacing ranging from 25 cm × 25 cm to 30 cm × 30 cm under non-flooded conditions. In this study, younger seedling and AWD irrigation were the main synergistic factors that increased the tillering ability (per hill and per area), panicle length, number of filled grains, and finally high yield for the 80% SRI treatment followed by the SRI compared to the CF.

Rice yield for the wet season was low with small differences among the treatments due to heavy and frequent rainfall at the beginning of the wet season (March and May) which resulted in sustained saturation/flooding during the wet period and prevented implementation of the SRI treatments.

Similar results were observed by Stoop et al. (2002) [22], who noted that it was not possible to attain higher yields with SRI compared to CF due to frequent rainfall events. Prolonged root zone saturation due to frequent rainfall events restrict root zone aeration under SRI thereby negatively affecting root and tiller growth. For the dry season, higher yield for the 80% SRI and SRI were due to: (1) adequate soil moisture required by the plant between transplanting and panicle initiation stages which enhances nutrients uptake; (2) reduced plant stress due to non-waterlogged conditions in the root zone which promotes healthier root growth; and (3) improved soil aeration which increases microbial metabolism activity. These factors resulted in better number of tillers and yield observed in this study for 80% SRI (Table 2).

3.3. Water Use and Productivity

3.3.1. Irrigation and Soil Moisture

Measured soil moisture within (30 cm) and below root zone (60 cm) indicates the success of treatment implementation. During the dry season, soil moisture of all the SRI treatments fluctuated around soil field capacity from the transplanting to the panicle initiation stage after which soil moisture for all treatments was similar due to sustained flooding (Figures 5 and 6). During the dry season, the only treatment that was allowed to fall below field capacity and, at times, close to wilting point during the tillering stage was 50% SRI (Figure 5). After panicle initiation stage, the soil moisture at 60 cm depth for all the SRI treatments for the dry season was similar to the CF because of continuous ponding maintained for all the treatments. Unlike dry season, soil in all the SRI treatments was near saturation during the initial part of the wet season due to frequent rainfall events (Figures 5 and 6). The desired SRI irrigation cycle (wetting for 3 consecutive days and drying for 5 days) was interrupted by heavy tropical rainfall at 30 days after transplanting (DAT) which negated the effects of the irrigation management and led to similar soil moisture in the root zone for all treatments resulting in reduced yield for the SRI treatments. Similar observations were made by Kombe (2012) [48]. Given that the 80% SRI resulted in maximum yield, it can be inferred that the desired soil moisture from transplanting to panicle initiation stage within the root zone (0–30 cm) is 44–48% (vol.) which falls between field capacity to saturation.

There was a large difference in the amount of water applied between the two seasons. In the dry season, total rainfall was 489 mm and irrigation volumes applied to the 80% SRI and CF treatments were 830 mm and 1286 mm, respectively (Figure 7A). On the other hand, during the wet season, there was a total of 1379 mm rainfall and the irrigation volume applied to the 80% SRI and CF were 554 mm and 787 mm, respectively (Figure 7). Frequent rainfall and variability in deep percolation losses resulted in similar water application for the 80% SRI and 50% SRI treatments (Figure 7). The irrigation volume varied depending on the rainfall, ET and deep percolation losses.

Results show that the 80% SRI can save 35% water compared to the CF during the dry season (Figure 7B). The water savings during the wet season will vary depending on rainfall. Tabbal et al. (2002) also reported that maintaining the soil moisture by alternate wetting and drying reduced irrigation volume by about 40–70%, compared with the traditional CF, without any significant loss in yield. In Ruaha basin of Tanzania, water abstracted from rivers for irrigation accounts for 56% of the wet season river flows and 93% of the dry season river flows [49]. Results from this study indicate that practicing the 80% SRI system for both rice growing seasons can achieve significant water savings which can be used for other purposes and/or help maintain environmental flows.

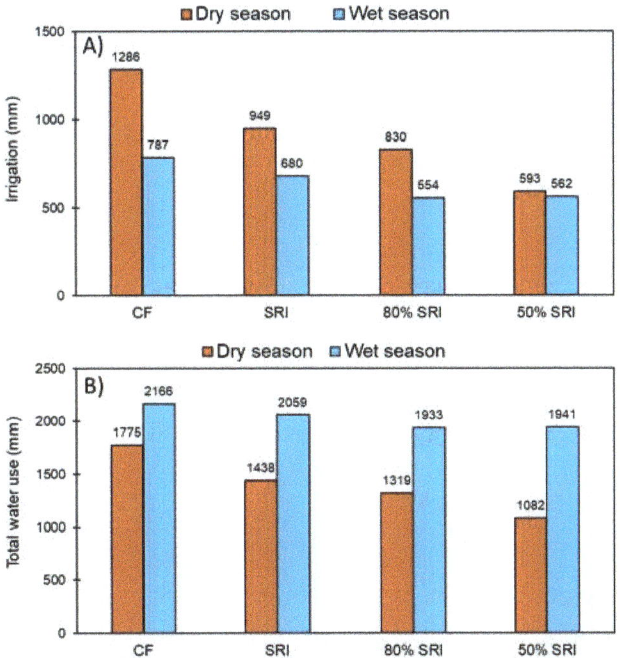

Figure 7. (**A**) Irrigation applied and (**B**) total water (irrigation + rainfall) input for continuous flood (CF), system of rice intensification (SRI), 80% SRI and 50% SRI treatments during the dry and wet seasons. 80% SRI and 50% SRI refers to 80% and 50% of the SRI ponding depth, respectively.

3.3.2. Economic Productivity of Water

All three SRI treatments had higher economic productivity than the CF production system indicating higher "income per drop of water" under SRI. The 80% SRI had the highest economic productivity during both dry and wet seasons (Figure 8). Dry season economic productivity for the 80% SRI was almost twice that of the CF (Figure 8). The 80% SRI treatment was 29% more productive than the SRI during the dry season (Figure 8). As compared to the dry season, wet season economic productivity showed little differences between the treatments. Despite 14–30% (107–233 mm) of difference in irrigation applied between the CF and SRI treatments, high wet season rainfall (1379 mm) masked the treatment effects on economic productivity (Figure 8). However, the 80% SRI was still 19% more beneficial than the CF during the wet season (Figure 8). Considering the yield advantage from the 80% SRI for both seasons, it is a better irrigation management strategy compared to the SRI and the CF (Figure 8). Although the economic productivity for the 50% SRI was higher than the CF for the dry season, the yield loss from this treatment is not likely to result in its acceptance over the 80% SRI. However, for areas with limited water supply the 50% SRI is still a viable option because yield reductions from the 50% SRI were only 13% compared to the CF over the two seasons.

In 2014, 957,218 ha of paddy area was harvested in Tanzania [8] almost all utilizing the CF production system. The yield advantage and water savings from the 80% SRI are likely to vary depending on rainfall amount and distribution, soil properties, water availability, and management strategies. For example, light textured soils (sandy loam) typically require higher irrigation volume than heavy texture soils (e.g., clay) mainly due to higher soil hydraulic conductivity that results in higher deep percolation losses. Assuming that the results from this study are applicable to the entire rice production area in Tanzania, implementation of the 80% SRI will result in annual water savings of 3.29 billion cubic meters and achieve additional production of 1.5 million tons of rice. If used, this water saved from the CF to the 80% SRI conversion can support 50% increase in the current rice irrigated area in Tanzania. Achieving these water savings and yield benefits is likely to increase the sustainability of rice production system in Tanzania and create additional water supplies for industry, environment, and other users.

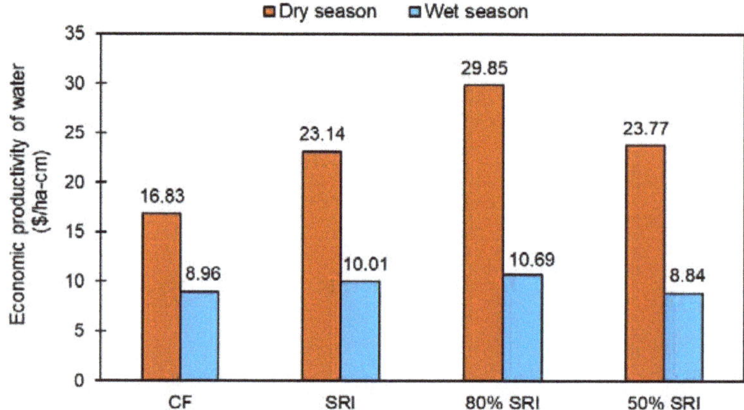

Figure 8. Economic productivity of water (US$/ha-cm) under different water management options during the dry and wet seasons. The CF is continuous flooding and SRI is system of rice intensification. 80% SRI and 50% SRI refers to 80% and 50% of the SRI ponding depth, respectively.

4. Conclusions

The traditional continuously flooded (CF) rice production system has no yield advantage over the SRI-AWD production system for both growing seasons in Tanzania. Consistently, highest yields were obtained from the 80% SRI system for both seasons which indicates that it is possible to increase yields while reducing the total irrigation volume. The 80% SRI system outperformed the CF system by 2762 kg/ha and 365 kg/ha with water savings of 456 mm and 233 mm during the dry and wet seasons, respectively. The water savings from the 80% SRI are 30–35% compared to the CF system. Given that 50% SRI system produced almost 90% yield compared to the CF during the wet season, the farmers in limited water supply regions are still likely to achieve a viable yield. For the farms that grow rice in both seasons, the annual water savings from the 80% SRI will be 689 mm with an additional production of 3127 kg/ha over the CF system. Considering that water savings from the 80% SRI accounts for 74% of annual rainfall (935 mm/year) in southern Tanzania, this irrigation management system has important implications for maintaining water supply and environmental flows in rivers.

To achieve large-scale yield and water saving benefits, there is a need to develop easy-to-understand water management recommendations for farmers in Tanzania. Maintaining 30 mm ponding depth (80% SRI) for three days followed by no irrigation for five days during transplanting to panicle initiation is an easy-to-follow recommendation. After panicle initiation, 20 mm ponding depth can be maintained to achieve increased yield with reduced irrigation. Farmers in the major river basins such as Ruaha are already experiencing water cuts which are likely to become more frequent or permanent in the future considering low reservoir levels for hydropower generation (e.g., Mtera and Kidatu power generation plants) [50]. Basin-scale implementation of the 80% SRI will not only help farmers sustain or improve the yields under current and future water cuts but also help maintain current power generation. One of the avenues for promoting large-scale implementation of the SRI system is to conduct on-farm demonstration studies. Furthermore, socio-economic factors including market prices, soil type, water availability, and existing irrigation infrastructure will have to be considered for wide-scale acceptance of the 80% SRI in Tanzania. Given the large-scale production of rice in Tanzania and projected water stress by 2025 [12], the 80% SRI has the potential to improve the well-being of farmers and contribute to food security in Tanzania.

Author Contributions: S.T.M., S.S., A.T., and S.D.T. conceived and designed the study, S.T.M. collected the data under the supervision of S.S., A.T. and S.D.T., S.T.M., S.S., and R.P.S. conducted the data analysis and wrote the manuscript.

Funding: This research was funded by the USAID, Feed the Future (iAGRI project) and Institute of Food and Agricultural Sciences, University of Florida.

Acknowledgments: The authors gratefully acknowledge D. N. Kimaro, Minde, and D. Kryabill for their helpful suggestions to improve this manuscript. We also thank Macha for data collection, Epignosis and Naza for their encouragement.

Conflicts of Interest: The authors declare no conflict of interest.

References

1. Rijsberman, F.R. Water scarcity: Fact or fiction? *Agric. Water Manag.* **2006**, *80*, 5–22. [CrossRef]
2. Hanjra, M.A.; Qureshi, M.E. Global water crisis and future food security in an era of climate change. *Food Policy* **2010**, *35*, 365–377. [CrossRef]
3. AQUASTAT. *Water Resources Development and Management Service*; Food and Agriculture Organization of the United Nations: Rome, Italy, 2010; Available online: http://www.fao.org/nr/water/aquastat/main/index.stm (accessed on 30 June 2016).
4. Siebert, S.; Burke, J.; Faures, J.M.; Frenken, K.; Hoogeveen, J.; Döll, P.; Portmann, F.T. Groundwater use for irrigation—A global inventory. *Hydrol. Earth Syst. Sci.* **2010**, *14*, 1863–1880. [CrossRef]
5. Nguyen, N.V. (Ed.) *Global Climate Changes and Rice Food Security*; FAO: Rome, Italy, 2002; pp. 24–30.
6. Geerts, S.; Raes, D. Deficit irrigation as an on-farm strategy to maximize crop water productivity in dry areas. *Agric. Water Manag.* **2009**, *96*, 1275–1284. [CrossRef]
7. Balasubramanian, V.; Sie, M.; Hijmans, R.J.; Otsuka, K. Increasing rice production in sub-Saharan Africa: Challenges and opportunities. *Adv. Agron.* **2007**, *94*, 55–133.
8. FAOSTAT. *Statistical Databases*; Food and Agriculture Organization of the United Nations: Rome, Italy, 2014.
9. Africa Rice Center. *Africa Rice Trends: Overview of Recent Developments in the Sub-Saharan Africa Rice Sector*; Africa Rice Center: Cotonou, Benin, 2007.
10. Nasrin, S.; Lodin, J.B.; Jirström, M.; Holmquist, B.; Djurfeldt, A.A.; Djurfeldt, G. Drivers of rice production: Evidence from five Sub-Saharan African countries. *Agric. Food Secur.* **2015**, *4*, 12. [CrossRef]
11. United Republic of Tanzania (URT). *State of the Environment Report—2006*; Division of Environment, Vice President's Office: Dar es Salaam, Tanzania, 2006; ISBN 9987-8990.
12. World Bank. *United Republic of Tanzania, Water Resources Assistance Stretagy, Improving Water Security for Sutaining Livelihoods and Growth*; Report No. 35327-TZ; Water and Urban Unit 1, Africa Region; World Bank: Washington, DC, USA, 2006; Available online: http://documents.worldbank.org/curated/en/378981468117562281/Tanzania-Water-resources-assistance-strategy-improving-water-security-for-sustaining-livelihoods-and-growth (accessesd on 6 July 2017).
13. Bouman, B.A.M.; Tuong, T.P. Field water management to save water and increase its productivity in irrigated rice. *Agric. Water Manag.* **2001**, *49*, 11–30. [CrossRef]
14. Guerra, L.C. *Producing More Rice with Less Water from Irrigated Systems*; International Water Management Institute (IWMI): Colombo, Sri Lanka, 1998; Volume 5.
15. Tuong, T.P.; Bouman, B.A.M. Rice production in water-scarce environments. *Water Prod. Agric. Limits Oppor. Improv.* **2003**, *1*, 13–42.
16. Bouman, B.A.M.; Peng, S.; Castaneda, A.R.; Visperas, R.M. Yield and water use of irrigated tropical aerobic rice systems. *Agric. Water Manag.* **2005**, *74*, 87–105. [CrossRef]
17. Nie, L.; Peng, S.; Chen, M.; Shah, F.; Huang, J.; Cui, K.; Xiang, J. Aerobic rice for water-saving agriculture. A review. *Agron. Sustain. Dev.* **2012**, *32*, 411–418. [CrossRef]
18. Tabbal, D.F.; Bouman, B.A.M.; Bhuiyan, S.I.; Sibayan, E.B.; Sattar, M.A. On-farm strategies for reducing water input in irrigated rice; case studies in the Philippines. *Agric. Water Manag.* **2002**, *56*, 93–112. [CrossRef]
19. Zhang, H.; Xue, Y.; Wang, Z.; Yang, J.; Zhang, J. An Alternate Wetting and Moderate Soil Drying Regime Improves Root and Shoot Growth in Rice. *Crop Sci.* **2009**, *49*, 2246–2260. [CrossRef]

20. Lampayan, R.M.; Rejesus, R.M.; Singleton, G.R.; Bouman, B.A. Adoption and economics of alternate wetting and drying water management for irrigated lowland rice. *Field Crops Res.* **2015**, *170*, 95–108. [CrossRef]

21. Bouman, B.A.M.; Feng, L.; Tuong, T.P.; Lu, G.; Wang, H.; Feng, Y. Exploring options to grow rice using less water in northern China using a modelling approach: II. Quantifying yield, water balance components, and water productivity. *Agric. Water Manag.* **2007**, *88*, 23–33. [CrossRef]

22. Stoop, W.A.; Uphoff, N.; Kassam, A. A review of agricultural research issues raised by the system of rice intensification (SRI) from Madagascar: Opportunities for improving farming systems for resource-poor farmers. *Agric. Syst.* **2002**, *71*, 249–274. [CrossRef]

23. Kassam, A.; Stoop, W.; Uphoff, N. Review of SRI modifications in rice crop and water management and research issues for making further improvements in agricultural and water productivity. *Paddy Water Environ.* **2011**, *9*, 163–180. [CrossRef]

24. Mati, B.M.; Wanjogu, R.; Odongo, B.; Home, P.G. Introduction of the system of rice intensification in Kenya: Experiences from mwea irrigation scheme. *Paddy Water Environ.* **2011**, *9*, 145–154. [CrossRef]

25. Thakur, A.K.; Uphoff, N.T.; Stoop, W.A. Scientific underpinnings of the system of rice intensification (SRI): What is known so far? In *Advances in Agronomy*; Academic Press: Cambridge, MA, USA, 2016; Volume 135, pp. 147–179.

26. Uphoff, N. Agroecological implications of the system of rice intensification (SRI) in Madagascar. *Environ. Dev. Sustain.* **1999**, *1*, 297–313. [CrossRef]

27. Zhao, L.; Wu, L.; Li, Y.; Lu, X.; Zhu, D.; Uphoff, N. Influence of the system of rice intensification on rice yield and nitrogen and water use efficiency with different N application rates. *Exp. Agric.* **2009**, *45*, 275–286. [CrossRef]

28. Allen, R.G.; Pereira, L.S.; Raes, D.; Smith, M. *Crop Evapotranspiration-Guidelines for Computing Crop Water Requirements—FAO Irrigation and Drainage Paper 56*; FAO: Rome, Italy, 1998; Volume 300, D05109.

29. Schaap, M.G.; Leij, F.J.; Van Genuchten, M.T. Rosetta: A computer program for estimating soil hydraulic parameters with hierarchical pedotransfer functions. *J. Hydrol.* **2001**, *251*, 163–176. [CrossRef]

30. Cobos, D.R.; Chambers, C. *Calibrating ECH2O Soil Moisture Sensors*; Application Note; Decagon Devices: Pullman, WA, USA, 2010.

31. Leib, B.G.; Jabro, J.D.; Matthews, G.R. Field evaluation and performance comparison of soil moisture sensors. *Soil Sci.* **2003**, *168*, 396–408. [CrossRef]

32. Dukes, M.D.; Simonne, E.H.; Davis, W.E.; Studstill, D.W.; Hochmuth, R. Effect of sensor-based high frequency irrigation on bell pepper yield and water use. In Proceedings of the 2nd International Conference on Irrigation and Drainage, Phoenix, AZ, USA, 12–15 May 2003; pp. 12–15.

33. Zotarelli, L.; Dukes, M.D.; Scholberg, J.M.S.; Femminella, K.; Munoz-Carpena, R. Irrigation scheduling for green bell peppers using capacitance soil moisture sensors. *J. Irrig. Drain. Eng.* **2010**, *137*, 73–81. [CrossRef]

34. Yin, X.; Schapendonk, A.H.; Kropff, M.J.; van Oijen, M.; Bindraban, P.S. A generic equation for nitrogen-limited leaf area index and its application in crop growth models for predicting leaf senescence. *Ann. Bot.* **2000**, *85*, 579–585. [CrossRef]

35. SAS Institute. *SAS/IML 9.3 User's Guide*; SAS Institute: Cary, NC, USA, 2011.

36. Somerville, P.N. On the conservatism of the Tukey-Kramer multiple comparison procedure. *Stat. Probab. Lett.* **1993**, *16*, 343–345. [CrossRef]

37. Gleick, P.H.; Christian-Smith, J.; Cooley, H. Water-use efficiency and productivity: Rethinking the basin approach. *Water Int.* **2011**, *36*, 784–798. [CrossRef]

38. Wilson, R.T.; Lewis, I. *The Rice Value Chain in Tanzania, a Report from the Southern Highlands Food Systems Programme*; Food and Agriculture Organization of the United Nations: Rome, Italy, 2015; pp. 1–15.

39. Chaudhary, D.K. Effect of water regimes and NPK levels on mid duration rice. (*Oryza sativa* L). Master's Thesis, Rajendra Agricultural University, Pusa, Bihar, 2003.

40. Parihar, S.S. Effect of crop-establishment method, tillage, irrigation and nitrogen on production potential of rice (*Oryza sativa*)-wheat (*Triticum aestivum*) cropping system. *Indian J. Agron.* **2004**, *49*, 1–5.

41. Panda, S.C.; Rath, B.S.; Tripathy, R.K.; Dash, B. Effect of water management practices on yield and nutrient uptake in the dry season rice. *Oryza* **1997**, *34*, 51–53.

42. Gani, A.; Rahman, A.; Rustam, D.; Hengsdijk, H. Water management experiments in Indonesia. In Proceedings of the International Symposium on Water Wise Rice Production, IARI, New Delhi, India, 2–3 November 2003; pp. 29–37.

43. Thakur, A.K.; Rath, S.; Patil, D.U.; Kumar, A. Effects on rice plant morphology and physiology of water and associated management practices of the system of rice intensification and their implications for crop performance. *Paddy Water Environ.* **2011**, *9*, 13–24. [CrossRef]
44. Uphoff, N. Higher yields with fewer external inputs? The system of rice intensification and potential contributions to agricultural sustainability. *Int. J. Agric. Sustain.* **2003**, *1*, 38–50. [CrossRef]
45. Takai, T.; Matsuura, S.; Nishio, T.; Ohsumi, A.; Shiraiwa, T.; Horie, T. Rice yield potential is closely related to crop growth rate during late reproductive period. *Field Crops Res.* **2006**, *96*, 328–335. [CrossRef]
46. Vijayakumar, M.; Ramesh, S.; Chandrasekaran, B.; Thiyagarajan, T.M. Effect of system of rice intensification (SRI) practices on yield attributes yield and water productivity of rice (*Oryza sativa* L.). *Res. J. Agric. Biol. Sci.* **2006**, *2*, 236–242.
47. Krishna, A.; Biradarpatil, N.K.; Channappagoudar, B.B. Influence of system of rice intensification (SRI) cultivation on seed yield and quality. *Agric. Sci.* **2008**, *21*, 369–372.
48. Kombe, E. The System of Rice Intensification (SRI) as a Strategy for Adapting to the Effects of Climate Change and Variability: A Case Study of Mkindo Irrigation Scheme in Morogoro, Tanzania. Unpublished Master's Thesis, Department of Agricultural Engineering and Land Planning, Sokoine University of Agriculture, Morogoro, Tanzania, 2012.
49. Mwakalila, S. Water resource use in the Great Ruaha Basin of Tanzania. *Phys. Chem. Earth Parts A/B/C* **2005**, *30*, 903–912. [CrossRef]
50. Makoye, K. *Farmers to Lose Water Access as Tanzania's Hydropower Runs Dry*; Reuters: London, UK, 2015; Available online: http://www.reuters.com/article/tanzania-water-hydropower/farmers-to-lose-water-access-as-tanzanias-hydropower-runs-dry-idUSL8N0ZC0X320150626 (accessed on 19 June 2018).

 *water*

Article

Year-Round Irrigation Schedule for a Tomato–Maize Rotation System in Reservoir-Based Irrigation Schemes in Ghana

Ephraim Sekyi-Annan [1,2], Bernhard Tischbein [1], Bernd Diekkrüger [3] and Asia Khamzina [4,*]

[1] Department of Ecology and Natural Resources Management, Center for Development Research, University of Bonn, Genscherallee 3, 53113 Bonn, Germany; sekyiannan@yahoo.com (E.S.-A.); tischbein@uni-bonn.de (B.T.)

[2] CSIR-Soil Research Institute, Academy Post Office, Private Mail Bag, Kwadaso-Kumasi, Ghana; sekyiannan@yahoo.com

[3] Department of Geography, University of Bonn, Meckenheimer Allee 166, 53115 Bonn, Germany; b.diekkrueger@uni-bonn.de

[4] Division of Environmental Science and Ecological Engineering, College of Life Science and Biotechnology, Korea University, 145 Anam-Ro, Seongbuk-Gu, Seoul 02841, Korea

* Correspondence: asia_khamzina@korea.ac.kr; Tel.: +82-2-3290-3062

Received: 23 March 2018; Accepted: 8 May 2018; Published: 10 May 2018

Abstract: Improving irrigation management in semi-arid regions of Sub-Saharan Africa is crucial to respond to increasing variability in rainfall and overcome deficits in current irrigation schemes. In small-scale and medium-scale reservoir-based irrigation schemes in the Upper East region of Ghana, we explored options for improving the traditional, dry season irrigation practices and assessed the potential for supplemental irrigation in the rainy season. The AquaCrop model was used to (i) assess current water management in the typical tomato-maize rotational system; (ii) develop an improved irrigation schedule for dry season cultivation of tomato; and (iii) determine the requirement for supplemental irrigation of maize in the rainy season under different climate scenarios. The improved irrigation schedule for dry season tomato cultivation would result in a water saving of 130–1325 mm compared to traditional irrigation practices, accompanied by approximately a 4–14% increase in tomato yield. The supplemental irrigation of maize would require 107–126 mm of water in periods of low rainfall and frequent dry spells, and 88–105 mm in periods of high rainfall and rare dry spells. Therefore, year-round irrigated crop production may be feasible, using water saved during dry season tomato cultivation for supplemental irrigation of maize in the rainy season.

Keywords: AquaCrop model; capillary rise; climate change; rainfall variability; supplemental irrigation

1. Introduction

Insufficient water availability, owing to variability in rainfall patterns and frequent dry spells exacerbated by climate change [1,2], threatens food security and rural livelihoods in Sub-Saharan Africa (SSA) [3]. In SSA, more than 95% of arable land is under rainfed crop production, which contributes 81% to the regional food basket [4,5]. Because of variable rainfall and low-input cultivation [6,7], grain yields are only from 1 to 2 Mg ha^{-1}, whereas attainable yields range between 4 and 5 Mg ha^{-1} in SSA [5,8]. Furthermore, risks of crop failure in SSA have increased due to land degradation and soil nutrient depletion [9,10], signified by negative annual NPK balances with -26 kg ha^{-1} N, -7 kg ha^{-1} P$_2$O$_5$, and -23 kg ha^{-1} K$_2$O, as reported in [11]. On a continental scale, annual NPK losses averaged 54 kg ha^{-1} (and ranged between 9 kg ha^{-1} in Egypt and 88 kg ha^{-1} in Somalia), resulting in land degradation in more than 40% of Africa's total farmland [12,13]. These risks have further reduced the already insufficient financial capacity of farmers to invest in sustainable land management (SLM) strategies [3,5]. However, such strategies are key for optimizing trade-offs between food production and other agro-ecosystem services [12]. In water-scarce environments such as the Upper East region of Ghana (UER), sustainable soil-water management has been identified as the most influential among agricultural management practices, including soil fertility management, selection of crop varieties, and control of pests and diseases [5,14], for enhancing food security as well as improving the smallholders' livelihoods [5,15–17].

The reservoir-based irrigation schemes in SSA, which store water (i.e., mostly surface runoff) in the rainy season, were originally designed to supply water for dry season crop irrigation, the livestock sector, fish farming, and domestic use, excluding supplemental irrigation in the rainy season. However, increasing climate variability calls for exploring the feasibility of supplemental irrigation for crop cultivation in the rainy season [3]. Supplemental irrigation has considerable potential to increase grain yield, particularly if provided during the critical stages of the crop growing cycle (i.e., booting and grain filling) [18].

Because of increasing competition for stored water in the dry season, the extra water demand for supplemental irrigation to bridge dry spells is likely to result in a mismatch between water supply and demand in the reservoir-based irrigation schemes. Thus, the requirement for supplemental irrigation might be satisfied with water saved through increased irrigation efficiency as a result of improving dry season irrigation scheduling [19]. As long as increased irrigation efficiency is accompanied by yield increments, this provides incentives for irrigators to engage in SLM [5,15,18,20]. Consequently, crop–water–soil–atmosphere models will be useful to determine the most appropriate irrigation schedules for the prevalent cropping practices and for assessing possible alternative scenarios [21–23]. Among the common models capable of simulating irrigated crop growth, those requiring large inputs of primary data, for instance APSIM [24] and CropSyst [25], and that are not available for free, such as the irrigation scheduling model ISAREG [26], might not be favorable for applications in SSA. The DSSAT model [27] has been commonly used to assess the impact of agronomic inputs on irrigated crop yield but at present is not suitable to evaluate the effectiveness of irrigation practices. Some other models, such as CROPWAT [28] do not distinguish between evaporation (i.e., non-beneficial water consumption) and crop transpiration, and do not provide an estimation of yield or, as with EPIC [29], apply simplified routines to evaluate the groundwater contribution to crop water use. Due to relatively modest data requirements, consideration of all major agro-hydrological processes, and its free availability, the AquaCrop model developed by the Food and Agriculture Organization of the Unites Nations (FAO) [22] has found many applications worldwide, including in SSA [30–32].

Current irrigation schedules in reservoir-based irrigation schemes in SSA are based on locally established rules governing access to water for irrigation, but with little consideration of crop- and site-specific water demands in terms of quantity and timing, resulting in the over-irrigation of crops [19]. For instance, in reservoir-based irrigation schemes in onion fields in the UER, the ratio of total water supply to gross irrigation demand ranged between 2.4 and 5.7 during dry season crop irrigation [33]. The problem of over-irrigation in reservoir-based irrigation schemes was further confirmed by gross irrigation amounts (GIAs) ranging from 380 to 852 mm for dry season tomato production in the UER [34], and between 274 and 838 mm for tomato cropping under groundwater irrigation in the same region [35]. Simulations have suggested that the net irrigation requirement (NIR) for dry season tomatoes ranges from 359 to 372 mm in the reservoir-based Koga irrigation scheme in Ethiopia [36], emphasizing the need as well as the potential to improve water management through irrigation scheduling to reduce water losses and increase productivity.

To the best of our knowledge, no study has attempted to develop an irrigation schedule for dry season cropping systems in the UER. Moreover, the limited number of studies on supplemental irrigation in SSA have not explored the feasibility of using dry season water savings in reservoir-based irrigation schemes. For example, Sanfo et al. [3] investigated the economic value of supplemental irrigation of grain crops using farm ponds of 300 m^3 capacity in south-western Burkina Faso, and reported that in years of low rainfall, supplemental irrigation could be a cost-effective intervention to reduce risks of crop failure and increase farmers' incomes. Fox and Rockström [37] also assessed the effect of supplemental irrigation, based on 150 m^3 capacity farm ponds, on the grain yield of sorghum in northern Burkina Faso and found that supplemental irrigation alone resulted in an approximately 56% increase in grain yield, making it a useful technology to mitigate dry spells and shorten the yield gap. Similarly, Mustapha [16] studied the water productivity of pearl millet under supplemental irrigation applied at five different crop growth stages in Nigeria and reported that the supplemental irrigation amount of 84 mm applied at booting and grain filling stages could result in a 69% increase in yields.

This study aims to improve the traditional dry season irrigation practices in reservoir-based irrigation schemes in the UER, and to assess the potential for introducing supplemental irrigation in the rainy season as an adaptation to climate change. To this end, we (i) parameterized and validated the AquaCrop model to render applications for irrigated crop production in the EUR of Ghana (ii) assessed the appropriateness of current water management in the typical tomato–maize rotational system; (iii) developed an improved irrigation schedule for dry season cultivation of tomato; and (iv) determined the requirement for supplemental irrigation of maize in the rainy season under different climate scenarios.

2. Materials and Methods

2.1. Study Area

The study was conducted between May 2014 and April 2016 in the medium-scale Vea irrigation scheme (VIS, 136 km^2) and the small-scale Bongo irrigation scheme (BIS, 0.98 km^2) in the UER, located between latitudes 10°30′ N and 11°15′ N and longitudes 0° W and 1°30′ W (Figure 1). The UER belongs to the Guinea–Sudano–Savanna agro-ecological zone characterized by a single rainy season starting in April/May and ending in September/October, followed by a dry season from November until April/May. The mean annual rainfall is 970 mm with high intra- and inter-seasonal variability, and the mean annual temperature is 29 °C (Figure 2). The annual potential evapotranspiration (ET$_0$) is twice as much as the annual precipitation, but evapotranspiration is exceeded by rainfall in the rainy season [34]. Soil types in the UER include Gleyic Lixisols, Ferric Lixisols, Haplic Lixisols, Lithic Leptosols, and Eutric Fluvisols, with loam and sandy loam as the dominating soil textures.

Two schemes were selected to capture the typical scale of irrigation schemes in the region, as well as the differing institutional settings in operations by the parastatal Irrigation Company of the Upper Region (ICOUR, Navrongo, Ghana) in Vea, and a community-based operation in Bongo (Figure 1). Furthermore, water allocation in the VIS is supply-driven, and thus a technician implements water supply schedules for 4–5 days continuously with 3–4 days interval between schedules. However, in the BIS, where water allocation is demand-driven, water can flow for the whole week (8 h per day on average) except on market days which occur twice a week. Irrigators in the UER tend to water their crops with as much water as is available resulting in over-irrigation, hence there is an urgent need for improved schedules which are crop- and site-specific [19]. On average, the total irrigation events for the dry season production of tomato ranges between 20 and 29 in both VIS and BIS.

The storage capacity, the elevation of the reservoir's spillway, and the irrigable area of the BIS are 0.43 MCM, 231 m and 12 ha, respectively, while the values for the VIS are 17.27 MCM, 189 m and 850 ha, respectively. The irrigable area in both schemes is equipped with lined trapezoidal primary canals which convey water by gravity to the cropping fields. Farm sizes range between 0.01 and 0.10 ha in the dry season, and up to 0.31 ha in the rainy season, in both irrigation schemes.

Rainfed crops include maize (*Zea mays*), pearl millet (*Pennisetum glaucum*), sorghum (*Sorghum bicolor*), and rice (*Oryza sativa*; cultivated also in the dry season under irrigation). Tomato (*Solanum lycopersicum*) and leafy vegetables such as roselle (*Hibiscus sabdariffa*), lettuce (*Latuca sativa*) and cowpea (*Vigna unguiculata*; grown primarily for the leaves) are irrigated in the dry season only. Currently, irrigation is not practiced in the rainy season. There are no soil bunds constructed on the cropping fields, except around rice fields. Furthermore, furrows are not blocked during irrigation, resulting in the surface runoff of irrigation water.

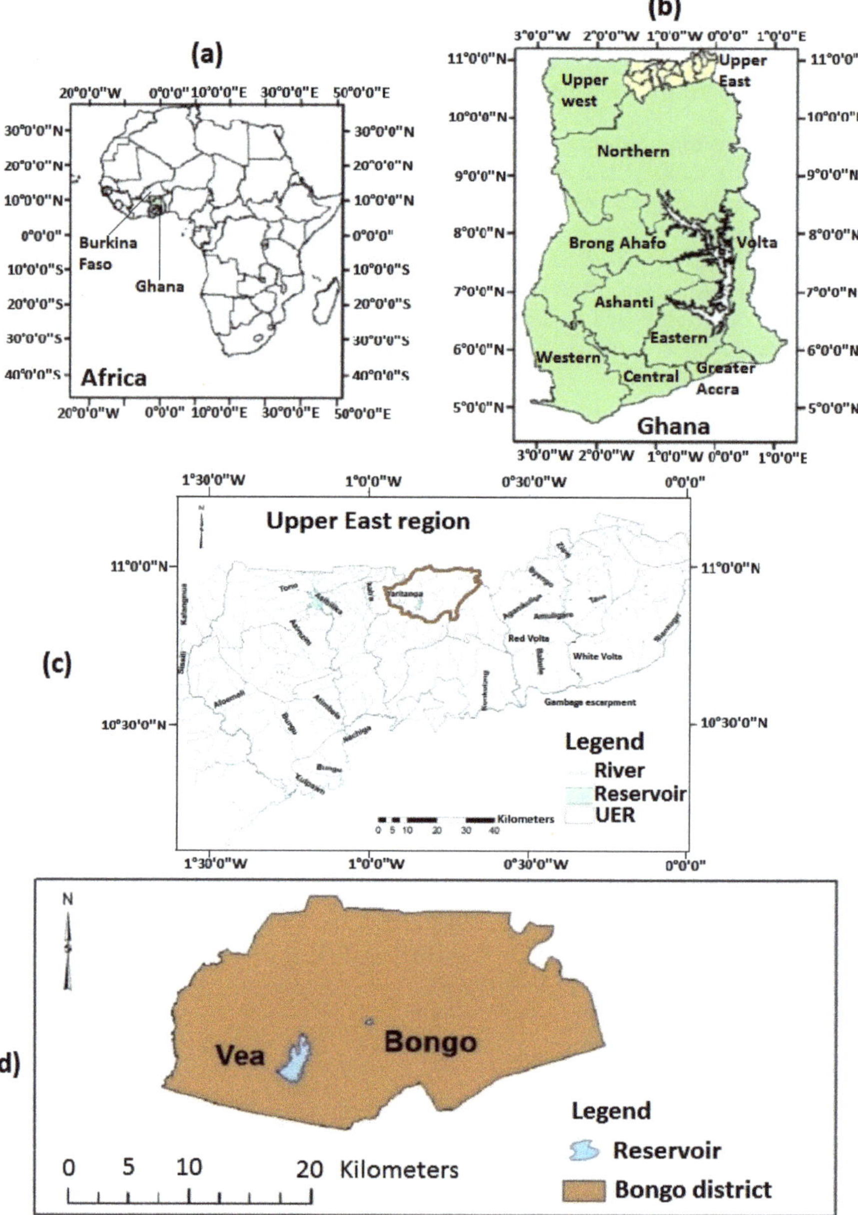

Figure 1. Location of Ghana in West Africa (**a**); the Upper East region (UER) of Ghana (**b**); hydrological network of the UER and location of the study area (**c**); and the study area including the Vea and Bongo reservoirs (**d**).

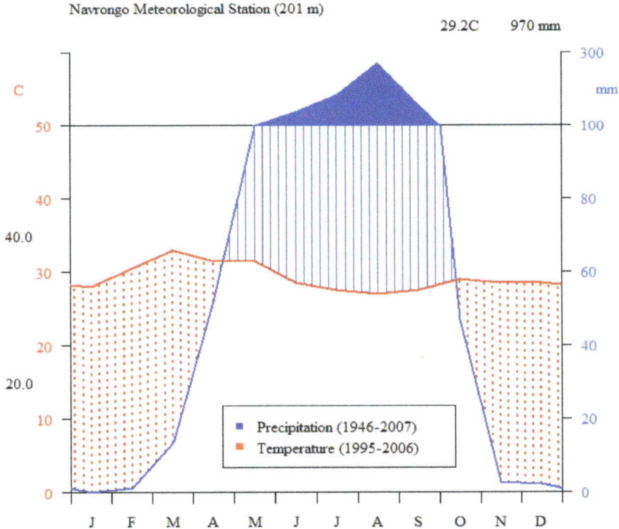

Figure 2. Walter–Lieth climate diagram for the Upper East region of Ghana based on data collected at Navrongo Meteorological Station (latitude 10°54′0″ N and longitude 1°06′0″ W; elevation 201 m above sea level). Precipitation data covered the period 1946–2007, and temperature data were measured between 1995 and 2006. Top of the graph shows the long-term mean annual temperature and rainfall. The value at the top-left of the temperature axis is the mean of the average daily maximum temperature of the hottest month; the value at the bottom of the same axis is the mean of the average daily minimum temperature of the coldest month. Area shaded in blue indicates the moist period and area shaded in red show the arid period. Area filled in blue indicates the period of excess water.

The principal cropping systems include tomato–maize rotation, millet/sorghum–leafy vegetable rotation, and rice mono-cropping under alternate wet–dry irrigation. In the VIS, the shares of the irrigable area in the dry season were 48%, 37%, and 12% for tomato, rice, and leafy vegetables, respectively, and 40%, 5%, and 55% for the same crops in the BIS. In the rainy season, the shares were 50%, 40%, and 10% for millet/sorghum, rice, and maize, respectively, in the BIS, while in the VIS these shares were 34%, 59% and 3%. In this study, the cropping system of dry season irrigated tomato in rotation with maize in the rainy season was selected for detailed analysis due to the socio-economic significance of these crops in the UER and SSA. Tomato was cropped once in the dry season in both the BIS and the VIS. The growing and irrigation period lasted for 113–123 days. In this period, mature tomato fruits were harvested 2–3 times. The duration of tomato seedling development was about 14 days. The growing period of maize sown directly in the field ranged between 84 and 113 days.

The application of cow manure (1 Mg ha^{-1}), NPK (0.21–0.7 Mg ha^{-1}) and ammonium sulfate fertilizer (0.1–0.34 Mg ha^{-1}) for tomato and maize, and Karate (i.e., lambda-cyhalothrin) and DDT insecticides for tomato only was observed in both schemes. On maize fields, manure was applied at ploughing, and mineral fertilizer at a later growth stage. Insufficient application of mineral fertilizer is common due to the high cost involved [9,38]. The fertilizers were applied twice in tomato and maize fields at 2–3 weeks after planting and later at 4–5 weeks after planting.

2.2. Model Description

AquaCrop is a crop water productivity model that simulates the response of crop yield to water supply and is particularly useful where water limits crop production. The model runs in daily time-steps which provides the basis for investigating the appropriateness of irrigation schedules to meet crop-specific demands in practical scheme operation. Consequently, the AquaCrop-based schedules have a high potential to increase crop water productivity [22]. The AquaCrop model can also simulate the effect of climate variability (including variations in temperature, atmospheric carbon dioxide and available water/rainfall) on crop production [22]. Additional useful features of the model are the ability to separate soil evaporation from crop transpiration and to quantify the capillary rise from shallow groundwater.

2.3. Data Collection and Preparation

The input data required for running AquaCrop were collected from two fields under a tomato–maize rotation system in each irrigation scheme (Figure 3).

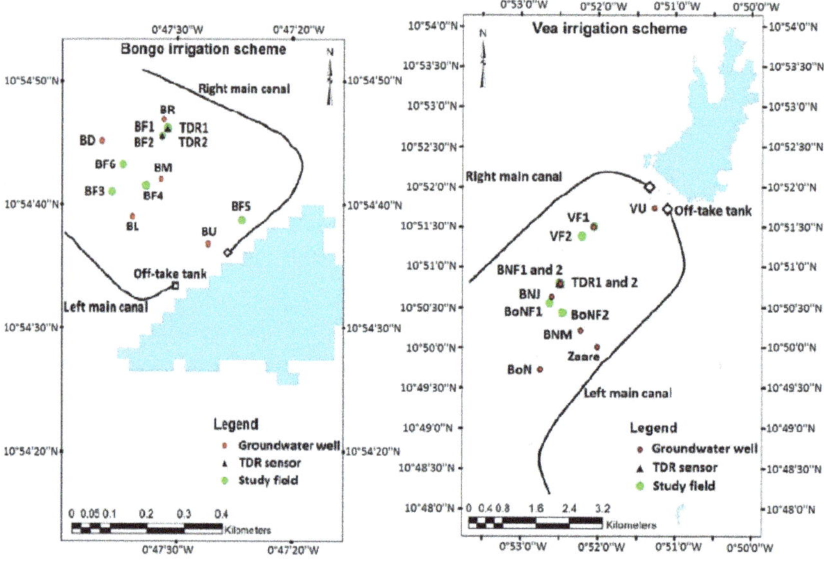

Figure 3. Layout of the irrigation schemes (reproduced from Sekyi-Annan et al. [19]). BF1–6 = Bongo fields, BNF1, 2 = Bongo Nyariga fields, BoNF1, 2 = Bolga Nyariga fields, BR = Bongo right well, BL = Bongo left well, BM = Bongo middle well, BD = Bongo downslope well, BU = Bongo upslope well, VF1, 2 = Vea fields, VU = Vea upslope well, BNM = Bongo Nyariga middle well, BNJ = Bongo Nyariga junction well, TDR1, 2 = Time domain reflectometers.

The model performance, based on the simulation of aboveground dry matter (DM), was assessed with multiple inbuilt statistical indicators including the coefficient of determination (R^2), normalized root mean square error (NRMSE), Nash–Sutcliffe model efficiency coefficient (EF), and Willmott's index of agreement (d). The R^2 indicates the fraction of the variance in observed data explained by the model and ranges from 0 (no agreement) to 1 (perfect agreement) between simulated and observed data. Typically, $R^2 > 0.5$ is acceptable for watershed simulations [39]. The NRMSE signifies the relative difference between the simulated results and the measured data, with NRMSE <10%, 10–20%, 20–30%, and >30% showing excellent, good, fair, and poor model performance, respectively. The EF quantifies the relative magnitude of the residual variance in comparison to the variance of the observed data. EF ranges between 1 and $-\infty$, where 1 signifies a perfect match between predictions and observations, 0 indicates that predictions are as accurate as the observed means, and a negative value indicates poor predictability. The d quantifies the extent to which the measured data are approached by the predictions and ranges from 0 (no agreement) to 1 (perfect agreement).

2.3.1. Estimation of Potential Evapotranspiration and Net Irrigation Requirement

Maximum and minimum air temperatures (T_{max} and T_{min}, °C), average relative humidity (RH, %), wind speed (U, m s^{-1}) and solar radiation (R_s, W m^{-2}) were measured by weather stations located near the study schemes, 10°54′54.1″ N and 0°49′35.3″ W in the BIS, and 10°50′44.6″ N and 0°54′43.9″ W in the VIS. Potential evapotranspiration was calculated based on the Penman–Monteith equation [40]:

$$ET_0 = \frac{1}{\lambda_w} \frac{\Delta(R_n - G) + \rho_a C_p (e_s - e_a)}{\Delta + \gamma_a \left(1 + \frac{r_c}{r_a}\right)} \tag{1}$$

where ET_0 is reference evapotranspiration (mm day^{-1}), R_n is net radiation (W m^{-2}), G is soil heat flux (W m^{-2}), ($e_s - e_a$) is the vapor pressure deficit of the air (kPa), ρ_a is mean air density at constant pressure (kg m^{-3}), C_p is the specific heat of the air (MJ kg^{-1} °C^{-1}), Δ is the slope of the saturation vapor pressure–temperature relationship (kPa °C^{-1}), λ_w is latent heat of vaporization (MJ kg^{-1}), γ_a is psychrometric constant (kPa °C^{-1}), r_c is crop resistance (s m^{-1}), and r_a is aerodynamic resistance (s m^{-1}).

Next, the net irrigation requirement was calculated based on the actual evapotranspiration simulated in AquaCrop as follows [22,41]:

$$NIR = \sum_{i=1}^{n} [(K_{cb} + K_e)ET_{0_i} - P_{e_i} - CR_i - W_{b_i}] \tag{2}$$

where NIR is the net irrigation requirement (mm), n is the number of days in the crop cycle, K_{cb} is the basal crop coefficient, K_e is the evaporation coefficient, P_e is effective rainfall (mm), CR is capillary rise (mm), and W_b is stored soil water (mm).

2.3.2. Rainfall and Scenario Analyses

Rainfall data during the years 1998–2014 were obtained for each scheme from the Tropical Rainfall Measuring Mission (TRMM) database. The total annual rainfall and total number and duration of dry spells were determined by the following conditions: (i) onset of rainfall is the beginning of a 10 day period between the second dekad of April and the first dekad of May during which the cumulative rainfall is ≥25 mm, and a dry spell ensuing within 30 days from the start of the 10 day period is ≤8 days [1,42]; (ii) cessation of rainfall is the last rainfall event between the third dekad of September and the second dekad of October [42]; (iii) dry spell is two or more consecutive non-rainy days [7], as even a period of two days without rainfall at critical growth stages is detrimental to crop production in savannah environments, particularly during periods of low rainfall; (iv) frequency of dry spells is the number of dry spells during the rainy season in the particular year under focus.

Additionally, the inter- and intra-seasonal variability of rainfall was expressed in the coefficient of variation based on the annual and monthly rainfall data, respectively:

$$CV = \frac{\sigma}{\mu} * 100 \; [\%] \tag{3}$$

where CV is the coefficient of variation, σ is the standard deviation, and μ the mean of the rainfall data.

For the estimation of the supplemental irrigation requirement for maize, two climate scenarios (i.e., wet and dry rainfall regimes) were formulated based on rainfall amount and the frequency of dry spells. The first scenario (S1) was a wet year characterized by ≤20% probability of exceedance (i.e., the likelihood of the occurrence of rainfall ≥1057 mm) and by less frequent dry spells [43]. The second scenario (S2) was a dry year characterized by ≥80% probability of rainfall occurrence exceeding 796 mm and by frequent dry spells [43] (Figure 4).

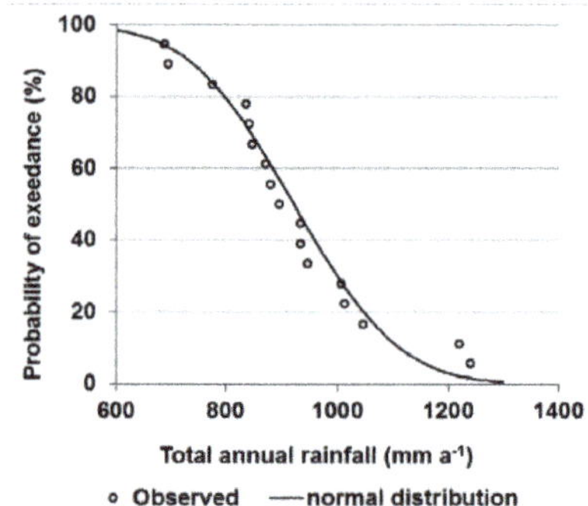

Figure 4. Probability plot of the total annual rainfall for the Vea and Bongo irrigation schemes for 1998–2014.

2.3.3. Soil Characteristics

Three soil profile pits were dug (at BF1 and BF4 in the BIS and at BNF2 in the VIS) to depths of 1.2–1.3 m in the Akrubu and Yaratanga soil series identified in the irrigation schemes. Soil samples were collected from the morphological soil horizons. The soil chemical properties, bulk density, soil moisture at saturation, field capacity, and permanent wilting point were determined in the laboratory [44] (Appendix A). Saturated hydraulic conductivity (K_{sat}) was determined in the laboratory by the falling head method [45] using undisturbed soil cores from the three soil pits. Comparison of the measured K_{sat} values with those determined from the pedo-transfer function based on soil texture and organic matter content [39,46] revealed significantly lower values from the laboratory measurements. This is likely caused by incomplete saturation of the undisturbed soil samples (especially samples with high clay content) before the test, leakage along the metal cylinder during the test, and the impact of soil structure and macropores. Consequently, the K_{sat} values determined from the pedo-transfer functions were used in the analysis.

2.3.4. Crop Growth and Yield Parameters

Sampling areas were demarcated within all selected fields for the collection of total aboveground biomass (AGB) [47]. Three rows were defined for bi-weekly AGB sampling, i.e., four times during the vegetative and reproduction stages. On each sampling day, three samples were collected per field from each defined row by cutting all plants along a 1 m rod. At harvest, two 8 m row sections in each of the selected fields were demarcated for AGB measurement. AGB was sampled and weighed as the yield components, i.e., maize grains and tomato fruits. The samples were weighed, oven-dried at 70–90 °C until a constant weight for at least 72 h, and subsequently, the DM and the yield components were weighed. The planting dates differed from one farmer to another, hence the growth stages of the crops at the time of sampling were not the same. Due to the late start of the field data collection in 2014, the AGB of maize during the vegetative stage was not measured. The harvest index (HI) was estimated as the ratio of the dry yield component to total aboveground DM. The tomato yields in BF1, VF1 and BNF1 could not be assessed during the 2015–2016 season owing to the early onset of the rainy season in 2016, leading to waterlogging and failed tomato yields. Tomato yield measurements were therefore conducted in the neighboring fields characterized by similar soil conditions and farming practices. In the 2014 rainy season, only the BF1 maize field was monitored in the BIS, as the BF6 maize field was not cropped by the farmer. Crop data could not be collected in 2015 in both schemes owing to technical challenges.

Plant density (PD) was determined in all sampling fields. Row spacing was measured as the average distance between two adjacent rows at five random locations in the field [47]. Leaf area index (LAI) was measured bi-weekly using the SunScan probe (SS1-UM-2.0) at five random locations at each field. LAI was converted into canopy cover (CC) using Equation (4), which was developed for maize and soybean but is also applicable to other crops with a similar leaf shape [21,22]:

$$CC = 1.005[1 - \exp(-0.6LAI)]^{1.2}$$ (4)

LAI measurements were interrupted in the 2014 rainy season and in the 2015–2016 dry season because of technical challenges. Maximum rooting depth (RD) was measured by manual excavations of at least three plants per crop at harvest time. A summary of all crop growth and yield parameters measured and details of their measurements each season is provided in Appendix B.

2.3.5. Gross Irrigation Amount

At the inlet of each of the selected fields, a Cipoletti weir, or a PVC pipe and a metallic staff gauge (50 cm) with metric graduation, was installed in the canal to measure the water inflow. A discharge equation for flow through the pipes during irrigation events was developed from in-situ measurements through a 'volumetric approach', using a bucket of a known volume (17.5 L) and a stopwatch. The time required to fill the bucket was recorded for seven different water depths read from the staff gauge. Discharges corresponding to the seven measured water depths were computed and, subsequently, discharge (Q, m^3 s^{-1}) was related to water depth (h, m) as follows:

$$Q = 0.073h^{1.334}$$ (5)

with R^2 = 0.972 and the standard error = 0.001.

The actual water abstraction rate was measured using the volumetric approach in VF1, where pump irrigation was practiced. The discharge was summed up over the irrigation event for the estimation of the gross irrigation amount (GIA) per event.

2.3.6. Groundwater and Capillary Rise

Groundwater was monitored from 1 October, 2014 to 11 May, 2016 to analyze the impact of the groundwater table on water fluxes. Seven georeferenced wells were installed in the irrigable area of the VIS and five in the BIS (Figure 2) at characteristic locations, such as valley bottoms, lateral sites, sites near the dam, and in the middle of the schemes. PVC pipes perforated up to 1 m from the base were used. The depths of the wells ranged from 2.7 to 5.5 m in the VIS and from 2 to 4.9 m in the BIS. An electric contact meter (Seba KLL 077) was used to measure the depth to groundwater table weekly throughout the 2014–2016 observation period. However, measurements could not be carried out between 3 June, 2015 and 15 July, 2015 owing to technical challenges. Because of the late start of the groundwater monitoring in 2014, measurements from the 2015 rainy season were used for the simulation of rainfed maize for 2014.

Capillary rise was estimated in AquaCrop based on soil type and hydraulic characteristics [48] as follows:

$$CR = \exp\left(\frac{\ln(z) - b}{a}\right)$$ (6)

where CR is the expected capillary rise in mm day^{-1}, z is the depth to groundwater table in m, and a and b are coefficients specific to the soil type and the hydraulic characteristics.

2.4. Model Parameterization and Validation

The 2014 rainy season dataset from VF1 was used to parameterize the AquaCrop model for maize and the 2014 maize dataset from BF1 was used to validate the model (i.e., inter-farm validation). The 2014 maize crop data from BNF1 were found to be unreliable owing to the effects of waterlogging, and thus were excluded from the analysis. For tomato, the 2014–2015 dry season data from BF1 were used for the parameterization, and inter-farm model validation was performed using 2014–2015 data from BF6. The inter-seasonal validation employed datasets from tomato BF1 collected in 2015–2016. Data from the other tomato fields (VF1 and BNF1) were either unavailable or incomplete owing to technical and environmental (i.e., crop disease attack) challenges in 2014–2015, and the early onset of rainfall destroying the crops in 2016.

The parameters modified in the model were climate, soil characteristics, and agronomic practice (Tables 1 and 2). All the default crop-specific parameters (i.e., yield response factor) for the study crops were used. The climate file in daily time-steps for the period 21 May, 2014 (i.e., beginning of the rainfed farming season in 2014) to 24 May, 2016 (i.e., end of the 2015–2016 dry season farming) was created using the AquaCrop ET_0 file, maximum and minimum temperature file, and a rainfall file.

Table 1. Modified parameters and field data used for the parameterization and validation of the AquaCrop model for maize.

Data Required	Model Parameterization	Inter-Farm Model Validation
Site conditions		
Cropping field	VF1 (2014)	BF1 (2014)
Crop variety	'Obatanpa'	'Obatanpa'
Growing cycle	3 July, 2014–25 August, 2014	24 May, 2014–14 August, 2014
Planting method	Direct sowing	Direct sowing
Soil fertility in relation to biomass	Poor	Poor
Initial canopy cover	High canopy cover	High canopy cover
Maximum canopy cover	Fairly covered	Fairly covered
Maximum rooting depth	0.30 m	0.36 m
Harvest index	0.51	0.53
Crop development	In growing degree days	In growing degree days
Field management		
Soil surface cover	No mulch	No mulch
Soil physical characteristics	Field capacity, wilting point, soil moisture, texture, and thickness of soil layer from soil pit 3 in BNF1	Field capacity, wilting point, soil moisture, texture, and thickness of soil layer from soil pit 1 in BF1
Groundwater level	Weekly depth to groundwater table from VF1 well	Weekly depth to groundwater table from BR well
Simulation period	Calendar of growing cycle	Calendar of growing cycle
Field data file	Aboveground dry matter from VF1	Aboveground dry matter from BF1

Table 2. Modified parameters and field data for the parameterization and validation of the AquaCrop model for tomato.

Data Required	Model Parameterization	Inter-Farm Model Validation	Inter-Seasonal Validation
Site conditions			
Cropping field	BF1 (2014–2015)	BF6 (2014–2015)	BF1 (2015–2016)
Crop variety	'Buffalo'	'Buffalo'	'Buffalo'
Growing cycle	22 October 22, 2014–11 February, 2015	11 November, 2014–6 March, 2015	23 November, 2015–18 March, 2016
Planting method	Transplanting	Transplanting	Transplanting
Soil fertility in relation to biomass	Moderate	Moderate	Moderate
Initial canopy cover	Very small cover	Very small cover	Very small cover
Maximum canopy cover	Fairly covered	Fairly covered	Fairly covered
Maximum rooting depth	0.35 m	0.37 m	0.28 m
Harvest index	0.29	0.29	0.21
Crop development	In growing degree days	In growing degree days	In growing degree days
Field management			
Soil surface cover	No mulch	No mulch	No mulch
Irrigation practice	Irrigation amount per event in mm from BF1	Irrigation amount per event in mm from BF6	Irrigation amount per event in mm from BF6
Soil physical characteristics	Field capacity, wilting point, soil moisture, texture, and thickness of soil layer from soil pit 1 in BF1	Field capacity, wilting point, soil moisture, texture, and thickness of soil layer field from soil pit 2 near BF6	Field capacity, wilting point, soil moisture, texture, and thickness of soil layer from soil pit 1 in BF1
Groundwater level	Weekly depth to groundwater table from BR well	Weekly depth to groundwater table from BD well	Weekly depth to groundwater table from BR well
Simulation period	Calendar of growing cycle	Calendar of growing cycle	Calendar of growing cycle
Field data file	Aboveground dry matter from BF1	Aboveground dry matter from BF6	Aboveground dry matter from BF1

2.5. Supplemental Irrigation Requirement for Maize

Irrigation scheduling was simulated for the maize fields (VF1 and BF1) by selecting the 'Net irrigation water requirement' option in AquaCrop, and 50% allowable root zone depletion. The simulation was run to determine the supplemental irrigation requirement under the two aforementioned climate scenarios.

2.6. Improved Irrigation Scheduling for Tomato

Datasets from tomato fields (BF1 and BF6) in 2014–2015 were used to optimize the irrigation schedule. Irrigation files for each of the fields were created for the furrow irrigation method. The time criterion selected was 'Allowable depletion of 80% of readily available water' and the irrigation depth criterion used was 'Back to field capacity'. The irrigation water quality was specified as 'excellent' assuming a negligible salinity of irrigation water.

3. Results

3.1. Rainfall Variability

Rainfall data revealed a high inter-seasonal variability of rainfall (i.e., 17%) and frequent dry spells lasting for 2–16 days (Figure 5). From 1998 to 2014, the frequency of dry spells in Vea and Bongo ranged between 18 and 28 occurrences. Furthermore, the analysis indicated increasing intra-seasonal rainfall variability in both schemes during the observation period, most likely due to climate change.

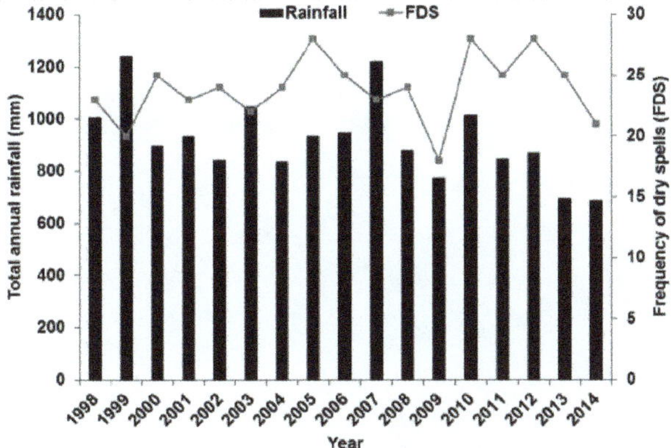

Figure 5. Total annual rainfall and frequency of dry spells (FDS) in the Vea and Bongo irrigation schemes during the years 1998–2014.

3.2. Crop Growth Parameters

The PD of maize ranged between 4.1 and 5.5 plants m^{-2} across the schemes (Table 3). The decline in the maize aboveground DM in VF1 and BNF1 in the VIS was attributed to the effects of late planting (i.e., 3 July, 2014) and waterlogging, respectively (Figure 6a). The PD of tomato was generally higher in the BIS (3.3–3.5 plants m^{-2} in 2014–2015, and 3.6–4.2 plants m^{-2} in 2015–2016) than in the VIS (2.6 plants m^{-2} in 2014–2015, and 3.3–3.5 plants m^{-2} in 2015–2016) (Table 3). The difference was partly due to the narrower inter-row spacing observed in the BIS (0.28–0.35 m) compared to that in the VIS (0.25–0.54 m). The remarkably low tomato DM in the Vea BNF1 field in 2014–2015 was due to the impact of plant root disease (Figure 6b).

Table 3. Crop growth and yield components in the Bongo and Vea irrigation schemes during the 2014–2016 observation period.

Crop Type	Farm Label	Plant Density (plants m^{-2})	Maximum Rooting Depth (m)	Fresh Yield (Mg ha^{-1})	Dry Yield (Mg ha^{-1})	Harvest Index
Bongo irrigation scheme				2014 rainy season		
Maize	BF1	4.4	0.36	n.d.	2.9	0.53
				2014–2015 dry season		
Tomato	BF1	3.5	0.35	49.2	2.3	0.29
Tomato	BF6	3.3	0.37	34.3	2.5	n.d.
				2015–2016 dry season		
Tomato	BF1	3.6	0.28	42.8	1.4	0.22
Tomato	BF6	4.2	n.d.	39.6	1.6	0.21
Vea irrigation scheme				2014 rainy season		
Maize	VF1	5.5	0.30	n.d.	2.6	0.51
Maize	BNF1	4.1	0.35	n.d.	1.2	0.41
				2015–2016 dry season		
Tomato	VF1	3.3	0.24	35.3	1.6	0.29
Tomato	BNF1	3.5	0.29	51.3	2.2	0.30

n.d. = not determined/applicable.

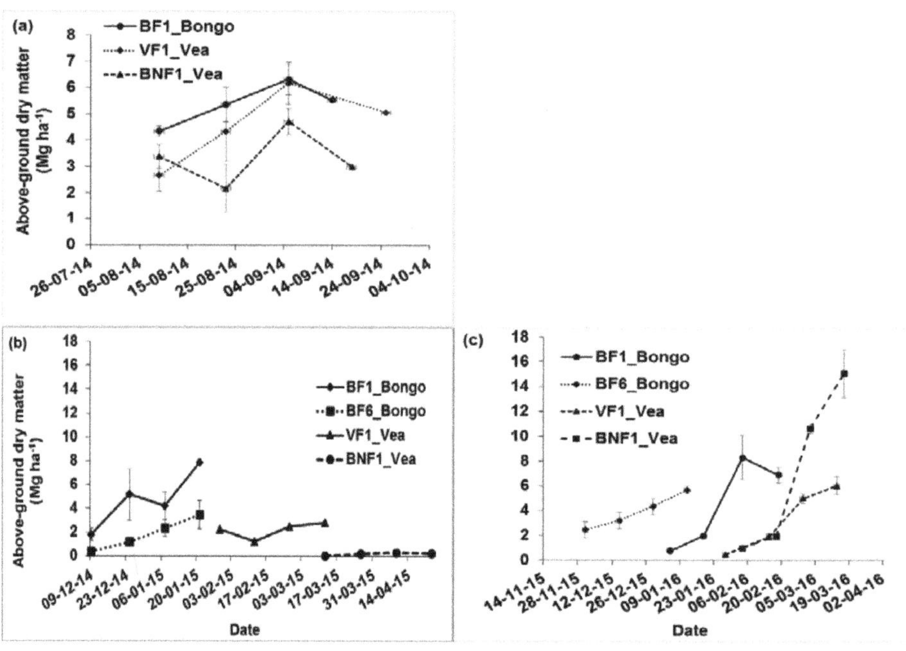

Figure 6. Aboveground dry matter of (**a**) maize in 2014; (**b**) tomato in 2014–2015 and (**c**) tomato in 2015–2016 in the Bongo and Vea irrigation schemes.

The higher tomato DM observed in the 2015–2016 dry season compared to the previous dry season could be due to excessive field-level water application in that season, when an increased water availability was recorded (Figure 6b,c). The downward trend of the LAI of tomato observed in the BIS in 2014–2015 might be due to an insufficient water supply in the later part of the dry season. In contrast, the upward trend of the LAI in the VIS reflects an adequate water supply (Figure 7). The maximum RD of tomato and maize ranged between 0.28 and 0.37 m (Table 1), a result of the shallow soil depth, not exceeding 0.4 m in the UER.

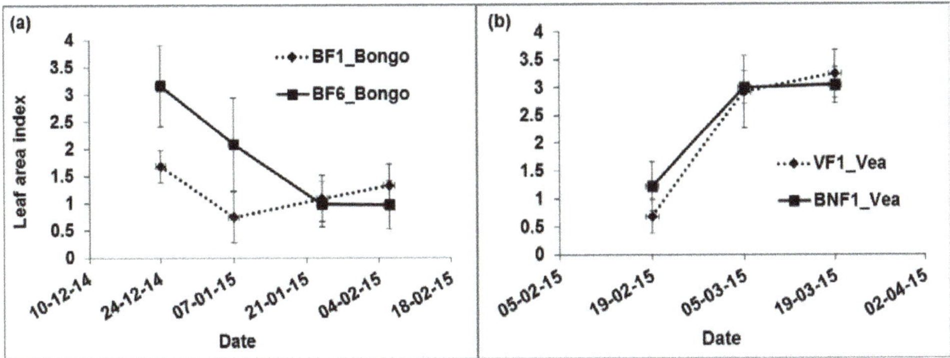

Figure 7. Leaf area index of tomato during the 2014–2015 dry season in the (**a**) Bongo and (**b**) Vea irrigation schemes.

3.3. Crop Yield Components

In 2014, the maize yield ranged between 1.2 and 2.9 Mg ha^{-1}, and the HI ranged from 0.41 to 0.53 across the irrigation schemes (Table 3). A relatively low yield was observed in the BNF1 maize field, in the VIS, possibly due to the combined effect of late planting and waterlogging that occurred in this farm. A relatively high maize yield was recorded in the BIS. The overall range for the annual fresh yields of tomato was 34.3–51.3 Mg ha^{-1} across the irrigation schemes and monitoring periods, and the tomato HI ranged between 0.21 and 0.3.

3.4. Groundwater Level and Capillary Rise

The average depth to groundwater table varied between 0.7 and 2.8 m in the VIS and between 0.6 and 1.3 m in the BIS during 2014–2016 (Figure 8). In the BIS, the soil waterlogging (detected in the BM well) occurred in August and the deepest level (3 m measured in the BD well) was observed in May. The BNF2 well in the VIS recorded the shallowest groundwater level (0.1 m) in August, while the VF1 well measured the deepest groundwater level (3.4 m) in May. The rise in the groundwater table in August most likely resulted from rainfall recharge.

Furthermore, the groundwater level was influenced by nearby streams, reservoirs, and fish ponds. For example, the BU well in the BIS and the VU, BNM and BoN wells in the VIS exhibited stable and relatively shallow groundwater levels due to their proximity to the Bongo reservoir, Vea fish ponds, and streams, even when deep groundwater levels were recorded at other wells (Figure 8). Irrigation events also impacted on the water table. For instance, the groundwater level in the BF1 well in the tomato field increased steadily from the beginning of the dry season and declined from 4 March, 2015, when 2014–2015 dry season irrigation was over. However, the VF1 and BNF2 wells in the VIS in the tomato and leafy vegetable fields, respectively, exhibited rather variable groundwater levels even during the irrigation period and a downward trend after the end of the irrigation period.

The simulated capillary rise into the root-zone of maize was 43–147 mm in 2014, while in the tomato fields it was 18–157 mm in 2014–2015, and 27–263 mm in the 2015–2016 across the irrigation schemes (Figure 9).

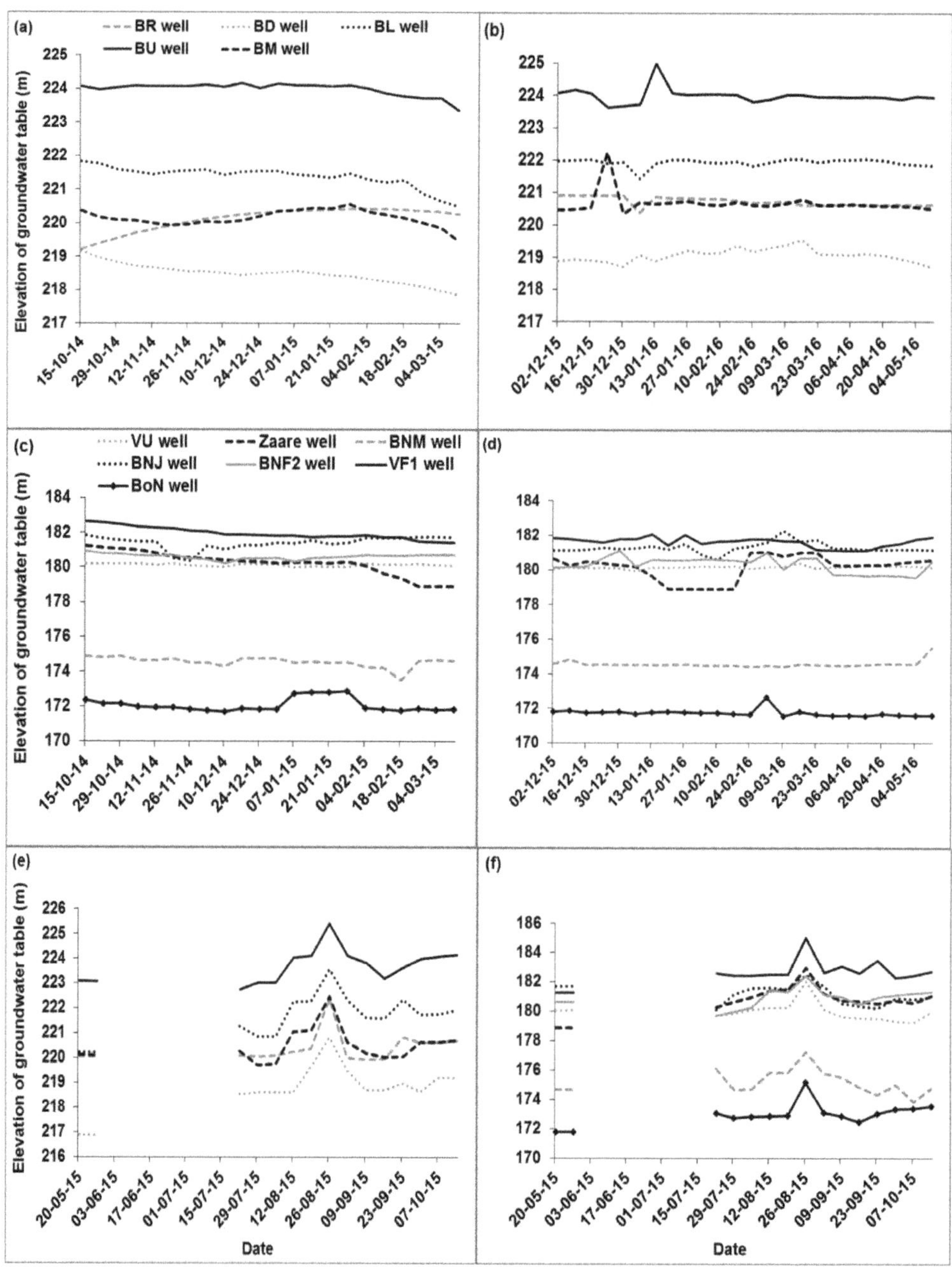

Figure 8. Elevation of the groundwater table during (**a**) 2014–2015 dry season in Bongo (**b**) 2015–2016 dry season in Bongo (**c**) 2014–2015 dry season in Vea (**d**) 2015–2016 dry season in Vea (**e**) 2015 rainy season in Bongo and (**f**) 2015 rainy season in Vea.

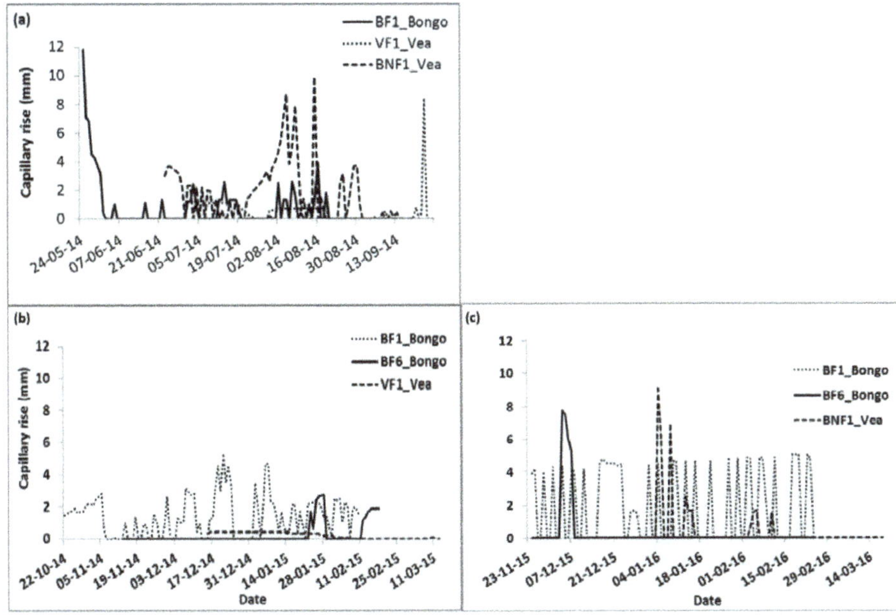

Figure 9. Daily groundwater contribution to the root-zone soil moisture of (**a**) maize in 2014; (**b**) tomato in 2014–2015 and (**c**) 2015–2016 computed by AquaCrop.

3.5. Traditional Irrigation Scheduling

The observed GIA for dry season tomato was lower in 2014–2015 than in 2015–2016 in both schemes (Table 4). Across both irrigation schemes and both dry seasons, the overall range of GIA was 21–67 mm per irrigation event and 584–2559 mm per season. The number of irrigation events for tomato ranged between 20 and 29 in both dry seasons. Particularly in the BIS, the irrigation interval in the tomato fields was generally shorter in 2015–2016 than in the previous dry season, owing to the increased availability of water in the Bongo reservoir.

Table 4. Observed field-level irrigation practices and water productivity for tomato in the Bongo and Vea irrigation schemes during the dry seasons.

Field Label	Gross Irrigation Amount Per Season (mm)	Gross Irrigation Amount Per Event (mm)	Average Irrigation Interval (day)	Water Productivity (kg m^{-3})
Bongo irrigation scheme		2014–2015		
BF1	586	19–50	4	8.4
BF6	1247	17–137	5	2.7
		2015–2016		
BF1	1719	20–93	3	2.5
BF6	2559	14–133	2	1.5
Vea irrigation scheme		2014–2015		
VF1	615	13–35	5	n.d.
BNF1	584	21–42	5	n.d.
		2015–2016		
BNF1	1137	33–79	4	4.5

n.d. = not determined.

3.6. Model Performance

The results of model evaluation for tomato DM indicated good agreement (EF = 0.65–0.83, and d = 0.87–0.96) and acceptable error margins (NRMSE = 17.7–42%) (Figure 10c–e).

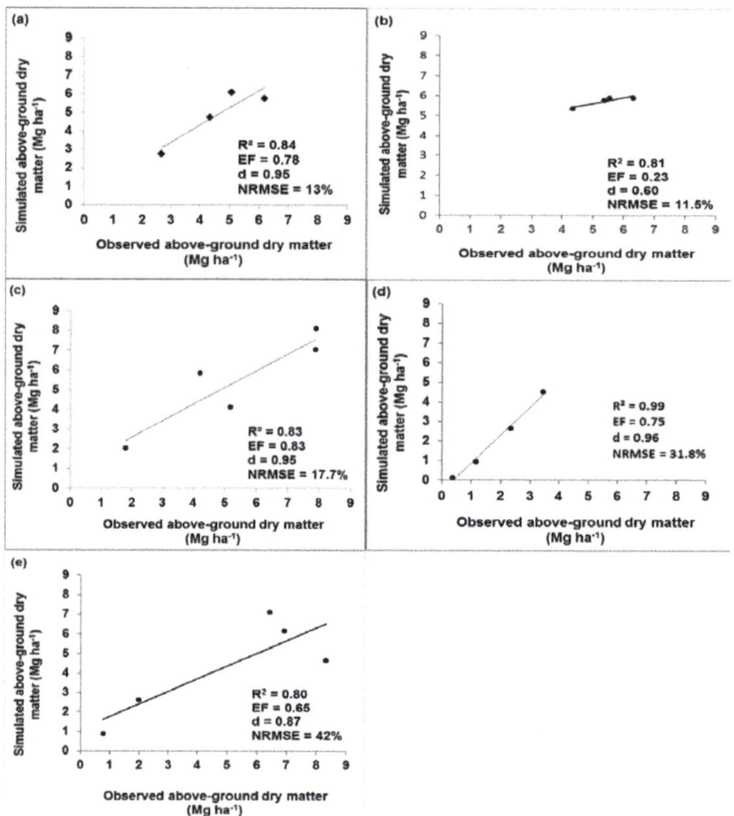

Figure 10. Evaluation of simulated and observed aboveground dry matter of (**a**) maize in VF1 in 2014; (**b**) maize in BF1 in 2014; (**c**) tomato in BF1 in 2014–2015; (**d**) tomato in BF6 in 2014–2015; and (**e**) tomato in BF1 in 2015–2016, in the Vea and Bongo irrigation schemes.

The model evaluation for maize DM also suggested a good agreement (EF = 0.23–0.78, and d = 0.60–0.95) and an acceptable error margin (NRMSE = 11.5–13%). However, the low EF (0.23) for maize in the BF1 field (Figure 10b) could be due to the missing biomass data for the vegetative stage due to the late start of data collection.

3.7. Improved Irrigation Schedule for Tomato

The optimized irrigation schedule for the dry season tomato cropping indicated the need for longer irrigation intervals (6–13 days) in the early crop growth stage and during ripening. In contrast, irrigation intervals should be shorter (2–8 days) in the flowering and yield formation stages (Figure 11). The simulated NIR for tomato ranged from 21 to 29 mm per irrigation event and from 311 to 495 mm per season. The GIA for tomato was estimated as 38–52 mm per irrigation event and 566–900 mm per season, assuming a 55% application efficiency.

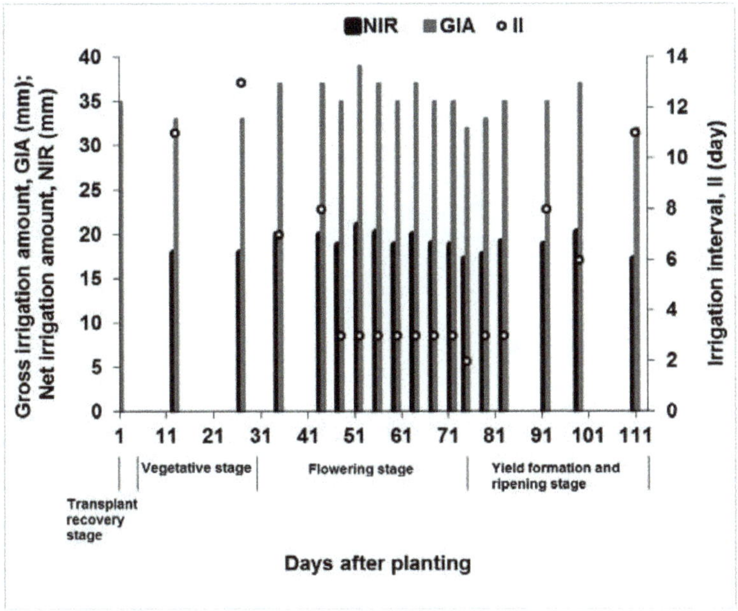

Figure 11. Improved irrigation schedule for tomato cultivation, based on the example of BF1 in the Bongo irrigation scheme. Gross irrigation amount was estimated using a field application efficiency of 55% [19].

The improved irrigation schedule would result in 4–14% yield increment while saving 130–1325 mm (22–52% of GIA) of water, which is otherwise lost through percolation beyond the root zone under the traditional irrigation practice in either scheme (Table 5).

Table 5. Potential water saving and yield increase under the improved irrigation schedule as simulated in AquaCrop.

Field Label	Potential Water Saving (mm)	Tomato Yield under Traditional Irrigation (Mg ha^{-1})	Tomato Yield under Improved Irrigation (Mg ha^{-1})	Potential Yield Increase (%)
BF1 (2014–2015)	130	2.30	2.40	4
BF6 (2014–2015)	775	2.01	2.30	14
BF1 (2015–2016)	1,325	1.58	1.79	14

3.8. Supplemental Irrigation Requirement for Maize

S1 was observed in 1999 when 1240 mm of rainfall and 20 dry spells were recorded in the rainy season, and S2 was observed in 2012 with 871 mm of rainfall and 28 dry spells, the highest frequency of dry spells during the 17 year observation period (Figure 5). Notably, although 2014 recorded the lowest rainfall (687 mm), it was not considered the driest year due to the lower frequency of dry spells (21) compared with 2012. The supplemental irrigation requirement for rainfed maize in the favorable climate scenario S1 was predicted in the range of 88–105 mm (i.e., 25–29% of NIR of maize). The values predicted for S2, the scenario of low rainfall and frequent dry spells, ranged between 107 and 126 mm (i.e., 30–35% of NIR of maize) (Figure 12). The simulated increase in maize yield under supplemental irrigation ranged between 5% and 14%.

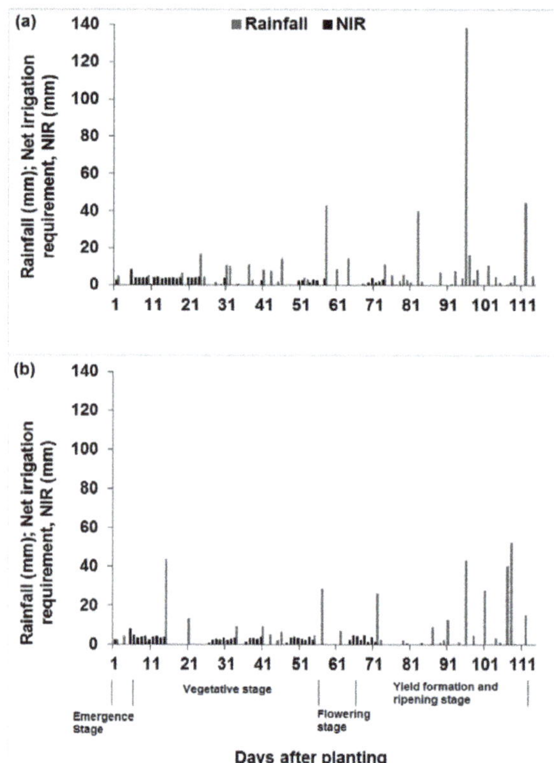

Figure 12. Daily rainfall and net irrigation requirement for supplemental irrigation of maize in the (**a**) 1999 wet year and (**b**) 2012 dry year, as simulated in AquaCrop based on the example of maize BF1 in the Bongo irrigation scheme.

4. Discussion

4.1. Crop Yields

The yields of fresh tomato fruits were similar across schemes but varied remarkably between fields owing to differences in field-level agronomic and irrigation practices and constraints. For instance, the tomato root disease in the VIS field impacted negatively on yield, as the application of insecticides protected only the aboveground biomass. Over-irrigated tomato in 2015–2016 showed high fresh yields and the higher water content of these fruits (i.e., 95% vs. 93% in water-scarce 2014–2015). The values of tomato fresh yield corresponded to the upper ranges measured by [35], reporting 20–36.8 Mg ha^{-1} of fresh tomato yields in the UER, and were greater than the 18 Mg ha^{-1} reported by [49] for rainfed tomato production in the Ashanti and Brong Ahafo regions of Ghana, reflecting the positive impact of irrigation. However, the lower HI of tomato observed in our study (0.21–0.3), compared to values (0.5–0.65) reported by [22] for rainfed tomato in drylands, could be partly due to over-irrigation. The excessive water use was reflected in the low field application efficiencies (30–59%) characteristic of almost all the examined fields [19].

Late planting and waterlogging, due to the lack of drainage facilities, reduced maize grain yield to only 1.2 Mg ha^{-1} in the affected field in the VIS. This observation confirms late planting as one of the causes of sub-optimal yield levels of rainfed maize as rainfall typically declines towards the end of the rainy season. Sallah et al. [50] reported a 30% loss in maize yields due to late planting in northern

Ghana. The observed range of maize grain yields in our study (1.2–2.9 Mg ha^{-1}) was similar to that reported for the fertilized 'Obatanpa' maize variety in Ghana (1.3 to 2.7 Mg ha^{-1}; [38]). The values of [8] for SSA (1.3–1.4 Mg ha^{-1}) were within the lower range of our results. However, Sugri et al. [51] reported the yield potential of the 'Obatanpa' maize variety to be 5.5 Mg ha^{-1} in Ghana. Variations in practices of soil nutrient management and often insufficient applications of fertilizer in the examined fields could also have contributed to the variability in yields. Folberth et al. [9] emphasized that even modest additions of N and P fertilizer might double maize production in most of SSA.

4.2. Irrigation Practice

The examination of field-level irrigation practices during the dry season revealed inappropriate, and in turn, ineffective water application for crop production resulting in over-irrigation in both schemes, mainly due to the lack of consideration of the crop growth stages and water storage characteristics of the soil. Over-irrigation was signified by the high GIA in the water-abundant 2015–2016 season, when farmers in both schemes used more water by shortening irrigation intervals (Table 4), leading to lower water productivity than in the previous, water-scarce season. Because of the lack of appropriate irrigation scheduling and the absence of flow measuring devices in the canals, farmers applied as much water as possible to the tomato crop, and further increased the water application rate with increasing water availability in the reservoir. Faulkner et al. [33] also observed the tendency for excessive water use in response to increasing water availability and attributed this phenomenon to the lack of knowledge of efficient and effective water application at field level. Moreover, the GIAs of tomato in our study were 100–400% larger than the range of values, 274 and 852 mm, previously reported for the UER [34,35], confirming the need for water saving.

4.3. Improved Irrigation Scheduling

The need to adjust irrigation schedules to local hydro-geological conditions is suggested by the modelling analysis, for example, a significant contribution of capillary rise from the groundwater was shown to satisfy the NIR of crops. The groundwater contribution to the NIR of the study crops was highly variable, reflecting the spatial variability in hydro-geological characteristics of the cropping fields. The need to account for this variability complicates the development and application of improved farmer irrigation scheduling in the UER. According to [52], there could be varying contributions of shallow groundwater ($\leq$3 m) to the root-zone soil moisture in fine-textured soils such as those mostly found in the Bongo and Vea irrigation schemes.

The observed increases in tomato yield (i.e., 4–14%) under the improved irrigation schedule most likely resulted from the reduction of the negative effect of over-irrigation on crop yield, as the over-irrigated cropping fields showed the highest potential (14%) to increase yields under the improved irrigation schedule. The simulated magnitude of water saving in the reservoir-based irrigation schemes, which was 22–52% of the GIA under the current irrigation practices, indicates that improving irrigation schedules offers considerable potential for water saving in the dry season in the UER irrigation systems. Overall, however, the improvement of field-level irrigation scheduling alone might not be sufficient for optimizing water productivity and availability in the schemes [16]. To achieve full benefits, equipping irrigation infrastructure with discharge-measuring and dosage structures, and reparation of the decaying water conveyance and distribution sub-systems in the UER would be necessary [18–20]. These interventions to upgrade infrastructure would need to be accompanied by the training of irrigators to handle these facilities, and by further development of water management institutions towards reliably implementing advanced irrigation schedules in order to utilize the full potential of improvements.

4.4. Feasibility of Supplemental Irrigation

The observed temporal variability in rainfall across the irrigation schemes highlights the urgent need for water management strategies to ensure a reduction of the associated risks in rainfed crop production. The intra-seasonal variability of rainfall revealed by the frequency of dry spells was found to be more influential on water demand for crop growth than the total rainfall over the growing season. In the VIS, for example, the supplemental irrigation requirement for maize simulated with the 2014 rainfall was 29 mm, whereas in the wetter year 2012, this value was 107 mm due to the higher frequency of dry spells (28). Similarly, although 1999 was recognized as the wet year for S1 according to the aforementioned criteria, the simulated NIR for supplemental irrigation of maize was 88 mm due to the higher intra-seasonal variability (69%) in that year than in 2014 (64%).

The supplemental irrigation requirement for maize estimated by AquaCrop (29–126 mm) was within the range of values, 20–240 mm, determined by [5] in semi-arid Mwala in Kenya. Furthermore, the temporal rainfall variability was consistent with the findings of [37], who observed high rainfall variability of >25% during the years 1923–1995 in the Sahelian region. Likewise, [7] estimated dry spells lasting for 2–13 days in the Savanna agro-ecological zone of Ghana.

Overall, considering only the crop irrigation sector, the quantity of water saved through improved irrigation scheduling of dry season tomato is largely sufficient to accommodate supplemental irrigation of maize in the rainy season, and thus adapt to rainfall variability and recurrent dry spells. Even for the dry climate scenario of low rainfall coupled with frequent dry spells, about 126 mm of water at field level would be required for the supplemental irrigation of maize during the rainy season. Furthermore, the simulated increase in maize yield upon the introduction of supplemental irrigation offers an incentive for managers of the Bongo and Vea schemes to explore this strategy. Notably, due to the reservoir losses through evaporation and seepage, some of the water saved in the dry season might not be available for supplemental irrigation in the rainy season. Hence, an effective year-round irrigation schedule is required so that supplemental irrigation in the rainy season does not compromise water availability for dry season crop production.

5. Conclusions

High temporal variability in rainfall and frequent dry spells lasting for 2–16 days are common in the UER, requiring adaptive measures to enhance rainfed crop production. The supplemental irrigation requirement for maize under the dry climate scenario of low rainfall and frequent dry spells was estimated between 107 and 126 mm, whereas for periods of high rainfall and rare dry spells, between 88 and 105 mm would be required. These demands can be satisfied via improved irrigation scheduling for dry season tomato that can potentially save 130–1325 mm of water, which would otherwise be lost through percolation and evaporation. Tomato and maize yield increments in the range of 4–14% and 5–14%, respectively, are predicted under the improved irrigation schedule and supplemental irrigation. The AquaCrop model, parameterized using field data collected in the small- and medium-scale reservoir-based irrigation schemes in the Upper East region of Ghana, can be further utilized to improve the irrigation schedule of other cropping systems in the UER. Given the sub-optimal nutrient management practices observed across the study sites, further research should investigate the potential of both soil fertility and water management practices combined for improving crop yields and year-round food security in sub-Saharan Africa.

Author Contributions: E.S.-A., B.T., B.D. and A.K. conceived and designed the field surveys; E.S.-A. performed the surveys; E.S.-A. and B.T. analyzed the data; E.S.-A., B.T., B.D. and A.K. wrote the paper.

Acknowledgments: This study was supported by the German Federal Ministry of Education and Research (BMBF) under the program WASCAL (West African Science Service Center on Climate Change and Adapted Land Use, project No. 00100218). The additional support by the German Academic Exchange Service (DAAD) and Korea University Grant (No. K1608421) is gratefully acknowledged. This paper was presented at the 23rd ICID Congress on Irrigation and Drainage in Mexico City, Mexico.

Conflicts of Interest: The authors declare no conflict of interest. The founding sponsors had no role in the design of the study; in the collection, analyses, or interpretation of data; in the writing of the manuscript, and in the decision to publish the results.

Appendix A

Table A1. Chemical properties of soils in the Bongo and Vea irrigation schemes.

Morphological Horizon (cm)	pH (1:1 H_2O)	OC (%)	Total N (%)	OM (%)	Exchangeable Cations (cmol kg^{-1})				TEB (cmol kg^{-1})	EA (cmol kg^{-1})	CEC (cmol kg^{-1})	BS (%)	Available-Brays	
					Ca^{2+}	Mg^{2+}	K^+	Na^+					ppmP	ppmK
Pit 1														
0–12	6.5	1.44	0.16	2.48	3.20	0.80	0.08	0.04	4.12	0.15	4.27	96.49	26.95	31.19
12–30	7.0	0.41	0.06	0.71	4.01	1.07	0.04	0.03	5.15	0.15	5.30	97.17	5.50	13.25
30–56	7.2	0.34	0.04	0.59	2.94	1.07	0.03	0.03	4.07	0.05	4.12	98.79	2.55	10.56
56–75	7.1	0.21	0.04	0.36	1.34	1.07	0.02	0.02	2.45	0.50	2.95	83.05	1.99	8.23
75–100	7.9	0.14	0.03	0.24	2.40	1.07	0.04	0.30	3.81	0.03	3.84	99.22	3.11	11.03
100–125	7.9	0.07	0.02	0.12	2.14	1.60	0.03	0.03	3.80	0.03	3.83	99.22	0.88	12.98
Pit 2														
0–20	7.8	1.75	0.16	3.02	15.49	7.34	0.09	0.06	22.98	0.05	23.03	99.78	34.12	35.14
20–44	8.4	0.51	0.07	0.88	9.08	5.07	0.10	0.06	14.31	0.03	14.34	99.79	2.79	37.12
44–80	8.7	0.34	0.04	0.59	7.08	4.01	0.10	0.06	11.25	0.03	11.28	99.73	0.24	38.62
80–120	8.8	0.31	0.04	0.53	8.28	11.21	0.10	0.06	19.65	0.03	19.68	99.85	1.59	34.21
Pit 3														
0–20	7.3	0.72	0.09	1.24	3.74	1.87	0.15	0.08	5.84	0.05	5.89	99.15	13.87	52.48
20–32	7.7	0.48	0.07	0.83	4.81	3.07	0.07	0.04	7.99	0.05	8.04	99.38	28.70	24.31
32–57	8.5	0.45	0.07	0.78	8.01	6.14	0.08	0.04	14.27	0.05	14.32	99.65	0.40	29.67
57–76	8.2	0.41	0.06	0.71	10.68	8.41	0.10	0.06	19.25	0.03	19.28	99.84	0.48	35.29
76–120/128	8.2	0.31	0.05	0.53	18.16	11.35	0.19	0.08	29.78	0.05	29.83	99.83	0.48	70.13

OC = Organic carbon, OM = Organic matter, TEB = Total exchangeable bases, EA = Exchangeable acidity, CEC = Cation exchange capacity, BS = Base saturation.

Table A2. Physical and hydraulic properties of soils in the Bongo and Vea irrigation schemes.

Morphological Horizon (cm)	Soil Texture	Bulk Density (g cm^{-3})	SAT (%)	FC (%)	PWP (%)	TAW (%)	Ksat (mm day^{-1})
Pit 1 (Bongo irrigation scheme)							
0–12	Sandy loam	1.10	49.7	16.9	6.2	10.6	1744
12–30	Loamy sand	1.27	47.5	19.1	4.3	14.9	1816
30–56	Sandy loam	1.26	45.8	19.1	5.8	13.3	1318
56–75	Sandy loam	1.28	46.7	14.1	4.2	9.9	1641
75–100	Sandy loam	1.36	45.2	18.3	6.4	11.9	1109
100–125	Sandy loam	1.44	44.2	17.8	6.3	11.6	885
Pit 2 (Bongo irrigation scheme)							
0–20	Silt loam	1.07	51.5	34.2	11.0	23.2	632
20–44	Loam	1.32	44.6	31.1	12.6	18.5	363
44–80	Loam	1.53	45.1	40.7	18.7	22.0	261
80–120	Loam	1.40	45.7	44.2	14.8	29.4	192
Pit 3 (Vea irrigation scheme)							
0–20	Sandy loam	1.37	47.3	20.3	5.4	14.9	1473
20–32	Sandy loam	1.59	45.3	22.0	8.8	13.2	625
32–57	Loam	1.57	45.1	32.9	14.8	18.1	226
57–76	Loam	1.56	47.0	38.6	14.0	24.6	159
76–128	Clay loam	1.52	49.7	47.3	17.5	29.8	86

SAT = Water content at saturation, FC = Field capacity, PWP = Permanent wilting point, TAW = Total available water, K_{sat} = Saturated hydraulic conductivity determined from pedo-transfer functions.

Appendix B

Table A3. Summary of crop growth and yield parameters and details of their measurements for each study season.

Parameter	Method of Data Collection	Frequency of Data Collection	Cropping Field (Figure 3)
Maize, 2014 rainy season			
Above-ground biomass	Destructive biomass sampling along a 1 m rod on three selected rows Destructive biomass sampling in two 8 m row sections at harvest	Three times during the crop reproduction stage at two weeks interval, and once at harvest time	BF1, VF1, BNF1
Plant density	Counting of total number of plants along the 1 m rod on the three selected rows Estimation of the sampling area	Three times during the crop reproduction stage at two weeks interval	BF1, VF1, BNF1
Row spacing	The average distance between two adjacent rows at five random locations	Once during the reproduction stage	BF1, VF1, BNF1
Maximum rooting depth	Manual excavations of at least three plants per crop	Once at harvest time	BF1, VF1, BNF1
Crop yield	Harvesting and weighing of total maize grain yield from two 8 m row sections	Once at harvest time	BF1, VF1, BNF1
Tomato, 2014–2015 dry season			
Above-ground biomass	Destructive biomass sampling along a 1 m rod on three selected rows Destructive biomass sampling in two 8 m row sections at harvest	Four times during the vegetative and reproduction stages, and once at harvest time	BF1, BF6, VF1, BNF1
Plant density	Counting of total number of plants along the 1 m rod on the three selected rows Estimation of the sampling area	Four times during the vegetative and reproduction stages	BF1, BF6, VF1, BNF1
Leaf area index	Measurements with the SunScan probe (SS1-UM-2.0) at five random locations	Four times during the vegetative and reproduction stages	BF1, BF6, VF1, BNF1
Row spacing	The average distance between two adjacent rows at five random locations	Once at harvest time	BF1, BF6, VF1, BNF1
Maximum rooting depth	Manual excavations of at least three plants per crop	Once at harvest time	BF1, BF6, VF1
Crop yield	Harvesting and weighing of total tomato fruits from two 8 m row sections	Once at harvest time	BF1, BF6

Table A3. *Cont.*

Parameter	Method of Data Collection	Frequency of Data Collection	Cropping Field (Figure 3)
	Tomato, 2015–2016 dry season		
Above-ground biomass	Destructive biomass sampling along a 1 m rod on three selected rows Destructive biomass sampling in two 8 m row sections at harvest	Four times during the vegetative and reproduction stages	BF1, BF6, VF1, BNF1
Plant density	Counting of total number of plants along the 1 m rod on the three selected rows Estimation of the sampling area	Four times during the vegetative and reproduction stages	BF1, BF6, VF1, BNF1
Row spacing	The average distance between two adjacent rows at five random locations	Once during the reproduction stage	BF1, BF6, VF1, BNF1
Maximum rooting depth	Manual excavations of at least three plants per crop	Once at harvest time	BF1, BF6, VF1, BNF1
Crop yield	Harvesting and weighing of total tomato fruits from two 8 m row sections	Once at harvest time	BF6, and fields close to BF1, VF1, BNF1

References

1. Cook, K.H.; Vizy, E.K. Impact of climate change on mid-twenty-first century growing seasons in Africa. *Clim. Dyn.* **2012**, *39*, 2937–2955. [CrossRef]
2. Sylla, M.B.; Nikiema, P.M.; Gibba, P.; Kebe, I.; Klutse, N.A.B. Climate Change over West Africa: Recent Trends and Future Projections. In *Adaptation to Climate Change and Variability in Rural West Africa*; Springer: Cham, Switzerland, 2016; pp. 25–40.
3. Sanfo, S.; Barbier, B.; Dabiré, I.W.P.; Vlek, P.L.G.; Fonta, W.M.; Ibrahim, B.; Barry, B. Rainfall variability adaptation strategies: An ex-ante assessment of supplemental irrigation from farm ponds in southern Burkina Faso. *Agric. Syst.* **2017**, *152*, 80–89. [CrossRef]
4. McCartney, M.; Smakhtin, V. *Water Storage in An Era of Climate Change: Addressing the Challenges of Increasing Rainfall Variability*; International Water Management Institute Blue Paper; IWMI: Colombo, Sri Lanka, 2010; pp. 1–24. Available online: https://www.agriskmanagementforum.org/sites/agriskmanagementforum.org/files/Documents/water%20storage%20in%20era%20of%20climate%20change%20IWMI%20Blue_Paper_2010-final.pdf (accessed on 20 November 2016).
5. Rockström, J.; Barron, J. Water productivity in rainfed systems: Overview of challenges and analysis of opportunities in water scarcity prone savannahs. *Irrig. Sci.* **2007**, *25*, 299–311. [CrossRef]
6. Adwubi, A.; Amegashie, B.K.; Agyare, W.A.; Tamene, L.; Odai, S.N.; Quansah, C.; Vlek, P. Assessing sediment inputs to small reservoirs in Upper East Region, Ghana. *Lakes Reserv. Res. Manag.* **2009**, *14*, 279–287. [CrossRef]
7. Kranjac-Berisavljevic, G.; Abdul-Ghanyu, S.; Gandaa, B.Z.; Abagale, F.K. Dry Spells Occurrence in Tamale, Northern Ghana–Review of Available Information. *J. Disaster Res.* **2014**, *9*, 468–474. [CrossRef]
8. Dzanku, F.M.; Jirström, M.; Marstorp, H. Yield Gap-Based Poverty Gaps in Rural Sub-Saharan Africa. *World Dev.* **2015**, *67*, 336–362. [CrossRef]
9. Folberth, C.; Yang, H.; Gaiser, T.; Abbaspour, K.C.; Schulin, R. Modeling maize yield responses to improvement in nutrient, water and cultivar inputs in sub-Saharan Africa. *Agric. Syst.* **2013**, *119*, 22–34. [CrossRef]
10. Vlek, P.L.G.; Khamzina, A.; Tamene, L. *Land Degradation and the Sustainable Development Goals: Threats and Potential Remedies*; CIAT Publication No. 440; International Center for Tropical Agriculture (CIAT): Nairobi, Kenya, 2017. Available online: http://hdl.handle.net/10568/81313 (accessed on 9 May 2018).
11. Drechsel, P.; Gyiele, L.; Kunze, D.; Cofie, O. Population density, soil nutrient depletion, and economic growth in sub-Saharan Africa. *Ecol. Econ.* **2001**, *38*, 251–258. [CrossRef]
12. Vlek, P.L.G.; Khamzina, A.; Azadi, H.; Bhaduri, A.; Bharati, L.; Braimoh, A.; Martius, C.; Sunderland, T.; Taheri, F. Trade-offs in multi-purpose land use under land degradation. *Sustainability* **2017**, *9*, 2196. [CrossRef]
13. Henao, J.; Baanante, C. *Agricultural Production and Soil Nutrient Mining in Africa: Implications for Resource Conservation and Policy Development*; International Center for Soil Fertility and Agricultural Development: Muscle Shoals, AL, USA, 2006.

14. Droogers, P.; Aerts, J. Adaptation strategies to climate change and climate variability: A comparative study between seven contrasting river basins. *Phys. Chem. Earth Parts ABC* **2005**, *30*, 339–346. [CrossRef]

15. Molden, D.; Oweis, T.; Steduto, P.; Bindraban, P.; Hanjra, M.A.; Kijne, J. Improving agricultural water productivity: Between optimism and caution. *Agric. Water Manag.* **2010**, *97*, 528–535. [CrossRef]

16. Mustapha, A.B. Effect of Dryspell Mitigation with Supplemental Irrigation on Yield and Water Use Efficiency of Pearl Millet in Dry Sub-Humid Agroecological Condition of Maiduguri. In *Proceedings of the 2nd International Conference on Environment Science and Biotechnology;* IACSIT Press: Singapore, 2012; pp. 46–49.

17. Zwart, S.J.; Bastiaanssen, W.G.M. Review of measured crop water productivity values for irrigated wheat, rice, cotton and maize. *Agric. Water Manag.* **2004**, *69*, 115–133. [CrossRef]

18. Ali, M.H.; Talukder, M.S.U. Increasing water productivity in crop production—A synthesis. *Agric. Water Manag.* **2008**, *95*, 1201–1213. [CrossRef]

19. Sekyi-Annan, E.; Tischbein, B.; Diekkrüger, B.; Khamzina, A. Performance evaluation of reservoir-based irrigation schemes in the Upper East region of Ghana. *Agric. Water Manag.* **2018**, *202*, 134–145. [CrossRef]

20. Pereira, L.S. Relating water productivity and crop evapotranspiration. *Options Méditerr. Ser. B* **2007**, *57*, 31–49.

21. Greaves, G.E.; Wang, Y.-M. Assessment of FAO AquaCrop Model for Simulating Maize Growth and Productivity under Deficit Irrigation in a Tropical Environment. *Water* **2016**, *8*, 557. [CrossRef]

22. Steduto, P.; Hsiao, T.C.; Fereres, E.; Reas, D. *Crop Yield Response to Water;* Food and Agriculture Organization of the United Nations: Rome, Italy, 2012. Available online: http://www.fao.org/docrep/016/i2800e/i2800e.pdf (accessed on 13 November 2016).

23. Sekyi-Annan, E.; Acheampong, E.N.; Ozor, N. Modeling the Impact of Climate Variability on Crops in Sub-Saharan Africa. In *Quantification of Climate Variability, Adaptation and Mitigation for Agricultural Sustainability;* Springer International Publishing: Cham, Switzerland, 2017; pp. 39–70.

24. Gaydon, D.S.; Balwinder-Singh; Wang, E.; Poulton, P.L.; Ahmad, B.; Ahmed, F.; Akhter, S.; Ali, I.; Amarasingha, R.; Chaki, A.K.; et al. Evaluation of the APSIM model in cropping systems of Asia. *Field Crops Res.* **2017**, *204*, 52–75. [CrossRef]

25. Sommer, R.; Kienzler, K.; Christopher, C.; Ibragimov, N.; Lamers, J.; Martius, C.; Vlek, P. Evaluation of the CropSyst model for simulating the potential yield of cotton. *Agron. Sustain. Dev.* **2008**, *28*, 345–354. [CrossRef]

26. Fortes, P.S.; Teodoro, P.R.; Campos, A.A.; Mateus, P.M.; Pereira, L.S. Model tools for irrigation scheduling simulation: WINISAREG and GISAREG. In *Irrigation Management for Combating Desertification in the Aral Sea Basin. Assessment and Tools;* Vita Color Publication: Tashkent, Uzbekistan, 2005; pp. 81–96.

27. Jones, J.W.; Hoogenboom, G.; Porter, C.H.; Boote, K.J.; Batchelor, W.D.; Hunt, L.A.; Wilkens, P.W.; Singh, U.; Gijsman, A.J.; Ritchi, J.T. The DSSAT cropping system model. *Eur. J. Agron.* **2003**, *18*, 235–265. [CrossRef]

28. Surendran, U.; Sushanth, C.M.; Mammen, G.; Joseph, E.J. Modelling the Crop Water Requirement Using FAO-CROPWAT and Assessment of Water Resources for Sustainable Water Resource Management: A Case Study in Palakkad District of Humid Tropical Kerala, India. *Aquat. Procedia* **2015**, *4*, 1211–1219. [CrossRef]

29. Wang, X.C.; Li, J. Evaluation of crop yield and soil water estimates using the EPIC model for the Loess Plateau of China. *Math. Comput. Model.* **2010**, *51*, 1390–1397. [CrossRef]

30. Wellens, J.; Raes, D.; Traore, F.; Denis, A.; Djaby, B.; Tychon, B. Performance assessment of the FAO AquaCrop model for irrigated cabbage on farmer plots in a semi-arid environment. *Agric. Water Manag.* **2013**, *127*, 40–47. [CrossRef]

31. Walker, S.; Bello, Z.A.; Mabhaudhi, T.; Modi, A.T.; Beletse, Y.G.; Zuma-Netshiukhwi, G. Calibration of AquaCrop Model to predict water requirements of African vegetables. *Acta Hortic.* **2013**, *1007*, 943–949. [CrossRef]

32. Mabhaudhi, T.; Modi, A.T.; Beletse, Y.G. Parameterisation and evaluation of the FAO-AquaCrop model for a South African taro (*Colocasia esculenta* L. Schott) landrace. *Agric. For. Meteorol.* **2014**, *192–193*, 132–139. [CrossRef]

33. Faulkner, J.W.; Steenhuis, T.; van de Giesen, N.; Andreini, M.; Liebe, J.R. Water use and productivity of two small reservoir irrigation schemes in Ghana's upper east region. *Irrig. Drain.* **2008**, *57*, 151–163. [CrossRef]

34. Mdemu, M.V. Water Productivity in Medium and Small Reservoirs in the Upper East Region (UER) of Ghana. Ph.D. Thesis, University of Bonn, Bonn, Germany, 2008. Available online: http://hss.ulb.uni-bonn.de/2008/1362/1362.pdf (accessed on 21 October 2016).

35. Barry, B.; Forkuor, G. *Contribution of Informal Shallow Groundwater Irrigation to Livelihoods Security and Poverty Reduction in the White Volta Basin (WVB): Current Status and Future Sustainability*; International Water Management Institute: Colombo, Sri Lanka, 2010. Available online: https://cgspace.cgiar.org/bitstream/handle/10568/3940/PN65_IWMI_Project%20Report_May10_final.pdf?sequence=1&isAllowed=y (accessed on 22 February 2017).

36. Asres, S.B. Evaluating and enhancing irrigation water management in the upper Blue Nile basin, Ethiopia: The case of Koga large scale irrigation scheme. *Agric. Water Manag.* **2016**, *170*, 26–35. [CrossRef]

37. Fox, P.; Rockström, J. Supplemental irrigation for dry-spell mitigation of rainfed agriculture in the Sahel. *Agric. Water Manag.* **2003**, *61*, 29–50. [CrossRef]

38. Srivastava, A.K.; Mboh, C.M.; Gaiser, T.; Ewert, F. Impact of climatic variables on the spatial and temporal variability of crop yield and biomass gap in Sub-Saharan Africa—A case study in Central Ghana. *Field Crops Res.* **2017**, *203*, 33–46. [CrossRef]

39. Raes, D.; Steduto, P.; Hsiao, T.C.; Fereres, E. Chapter 2—Users guide. In *Reference Manual: AquaCrop, Version 4.0*; FAO, Land and Water Division: Rome, Italy, 2012; pp. 1–164.

40. Allen, R.G.; Pereira, L.S.; Raes, D.; Smith, M. *Crop Evapotranspiration: Guidelines for Computing Crop Water Requirements*; Food and Agricultural Organization of the United Nations: Rome, Italy, 1998. Available online: https://www.unirc.it/documentazione/materiale_didattico/1462_2016_412_24101.pdf (accessed on 15 November 2016).

41. Doorenbos, J.; Pruitt, W.O.; Aboukhaled, A.; Damagnez, J.; Dastane, N.G.; Van Den Berg, C.; Rijtema, P.E.; Ashford, O.M.; Frère, M. *Crop Water Requirements*; FAO Irrigation and Drainage Paper No. 24; Food and Agriculture Organization of the United Nations: Rome, Italy, 1992; pp. 70–72.

42. Amekudzi, L.; Yamba, E.; Preko, K.; Asare, E.; Aryee, J.; Baidu, M.; Codjoe, S. Variabilities in Rainfall Onset, Cessation and Length of Rainy Season for the Various Agro-Ecological Zones of Ghana. *Climate* **2015**, *3*, 416–434. [CrossRef]

43. Raes, D. *Frequency Analysis of Rainfall Data*; International Centre for Theoretical Physics (ICTP): Leuven, Belgium, 2004. Available online: http://indico.ictp.it/event/a12165/session/21/contribution/16/material/0/0.pdf (accessed on 3 January 2017).

44. Mbah, C.N. Determining the field capacity, wilting point and available water capacity of some Southeast Nigerian soils using soil saturation from capillary rise. *Niger. J. Biotechnol.* **2012**, *24*, 41–47.

45. Pedescoll, A.; Samsó, R.; Romero, E.; Puigagut, J.; García, J. Reliability, repeatability and accuracy of the falling head method for hydraulic conductivity measurements under laboratory conditions. *Ecol. Eng.* **2011**, *37*, 754–757. [CrossRef]

46. Saxton, K.E.; Rawls, W.J. Soil Water Characteristic Estimates by Texture and Organic Matter for Hydrologic Solutions. *Soil Sci. Soc. Am. J.* **2006**, *70*, 1569–1578. [CrossRef]

47. Bell, M.A.; Fischer, R.A. *Guide to Plant and Crops Sampling: Measurements and Observations for Agronomic and Physiological Research in Small Grain Cereals*; CIMMYT: Mexico City, Mexico, 1994. Available online: http://libcatalog.cimmyt.org/download/cim/53067.pdf (accessed on 11 November 2016).

48. Raes, D.; Steduto, P.; Hsiao, T.C.; Fereres, E. Chapter 3: Calculation procedures. In *Reference Manual: AquaCrop, Version 4.0*; FAO, Land and Water Division: Rome, Italy, 2012; pp. 1–130.

49. Adu-Dapaah, H.K.; Oppong-Konadu, E.Y. Tomato production in four major tomato-growing districts in Ghana: Farming practices and production constraints. *Ghana J. Agric. Sci.* **2002**, *35*, 11–22. [CrossRef]

50. Sallah, P.Y.K.; Twumasi-Afriyie, S.; Kasei, C. Optimum planting dates for four maturity groups of maize varieties grown in the Guinea savanna zone. *Ghana J. Agric. Sci.* **1997**, *30*, 63–69. [CrossRef]

51. Sugri, I.; Kanton, R.A.L.; Kusi, F.; Nutsugah, S.K.; Buah, S.S.J.; Zakaria, M. Influence of Current Seed Programme of Ghana on Maize (*Zea mays*) Seed Security. *Res. J. Seed Sci.* **2013**, *6*, 29–39. [CrossRef]

52. Bos, M.G.; Kselik, R.A.; Allen, R.G.; Molden, D. Capillary Rise. In *Water Requirements for Irrigation and the Environment*; Springer Netherlands: Dordrecht, The Netherlands, 2009; pp. 103–118.

Article

Groundwater Table Effects on the Yield, Growth, and Water Use of Canola (*Brassica napus* L.) Plant

Hakan Kadioglu [1], Harlene Hatterman-Valenti [2], Xinhua Jia [3], Xuefeng Chu [4], Hakan Aslan [5] and Halis Simsek [2,*]

1   School of Natural Resource Sciences, North Dakota State University, Fargo, ND 58105, USA
2   Agricultural & Biosystems Engineering, North Dakota State University, Fargo, ND 58105, USA
3   Plant Science, North Dakota State University, Fargo, ND 58105, USA
4   Civil & Environmental Engineering, North Dakota State University, Fargo, ND 58105, USA
5   Farm Structure and Irrigation, Ondokuz Mayis University, 55270 Samsun, Turkey
*   Correspondence: halis.simsek@ndsu.edu; Tel.: +1-701-231-6107

Received: 3 July 2019; Accepted: 17 August 2019; Published: 20 August 2019

Abstract: Lysimeter experiments were conducted under greenhouse conditions to investigate canola (*Brassica napus* L.) plant water use, growth, and yield parameters for three different water table depths of 30, 60, and 90 cm. Additionally, control experiments were conducted, and only irrigation was applied to these lysimeters without water table limitations. The canola plant's tolerance level to shallow groundwater was determined. Results showed that groundwater contributions to canola plant for the treatments at 30, 60, and 90 cm water table depths were 97%, 71%, and 68%, respectively, while the average grain yields of canola were 4.5, 5.3, and 6.3 gr, respectively. These results demonstrate that a 90 cm water table depth is the optimum depth for canola plants to produce a high yield with the least amount of water utilization.

Keywords: lysimeter; canola; water table; water use efficiency; root distribution; evapotranspiration

---

1. Introduction

As the global population grows, the demand for fresh water in many regions has increased dramatically. These population increases have caused more water stress for agriculture, the production of energy, industrial uses, and human consumption. Even though many countries currently have not faced a lack of water, water can no longer be considered an infinite source. Numerous regions have water use restrictions, so additional strategies to decrease the impact of water crises across the globe are needed [1–3].

One strategy for agricultural water management would encourage farmers to use shallow groundwater. Approximately 80% of available water resources in the world are being used in agricultural applications, and, therefore, the gap between adequate water availability and water needs is increasing [2]. Hence, the management of groundwater utilization in agriculture may be an acceptable alternative strategy to reduce freshwater demand. Therefore, surface water and shallow groundwater resources have become important for water demands.

Water use efficiency (WUE) is defined as a grain crop yield or total crop biomass per unit of water use [4]. Improved and well-managed WUE in agricultural water management systems is an important strategy to increase the productivity and reliability of crop yields. The consumption of groundwater is an extremely significant part of WUE. However, describing WUE for irrigation is complicated [5].

Good quality groundwater is a supplemental irrigation water source that can supply crops' water demands. When managed correctly, shallow groundwater can reduce both drainage and irrigation requirements. Some crops, such as canola (*Brassica napus* L.), soybean (*Glycine max*), and safflower (*Carthamus tinctorius*), are able to use moderate saline groundwater and could help to increase the utilization of groundwater and decrease the utilization of surface irrigation water [1,6,7]. In addition, there are obvious relationships between water table management (WTM), crop productivity, and environmental pollution. The environmental and economic benefits of WTM could decrease environmental pollution and increase crop productivity and irrigation intervals. However, WTM must be utilized correctly to supply sufficient soil moisture content to the crops [8].

The consumption of shallow groundwater as a crop water supply depends on several factors, such as groundwater table depths, groundwater availability and quality, crop species, distribution of the plant root system, weather conditions, and soil types [7,9]. The quantity and quality of groundwater are also affected by the irrigation method and management practices, as an excessive amount of irrigation water will increase groundwater utilization. It is impossible to control all these factors under field conditions because groundwater contributions are highly variable and difficult to estimate. Therefore, lysimeters are often used to evaluate a single parameter at a time [10].

Mejia et al. [8] utilized lysimeters to determine the effect of two different water table depths (50 and 75 cm) on corn and soybean grain yields. A free drainage system was installed 100 cm below the soil surface for both treatments. In the first year, corn yield was determined to be 13.8% higher with the free drainage treatment compared to the treatment without drainage at the 50 cm water table depth. However, only a 2.8% corn yield increase was observed at the 75 cm water table depth. In the second year, corn yield increases with the free drainage treatment compared to no drainage were measured as 6.6% at the 50 cm water table depth and 6.9% at the 75 cm water table depth. Similar results were observed for soybean. The authors concluded that the 75 cm water table depth with a free drainage system for corn and soybean was the most efficient water table depth.

Luo and Sophocleous [10] used lysimeters to evaluate the influence of the groundwater evaporation's contribution to winter wheat crop water use. Different water table depths, climates, and irrigation conditions were used to determine the amount of crop water use from the desired groundwater table levels. The relationship between wheat crop water use and water table depth varied. Winter wheat was supplied with 75% of crop water-use from a 100 cm groundwater depth without an irrigation application, while 3% of crop water use was supplied from the 300 cm groundwater level with three irrigation applications. The results showed that the water table contribution was affected not only by the water table depth, but also by the soil profile, rainfall, irrigation, and climatic variations.

Plant water uptake from shallow groundwater is affected by water table depth, plant salt tolerance, and plant root characteristics, the soil's hydraulic properties, the salinity level of the groundwater, and the presence of irrigation and drainage systems. Plant salt tolerance is the leading factor affecting water extraction from shallow groundwater. Each plant has a different tolerance to salinity, and plant tolerance differs in each growth stage. All the plants tend to be more susceptible to salinity in their early stages [11,12].

Fidantemiz et al. [13] used lysimeters under a controlled environment condition to determine the effect of different groundwater table levels (30, 50, 70, and 90 cm) on soybean growth. The highest grain yield and WUE results were obtained from 90 cm water table depth with 17.2 g/lys and 0.31 g/lys./c, respectively. In terms of WUE, grain yield and root distribution, both 70 and 90 cm water table depths were optimum for soybean yield in the experiments conducted without surface irrigation.

In this current study, canola plants are grown in the lysimeters. Canola can be grown with inadequate irrigation and weather conditions and, therefore, is highly adapted to cold weather conditions with insufficient water availability. High temperatures may cause abiotic stress on canola plant and influences its growth. Canola's sensitivity to high temperatures is higher in the flowering period than the podding period. During the blooming season of the canola plant, heat stress may shorten the flowering period. Two common types of canola, winter (*B. rapa*) and spring (*B. napus*) canola, can be grown in North Dakota. Although winter canola can be produced in ND and northwestern Minnesota, ND farmers mainly prefer to plant spring canola since spring canola can survive under the harsh winter condition, and its yield growth is higher than that of winter canola [14–16].

The main scope of this study was to determine an optimum shallow groundwater depth to achieve a high yield for canola plants. The lysimeter experiment was conducted to: (1) determine the optimum groundwater depth for canola growth and yield parameters for water table depths of 30, 60, and 90 cm without irrigation, (2) to quantify the amount of water consumption for water table depths of 30, 60, and 90 cm during canola growth, and (3) to determine the canola plant root distribution at water table depths of 30, 60, and 90 cm.

2. Materials and Methods

2.1. Lysimeter Design and Preparation

A greenhouse located in the North Dakota State University campus, Fargo, ND was used for the lysimeter study. Four treatments at 30 cm (T_{30}), 60 cm (T_{60}), and 90 cm (T_{90}) water table depths with no irrigation application and a control treatment ($T_{control}$, no water table) with a surface irrigation application were used. These three different water table depths were selected because they represent the elevated water table conditions in the fields where canola is normally grown. Each treatment had eight replications, so a total of 32 lysimeters were used. For the control treatment, 50% of the total available moisture (TAM) was considered as readily available moisture (RAM) in the soil profile. RAM is defined as the portion of the available water (field capacity minus permanent wilting point) before growth and yield are affected. RAM varies with crop and the evapotranspiration (ET) rates. According to Huffman et al. [17], 50% of the TAM for canola and a maximal ET rate of 6 mm/day was recommended. Tap water was used for both the groundwater and irrigation water sources. All the lysimeters in the greenhouse were distributed using a randomized complete block design method with eight replications.

Amber colored class bottles were used as Mariotte bottles to prevent algal growth and connected to the 24 lysimeters used for the water table depth treatments. The volume of the Mariotte bottles were 4 L, and four adjustable shelves were used to adjust the desired water table depth. The variation of the water volume in the Mariotte bottles was measured to determine the water consumption of canola. The Mariotte bottles were connected to the lysimeters from the bottom and continuously fed the lysimeters with a constant flow rate (Figure 1). The water reduction on the Mariotte bottles was monitored, and the difference was considered as the canola water consumption that supplied from the groundwater. Graduated cylinders were used for replenishment in the Mariotte bottles to obtain reliable measured water use.

2.2. Soil Packing and Sensor Installation

The loam soil was used to pack all the lysimeters. Bulk soil samples were obtained from an agricultural field in Fergus Falls, MN. The soil texture was classified as a loam soil based on the USDA/FAO texture classification system. The soil was then air-dried and sieved through a 2 mm screen and packed into the lysimeters. At the beginning of the study, the soil compaction problem in the lysimeters was observed, and 300 g of sand was added to 1.0 kg of the soil to deal with this problem. According to the laboratory analysis, the packed soil field capacity, readily available water, permanent wilting point, and bulk density were 0.32 cm³/cm³, 0.27 cm³/cm³, 0.21 cm³/cm³, and 1.14 Mg/m³, respectively. All these parameters were measured using the combined HYPROP (Data Evaluation Software) and WP4 method [18]. Gravel (8 cm) was packed at the bottom of the lysimeters, sand (8 cm) was then packed, and finally the processed loam soil (100 cm) was used to fill the lysimeters (Figure 1). All lysimeters were packed identically. Each lysimeter's diameter, wall thickness, and height were 152.4 mm, 5 mm, and 1260 mm, respectively. The lysimeters were made of Schedule-40 PVC material. The bottoms of the lysimeters were closed with a cap and glued to prevent leaking.

In the control treatment lysimeters ($T_{control}$), three soil water potential sensors (TEROS-21, METER Group, Inc., Pullman, WA, USA) were used to determine (i) the irrigation timing and (ii) the water needed for irrigation. Water potential sensors were installed at depths of 15, 45, and 75 cm in the lysimeters. For the remaining treatments (T_{30}, T_{60}, and T_{90}), six soil water potential sensors were used and placed at the appropriate depths. One soil water potential sensor was placed at a depth of 15 cm from the top of the soil surface in the T_{30} lysimeter. Two soil water potential sensors were placed at depths of 15 and 45 cm in the T_{60} lysimeter, and three soil water potential sensors were placed at the depths of 15, 45, and 75 cm in the T_{90} lysimeter [13]. To ensure hydraulic contact between sensors and moisture in the soil, all the sensors were placed horizontally in the lysimeters. All 9 water potential sensors were plugged into two Em50G (Decagon Inc.) dataloggers, and the data recording time interval was selected as 10 min.

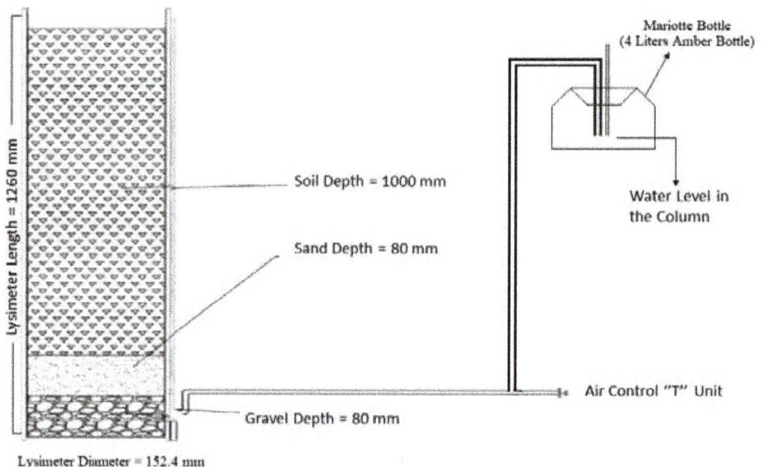

Figure 1. Schematic diagram of a lysimeter and Mariotte bottle system.

Two ETgage model E atmometers (C&M Meteorological Supply, Colorado Springs, CO, USA) were used to measure reference crop evapotranspiration (ET_0) in the greenhouse and recorded daily using HOBO Pendant Event Data Loggers (Onset Computer, Bourne, MA, USA) between 4 November 2018 (planting) and 4 February 2019 (harvesting). In addition, air temperature, barometric pressure, relative humidity, and vapor pressure were measured using an Atmos 14 sensor (Decagon Devices, Inc., Pullman, WA, USA). The device was connected to an Em50G datalogger to transfer the data to a computer.

2.3. Planting and Harvesting of Canola

Canola seeds (NDOLA-01) were planted on 4 November 2018 and harvested from 4 February to 10 February 2019, according to the plant harvest stages (Table 1). Ten seeds were sowed from 1 to 3 cm soil depths and thinned so that the three healthiest seedlings remained in each lysimeter. All the planted seeds were germinated in eight days. Although iron deficiency was observed at the beginning of the experiment, beneficial nematodes, supplements, and chemicals were not applied during the experiment. Similarly, no fertilizer was used in this study.

Table 1. Canola harvesting dates.

Treatments	Replications							
	1	2	3	4	5	6	7	8
$T_{control}$	5 February	5 February	6 February	6 February	5 February	5 February	5 February	5 February
T_{30}	9 February	10 February	10 February	9 February	10 February	9 February	9 February	10 February
T_{60}	7 February	6 February	6 February	7 February	6 February	7 February	7 February	7 February
T_{90}	5 February	4 February	5 February	4 February	5 February	4 February	4 February	4 February

To provide identical water curve conditions at the beginning of the experiment, all the lysimeters were filled with water to the soil surface in the lysimeters. Then, the valves at the bottom of the lysimeters were opened, and water in the lysimeters was drained. Approximately 30 h later, the valves were closed to maintain adequate moisture in the lysimeters for germination and the Mariotte bottles were connected and adjusted for the desired water table level for each lysimeter. For the control experiments, surface irrigation was applied based on the data obtained from the sensors, regardless of the germination stage in the lysimeters. Thus, starting from the first irrigation application, the irrigation timing and the amount of water needed for irrigation were determined by considering only the sensors' outcomes. Therefore, the germination stages in the columns were not monitored. As explained earlier, our goal was to maintain the packed soil field capacity at 0.32 cm^3/cm^3 and readily available water content at 0.27 cm^3/cm^3.

2.4. Calculation of Crop Water Use from Groundwater and Irrigation Water

After plant harvest, four randomly selected lysimeters from each treatment were cut vertically to determine the canola plant root's dry mass. In order to analyze the entire root distribution in each treatment, lysimeters were cut from the top through the bottom using electric saw. During the soil extraction process, three plant root depth intervals (0–30, 30–60, and 60–90 cm) were selected based on three water table depths. The soil in the lysimeters was washed, and the roots were separated gently from the soil. The roots were air-dried for 24 h before weighing to determine the root distribution and dry mass at each depth interval. Evapotranspiration in each lysimeter was calculated using Equation (1):

$$(\Delta S) = (I + Cr) - (Dp + ET) \tag{1}$$

where Cr is the water inflow due to capillarity, I is the irrigation, Dp is the deep percolation, ET is the evapotranspiration, and ΔS is the change in soil water content. Precipitation, runoff, and deep percolation were not applicable in this study since the experiments was performed in a controlled greenhouse. Irrigation was only applied to the control experiments. After evaluation of the controlled environment's conditions, the soil water balance equation was used to determine ET for each treatment (Equation (2)):

$$ET = Cr + S_1 - S_2 \tag{2}$$

where S_1 is the initial soil water storage (soil moisture) and S_2 is the final soil water storage in the lysimeters. Water reduction in the Mariotte bottles was measured every 15 days to determine the capillary water inflow in the lysimeters. The amount of water used by the canola was calculated using the soil water balance equation (Equation (1)) [19].

To determine the initial moisture conditions of the lysimeters at the beginning of the experiment, soil water potential sensors were used. After cutting the sixteen lysimeters, soil water content was measured, and the final moisture conditions of the sixteen lysimeters were determined. The soil water release curve was used to consider 50% of the total available moisture as the RAM in the soil profile of control treatment. The irrigation water depths for the lysimeters were calculated by using Equation (3) [20]:

$$d = \sum_{i=1}^{n} \frac{F_{ci} - M_{bi}}{100} \times A_{si} \times D_i \tag{3}$$

where d is the equivalent depth of water in cm, F_{ci} is the field capacity of the soil layer in percent by weight, M_{bi} is the current water content of the soil layer in percent by weight, A_{si} is the apparent specific gravity (bulk density), D_i is the depth of each soil layer, and n is the total number of soil layers.

To determine the soil water retention curve, the water in each lysimeter was drained out through a valve at the bottom of the lysimeter until 50% readily available soil moisture content was obtained in the lysimeter. For the control treatment, supplemental water was applied at the surface of the lysimeters to maintain the soil field capacity at 0.32 cm^3/cm^3.

WUE was calculated for both grain yield (harvested seed weight) and total biomass (harvested total dry matter). Since sixteen lysimeters were cut, the grain yield and total biomass values of sixteen lysimeters were used for grain yield and biomass WUE calculations. The same statistical difference in the grain yield and total biomass WUE results of thirty-two lysimeters was extrapolated by using the data of sixteen lysimeters in response to different WTDs.

2.5. Statistical Analysis

A randomized complete block design method was used in this study. The effect of different groundwater levels on canola growth and yield parameters (crop water use, plant height, seed weight, pod weight, total biomass, root-shoot ratio, and root distribution) were analyzed by using a one-way analysis of variance (ANOVA) with a $P \leq 0.05$ level of significance. The Statistical Package for the Social Sciences version 25 (SPSS) and Duncan homogeneous test comparisons with the $P = 0.05$ probability level were used to conduct mean separation tests, when appropriate.

3. Results and Discussions

3.1. Evapotranspiration and Climate Conditions in the Greenhouse

To determine the relationship between evapotranspiration and temperature in the greenhouse and interpret the temperature and ET_0 changes during the different canola growing stages (germination, growing, and harvesting), daily average ET_0 rates and temperature data were collected between 4 November 2018 (planting) and 4 February 2019 (first harvesting). According to the result obtained from ETgages, the lowest and highest temperature in the greenhouse were determined as 15.5 °C and 29.5 °C, respectively (Figure 2). The lowest temperatures were observed during the first 10 days after planting because of extreme cold ambient temperatures. The temperatures in the greenhouse were 25 ± 5 °C from 14 November 2018 to 4 February 2019.

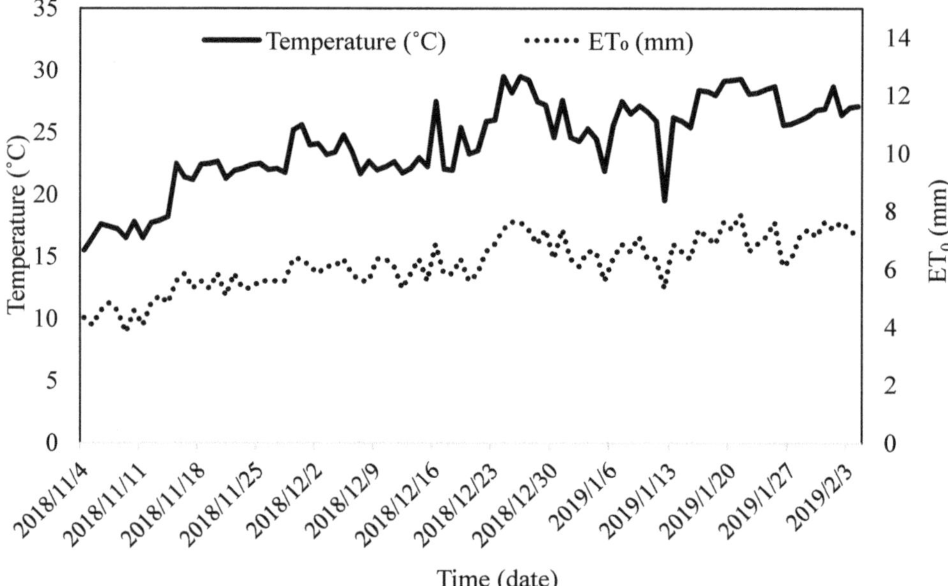

Figure 2. Measured daily air temperature (°C) and ET_0 values in the greenhouse. (ET_0 is reference crop evapotranspiration).

The lowest daily ET_0 was measured as 3.80 mm during the germination stage of canola (Figure 2). After 10 days of planting (when the canola was germinating and emerging), ET_0 rates fluctuated, with the highest ET_0 measured at 7.80 mm. Cumulative ET_0 was calculated as 577 mm during the entire experimental period (92 days). Fluctuations in air temperature influenced evapotranspiration. When the greenhouse temperature dropped from time to time, ET_0 also decreased accordingly.

3.2. Canola Irrigation Water Use

In the control treatment, the water content was kept between the field capacity (0.32 cm^3/cm^3) and the RAM (0.27 cm^3/cm^3) in order to prevent water stress and the application of an excessive amount of irrigation water. Three different plant root depths were considered for water requirement calculations. The canola root depth was projected 30 cm between 4 November and 4 December 2018 for the calculation of crop water requirements. Between 4 December 2018 and 4 January 2019, the control plants were irrigated up to a 60 cm root depth. After 4 January 2019, the crop water requirement was calculated for a 90 cm root depth. The volumetric water content for the specified root depths of the control plants was always maintained between field capacity and RAM with supplemental watering (Figure 3).

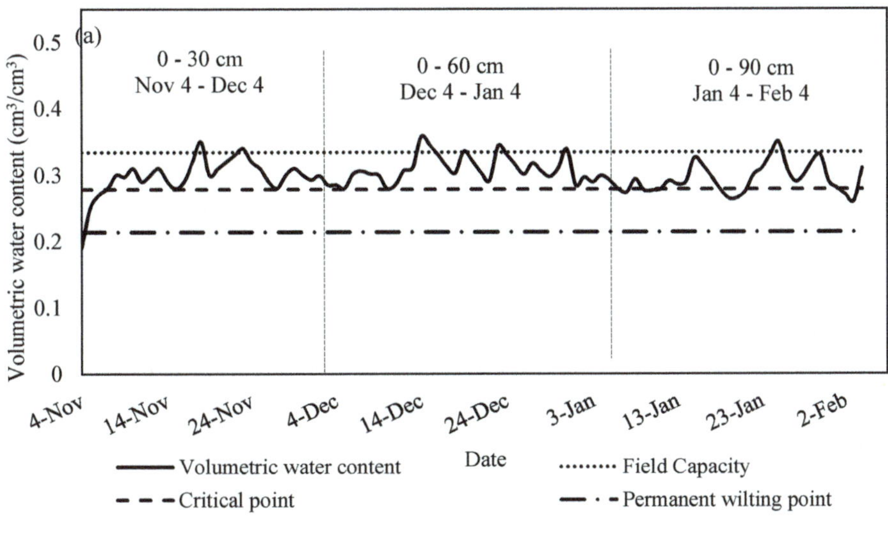

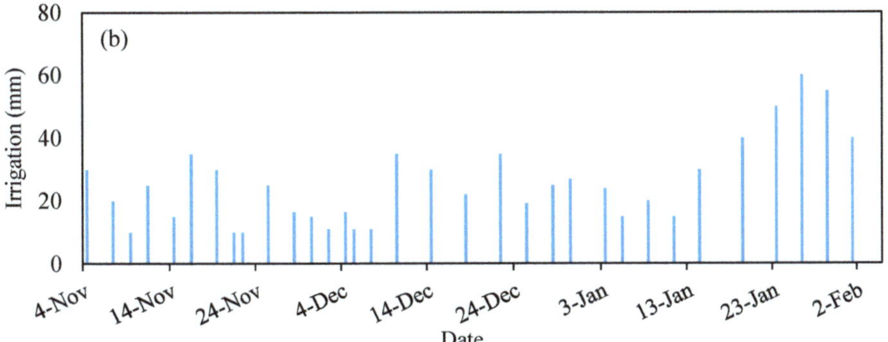

Figure 3. (**a**) Soil moisture content measurements for the control treatment at the desired depth of soil profile and (**b**) irrigation water applied to the lysimeters.

Control lysimeters received 752 mm of cumulative irrigation water. The calculated cumulative canola plant water use varied from 733 to 749 mm, with an average ET_c of 740 mm for the control treatments (Table 2).

Table 2. Total canola water-use for the control treatments.

Lysimeter Number	Initial Condition	Cumulative Irrigation Water	Final Condition	Cumulative ET_c	Mean ET_c
#	mm	mm	mm	mm	mm
R_3-$T_{control}$	162	752	181	733	
R_4-$T_{control}$	162	752	174	740	740
R_6-$T_{control}$	162	752	165	749	
R_7-$T_{control}$	162	752	176	738	

3.3. Canola Groundwater Use

The total canola plant water use and the groundwater contribution for the 12 lysimeters (3 lysimeters from each treatment) are presented in Table 3. Data collected from the water potential sensors were used to calculate soil water content. Initial soil water content was 350 mm in all the lysimeters. Similar to the control treatments, the canola evapotranspiration values in the water table treatments were measured. Each root depth had different ET_c values because evapotranspiration was influenced by WTD (Table 3). According to the ET_c value comparisons from different WTDs, the 30 cm soil profile had the highest ET_c, 717 mm. A significant difference in ET_c was observed between the 30 and 90 cm soil profiles. These results showed an inverse relationship between WTD and evapotranspiration. Additionally, the same inverse relationship was obtained between WTD and groundwater contribution. When the water table depth increased from 30 to 90 cm, the amount of water use from the groundwater also increased.

Table 3. Total canola water use from the groundwater for different depths.

Lysimeter Number	Depth	Initial Condition	Water Use from GW	Final Condition	ET_c	Mean ET_c	Mean ET_c
#	cm	mm	mm	mm	mm	mm	% $T_{control}$
R_3	30	350	632	241	741		
R_4	30	350	615	268	697	717	97
R_6	30	350	643	245	748		
R_7	30	350	607	275	682		
R_3	60	350	485	279	556		
R_4	60	350	426	289	487	527	71
R_6	60	350	433	271	512		
R_7	60	350	454	251	553		
R_3	90	350	402	235	517		
R_4	90	350	355	225	480	501	68
R_6	90	350	379	214	515		
R_7	90	350	341	198	493		

Note: R denotes replication. The initial condition is assumed to be identical for all lysimeters.

3.4. Growth and Yield Parameters

Plants in the T_{90} treatments were taller than the plants in the $T_{control}$, T_{30}, and T_{60} treatments, with a mean plant height of 134.6 cm for plants in the T_{90} treatment, while the shortest plants (113.3 cm) were in the T_{30} treatment. An inverse relationship was observed between the mean plant height and WTD, which was similar to the relationship between the WTD and the groundwater contribution.

When the water level was increased from 90 to 60 cm (measured from the soil surface) in the lysimeters, the canola plant height decreased from 134.6 to 113.3 cm.

Treatment differences were significant for total biomass, pod weight, and seed weight per plant (Table 4). The highest mean total biomass, pod weight, and seed weight were 22.1, 12.6, and 6.3 gr, respectively, for the T_{90} treatment. The lowest mean total biomass, pod weight, and seed weight results were 15.1, 9.5, and 4.8 gr, respectively, for the T_{30} treatment. These results indicate that the shallower water table depth decreased canola harvesting. Overall, the statistical results suggest a negative correlation between the canola plant growth and the yield parameters and WTD. A similar correlation between the WTD and crop harvesting results was reported by Mejia et al. [8]. According to the 2 year lysimeter experiment results, corn and soybean grain yield weights increased when the WTD decreased from 50 cm to 75 cm [8].

Plants in the control treatment consumed the most water (Tables 2 and 3), while plants in the T_{90} treatment had the highest yield values compared to other treatmnets (Table 4). As explained earlier, plants in $T_{control}$ used the optimum amount of irrigation water, 740 mm, while plants in T_{90} consumed 501 mm of water from the groundwater. However, better growth and yield results were obtained from plants in the T_{90} treatment.

Table 4. Statistical analysis of the canola growth and yield parameters.

Treatment #	Plant Height cm/plant	Total Biomass g/plant	Pod Weight g/plant	Seed Weight g/plant
$T_{control}$	118.0 [a]	19.4 [c]	11.7 [c]	5.8 [c]
T_{30}	113.2 [a]	15.1 [a]	9.5 [a]	4.7 [a]
T_{60}	113.8 [a]	17.8 [b]	10.8 [b]	5.3 [b]
T_{90}	134.6 [b]	22.0 [d]	12.5 [d]	6.3 [d]

Note: Lowercase letters; [a], [b], [c], and [d] show statistical significance at $p = 0.05$. The values followed by the same letter in each column are not significantly different.

3.5. Water Use Efficiency (WUE)

The relationship between WUE for canola total biomass and grain yield was significant. The correlation of both WUE values was also determined (Figure 4). The effects of different WTD levels on both the grain yield and total biomass WUEs were also significant. The highest grain yield and biomass WUE values for T_{90} were 0.0126 and 0.0449 g lys^{-1} mm^{-1}, respectively. The lowest WUE values for both parameters occurred with the T_{30} treatment (Table 5).

After cutting the 16 lysimeters as mentioned earlier, each soil profile was divided into three different layers: 0–30, 30–60, and 60–90 cm (measured from the top) to determine the percentage of the root mass distribution in terms of WTD (Table 6). Overall, the highest root–mass ratio was found with plants in $T_{control}$ at the 0–30 cm soil layer (4.52 g and 54.6%). There was an inverse relationship between soil depth and root mass distribution for $T_{control}$ plants. When WTD changed from 90 to 30 cm, the root weight increased from 0.73 to 4.52 g. Since $T_{control}$ did not have WTD, a lower amount of roots was found in deeper soil layers. Significant differences were observed in 0–30 cm between T_{90} and other treatments, and a greater root mass was observed at 60–90 cm in treatment T_{90}. The highest average root weight was 7.97 g in the third layer for plants in the T_{90} treatment (Table 6). Similar to the inverse relationship between the soil depth and root mass for $T_{control}$, an inverse correlation was observed between WTD and the root mass for T_{90}.

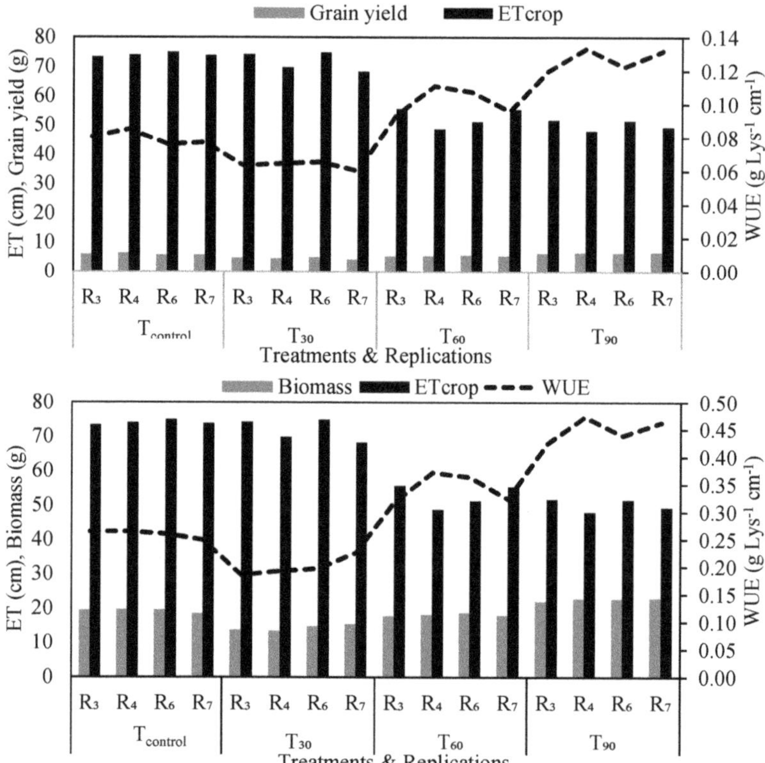

Figure 4. Canola water use efficiency with treatments: (**a**) grain yield water use efficiency (WUE) and (**b**) total biomass WUE.

Table 5. Statistical analysis results of the canola's mean grain yield, total biomass, water use, and water use efficiency.

Treatment	Mean Grain Yield (g/lys)	Mean Total Biomass (g/lys)	Mean Crop Water Use (mm)	Mean Grain Yield WUE (g/lys·mm)	Mean Total Biomass WUE (g/lys·mm)
$T_{control}$	5.9 [c]	19.2 [c]	740.0 [b]	0.0079 [b]	0.0259 [b]
T_{30}	4.5 [a]	14.3 [a]	717.0 [b]	0.0063 [a]	0.0199 [a]
T_{60}	5.3 [b]	18.0 [b]	527.0 [a]	0.0101 [c]	0.0344 [c]
T_{90}	6.3 [d]	22.5 [d]	501.2 [a]	0.0126 [d]	0.0449 [d]

Note: Lowercase letters; [a], [b], [c], and [d] show statistical significance at $p = 0.05$. The values followed by the same letter in each column are not significantly different.

Table 6. Average root mass and proportions of roots.

Layers	Depth	Average Root Mass and Percentage							
		$T_{control}$		T_{30}		T_{60}		T_{90}	
	cm	g	%	g	%	g	%	g	%
1th	0–30	4.52 [b]	54.6	4.4 [b]	47.8	4.60 [b]	42.9	3.13 [a]	19.1
2nd	30–60	3.02 [a]	36.7	2.95 [a]	31.9	4.02 [b]	37.6	5.22 [c]	32.1
3rd	60–90	0.73 [a]	8.7	1.87 [b]	20.3	2.08 [b]	19.5	7.97 [c]	48.8
Total		8.27 [a]	100	9.22 [a]	100	10.7 [b]	100	16.32 [c]	100

Note: Lowercase letters; [a], [b], [c], and [d] show statistical significance at $p = 0.05$. The values followed by the same letter in each column are not significantly different.

3.6. Root Mass Distribution

There is no significant total root mass difference between plants in the T_{30} and $T_{control}$ treatments. However, plants in the T_{30} and $T_{control}$ treatments had lower mean total root weights compared to plants in the T_{60} and T_{90} treatments. The mean total root mass for plants in the T_{90} treatment was always two-fold that for $T_{control}$. A relatively lower dry root mass was measured at the second and third layers of $T_{control}$, T_{30}, and T_{60} but a higher dry root mass was found at the second and third layers of T_{90}. Similar to the results of the grain yield, the plant height, total biomass, pod weight, and WUE, the best root mass results were obtained from T_{90}. Fidantemiz et al. [13] found similar results for the T90 treatment. Their results also showed the inverse relationship between the WTD and root mass distribution.

4. Conclusions

In this study, the canola plant height, water use from different groundwater levels, total biomass WUE, grain yield WUE, root mass, root-shoot ratio, and harvesting results (total biomass, pod and seed weight) were determined and compared with three different water table depths in a greenhouse. In addition, the effects of the optimum amount of irrigation water and different WTDs on canola were exanimated. Results suggest that the canola plant was affected by different water table levels since inverse linear relationships were found with the different WTDs.

The highest measured pod weight, total biomass, and seed weight were found for plants with the T_{90} treatment, although plants from this treatment consumed the lowest amount of water from the groundwater. Plants with the greatest harvest results and the lowest amount of water utilization also had the greatest total biomass and grain yield WUEs. On the other hand, plants with the lowest harvest results and the highest crop water use occurred when plants were at the 30 cm water table depth. As a result, a high WTD level (30 cm) negatively impacted the canola growth.

Significant statistical differences were found between the root distribution and soil layers. In addition, stronger and heavier roots were found near the water table level. In contrast, the total root weight was affected by WTD, and significant statistical differences were observed among the treatments. The total root weight of the 90 cm lysimeter was significantly higher than that of the other treatments. It was projected that canola in a drier lysimeter developed its root structure very well, since canola plants have a tendency to reach the water. Overall, the results from this study can be used to guide water management through drainage water management, in order to achieve the best yield potential.

Author Contributions: Conceptualization, H.S., X.C. and X.J.; methodology, H.S., X.J. and H.A.; investigation, H.S.; resources, H.H.-V., H.A.; writing—original draft preparation, H.K.; writing—review and editing, H.H.-V., X.J., X.C., H.S.; supervision, H.S.; funding acquisition, H.S.

Funding: This research was funded by USDA-NIFA North Central Region Canola Research (grant number FAR0029510), North Dakota Water Resources Research Institute (Grant number FAR0025807), and North Dakota Agricultural Extension Station. Additionally, the Turkish Government Ministry of Education (YLSY program) and General Directorate of State Hydraulic Works (DSI) provided stipend and tuition funding for the M.S. student, Hakan Kadioglu.

Acknowledgments: We thank our colleagues Hans Kendal and Mukhlesur Rahman from Plant Science Department and Aaron L.M. Daigh from Soil Science Department at North Dakota State University for their valuable suggestions throughout the study. We are also immensely grateful to Izzet Kadioglu from Plant Protection Department at Gaziosmanpasa University, Tokat, Turkey for his support and encouragement during our experiments. Any opinions, findings, conclusions, or recommendations expressed in this material are those of the author(s) and do not necessarily reflect the views of the funding institutions.

Conflicts of Interest: The authors declare no conflict of interest.

References

1. Hamdy, A.; Ragab, R.; Scarascia-Mugnozza, E. Coping with water scarcity: Water saving and increasing water productivity. *Irrig. Drain.* **2003**, *52*, 3–20. [CrossRef]

2.   Condon, A.G.; Richards, R.; Rebetzke, G.; Farquhar, G. Breeding for high water-use efficiency. *J. Exp. Bot.* **2004**, *55*, 2447–2460. [CrossRef] [PubMed]

3.   Ripoll, J.; Urban, L.; Staudt, M.; Lopez-Lauri, F.; Bidel, L.P.; Bertin, N. Water shortage and quality of fleshy fruits—Making the most of the unavoidable. *J. Exp. Bot.* **2014**, *65*, 4097–4117. [CrossRef] [PubMed]

4.   Sinclair, T.R.; Tanner, C.; Bennett, J. Water-use efficiency in crop production. *Bioscience* **1984**, *34*, 36–40. [CrossRef]

5.   Howell, T.A. Enhancing water use efficiency in irrigated agriculture. *Agron. J.* **2001**, *93*, 281–289. [CrossRef]

6.   Yang, J.; Wan, S.; Deng, W.; Zhang, G. Water fluxes at a fluctuating water table and groundwater contributions to wheat water use in the lower Yellow River flood plain, China. *Hydrol. Process. Int. J.* **2007**, *21*, 717–724. [CrossRef]

7.   Ghamarnia, H.; Golamian, M.; Sepehri, S.; Arji, I.; Rezvani, V. Groundwater contribution by safflower (*Carthamus tinctorius* L.) under high salinity, different water table levels, with and without irrigation. *J. Irrig. Drain. Eng.* **2011**, *138*, 156–165. [CrossRef]

8.   Mejia, M.; Madramootoo, C.; Broughton, R. Influence of water table management on corn and soybean yields. *Agric. Water Manag.* **2000**, *46*, 73–89. [CrossRef]

9.   Huo, Z.; Feng, S.; Huang, G.; Zheng, Y.; Wang, Y.; Guo, P. Effect of groundwater level depth and irrigation amount on water fluxes at the groundwater table and water use of wheat. *Irrig. Drain.* **2012**, *61*, 348–356. [CrossRef]

10.   Luo, Y.; Sophocleous, M. Seasonal groundwater contribution to crop-water use assessed with lysimeter observations and model simulations. *J. Hydrol.* **2010**, *389*, 325–335. [CrossRef]

11.   Kruse, E.; Champion, D.; Cuevas, D.; Yoder, R.; Young, D. Crop water use from shallow, saline water tables. *Trans. ASAE* **1993**, *36*, 697–707. [CrossRef]

12.   Talebnejad, R.; Sepaskhah, A. Effect of different saline groundwater depths and irrigation water salinities on yield and water use of quinoa in lysimeter. *Agric. Water Manag.* **2015**, *148*, 177–188. [CrossRef]

13.   Fidantemiz, Y.F.; Jia, X.; Daigh, A.L.; Hatterman-Valenti, H.; Steele, D.D.; Niaghi, A.R.; Simsek, H. Effect of Water Table Depth on Soybean Water Use, Growth, and Yield Parameters. *Water* **2019**, *11*, 931. [CrossRef]

14.   Johnston, A.M.; Tanaka, D.L.; Miller, P.R.; Brandt, S.A.; Nielsen, D.C.; Lafond, G.P.; Riveland, N.R. Oilseed crops for semiarid cropping systems in the northern Great Plains. *Agron. J.* **2002**, *94*, 231–240. [CrossRef]

15.   Kandel, H. *Soybean Production Field Guide for North Dakota and Northwestern Minnesota*; North Dakota State University: Fargo, ND, USA, 2010.

16.   Kutcher, H.; Warland, J.; Brandt, S. Temperature and precipitation effects on canola yields in Saskatchewan, Canada. *Agric. For. Meteorol.* **2010**, *150*, 161–165. [CrossRef]

17.   Huffman, R.L.; Fangmeier, D.D.; Elliot, W.J.; Workman, S.R. Conservation and the Environment. In *Soil and Water Conservation Engineering*, 7th ed.; Cengage Learning: Boston, MA, USA, 2012; pp. 1–7.

18.   Roy, D.; Jia, X.; Steele, D.D.; Lin, D. Development and comparison of soil water release curves for three soils in the red river valley. *Soil Sci. Soc. Am. J.* **2018**, *82*, 568–577. [CrossRef]

19.   Hillel, D. *Environmental Soil Physics: Fundamentals, Applications, and Environmental Considerations*; Elsevier: Amsterdam, The Netherlands, 1998.

20.   Majumdar, D.K. *Irrigation Water Management: Principles and Practice*; Prentice-Hall India Pvt. Ltd.: Delhi, India, 2004; p. 500.

Article

Spatial Distribution of Salinity and Sodicity in Arid Climate Following Long Term Brackish Water Drip Irrigated Olive Orchard

John Rohit Katuri [1,2], Pavel Trifonov [1] and Gilboa Arye [1,*

[1] French Associates Institute for Agriculture and Biotechnology of Drylands, Jacob Blaustein Institutes for Desert Research, Ben Gurion University of the Negev, Sede Boqer Campus, Midreshet Ben-Gurion 84990, Israel; john.katuri@gmail.com (J.R.K.); paveltri@post.bgu.ac.il (P.T.)

[2] Department of Agronomy, Directorate of Crop Management, Tamil Nadu Agricultural University, Coimbatore, Tamil Nadu 641003, India

* Correspondence: aryeg@bgu.ac.il

Received: 7 November 2019; Accepted: 28 November 2019; Published: 3 December 2019

Abstract: The availability of brackish groundwater in the Negev Desert, Israel has motivated the cultivation of various salinity tolerant crops, such as olives trees. The long term suitability of surface drip irrigation (DI) or subsurface drip irrigation (SDI) in arid regions is questionable, due to salinity concerns, in particular, when brackish irrigation water is employed. Nevertheless, DI and SDI have been adopted as the main irrigation methods in olive orchards, located in the Negev Desert. Reports on continued reduction in olive yields and, essentially, olive orchard uprooting are the motivation for this study. Specifically, the main objective is to quantify the spatial distribution of salinity and sodicity in the active root-zone of olive orchards, irrigated with brackish water (electrical conductivity; EC = 4.4 dS m^{-1}) for two decades using DI and subsequently SDI. Sum 246 soil samples, representing 2 m^2 area and depths of 60 cm, in line and perpendicular to the drip line, were analyzed for salinity and sodicity quantities. A relatively small leaching-zone was observed below the emitters depth (20 cm), with EC values similar to the irrigation water. However, high to extreme EC values were observed between nearby emitters, above and below the dripline. Specifically, in line with the dripline, EC values ranged from 10 to 40 dS m^{-1} and perpendicular to it, from 40 to 120 dS m^{-1}. The spatial distribution of sodicity quantities, namely, the sodium adsorption ratio (SAR, (meq L^{-1})$^{0.5}$) and exchangeable sodium percentage (ESP) resembled the one obtained for the EC. In line with the dripline, from 15 to 30 (meq L^{-1})$^{0.5}$ and up to 27%, in perpendicular to the drip line from 30 to 60 (meq L^{-1})$^{0.5}$ and up to 33%. This study demonstrates the importance of long terms sustainable irrigation regime in arid regions in particular under DI or SDI. Reclamation of these soils with gypsum, for example, is essential. Any alternative practices, such as replacing olive trees and the further introduction of even high salinity tolerant plants (e.g., jojoba) in this region will intensify the salt buildup without leaving any option for soil reclamation in the future.

Keywords: arid region; brackish water; sub surface drip irrigation (SDI); salinity; sodicity; olives trees

---

1. Introduction

Salinity and drought are the major abiotic stress factors limiting yield in arid regions [1]. To counteract these limitations, advanced irrigation management practices, such as drip irrigation (DI), were introduced and soon hailed as a breakthrough in agricultural efficiency [2]. Additionally, advanced breeding methods and genetic engineering tools have been developed to confer abiotic stress tolerance in different crops, with emphasis on enhanced tolerance to drought and high soil salinity [3]. With the advent of these technologies, saline water agriculture has gained importance and facilitated cultivation in arid environments. Due to drought conditions (low precipitation) in arid regions soil, salinity often increases, impeding plant water uptake. The initial plant responses to salinity and drought stress are fundamentally identical across species and are often complex [4,5]. Plant root adaptations play a key role in coping with these stresses [6]. For the successful management of arid agriculture choice of crop, cultivar and irrigation management regimes play a key role.

In the late 1970s, the introduction and cultivation of various saline tolerant crops with brackish water started in the Negev Desert of Israel [7]. Today farmers in the Negev region grow olives using DI or sub-surface drip irrigation (SDI) with brackish ground water (EC ~4.5 dS m^{-1}) from the local aquifer, as they have no alternative for other economical irrigation water source [8]. Olive trees are generally tolerant of drought and salinity [9,10]. However, salinity tolerance in olives is a cultivar specific trait. The main active root zone distribution in olives trees is at a depth of 30 to 60 cm [11,12] and various studies have reported that the upper critical limit of soil EC for normal olive development is 4 to 6 dS m^{-1} [12–15]. In olives trees, the maximum root growth rate can be achieved under fresh water irrigation and the high root mortality rate and root growth restriction occurs under moderately saline irrigation (4.2 dS m^{-1}) [16–19]. Irrigation water salinity of 4 dS m^{-1} limits significant production of the potential yield possible with good quality water [15] and there is a gradual buildup of soil salinity over the years in the root zone [16]. Therefore, an appropriate management of irrigation regime and salinity in root zone is necessary to optimize yield and oil quality in olive orchards irrigated with saline water [15,20].

In the long term, the commitment to utilizing marginal irrigation water sources, such as brackish water, may be fundamentally unsustainable, in particular, in arid lands where precipitation is too low to leach the accumulated salts from the active root zone [21]. There is a higher risk of soil salinization if rainfall is lower than 250 mm and the salts are not leached from the upper 60 cm depth [22–24]. The Negev region has an arid climate with high rates of evapotranspiration (about 2600 mm year^{-1}) and low rainfall (70 to 125 mm/year) [8,25]. When SDI was employed it reduced evaporation and improved irrigation water-use efficiency with olive yield similar to DI irrigation [26,27]. However, in SDI systems, salt accumulation above the dripper is high and does not offer an advantage over DI in regard to soil salt distribution under conditions of high evaporative demand [28,29]. In arid and semiarid areas, using SDI placed at shallow depths (about 20 cm) resulted in large amounts of salt accumulation near the soil surface [30], specifically located above the dripline [31,32]. When salts accumulate in soil surface layers, sprinkler irrigation is commonly used in SDI plots to leach salts below the drip tapes, but, in the long term it affects the economic sustainability of SDI [30]. Nevertheless, it was recently demonstrated [33,34] that a sequential practice of sprinkler irrigation for potato germination, followed by low discharge shallow SDI with brackish irrigation water, can result in similar potato yields to traditional methods that utilize sprinkler irrigation with fresh water.

There is high transient salinity and sodicity risk associated with saline water SDI in orchards [35] and they change with the amount and quality of infiltrated water, evapotranspiration rates, and rainfall [36]. When water quality of EC >2.5 dS m^{-1} and SAR >4 was used in olive and other orchards with SDI, there was a significant increase in soil salinity and sodicity values at 0–60 cm soil depths [37–40]. Most studies which examined the salinity and/or sodicity effect on olive growth and yield are short term (<8 years) studies [19,24,41,42] and, consequently, a severe accumulation of salts in the soil profile was not reported.

As mentioned, the introduction and cultivation of salt-tolerant crops in the arid regions in conjunction with brackish irrigation water for the past few decades has resulted in increasing soil salinity. In the current study, we quantify the salinity and sodicity spatial distribution in an olive orchard following twenty years of irrigation with brackish water. The motivation for this study stems from recent reports on continues decrease in yields (Figure 1) and the eventual uprooting of some olive orchards due to unprofitability. Therefore, it is necessary to understand the sustainability of olive cultivation under saline brackish water with SDI, so that secondary salinization is prevented and the soil can be reclaimed for agriculture in future years. The main objective of this study is to fill the knowledge gap regarding the spatial distribution of salinity and sodicity in long term sub-surface drip irrigated soils with brackish irrigation water. Given the relatively high distance (1 m) between drippers, we hypothesized that a high level of salinity and sodicity will be established between nearby drippers.

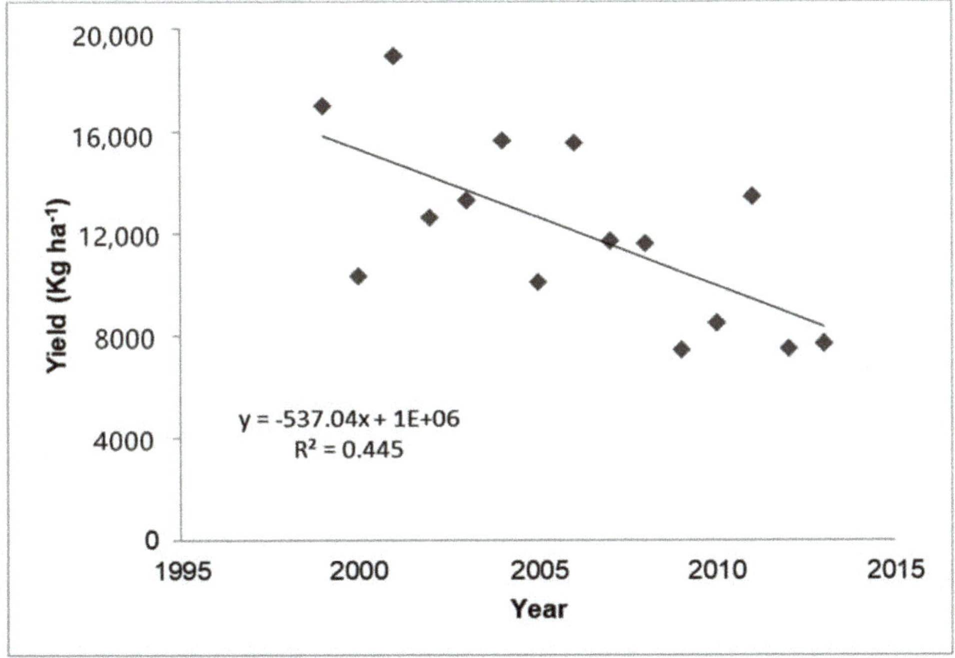

Figure 1. Yield trend for 15 years of the Barnea olive variety grown in the Revivim orchard.

2. Materials and Methods

2.1. Site Description

A field investigation was carried out in the olive (*Olea europaea*) cultivar 'Barnea' orchard of Kibbutz Revivim (31.0436° N, 34.7212° E), located in the central Negev Desert, Israel. The climatic conditions of the location are of the typical arid type, with cooler night temperatures and hot, dry summers (Figure 2). The mean annual rainfall ranges from 75 to 125 mm [25]. During the 2014/2015 season, the total precipitation was 105 mm and cumulative potential evapotranspiration was 2500 mm (Figure 3). The olive orchard was planted in 1995 and has been irrigated with brackish groundwater (EC = 4.4 dS m^{-1}) since then, for approximately 20 years. Water quality parameters of the brackish irrigation water are presented in Table 1.

During the first 15 years, DI was used but later converted to SDI by placing drip laterals at 20 cm soil depth and about 1 m distance from the tree line with an emitter flow-rate of 4 L h^{-1} and 1 m distance between nearby emitters. An initial tree spacing of about 3 × 7 m was first established and after ten years, each alternate tree within a row was uprooted, giving the current spacing of 6 × 7 m. Irrigation was scheduled according to class evaporation pan located nearby the orchard. Specifically, in average, a factor of 35% to 60 % was used to calculate the irrigation amounts from the predetermined cumulative pan evaporation (class A pan) [8]. Accordantly, irrigation intervals were scheduled every 3 days during summer and every 7 days during winter. Approximately, 800 mm plus an excess of 100 mm, as the leaching requirement of irrigation water, was applied annually.

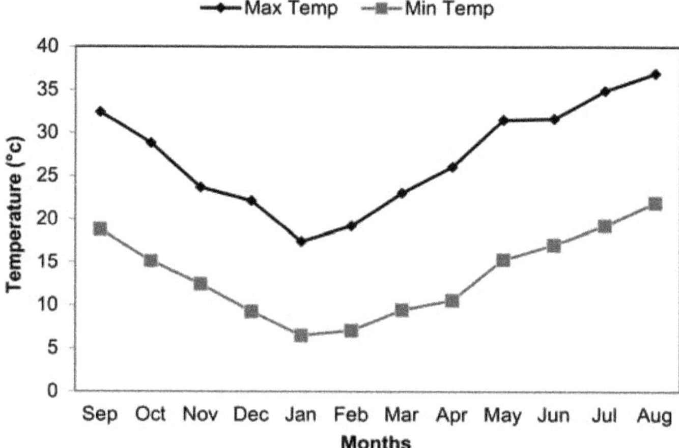

Figure 2. Minimum and maximum temperature in Revivim during 2014–2015.

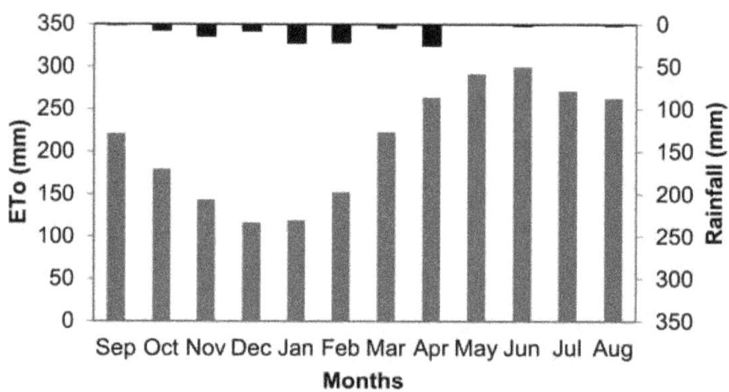

Figure 3. Rainfall and potential evapotranspiration (ETo) in Revivim during 2014–2015.

Table 1. Irrigation water quality parameters (Mekerot Water Company).

Parameter	Value
Boron (mg L^{-1})	1.200
Calcium (mg L^{-1})	171.000
Chloride (mg L^{-1})	1120.000
Electrical Conductivity dS m^{-1}	4.400
CaCO$_3$ (mg L^{-1})	748
HCO$_3$ (mg L^{-1})	301
Potassium (mg L^{-1})	1900
Magnesium (mg L^{-1})	78.000
Sodium (mg L^{-1})	684.000
pH	7.000
Total organic carbon (mg L^{-1})	<0.200
Total dissolved matter (mg L^{-1})	2697.000
SAR (meq L^{-1})$^{0.5}$	10.900

2.2. Soil Sampling and Analysis

Comprehensive soil sampling was carried out to explore the spatial distribution of salinity and sodicity along and perpendicular to the drip-line, representing a total area of 2 m^2 (Figure 4). Soil samples were collected from 41 locations along the drip-line between three nearby emitters that were perpendicular and diagonal to the central emitter. At each sampling location, disturbed soil samples (n = 246) were taken from six depths: 0–5, 5–10, 10–15, 15–30, 30–45, and 45–60 cm. In addition, representative intake soil samples were taken near the sampling locations mentioned above, from which the bulk density of each layer was calculated (Table 2). The gravimetric water content (WC) was measured shortly after the sampling event from the differences in weight before and after drying at 105 °C for 24 h. The rest of the soil samples were air-dried and thereafter passed through a 2-mm sieve.

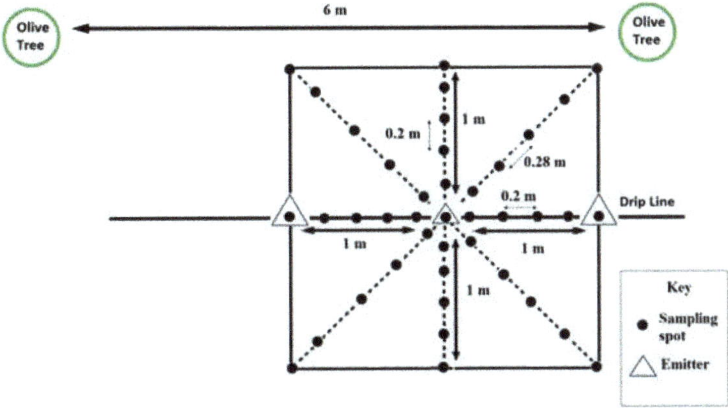

Figure 4. Schematic view of soil sampling spots in an olive orchard under sub surface drip irrigation.

Table 2. Revivim olive orchard soil properties.

Depth	Texture	Sand	Silt	Clay	Organic Matter	Bulk Density	CEC	SSA*	CaCO$_3$
(cm)		(%)	(%)	(%)	(%)	(g/cm^3)	(meq/100g)	(g/m^2)	(%)
0–5	Loamy Sand	80.83 ± 5.77	11.67 ± 5.77	7.50 0.00	10.22 ± 0.05	0.96 ± 0.017	8.54 ± 1.13	28.29 0.00	8.0 ± 3.5
5–10	Sandy Loam	70.83 ± 1.44	16.67 ± 2.89	12.50 ± 2.5	5.41 ± 3.28	1.09 ± 0.006	9.27 ± 0.77	57.19 ± 14.45	8.9 ± 1.5
10–15	Sandy Loam	67.50 0.00	19.17 ± 2.89	13.33 ± 2.89	1.62 ± 0.17	1.46 ± 0.006	8.61 ± 0.22	62.00 ± 16.69	8.5 ± 2.3
15–30	Sandy Loam	65.83 ± 5.20	19.17 ± 5.20	15.00 0.00	1.60 ± 0.20	1.48 ± 0.006	8.59 ± 0.63	71.64 0.00	9.5 ± 1.6
30–45	Loam	50.83 ± 3.82	28.33 ± 1.44	20.83 ± 2.89	1.53 ± 0.21	1.45 ± 0.006	9.19 ± 0.94	105.35 ± 16.69	8.6 ± 1.2
45–60	Loam	54.17 ± 3.82	27.50 ± 2.50	18.33 ± 1.44	1.38 ± 0.11	1.47 ± 0.006	9.03 ± 0.19	90.90 ± 8.34	10.9 ± 1.9

± Standard deviation, SSA*—specific surface area (calculated according to [43]).

The concentration of the main cations (Na, K, Mg, and Ca) in the soil solution was obtained from the extraction of the soil to distilled water ratio of 1:1. Samples were shaken on an end-over shaker and then centrifuged at 4000 rpm for 10 min. The supernatant was analyzed for soluble cations, bicarbonate, and chloride concentration. The cation exchange capacity (CEC) was measured by the sodium acetate method [44] and the exchangeable cations concentrations (xNa, xK, xMg, and xCa) from the sodium acetate extraction [45]. The cations concentration was measured by atomic adsorption spectrophotometer (Analyst 400, ParkinElmer) and Chloride (Cl^{-1}) concentration by Chloride Analyzer (926, Sherwood).

The sodium adsorption ratio (SAR) and the exchangeable sodium ratio (ESR) were calculated according to the Gapon equation.

$$\frac{xNa}{xCa + xMg} = K_G \cdot \frac{Na}{\sqrt{0.5 \cdot (Ca + Mg)}} \equiv ESR = K_G \cdot SAR \tag{1}$$

where, the concentrations of the soluble and exchangeable cation are in meq L^{-1} and meq Kg^{-1}, respectively. The K_G is the Gapon selectivity coefficient.

Contours map of the spatial distribution of the WC, EC, Cl^{-1}, SAR, and ESP in the soil profile were established with Surfer software (version 8, Golden Software, Colorado, USA) using the Kriging regression.

3. Results and Discussion

From the soil properties (Table 2) it can be seen that the texture in the examined soil layers changes from loamy-sand in the top soil layer (0–5cm), sandy-loam in the middle ones (5–30 cm), and loam in the deeper layers (30–60 cm). A distinct difference in organic matter (OM) percentage could be observed from 10.2% (0–5 cm), 5.4% (5–10 cm) and similar values ranging from 1.62% (10–15 cm) to 1.38% (45–60 cm). The bulk density exhibited an inverse linear correlation to OM content (BD = 1.54 − 0.06 × OM, R^2 = 0.93) rather than any of size fractions; ranging from about 1 g cm^{-3} in the top soil layer (0–10 cm) and exhibited similar values of about 1.45 g cm^{-3} for the rest of the soil profile. The above observation may imply a higher water holding capacity in the top soil layers, due to water adsorption and/or structures formation induced by the level of soil OM.

In the followings, the spatial distribution obtained for water content, salinity, and sodicity quantities are presented for two 60 cm soil transect: (i) along the drip line and (ii) perpendicular to the drip line (crossing the middle dripper), (Figure 4). In addition, a three dimensional visualization is presented as a counter map calculated from all measured data points of the four transects for a given soil layer (Table 2).

3.1. Water Content Spatial Distribution

In Figure 5 the spatial distribution obtained for the WC is presented for the sampled transect along (Figure 5a) and perpendicular (Figure 5b) to the drip line. The WC distribution demonstrates that relatively higher WC can be found directly above and below the location of the emitters (i.e., 20 cm depth). A typical wetting bulb of relatively light-texture soil can be observed with the bulb radius; the horizontally wetted radius is less than the vertically wetted depth radius [46,47]. The near-saturation zone was located about 20 cm from the emitters from which a gradual reduction in WC can be observed up to 50 cm distance, which is located in the middle, between two nearby drippers. At this location, the WC above the emitters is the lowest one, suggesting that there is no significant overlap between nearby emitters. The relatively large distance between the emitters (i.e., 1 m) and the corresponding spatial distribution of the WC also affected the salinity and sodicity spatial distribution, as is demonstrated below. Regarding the perpendicular transect (Figure 5b), it should be noted that +100 cm on the x-axis is towards the tree-line and −100 cm is towards the road, i.e. away from the tree-line. Toward the tree-line, there is a gradual reduction in WC which is likely due to root water uptake. Away from the tree-line, the reduction in WC may stem from higher evaporation rates, due to less shading from the tree.

The three dimensional visualization (Figure 6) shows an entire 2 m^2 view for the spatial WC distribution at six individual depths (Table 2). It is clearly illustrated that down to 30 cm (the three top layers), the dryer zone prevails toward the tree line compared to the corresponding locations, away from the tree line. The dryer WC zone may indicate water uptake by the active root zone [11,12]. The relatively low overlap between the wetting fronts of the nearby emitters is also illustrated, suggesting that in the long-term, the solute fluxes, due to convection, dispersion, and diffusion might have reached the wetting front of individual emitter and accumulated at this location. Consequently, in the long-term, higher salinity can be expected between emitters and perpendicular to the emitter.

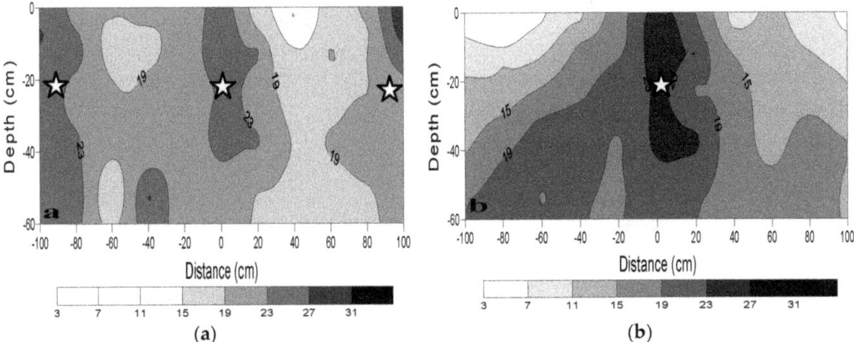

(a)

(b)

Figure 5. Gravimetric water content (%) distribution (**a**) along the drip line and (**b**) perpendicular to the drip line. The black and white stars indicate the location of the drippers.

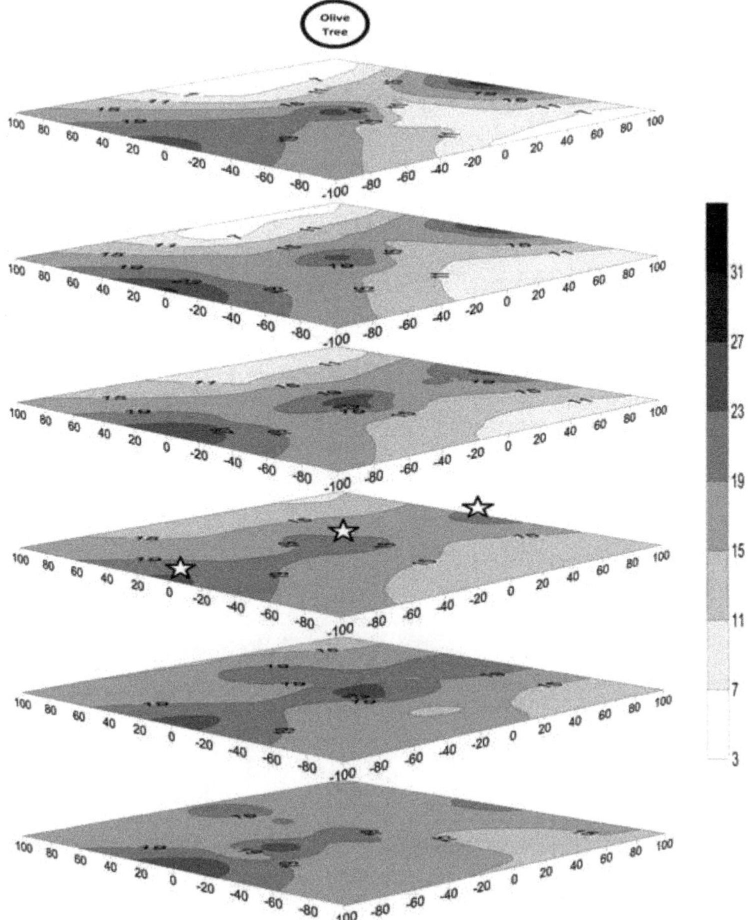

Figure 6. An overall gravimetric water content (%) distribution at all six depths, 0–5, 5–10, 10–15, 15–30, 30–45, and 45–60 cm. The black and white stars indicate the location of the drippers.

3.2. Salinity Spatial Distribution

Two quantities were used to describe the long-term accumulation of salinity: (i) electrical conductivity (EC)—representing the total salinity, and (ii) chloride concentration—as a soil native conservative tracer. The EC distribution is shown for the transect along (Figure 7a) and perpendicular (Figure 7b) to the drip line. The chloride distribution is shown in Figure 7c and d for transects along and perpendicular to the drip line, respectively. For both transects, salinity and chloride distribution exhibited similar patterns. Specifically, both quantities demonstrated a leaching zone above and below the emitters and salt accumulations zone between nearby emitters. For the transect along the drip-line, the highest salinity prevailed above the drip line in the middle of two nearby emitters. Nevertheless, the salinity values below the drip-line are also very high and may reduce water uptake by the olive trees' roots, due to the high osmotic pressure, even if a high water content is maintained. Regarding the perpendicular transect, a distinct, uneven distribution could be observed. Specifically, the salt accumulation away from the tree (−100 to 0 cm) is significantly higher than the one obtained toward the tree line (0 to 100 cm). The lowest salinity obtained near the tree line may be explained by a reduced evaporation and capillary rise toward the soil surface, due to the surface shading by the olive trees. However, the entire zone exhibited very-high to extreme values of EC, which indicates the salinization of the olive plantation, as clearly illustrated from the three dimensional visualization of the entire 2 m^2 view of the spatial EC distribution at six individual depths (Figure 8). The representation of the entire domain emphasizes the extreme values of salinity above the drip-line and away from the tree-line. A clear pattern could not be observed, due to the large salinity spectrum that was considered in this counter map. Nevertheless, the leaching zones above and below the emitter is clearly demonstrated, indicating moderate to high salinity levels.

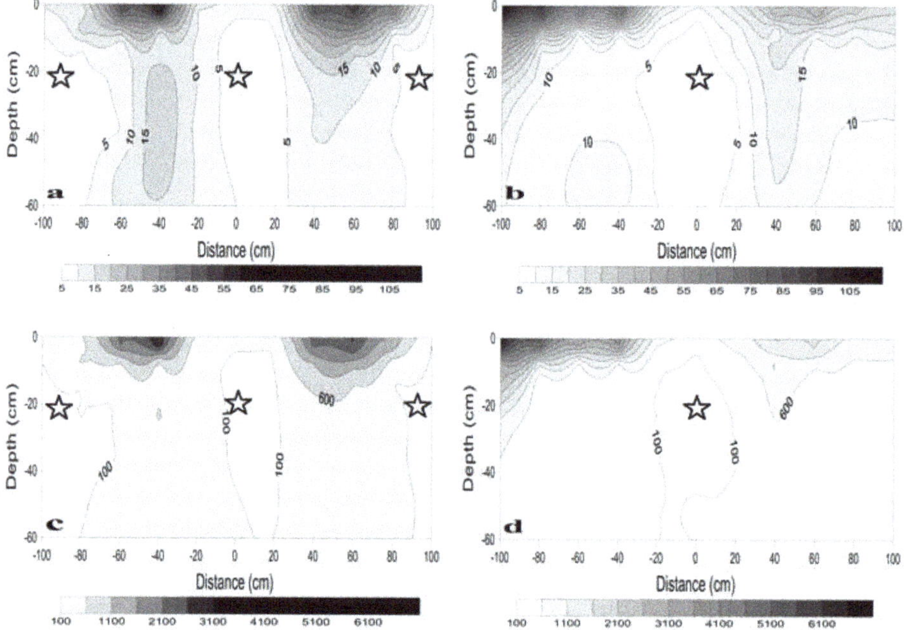

Figure 7. Electrical conductivity (dS m^{-1}) distribution (**a**) along the drip line, (**b**) perpendicular to the drip line, and soil chloride (mg L^{-1}) distribution (**c**) along the drip line and (**d**) perpendicular to the drip line. The black and white stars indicate the location of the drippers.

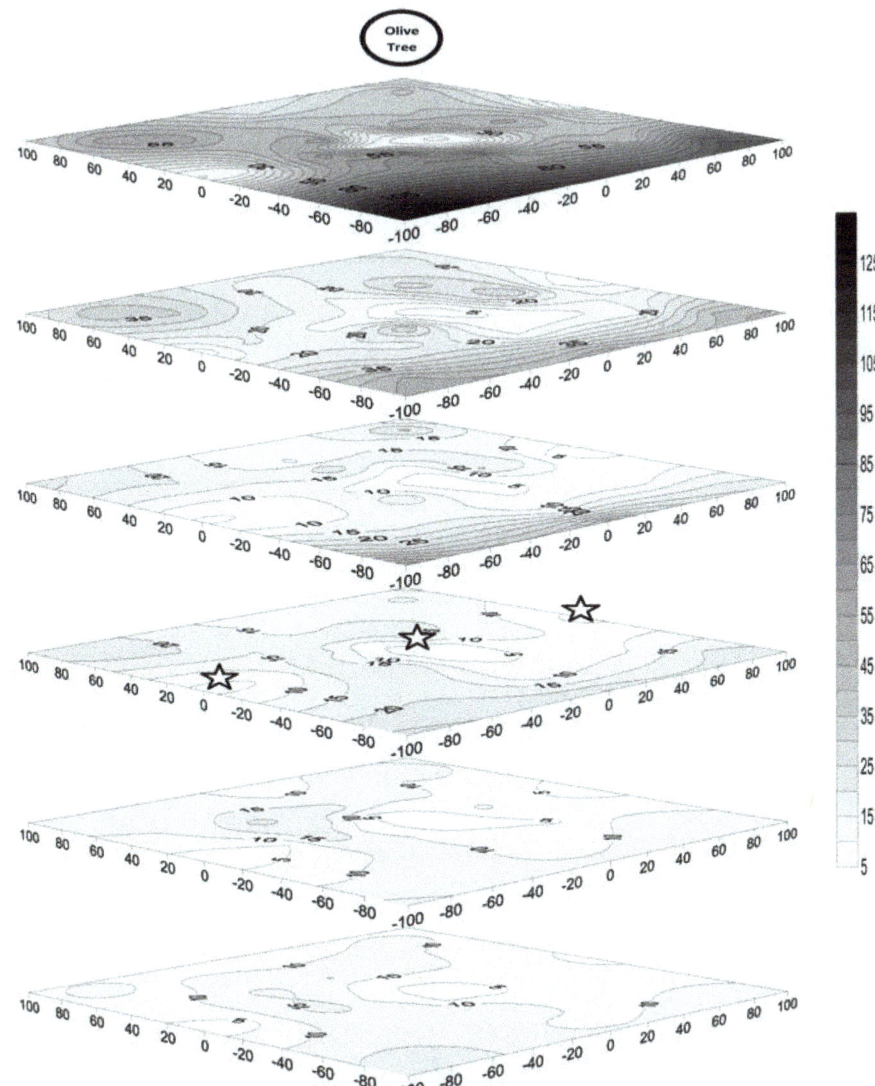

Figure 8. An overall distribution of the electrical conductivity (dS m^{-1}) at all six depths, 0–5, 5–10, 10–15, 15–30, 30–45, and 45–60 cm. The black and white stars indicate the location of the drippers.

As mentioned, the mean annual rainfall in this region is below 125 mm and may be distributed over ten low-rain events (Figure 3). Under these conditions, there is insufficient rain to leach the accumulated salts from the soil surface below the active root zone. This minimal rainfall can also exacerbate the salinity problem by bringing surface salts (0–15 cm) to the root zone (30–60 cm) after one or several rainfall events. The salinity observed in the orchard soils is far above the normal threshold salinity level for olive growth, i.e., a soil EC value of 4 to 6 dS m^{-1} is the accepted critical limit for normal olive growth [12–15]. To leach the excess salts from the root zone, high rainfall events >600 mm are required [48,49] or sprinkler irrigation has to be implemented in order to leach salts, but the long-term economic sustainability of this system is questionable [30].

3.3. Sodicity Spatial Distribution

The outcome of the long-term sodification is described by the calculated values of the SAR as a measure for the sodicity of the liquid phase and by the sodium adsorption percentage (ESP = 100 xNa/CEC) as a measure the sodicity of the solid phase. The spatial SAR distribution is shown for the transect along (Figure 9a) and perpendicular (Figure 9b) to the drip line. The spatial ESP distribution is shown in Figure 9c,d for transects along and perpendicular to the drip line, respectively. In general terms, the spatial distribution patterns obtained for the SAR and ESP resemble the one obtained for the salinity (Figure 7), demonstrating that a higher salinity in the soil solution resulted in higher SAR and consequently higher ESP. The SAR values obtained between nearby emitters, above a below the dripline, exhibited values >15% and reached values even higher than 30% at the top soil layers (Figure 10). Therefore, sodicity hazardous of soil structure degradation which can negatively affects soil hydraulic properties should be considered. Nevertheless, since high sodicity levels were accompanied by high salinity levels, the latter, may offset the negative sodicity effect on the stability of soil structure. The fact that sodicity levels increased with salinity implied a chemical equilibrium between the soil-solution and solid phase. In support of this argument is the linear correlation obtained from all samples between the ESR and SAR (Figure 11) with a slope of 0.0134, which is close to the commonly accepted value of the Gapon constant, K_G = 0.015, e.g., [50]. In addition, a positive linear correlation was obtained (data is not shown) between ESP and SAR (ESP = 0.77SAR + 3.34, R^2 = 0.73). It is well established [51] that if cation exchange reactions have reached equilibrium, the ESP values are similar to the SAR at the range of 0 to 40.

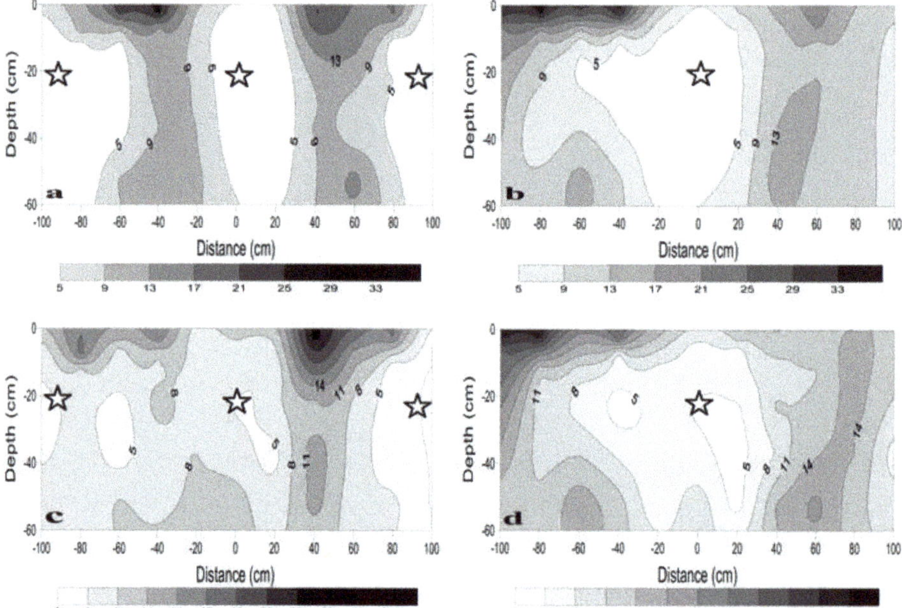

Figure 9. SAR (meq $L^{-1})^{0.5}$ distribution (**a**) along the drip line, (**b**) perpendicular to the drip line and, ESP (%) distribution, (**c**) along the drip line, and (**d**) perpendicular to the drip line. The black and white stars indicate the location of the drippers.

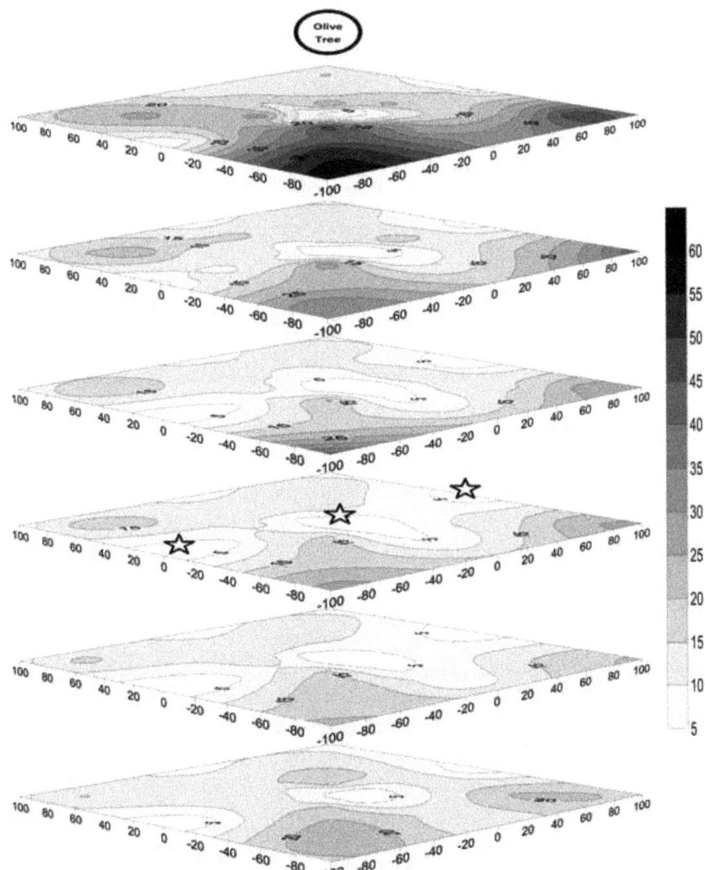

Figure 10. An overall distribution of the SAR (meq L^{-1})$^{0.5}$ at all six depths, 0–5, 5–10, 10–15, 15–30, 30–45, and 45–60 cm. The black and white stars indicate the location of the drippers.

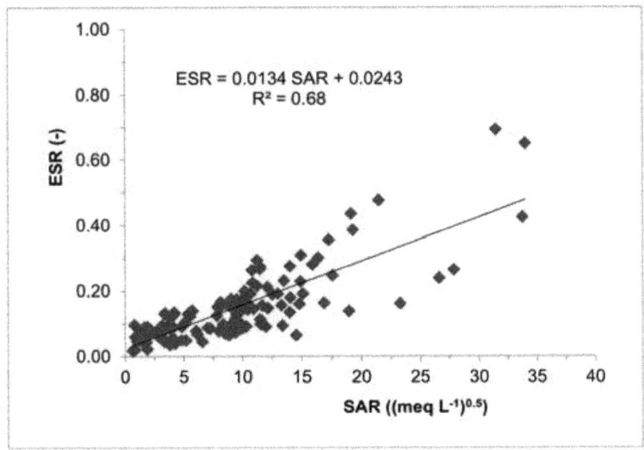

Figure 11. Exchangeable sodium ratio (ESR) as a function of sodium adsorption ratio (SAR).

4. Summary and Conclusions

The main goal of this study was to quantify the long term development of salinity and sodicity in an olive orchard grown in an arid region and irrigated with brackish water for two decades using DI and, subsequently, SDI. The study was motivated by reports on olive orchard uprooting in the Negev Desert, due to the continual reduction in olive yields. We assumed that under the climate conditions that prevail in this arid region, long term salinization and sodification at the active root zone is inevitable, in particular, under SDI with brackish irrigation water. The results of this study clearly demonstrate that following twenty years of irrigation with brackish irrigation water, salinization and sodification took place in the examined soil profile (0–60 cm), which represents the active root zone of the olive trees. The relatively large distance (1 m) between nearby drippers resulted in no significant overlaps between the wetting fronts of two nearby emitters. Consequently, a relatively small area of salt leaching could be observed below the emitters, with EC values close to the ones in the brackish irrigation water. However, moderate salt buildup took place above the emitters. The salinity buildup between nearby emitters were above the salinity threshold level for olive trees with extreme EC values above the drip line and high ones below it. The spatial distribution of the sodicity levels resembled the ones obtained for salinity, corresponding to high sodicity levels (in terms of SAR and ESP) where salinization took place. The linear correlation obtained between the sodicity quantities (i.e., ESR vs. SAR and ESP vs. SAR) implies that chemical equilibrium has been reached between the brackish irrigation water, soil solution, and the solid phase.

The results of this study show that in arid regions, the benefits of water saving, attributed to SDI, are masked by soil salinization and sodification that was induced by this irrigation method. The quantification of the long term suitability of brackish water irrigation with SDI may assist in improving the irrigation system design, for example, by significantly reducing the distance of nearby emitters and increasing the allocated leaching fraction. Finally, this study emphasizes the current necessity for salinity and sodicity reclamation in the studied region. Any alternative practices of replacing olives trees and further introduction of even higher salinity tolerant plants (e.g., jojoba) in this region will intensify the salt buildup without leaving any option for soil reclamation in the future.

Author Contributions: The study was conceived and designed by G.A., J.R.K. and P.T. performed the field experiment. All authors took a part in the data analysis, interpretation, and writing the paper. All authors have read and approved the final manuscript.

Funding: The study was partially funded by the Goldinger Trust, Jewish Federation of Delaware and the Frances and Elias Margolin Trust.

Acknowledgments: We thank E.N. and his team from Kibbutz Revivim for providing the agricultural facilities. We also appreciate the technical assistance and support by Y.M.

Conflicts of Interest: The authors declare no conflict of interest.

References

1. Ladeiro, B. Saline agriculture in the 21st century: Using salt contaminated resources to cope food requirements. *J. Bot.* **2012**, *2012*, 310705. [CrossRef]
2. Siegel, S.M. *Let There be Water: Israel's Solution for a Water-Starved World*; Macmillan: New York, NY, USA, 2015.
3. Fita, A.; Rodríguez-Burruezo, A.; Boscaiu, M.; Prohens, J.; Vicente, O. Breeding and domesticating crops adapted to drought and salinity: A new paradigm for increasing food production. *Front. Plant Sci.* **2015**, *6*, 978. [CrossRef] [PubMed]
4. De Oliveira, A.B.; Alencar, N.L.M.; Gomes-Filho, E. Comparison between the water and salt stress effects on plant growth and development. In *Responses of Organisms to Water Stress*; Akinci, S., Ed.; IntechOpen: London, UK, 2013.
5. Uddin, M.N.; Hossain, M.A.; Burritt, D.J. Salinity and drought stress: Similarities and differences in oxidative responses and cellular redox regulation. *Water Stress Crop Plants Sustain. Approach* **2016**, *1*, 86–101.
6. Muscolo, A.; Sidari, M.; Panuccio, M.R.; Santonoceto, C.; Orsini, F.; De Pascale, S. Plant responses in saline and arid environments: An overview. *Eur. J. Plant Sci. Biotechnol.* **2011**, *5*, 1–11.

7. Pasternak, D.; Aronson, J.A.; Ben-Dov, J.; Forti, M.; Mendlinger, S.; Nerd, A.; Sitton, D. Development of new arid zone crops for the Negev Desert of Israel. *J. Arid Environ.* **1986**, *11*, 37–59. [CrossRef]

8. Dag, A.; Tugendhaft, Y.; Yogev, U.; Shatzkin, N. Commercial cultivation of olive (*Olea europaea* L.) with saline water under extreme desert conditions. In Proceedings of the Fifth International Symposium on Olive Growing, Izmir, Turkey, 27 September 2004; pp. 279–284.

9. Gucci, R.; Tattini, M. Salinity tolerance in olive. *Hortic. Rev.* **1997**, *21*, 177–214.

10. Gucci, R.; Caruso, G. Environmental stresses and sustainable olive growing. In Proceedings of the XXVIII International Horticultural Congress on Science and Horticulture for People (IHC2010): Olive Trends Symposium, Lisbon, Portugal, 22 August 2010; pp. 19–30.

11. Fernández, J.E.; Moreno, F.; Cabrera, F.; Arrue, J.L.; Martín-Aranda, J. Drip irrigation, soil characteristics and the root distribution and root activity of olive trees. *Plant Soil* **1991**, *133*, 239–251. [CrossRef]

12. Weissbein, S.; Wiesman, Z.; Ephrath, Y.; Silberbush, M. Vegetative and reproductive response of olive cultivars to moderate saline water irrigation. *HortScience* **2008**, *43*, 320–327. [CrossRef]

13. Ayers, R.S.; Westcot, D.W. *Water Quality for Agriculture. Irrigation and Drainage Paper No. 29*; Food and Agriculture Organization of the United Nations: Rome, Italy, 1985.

14. Aragüés, R.; Puy, J.; Royo, A.; Espada, J.L. Three-year field response of young olive trees (*Olea europaea* L., cv. Arbequina) to soil salinity: Trunk growth and leaf ion accumulation. *Plant Soil* **2005**, *271*, 265–273. [CrossRef]

15. Ben-Gal, A. Salinity and olive: From physiological responses to orchard management. *Isr. J. Plant Sci.* **2011**, *59*, 15–28. [CrossRef]

16. Weissbein, S. *Characterization of New Olive (Olea Europea L.) Varieties Response to Irrigation with Saline Water in the Ramat Negev Area*; Ben Gurion University: Beersheba, Israel, 2006.

17. Rewald, B.; Rachmilevitch, S.; Ephrath, J.E. Salt stress effects on root systems of two mature olive cultivars. *Acta Hortic.* **2011**, *888*, 109–118. [CrossRef]

18. Hill, A.; Rewald, B.; Rachmilevitch, S. Belowground dynamics in two olive varieties as affected by saline irrigation. *Sci. Hortic.* **2013**, *162*, 313–319. [CrossRef]

19. Soda, N.; Ephrath, J.E.; Dag, A.; Beiersdorf, I.; Presnov, E.; Yermiyahu, U.; Ben-Gal, A. Root growth dynamics of olive (*Olea europaea* L.) affected by irrigation induced salinity. *Plant Soil* **2017**, *411*, 305–318. [CrossRef]

20. Wiesman, Z.; Itzhak, D.; Dom, N.B. Optimization of saline water level for sustainable Barnea olive and oil production in desert conditions. *Sci. Hortic.* **2004**, *100*, 257–266. [CrossRef]

21. Tal, A. Rethinking the sustainability of Israel's irrigation practices in the Drylands. *Water Res.* **2016**, *90*, 387–394. [CrossRef]

22. Keller, J.; Bliesner, R.D. *Sprinkle and Trickle Irrigation*; Springer: New York, NY, USA, 1990.

23. Metochis, C. Irrigation of 'Koroneiki' olives with saline water. *Olivae* **1999**, *76*, 22–24.

24. Melgar, J.C.; Mohamed, Y.; Serrano, N.; García-Galavís, P.A.; Navarro, C.; Parra, M.A.; Benlloch, M.; Fernández-Escobar, R. Long term responses of olive trees to salinity. *Agric. Water Manag.* **2009**, *96*, 1105–1113. [CrossRef]

25. Bruins, H.J. Ancient desert agriculture in the Negev and climate-zone boundary changes during average, wet and drought years. *J. Arid Environ.* **2012**, *86*, 28–42. [CrossRef]

26. Lamm, F.R.; Bordovsky, J.P.; Schwankl, L.J.; Grabow, G.L.; Enciso-Medina, J.; Peters, R.T.; Colaizzi, P.D.; Trooien, T.P.; Porter, D.O. Subsurface drip irrigation: Status of the technology in 2010. *Trans. ASABE* **2012**, *55*, 483–491. [CrossRef]

27. Martínez, J.; Reca, J. Water use efficiency of surface drip irrigation versus an alternative subsurface drip irrigation method. *J. Irrig. Drain. Eng.* **2014**, *140*. [CrossRef]

28. Dorta-Santos, M.; Tejedor, M.; Jiménez, C.; Hernández-Moreno, J.M.; Palacios-Díaz, M.P.; Díaz, F.J. Evaluating the sustainability of subsurface drip irrigation using recycled wastewater for a bioenergy crop on abandoned arid agricultural land. *Ecol. Eng.* **2015**, *79*, 60–68. [CrossRef]

29. Dorta-Santos, M.; Tejedor, M.; Jiménez, C.; Hernández-Moreno, J.M.; Díaz, F.J. Using marginal quality water for an energy crop in arid regions: Effect of salinity and boron distribution patterns. *Agric. Water Manag.* **2016**, *171*, 142–152. [CrossRef]

30. Roberts, T.L.; White, S.A.; Warrick, A.W.; Thompson, T.L. Tape depth and germination method influence patterns of salt accumulation with subsurface drip irrigation. *Agric. Water Manag.* **2008**, *95*, 669–677. [CrossRef]

31. Oron, G.; De Malach, Y.; Gillerman, L.; David, I.; Rao, V.P. Improved saline-water use under subsurface drip irrigation. *Agric. Water Manag.* **1999**, *39*, 19–33. [CrossRef]

32. Thompson, T.L.; Pang, H.C.; Li, Y.Y. The potential contribution of subsurface drip irrigation to water-saving agriculture in the western USA. *Agric. Sci. China* **2009**, *8*, 850–854.

33. Trifonov, P.; Lazarovitch, N.; Arye, G. Increasing water productivity in arid regions using low-discharge drip irrigation: A case study on potato growth. *Irrig. Sci.* **2017**, *35*, 287–295. [CrossRef]

34. Trifonov, P.; Lazarovitch, N.; Arye, G. Water and Nitrogen Productivity of Potato Growth in Desert Areas under Low-Discharge Drip Irrigation. *Water* **2018**, *10*, 970. [CrossRef]

35. Mounzer, O.; Pedrero-Salcedo, F.; Nortes, P.A.; Bayona, J.M.; Nicolás-Nicolás, E.; Alarcón, J.J. Transient soil salinity under the combined effect of reclaimed water and regulated deficit drip irrigation of Mandarin trees. *Agric. Water Manag.* **2013**, *120*, 23–29. [CrossRef]

36. Oster, J.D.; Shainberg, I. Soil responses to sodicity and salinity: Challenges and opportunities. *Soil Res.* **2001**, *39*, 1219–1224. [CrossRef]

37. Levy, G.J.; Fine, P.; Goldstein, D.; Azenkot, A.; Zilberman, A.; Chazan, A.; Grinhut, T. Long term irrigation with treated wastewater (TWW) and soil sodification. *Biosyst. Eng.* **2014**, *128*, 4–10. [CrossRef]

38. Bedbabis, S.; Trigui, D.; Ahmed, C.B.; Clodoveo, M.L.; Camposeo, S.; Vivaldi, G.A.; Rouina, B.B. Long-terms effects of irrigation with treated municipal wastewater on soil, yield and olive oil quality. *Agric. Water Manag.* **2015**, *160*, 14–21. [CrossRef]

39. Ayoub, S.; Al-Shdiefat, S.; Rawashdeh, H.; Bashabsheh, I. Utilization of reclaimed wastewater for olive irrigation: Effect on soil properties, tree growth, yield and oil content. *Agric. Water Manag.* **2016**, *176*, 163–169. [CrossRef]

40. Erel, R.; Eppel, A.; Yermiyahu, U.; Ben-Gal, A.; Levy, G.; Zipori, I.; Schaumann, G.E.; Mayer, E.; Dag, A. Long-term irrigation with reclaimed wastewater: Implications on nutrient management, soil chemistry and olive (*Olea europaea* L.) performance. *Agric. Water Manag.* **2019**, *213*, 324–335. [CrossRef]

41. Bader, B.; Aissaoui, F.; Kmicha, I.; Salem, A.B.; Chehab, H.; Gargouri, K.; Boujnah, D.; Chaieb, M. Effects of salinity stress on water desalination, olive tree (*Olea europaea* L. cvs 'Picholine','Meski'and 'Ascolana') growth and ion accumulation. *Desalination* **2015**, *364*, 46–52. [CrossRef]

42. Ben-Gal, A.; Beiersdorf, I.; Yermiyahu, U.; Soda, N.; Presnov, E.; Zipori, I.; Crisostomo, R.R.; Dag, A. Response of young bearing olive trees to irrigation-induced salinity. *Irrig. Sci.* **2017**, *35*, 99–109. [CrossRef]

43. Banin, A.; Amiel, A. A correlative study of the chemical and physical properties of a group of natural soils of Israel. *Geoderma* **1970**, *3*, 185–198. [CrossRef]

44. Rhoades, J.D. Cation exchange capacity. In *Agronomy Monograph No. 9 Methods of Soil Analysis: Part 2. Chemical and Microbiological Properties*, 2nd ed.; Page, A.L., Miller, R.H., Kearney, D.R., Eds.; ASA, SSSA: Madison, WI, USA, 1986; pp. 149–157.

45. Thomas, G.W. Exchangeable cations. In *Agronomy Monograph No 9 Methods of Soil Analysis. Part 2. Chemical and Microbiological Properties*, 2nd ed.; Page, A.L., Miller, R.H., Keerney, D.R., Eds.; ASA, SSSA: Madison, WI, USA, 1986; pp. 159–164.

46. Fares, A.; Parsons, L.R.; Wheaton, T.A.; Morgan, K.T.; Simunek, J.; Van Genuchten, M.T. Simulated drip irrigation with different soil types. *Proc. Fla. State Hortic. Soc.* **2001**, *114*, 22–24.

47. Sevostianova, E.; Leinauer, B.; Sallenave, R.; Karcher, D.; Maier, B. Soil salinity and quality of sprinkler and drip irrigated warm-season turfgrasses. *Agron. J.* **2011**, *103*, 1773–1784. [CrossRef]

48. Ben-Hur, M.; Li, F.H.; Keren, R.; Ravina, I.; Shalit, G. Water and salt distribution in a field irrigated with marginal water under high water table conditions. *Soil Sci. Soc. Am. J.* **2001**, *65*, 191–198. [CrossRef]

49. Lado, M.; Bar-Tal, A.; Azenkot, A.; Assouline, S.; Ravina, I.; Erner, Y.; Fine, P.; Dasberg, S.; Ben-Hur, M. Changes in chemical properties of semiarid soils under long-term secondary treated wastewater irrigation. *Soil Sci. Soc. Am. J.* **2012**, *76*, 1358–1369. [CrossRef]

50. Levy, R.; Hillel, D. Thermodynamic Equilibrium Constants of Sodium-Calcium Exchange in Some Israel SOILS1. *Soil Sci.* **1968**, *106*, 393–398. [CrossRef]

51. United States Department of Agriculture. Diagnosis and improvement of saline and alkali soils. In *Agriculture Handbook*; US Government Printing Office: Washington, DC, USA, 1954; Volume 60, pp. 83–100.

Article

Analysis of the Effects of High Precipitation in Texas on Rainfed Sorghum Yields

Om Prakash Sharma [1], Narayanan Kannan [2,*], Scott Cook [3], Bijay Kumar Pokhrel [2] and Cameron McKenzie [4]

[1] Department of Chemistry, Geosciences and Physics, Tarleton State University, Stephenville, TX 76402, USA; omprakash.sharma@go.tarleton.edu

[2] Texas Institute for Applied Environmental Research (TIAER), Tarleton State University, Stephenville, TX 76402, USA; pokhrel@tarleton.edu

[3] Department of Mathematics, Tarleton State University, Stephenville, TX 76402, USA; scook@tarleton.edu

[4] Department of History, Sociology, Geography and GIS, Tarleton State University, Stephenville, TX 76402, USA; cameron.mckenzie@tarleton.edu

* Correspondence: kannan@tarleton.edu

Received: 31 July 2019; Accepted: 10 September 2019; Published: 14 September 2019

Abstract: Most of the recent studies on the consequences of extreme weather events on crop yields are focused on droughts and warming climate. The knowledge of the consequences of excess precipitation on the crop yield is lacking. We attempted to fill this gap by estimating reductions in rainfed grain sorghum yields for excess precipitation. The historical grain sorghum yield and corresponding historical precipitation data are collected by county. These data are sorted based on length of the record and missing values and arranged for the period 1973–2003. Grain sorghum growing periods in the different parts of Texas is estimated based on the east-west precipitation gradient, north-south temperature gradient, and typical planting and harvesting dates in Texas. We estimated the growing season total precipitation and maximum 4-day total precipitation for each county growing rainfed grain sorghum. These two parameters were used as independent variables, and crop yields of sorghum was used as the dependent variable. We tried to find the relationships between excess precipitation and decreases in crop yields using both graphical and mathematical relationships. The result were analyzed in four different levels; 1. Storm by storm consequences on the crop yield; 2. Growing season total precipitation and crop yield; 3. Maximum 4-day precipitation and crop yield; and 4. Multiple linear regression of independent variables with and without a principal component analysis (to remove the correlations between independent variables) and the dependent variable. The graphical and mathematical results show decreases in rainfed sorghum yields in Texas for excess precipitation could be between 18% and 38%.

Keywords: grain sorghum; precipitation; rainfed; multiple linear regression; crop yield; principal component analysis

1. Introduction

Sorghum is a crop that can be grown as either a grain or cash crop. It is one of the top five cereal crops in the world. Sorghum is also required for the survival of humankind in different parts of the world, especially in Africa and Asia. The United States (US) is the largest producer of sorghum in the world [1]. In the US, sorghum usually grows throughout the sorghum belt from South Dakota to southern Texas [2]. The top five sorghum producing states are Kansas, Texas, Colorado, Oklahoma, and South Dakota. In the US, sorghum grain is primarily used for feeding of livestock and ethanol production, but it is becoming popular in the consumer food industry and other markets [3]. The livestock industry is one of the oldest standing marketplaces for sorghum grain in the US. Sorghum is utilized in feed rations for poultry, beef, dairy, and swine [3]. Also, a large portion of sorghum is used for biofuel production. It is also exported to the different parts of the world, including Mexico, China, and Japan.

Sorghum grain is highly resistant to drought and can withstand waterlogging better than any other cereal crop. Sorghum has a special fibrous root system, which can extend to a depth of 1.2 to 1.8 m (4 to 6 feet) deep in the soil. More than 75% of water and nutrients taken by root system are from the top 0.9 m (3 feet). Therefore, the deep extension of the root system helps sorghum withstand drought conditions better than any other cash crops [4]. Grain sorghum exhibits yield stability greater than maize. Drought resistance and heat tolerance make it a popular choice for marginal rainfall areas of semiarid zones of Africa where food shortages are common.

Total water use by a sorghum crop depends on the variety, maturation, planting date, and geographical and environmental conditions. It is estimated that the total use of 1750 mm/ha (28 inches of water/acre) water is needed for good sorghum yield of 783 kg/ha (700 lb/acre) [5]. The water use of sorghum depends on the growth stage of the sorghum plant (Table 1). During the early part of plant development, water use is relatively low but water stress during this time can affect plant growth and yield. Rainfall of 25 to 50 mm (1 to 2 inches) in the second week following sorghum pollination would result in the best yield if the period of pollination had adequate soil moisture [5,6]. The period from sorghum pollination to maturity is about 60 days. At the time of growth, a dry spell in the field from 14 to 60 days after pollination may have a small effect on the final harvesting yield of the sorghum crop. If no rain were to occur during the final period of 46 days, the yield of the sorghum crop would be greatly reduced [5,6]. Therefore, rainfall and its timing are important factors for the growth and yield of sorghum.

Table 1. Estimated grain sorghum water use by growth stage [5].

Days after Sorghum Planting	Water Requirement (Inches/Day)	Water Requirement (mm/day)
0–30 (early plant growth)	0.05–0.10	1.3–2.5
30–60 (rapid plant growth)	0.10–0.20	2.5–5.0
60–80 (boot and flowering)	0.25–0.30	6.3–7.5
80–120 (grain fill to maturity)	0.10–0.25	2.5–6.3

Although sorghum is tolerant of some waterlogging, it suffers damage under prolonged wetting of soil under very high rainfall [6]. Researchers from Australia, Germany, and the US have quantified the overall of extremes climate effects like drought, heat wave problems and precipitation on the crop yield variability of different staple crops around the world [7]. The year-to-year overall changes in the climatic factor in the growing season of maize, rice, sorghum, and wheat accounted the fluctuations of 20% to 49% of total yields [8]. Climatic extremes like hot and cold climates, drought, and heavy rainfall accounted for 18% to 43% of inter-annual variations in different crops yields [9]. Therefore, it is important to understand the consequences of climate extremes on crop yields to secure our food supply. A large body of literature already exists for drought. However, studies on the consequences of extremely high precipitation on crop yield are sparse, especially for grain sorghum. Therefore, an attempt is made in this study to analyze the consequences of high precipitation on rainfed grain sorghum yields.

Extreme precipitation events are producing more and more rain, and are now becoming one of the most common events since the beginning of the 1950s in many regions of the world, including the US. Scientists expect heavy rainfall as a consequence of a warming planet [10,11]. Warmer air mass can hold more water vapor content than cold air mass. For each degree of warming in the earth, the air mass capacity for holding water vapor goes up by about 7%. An atmosphere with more moist air can produce more heavy and continuous rainfall events, which is what has been observed all over the world since the 1950s [10,11].

An increase in continuous heavy rainfall events may not always show the increases in total rainfall over a season or year. Some studies show a small decrease in rainfall and show an increase of dry periods, which offsets rainfall increases falling during heavy events. The most immediate effect of heavy rainfall is the flooding. There are several recent examples of heavy rainfall events. In August 2017, Hurricane Harvey produced 1220 mm (48 inches) of heavy rainfall on Houston, Texas from a single event and was the biggest threat from tropical cyclones. In July 2016, more than 150 mm (6 inches) of heavy rainfall occurred in less than two hours in Ellicott City, Maryland, the estimated cost of the damage is more than $22 million dollars. In summary, we incur a huge economic loss because of heavy precipitation events, including some losses coming from a reduction in crop yields.

Precipitation is generally useful in recharging the soil profile, which is very important for crop growth. The precipitation efficiency in recharging the soil profile depends on intensity and rate at which precipitation occurs. Precipitation that falls on the soil at rates greater than 127 cm/h (0.5 inches an hour) are less efficient compared to lighter rain, because the water that runs off from the surface carries the fertile soil to the streams, lakes, and rivers which decreases future yields. The timing of rainfall while crops are growing is critical. During germination and stand establishment, either heavy rainfall or little rainfall can substantially affect the yield.

In general, the more precipitation during the crop growing season, the better the crop growth. However, too much precipitation will damage the crop by saturating the soil profile and removing air, which is also important for healthy plant growth. The majority of the previous studies relating extreme climate events and food production are focused on increasing temperatures and drought. The consequences of high precipitation on probable reduction in crop yields are often ignored. There is a big knowledge gap of understanding the consequences of extremely high precipitation on the yield of food crops and relating it to subsequent consequences in food production scenarios at different spatial scales. Addressing the knowledge gap and exploring the less-studied relationship between excess precipitation and rainfed food crop yields are the novelties of this study. Detailed analysis of the above-mentioned relationship using a combination of established mathematical principles and graphical tools are some of the unique aspects of this study. The results from our study and other similar studies have applications in crop insurance, parameterization of computer models (estimating crop yield reductions based on aeration stress), policy level decisions on rainfed crop selection, yield forecasting, estimating food production, and water footprint analysis.

The specific objectives of the study are to: (1) Identify historic extreme high precipitation events during the crop growth of rainfed sorghum in Texas, (2) Extract continuous serially complete crop yield information for rainfed sorghum by county, (3) Collect continuous records of daily average precipitation corresponding to the sorghum crop yield data, (4) Estimate the growing season total precipitation and 4-day maximum precipitation using the precipitation data, and (5) Relate items 3 and 4 above using visual patterns and statistical principles to quantify the consequences of high precipitation on crop yields.

2. Materials and Methods

2.1. Data Collection and Arrangement

2.1.1. GIS Data

The map of county boundaries was downloaded from the Texas Natural Resources Information System (TNRIS) website [12]. The cultivated area map of Texas was downloaded from the National Land Cover Dataset (NLCD) [13] and overlaid with county boundaries. A map showing the location of meteorological stations in Texas was developed using the latitude and longitude information that came with the precipitation data. It was overlaid with county boundaries to identify the list of weather stations within each county. Continuous records of Sorghum yield data (without gaps) are required for the analysis. In addition, the data availability period had to be consistent for different counties in Texas. The period from 1973 to 2000 satisfied the criteria of no data gaps and consistent availability of data for many counties. Therefore, only those counties with rainfed sorghum yield data (Figure 1, Table A1 in Appendix A) for the period 1973–2000 are included in the analysis and 26 United State Geological Survey (USGS) precipitation gaging data satisfied these criteria are considered for further analysis. Twenty-six meteorological stations (precipitation data from USGS) correspond to the counties having rainfed sorghum yield.

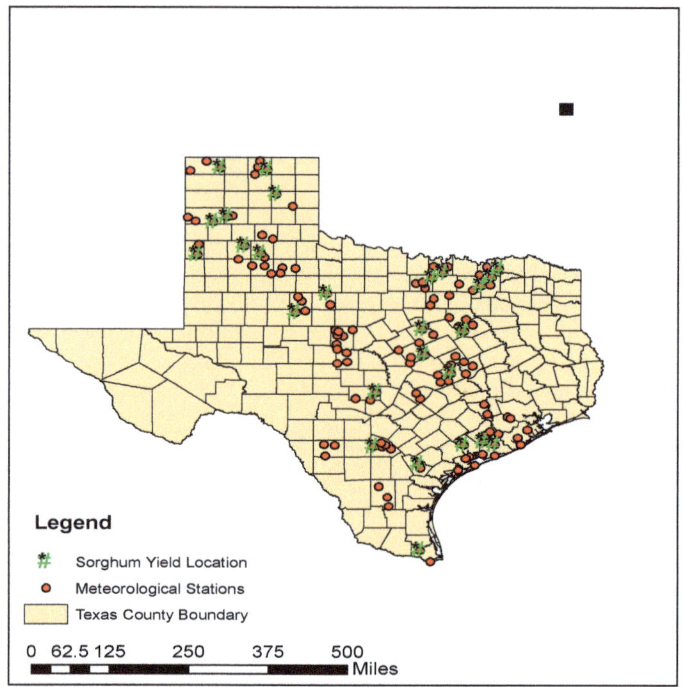

Figure 1. Map showing the location of Meteorological stations and rainfed sorghum cultivated location.

2.1.2. Estimation of Sorghum Growing Season for Different Counties in Texas

Grain sorghum is a hot season crop grown in most arid plain states that do not have enough moisture to grow other crops. Sorghum is planted once the soil temperature is consistent at about 15.5 °C (60 °F). This sometimes depends on the local condition so it can occur as early as late February in warmer climates or May in colder climates. This crop has longer maturity stages than other corn and cereal crops.

The planting dates of sorghum were estimated from USDA-ARS [14], taking into consideration the north-south temperature gradient. The harvest dates were estimated based on the planting date and the crop duration of 120 days (assumption). The detail of dates of planting and harvesting estimated for different counties in Texas are shown below in Table 2.

There is a north-south temperature gradient in Texas. Therefore, planting starts from the south and moves toward the northern region of Texas. Sorghum is planted in the southern region of Texas first around the last week of March and then towards the south-central region followed by the far eastern and eastern regions and finally ends toward the north in the last week of May. We selected a date from the range of dates in between the early and late planting dates of each county listed in the table above. That day is taken as a base for analysis with the precipitation data (Figure 2).

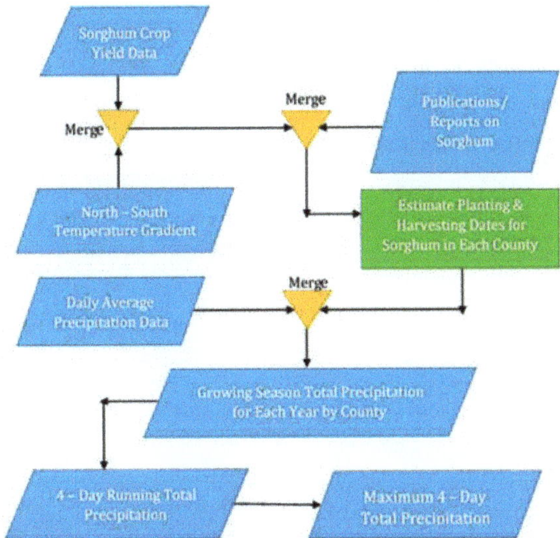

Figure 2. Estimation of growing season total precipitation and maximum 4-day total precipitation.

Table 2. Plant date of sorghum in different counties of Texas [14].

County	Planting Date			Harvest Date	Precipitation Start	Precipitation End
	Early	Late	Used			
Atascosa	3/10–3/15	3/15–3/25	15-March	3-July	5-March	25-June
Bailey	3/5–3/10	3/10–3/20	10-March	28-June	1-March	18-June
Bee	1/21–1/30	1/31–2/10	30-January	20-May	20-Jan	10-May
Bosque	3/15–3/25	3/26–4/5	25-March	13-July	15-March	3-July
Cameron	1/21–1/30	1/31–2/10	30-January	20-May	20-Jan	10-May
Collin	3/25–4/4	4/5–4/15	4-April	23-July	26-March	13-July
Cooke	3/15–3/25	3/26–4/5	25-March	13-July	15-March	3-July
Coryell	3/15–3/25	3/25–4/5	25-March	13-July	15-March	3-July
Dallam	3/5–3/10	3/10–3/20	10-March	28-June	1-March	18-June
Fannin	3/25–4/4	4/5–4/15	4-April	23-July	26-March	13-July
Floyd	3/5–3/10	3/10–3/20	10-March	28-June	1-March	18-June
Gillespie	3/10–3/15	3/15–3/25	15-March	3-July	5-March	25-June
Gray	3/5–3/10	3/10–3/20	10-March	28-June	1-March	18-June
Hale	3/5–3/10	3/10–3/20	10-March	28-June	1-March	18-June
Hansford	3/5–3/10	3/10–3/20	10-March	28-June	1-March	18-June
Hunt	3/25–4/4	4/5–4/15	4-April	23-July	26-March	13-July
Jackson	2/15–2/21	2/22–3/5	21-February	11-June	11-February	1-June
Jones	3/5–3/10	3/10–3/20	10-March	28-June	1-March	18-June
Matagorda	2/15–2/21	2/22–3/5	21-February	11-June	11-February	1-June
Milam	3/15–3/25	3/26–4/5	25-March	13-July	15-March	3-July
Navarro	3/25–4/4	4/5–4/15	4-April	23-July	26-March	13-July
Nolan	3/5–3/10	3/10–3/20	10-March	28-June	1-March	18-June
Randall	3/5–3/10	3/10–3/20	10-March	28-June	1-March	18-June
Wharton	2/15–2/21	2/22–3/5	21-February	11-June	11-February	1-June
Wise	3/15–3/25	3/26–4/5	25-March	13-July	15-March	3-July

2.1.3. Estimation of Growing Season Precipitation by County

The growing season is the number of consecutive days from the beginning of planting date to the harvesting date. It is calculated for every county. To obtain the growing season total precipitation, the precipitation of all daily values within the growing season is added together. The precipitation data used for analysis for each county was taken from 10 days before the planting and harvesting dates of each station from the base date. This is because farmers would use soil moisture from any precipitation event before planting the seeds. Also, they harvest the crop only when the crop is adequately dry, avoiding days for harvest soon after precipitation (Figure 2).

2.1.4. Estimation of Maximum 4-Day Running Total Precipitation

The 4-day running total is the cumulative value of continuous four days of precipitation data. Continuous four days of precipitation is added to get one value, and so on. In this way, it is calculated for every day in the growing season for each year and station considered for the analysis. Finally, the maximum of four days of total precipitation within the grain sorghum growing season each year is calculated for every station for further graphical analysis.

2.2. Data Analysis

2.2.1. Level 1: Historically Documented Extreme Precipitation Events and Sorghum Yield in Texas

The High Plains and Low Rolling Plains climatic regions of Texas received an extreme rainfall of 508 mm (20 inches) over 26 km^2 (10 square miles) area and 254 mm (10 inches) over 26,000 km^2 (10 thousand square miles) from 1 August to 4 August in 1978. The East Texas and Upper Coast climatic regions of Texas received an extreme rainfall of 1000 mm (40 inches) for about 26 km^2 (10 square miles) area and 254 mm (10 inches) for 26,000 km^2 (10 thousand square miles) from 24 July to 28 July in 1979. Randall County had a storm during 26 May to 27 May 1978. The rainfall amount during the period averaged 100 mm to 254 mm (4 in. to 10 in.) on the High Plains. Out of all the extreme precipitation events documented, only the May 1978 storm in Randall County fell within the sorghum-growing season. Therefore, only the details of the May 1978 storm will be included for further analysis under this category [15,16].

2.2.2. Level 2: Growing Season Precipitation and Rainfed Sorghum Crop Yields

The growing season's total precipitation and rainfed sorghum crop yield for different years is plotted to identify graphical relationships (Figure A1). The trends in data for every county were analyzed.

2.2.3. Level 3: Maximum 4-Day Running Total Precipitation and Crop Yield

The maximum 4-day running total precipitation and rainfed sorghum crop yield for different years is plotted to identify graphical relationships (Figure A2). The trends in data were studied for every county considered for the analysis.

2.2.4. Level 4: Generation of Mathematical Relationships between Rainfed Sorghum Yield and Excess Precipitation

Principal component analysis (PCA) is a commonly used mathematical tool used to display patterns in multivariate data. It removes correlation within a large set of variables and sorts them according to importance (explained variance) [17]. While PCA is commonly used for dimensionality reduction, it was not used for that purpose in this study. Total precipitation and max 4-day precipitation are somewhat correlated, which could affect the regression relationships. PCA transforms the input variables to remove such correlation. A downside of PCA is that while the original variables have clear interpretations (total growing season precipitation and max 4-day precipitation), the PCA-transformed variables do not. They are called "principal components" 1 and 2.

In our regression analysis, the dependent variable was taken as the rainfed grain sorghum yield data, and the independent variables were growing season total precipitation and maximum 4-day total precipitation (Figure 3). Multiple linear regression (MLR) analysis (Equation (1)) [18] was performed with the data analysis tool available in Microsoft Excel.

$$Y = A + B_1X_1 + B_2X_2,$$ (1)

where Y is crop yield, A is an intercept, X_1 and X_2 are growing season total precipitation and maximum 4-day total precipitation respectively, and B_1 and B_2 are partial regression coefficients [18].

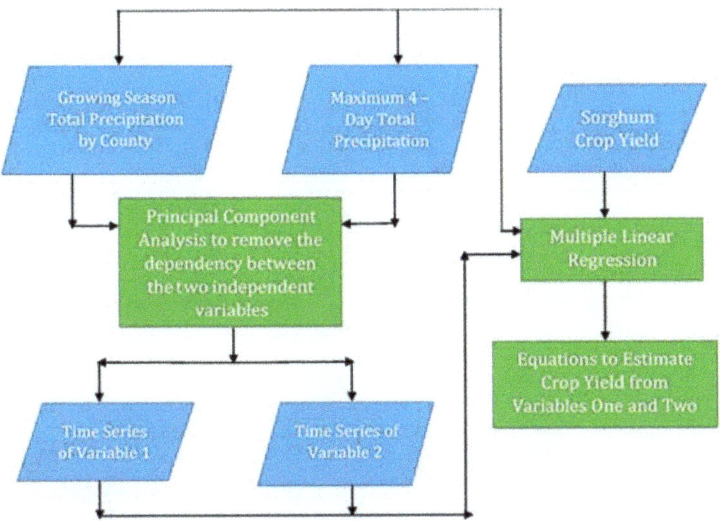

Figure 3. Schematic for the multiple linear regression (MLR) with and without a principal component analysis (PCA) (level 4 results).

3. Results

3.1. Level 1 Results

In the year 1978, Randall County encountered a storm event during the grain sorghum crop-growing period (26–27 May) (Figure 4). The 4-day maximum precipitation during the crop growing period was 182.9 mm (7.2 inches) which is 206% more than the average 4-day maximum precipitation (60.9 mm [2.4 inches]) that occurred during the sorghum crop growing period between 1973 and 2000. Also, the growing season total precipitation during the 1978 grain sorghum crop growing period was 271.8 mm (10.7 inches) which is 72.3% more than the average of the growing season total precipitation (157.5 mm (6.2 inches)) that occurred during the sorghum crop growing period between 1973 and 2000. The storm event could have brought down the rainfed sorghum yield by 27.5% (corresponding to the year 1978) when compared to the average rainfed sorghum yield from 1973 to 2000. This is evident from Figures 5 and 6, which show the sharp declines in crop yields based on 4-day maximum precipitation and growing season total precipitation separately.

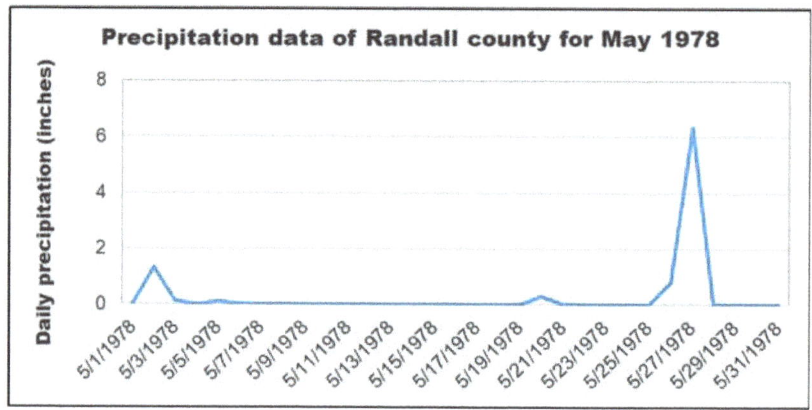

Figure 4. Storm event of May 1978 in Randall County.

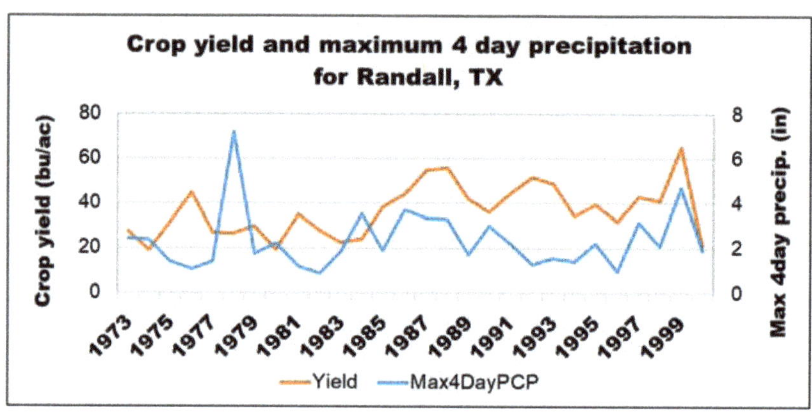

Figure 5. Sorghum yield reductions for Randall County in 1978 coming from maximum 4-day total precipitation.

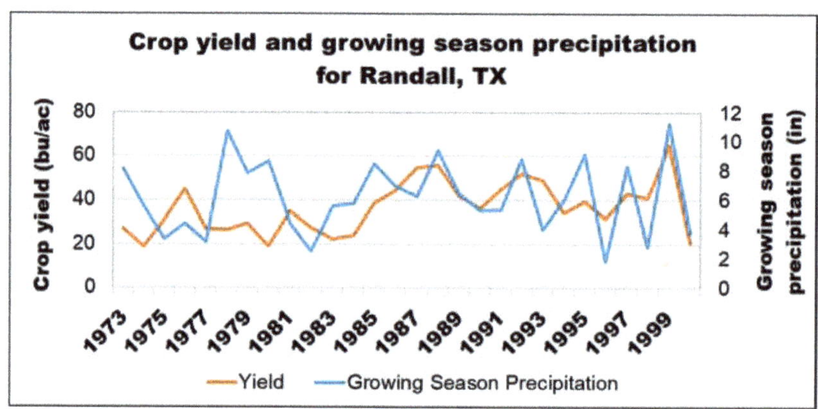

Figure 6. Sorghum yield reductions for Randall County in 1978 coming from excess growing season precipitation.

3.2. Level 2 Results

The graphical relationship between growing season total precipitation and rainfed sorghum crop yield was studied. Crop yield trends closely followed the growing season total precipitation for Texas counties Bailey, Bee, Cameron, Collin, Cooke, Dallam, Fannin, Hansford, Hunt, Jackson, and Wharton. When there was an increase in precipitation, there was a corresponding increase in the crop yield and vice versa (Figure 7). However, for some counties (e.g., Figure 8) there were declines in crop yield for excess precipitation. For Bosque County in 1976, growing season precipitation increased to 635 mm (25 inches) which resulted in a sharp decrease of crop yield. For Coryell County, when the annual growing season rainfall increased to 381 mm (15 inches) in 1976, it showed a decrease in crop yield. For Milam County in 1976, 1978, and 1994, increases in growing season total precipitation brought decreases in crop yield. Similar noticeable yield declines for excess precipitation results were observed for Atascosa, Gillespie, Hansford, Navarro, Randall, and Wise counties in Texas (Table 3) (graphs not shown in the manuscript for the sake of brevity).

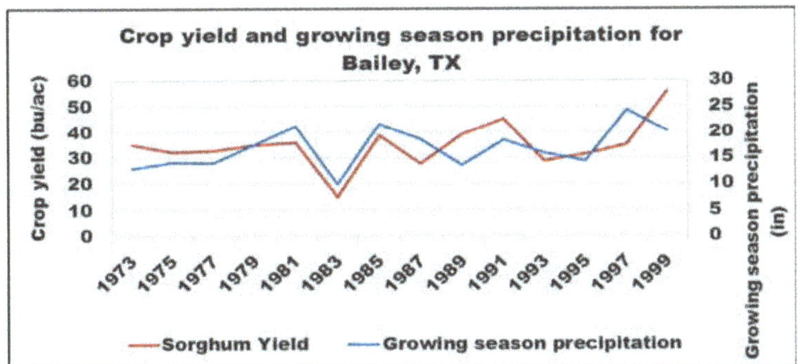

Figure 7. Example for crop yield trends closely following growing season total precipitation.

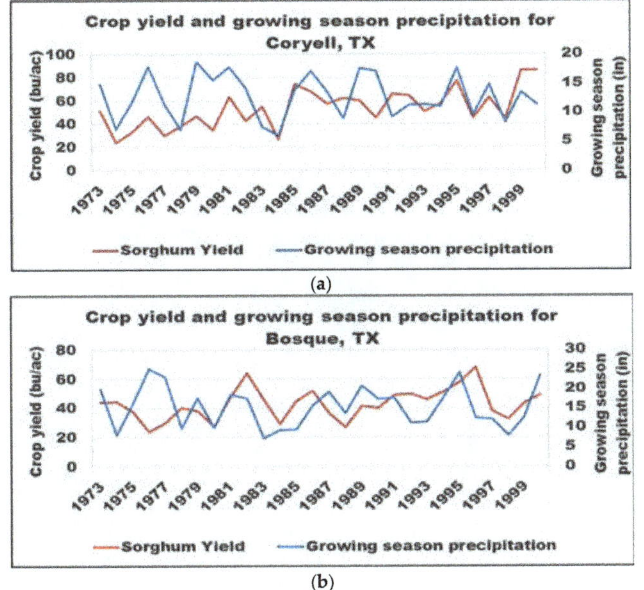

Figure 8. *Cont.*

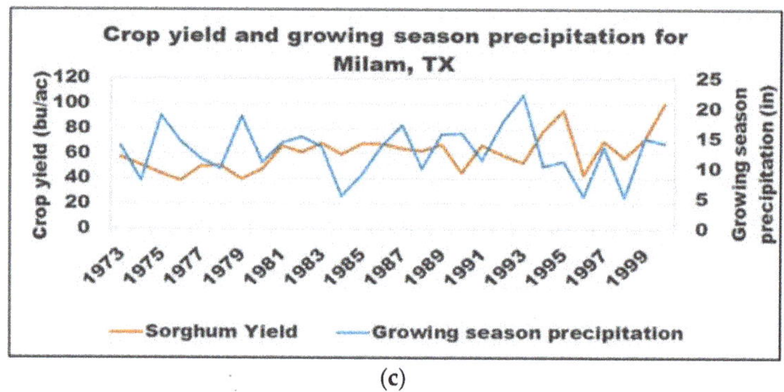

(c)

Figure 8. Relationship between growing season precipitation and crop yield for (**a**) Coryell County, (**b**) Bosque County, and (**c**) Milam County.

Table 3. Differences in sorghum yield between average growing season total precipitation (for years 1973–2000) and years showing high growing season total precipitation (column 2) and years nearby high growing season total precipitation (column 3).

County	% Differences in Sorghum Yield between the High Growing Season Precipitation and	
	Growing Season Precipitation for 1973–2000	Years Nearby High Growing Season Precipitation
Atascosa	40.5	9.5
Bosque	37.5	31.1
Coryell	33.4	27.3
Gillespie	4.88	−22.7
Hansford	34.59	28.5
Milam	27.87	21.9
Navarro	34.72	23.3
Randall	37.34	24.5
Wise	20.71	14.3
Average	30.2	17.5
95% CI	23 to 37	7 to 28

The numerical analysis of crop yields and growing season total precipitation are provided in Table 3. When compared to the average for the period 1973 to 2000, the decreases in crop yield corresponding to the year(s) with excess precipitation is about 30% (95% confidence intervals 23% to 37%). When compared to the nearby years (before and after the year with excess precipitation), the years with excess precipitation showed a decrease in crop yield of 17% (95% confidence intervals 7% to 28%) (Table 3).

In summary, the analysis of numerical and graphical crop yield trends with respect to growing season total precipitation highlighted decreases in rainfed sorghum crop yield when the precipitation received is higher than the average or what could probably be necessary for healthy crop growth.

3.3. Level 3 Results

The graphical relationships of maximum 4-day total precipitation with rainfed sorghum crop yields were analyzed. Some of the results are shown in Figure 9. Crop yield trends closely follow the maximum 4-day total precipitation for Bailey, Bee, Bosque, Fannin, Dallam, Hale, Hunt, Jones, Matagorda, Nolan, and Wise counties. Atascosa County shows four days maximum total of 8 inches and results in the sharp decrease in crop yield for the year 1980 while for the other years the crop yield trends follow precipitation. For Hunt County, the four days precipitation go above 254 mm (10 inches) and result in a decrease in crop yield comparing to other years. The decrease in crop yield was observed for Milam County as well when the maximum 4-day total precipitation reached 254 mm (10 inches).

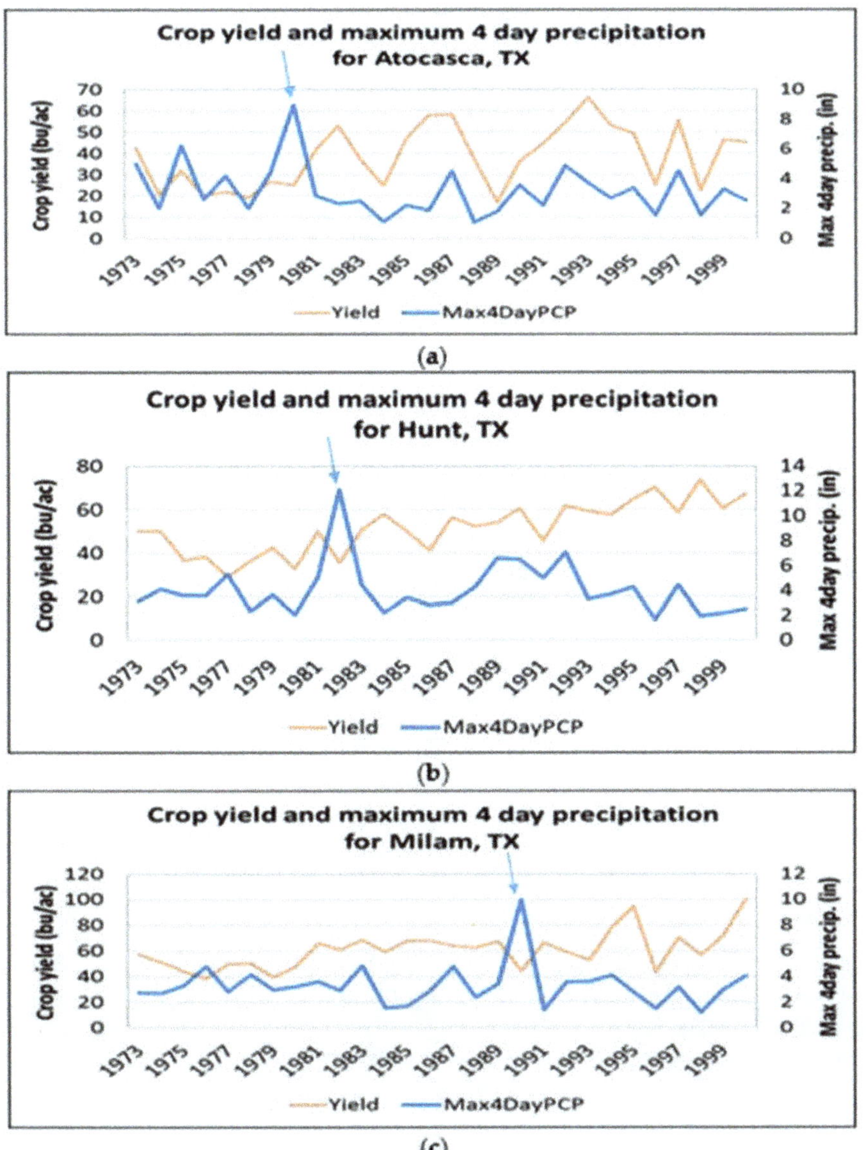

Figure 9. Relationship between maximum 4-day total precipitation and rainfed sorghum yield for (**a**) Atascosa (**b**) Hunt, and (**c**) Milam counties.

Similar rainfed sorghum yield declines were observed for high values of maximum 4-day running total precipitation for Coryell, Gillespie, Grey, Hansford, Matagorda, Navarro, Nolan, Randall, Wharton, and Wise counties. In summary, whenever the maximum four days running total precipitation is higher, that results in a decrease in crop yield of rainfed grain sorghum.

The numerical analysis of crop yields and maximum 4-day total precipitation are provided in Table 4. When compared to the average for the period 1973 to 2000 the decreases in crop yield corresponding to the year(s) with excess four-day precipitation is about 25% (95% confidence intervals 18% to 31%). When compared to the nearby years, the years with excess 4-day maximum precipitation showed a decrease in crop yield of 22% (95% confidence intervals 14% to 31%) (Table 4). In summary, the analysis of graphical and numerical crop yield trends with respect to maximum 4-day total precipitation pointed out decreases in rainfed sorghum crop yield when the precipitation received was much higher than the average or what could be necessary for healthy crop growth.

Table 4. Differences in sorghum yield between maximum 4-day precipitation (for years 1973–2000) and years showing high growing season total precipitation and years nearby high growing season total precipitation (Level 3 results).

County	% Differences in Sorghum Yield between the High Growing Season Precipitation and	
	Growing Season Precipitation for 1973–2000	Years Nearby High Growing Season Precipitation
Atascosa	35.0	26.8
Coryell	31.7	31.5
Hunt	30.6	29.0
Gillespie	10.2	−1.3
Gray	19.31	24.39
Hansford	22.29	24.95
Matagorda	7.16	5.80
Milam	26.52	33.23
Navarro	53.88	53.75
Nolan	20.14	32.52
Randall	27.54	10.13
Wharton	13.37	5.14
Wise	25.21	16.12
Average	24.8	22.5
95% CI	18 to 31	14 to 31

3.4. Level 4 Results

A multiple linear regression (MLR) analysis was performed with growing season total precipitation and maximum 4-day total precipitation as independent variables and rainfed sorghum yield as dependent variable the results of which are presented in Table 5. Although the R^2 values (Column 5 of Table 5) appear smaller, the regression relationships are significant, as evidenced by the F values of regression relationships presented in Table 6. Looking at the regression relationships by county, negative coefficients appear for growing season total precipitation for counties Bosque, Dallam, Hansford, and Milam only. Twenty-three out of 27 counties analyzed mathematically did not show declines in crop yield for excess precipitation when analyzed by growing season total precipitation. However, when analyzed by the maximum 4-day total precipitation, 21 out of 27 counties show negative coefficients substantiating the declines in crop yield for excess precipitation. The counties that do not show negative coefficients (with maximum 4-day total precipitation) are Bee, Bosque, Dallam, Deaf Smith, Floyd, and Hansford. Majority of the counties analyzed mathematically exhibit declining crop yields for excess precipitation showing negative coefficients mostly in maximum 4-day total precipitation and some in growing season total precipitation. Milam is the only county showing a negative coefficient for both the independent variables. Although Deaf Smith and Floyd showed some graphical relationships, they were the only counties that did not mathematically exhibit the regression relationship between the independent variables and the dependent variable.

Table 5. Results of multiple linear regression analysis (without principal component analysis) using annual growing season precipitation, 4-day maximum precipitation, and crop yield.

County	Coefficients for Independent Variables		Intercept	Regression Analysis without PCA
	Growing Season Precipitation	Maximum 4-Day Precipitation		Calculated (R^2)
Atascosa	1.922	−2.499	27.274	0.204
Bailey	1.595	−1.923	9.759	0.239
Bee	0.364	5.311	35.281	0.363
Bosque	−0.070	0.970	40.030	0.031
Cameron	1.346	−1.337	49.332	0.072
Collin	0.710	−0.481	46.576	0.086
Cooke	0.938	−1.453	48.932	0.151
Coryell	2.119	−2.830	36.120	0.119
Dallam	−0.476	1.762	31.160	0.024
Deaf Smith	0.506	2.239	31.146	0.053
Fannin	0.478	−2.411	56.630	0.094
Floyd	0.109	1.000	36.176	0.018
Gillespie	2.597	−3.674	26.381	0.350
Gray	1.568	−4.507	34.518	0.058
Hale	1.974	−3.905	32.998	0.056
Hansford	−1.161	3.661	42.566	0.028
Hunt	0.274	−1.941	55.071	0.064
Jackson	0.270	−2.128	78.370	0.103
Jones	2.003	−1.146	16.200	0.317
Matagorda	0.620	−0.750	69.784	0.075
Milam	−0.066	−0.977	64.811	0.014
Navarro	1.175	−4.179	50.711	0.070
Nolan	1.621	−1.206	21.611	0.268
Randall	1.730	−0.666	27.504	0.105
Wharton	0.177	−0.652	77.381	0.008
Wise	0.464	−1.952	41.285	0.054
All stations	2.523	−5.651	36.647	0.371

An MLR analysis like the one described above was performed with a PCA. The PCA was carried out to remove the relationship between the two independent variables. The results of the MLR are presented in Table 6; although the R^2 values (Column 5 of Table 7) appear smaller, the regression relationships are significant as evidenced by the F values of regression relationships presented in Table 6. Looking at the regression relationships (with PCA) by county, negative coefficients appear for growing season total precipitation for Fannin, Hansford, Hunt, Jackson, and Milam counties only. Twenty-two out of 27 counties analyzed did not show declines in crop yield for excess precipitation when analyzed mathematically using regression relationships with growing season total precipitation and crop yields. However, when analyzed by the maximum 4-day total precipitation, 21 out of 27 counties show negative coefficients substantiating the declines in crop yield for excess precipitation. The counties that do not show negative coefficients are Bee, Bosque, Dallam, Deaf Smith, Floyd, and Hansford. Like the MLR without a PCA, most of the counties analyzed mathematically exhibit declining crop yields for excess precipitation showing negative coefficients mostly in maximum 4-day total precipitation and some in growing season total precipitation. Milam is the only county showing a negative coefficient for both the independent variables. Although showing some graphical relationships, Deaf Smith and Floyd are the only counties that did not mathematically exhibit the regression relationship between the independent variables and the dependent variable.

Table 6. Relevance of regression relationships.

County	Significance of Regression without PCA		Significance of Regression with PCA	
	F	Significance F	F	Significance F
Atascosa	3.203	0.057	3.056	0.065
Bailey	3.933	0.032	4.251	0.026
Bee	6.822	0.004	7.333	0.003
Bosque	0.386	0.683	0.386	0.683
Cameron	0.933	0.406	0.933	0.406
Collin	0.841	0.447	0.468	0.631
Cooke	1.595	0.230	1.235	0.308
Coryell	1.617	0.219	1.617	0.219
Dallam	0.295	0.747	0.295	0.747
Deaf Smith	0.669	0.521	0.669	0.521
Fannin	1.244	0.306	1.244	0.306
Floyd	0.226	0.800	0.225	0.799
Gillespie	6.456	0.006	6.465	0.005
Gray	0.739	0.487	0.739	0.487
Hale	0.710	0.508	0.710	0.501
Hansford	0.346	0.710	0.346	0.710
Hunt	0.817	0.453	0.817	0.453
Jackson	1.439	0.255	1.324	0.284
Jones	5.795	0.008	6.100	0.007
Matagorda	1.007	0.379	1.013	0.377
Milam	0.179	0.836	0.182	0.834
Navarro	0.947	0.401	0.914	0.414
Nolan	4.572	0.020	3.109	0.062
Randall	1.473	0.248	1.473	0.248
Wharton	0.097	0.907	0.097	0.907
Wise	0.717	0.497	0.586	0.564
All stations	7.37	0.003	7.272	0.003

Table 7. Results of multiple linear regression analysis (with PCA for removing the relationship between the two independent variables) using annual growing season precipitation, 4-day maximum precipitation, and crop yield.

County	Coefficients for Independent Variables		Intercept	Regression Analysis with PCA
	Variable (X1)	Variable (X2)		Calculated (R^2)
Atascosa	1.156	−2.986	38.517	0.203
Bailey	1.345	−2.153	30.925	0.262
Bee	1.753	5.296	53.441	0.379
Bosque	0.093	0.968	42.508	0.031
Cameron	0.325	−1.869	54.574	0.072
Collin	0.389	−0.232	53.336	0.038
Cooke	0.418	−1.637	53.976	0.093
Coryell	1.331	−3.275	53.757	0.119
Dallam	0.185	1.816	31.922	0.024
Deaf Smith	1.068	2.032	38.246	0.053
Fannin	−0.036	−2.458	54.716	0.094
Floyd	0.381	0.931	39.602	0.018
Gillespie	1.526	−4.232	45.486	0.350
Gray	0.171	−4.769	37.397	0.058
Hale	0.463	−4.351	38.201	0.056
Hansford	−0.320	3.827	41.515	0.028
Hunt	−0.363	−1.926	51.713	0.064
Jackson	−0.150	−2.231	73.163	0.099
Jones	1.718	−1.643	30.740	0.337

Table 7. *Cont.*

| County | Coefficients for Independent Variables | | Intercept | Regression Analysis with PCA |
	Variable (X1)	Variable (X2)		Calculated (R^2)
Matagorda	0.448	−1.116	74.051	0.078
Milam	−0.253	−1.003	60.859	0.015
Navarro	0.167	−4.396	52.597	0.071
Nolan	1.041	−1.337	30.958	0.206
Randall	1.366	−1.252	36.707	0.105
Wharton	0.051	−0.674	76.854	0.008
Wise	0.133	−1.738	41.469	0.047
All stations	1.402	−6.034	47.693	0.377

A comparison of the R^2 values of regression relationships with and without PCA are presented in Table 8 which pointed out that the PCA did not offer a significant improvement in identifying relationships between excess precipitation and rainfed sorghum yield. However, there is some difference in the regression analysis results. In the regression without a PCA, only one county (Milam) did not mathematically show any declining crop yields with excess precipitation. In the regression with PCA, six out of 27 counties analyzed (Bee, Bosque, Dallam, Deaf Smith, Floyd, and Hansford) did not show declining crop yields with excess precipitation. However, the results analyzed in all four different levels point out the existence of crop yield declines with excess precipitation.

Table 8. R^2 with and without PCA.

County	(R^2) without PCA	(R^2) with PCA
Atascosa	0.204	0.203
Bailey	0.239	0.262
Bee	0.363	0.379
Bosque	0.031	0.031
Cameron	0.072	0.072
Collin	0.086	0.038
Cooke	0.151	0.093
Coryell	0.119	0.119
Dallam	0.024	0.024
Deaf Smith	0.053	0.053
Fannin	0.094	0.094
Floyd	0.018	0.018
Gillespie	0.350	0.350
Gray	0.058	0.058
Hale	0.056	0.056
Hansford	0.028	0.028
Hunt	0.064	0.064
Jackson	0.103	0.099
Jones	0.317	0.337
Matagorda	0.075	0.078
Milam	0.014	0.015
Navarro	0.070	0.071
Nolan	0.268	0.206
Randall	0.105	0.105
Wharton	0.008	0.008
Wise	0.054	0.047
All stations	0.371	0.377

3.5. Substantiation of Crop Yield Declines with Excess Precipitation

In the previous section, the existence of crop yield decline with excess precipitation was identified based on separate graphical relationships between crop yield and growing season total precipitation, and crop yield and maximum 4-day total precipitation. The presence of crop yield decline for excess precipitation are substantiated by the graphical plot of crop yield, growing season total precipitation, and maximum 4-day total precipitation together for Hunt county in TX. The thin green rectangle outlined in Figure 10 identifies the hotspots that substantiate our findings described in the previous section(s).

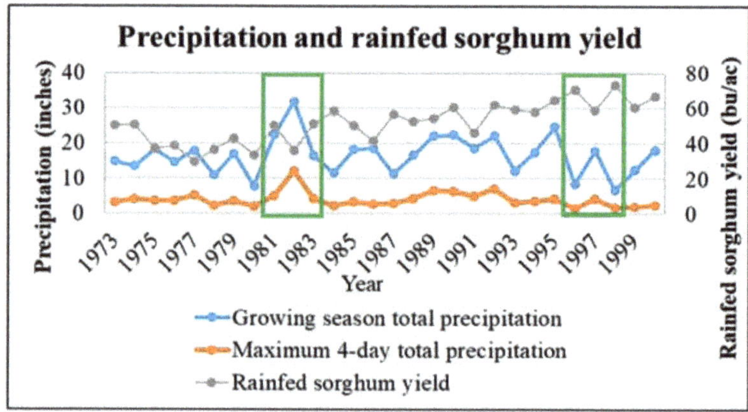

Figure 10. Graph showing rainfed sorghum yield, maximum 4-day total precipitation, and growing season total precipitation for Hunt County.

3.6. Spatial Variation of Declines in Crop Yield for Excess Precipitation

Counties and climate regions in Figures 11 and 12, respectively show the spatial variation of declines in yield of sorghum for excess precipitation. Based on both growing season total precipitation and maximum 4-day total precipitation, the North Central region of Texas appears to be more vulnerable to rainfed sorghum yield declines than other parts of Texas. The other regions showing some crop yield decline for excess precipitation are the High Plains and Southern regions. The large variation of precipitation within the region (Figure 13) and precipitation patterns appear to be the probable reason that can be attributed. However, we need more evidence to substantiate this finding.

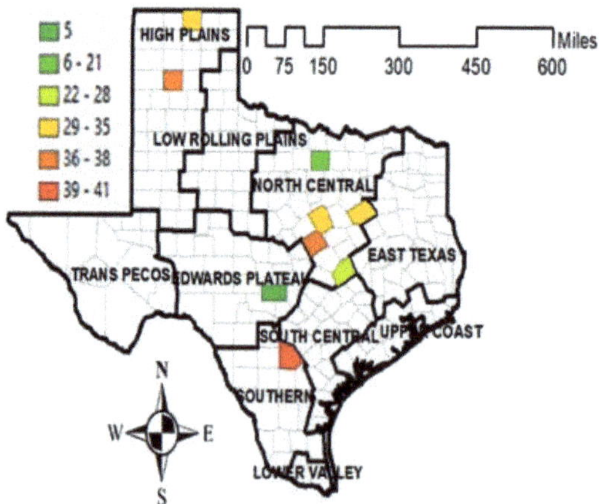

Figure 11. Percent reduction in rainfed sorghum yield between the year with excess precipitation and average crop yield from 1973 to 2000 (based on growing season total precipitation).

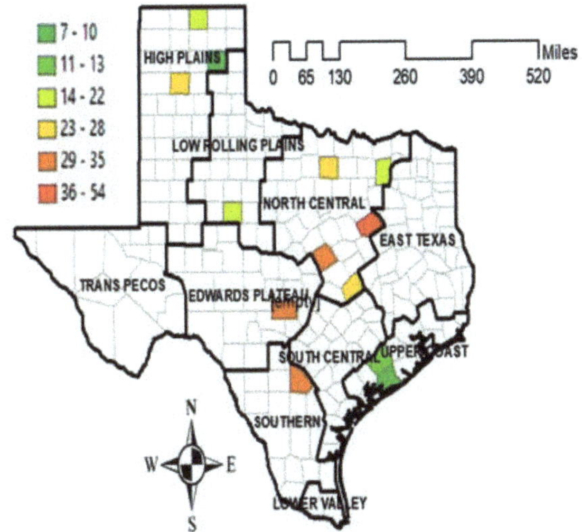

Figure 12. Percent reduction in rainfed sorghum yield between the year with excess precipitation and average crop yield from 1973 to 2000 (based on maximum 4-day total precipitation).

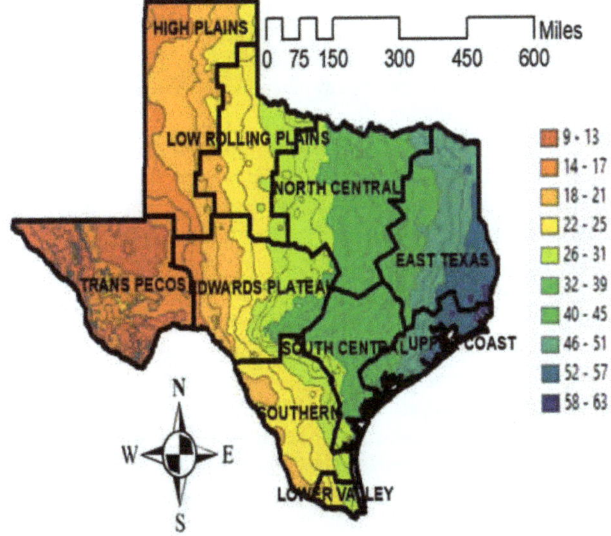

Figure 13. Variation of precipitation in different climate regions of Texas [19].

4. Discussion

For estimating crop yield losses, our study considered the quantity of precipitation alone leaving out another important aspect of precipitation, the timing with respect to the sorghum-growing season. In addition to excess precipitation, there are other contributing factors to yield losses such as high/low temperature (higher than optimum temperature for crop growth and lower than the crop base temperature), wind speed (high winds can dislodge the crop), humidity levels (excess would cause fungal problems), quality of soil (pH, drainage characteristics, depth), human decisions (e.g., whether or not going for pesticide application, irrigation, etc.), human errors in timing of land management operations (fertilizer or pesticide application, tillage, irrigation, and harvest). Therefore, care should be taken when interpreting the results of our study.

In addition to the approach used in this study, there are other ways of estimating crop yield losses by excess precipitation. The possibility of using remote sensing techniques to estimate crop yield losses by flooding was explored in Tapia-Silva et al. [20] using the August 2002 flooding event in Germany. In their approach, the flood crop loss is a function of crop value and a damage factor. The damage factor is a function of type of crop, timing of flood event, and inundation duration. When compared to field observations, they were able to estimate the crop losses with limited success. Their analysis dealt with flood inundation area of cropped fields rather than the proportion of yield loss.

There were a few other studies that explored the relationship between excess precipitation and crop yield reductions. Rosenzweig et al. [21] documented the extreme weather events that occurred in the US between 1977 and 1998; many of them include severe flooding events that resulted in reductions in crop yield. Increased moisture resulting from excess precipitation helps to spread epidemics and prevalence of leaf fungal pathogens, for example, fungal epidemics in corn, soybean, alfalfa, and wheat reported to have occurred in the US Midwest in 1993. The same period also saw incidences of soybean sudden death and mycotoxin increases [21]. Continuous soil saturation causing crazy top and common smut are also documented in the same study.

Corn yield reductions due to excess soil moisture (resulting from high precipitation) during current conditions and future conditions (under climate change) were estimated by Rosenzweig et al. [9] using CERES-maize model for the US Midwest. The current conditions showed a 3% reduction in corn yield ($600 million for the US corn production) because of aeration stress resulting from excess precipitation in the US Midwest. However, they have also estimated the increase in frequency of excess precipitation events in the future because of climate change. The same study also points out that when compared to the present, 90% more decreases in crop yield losses by 2030 and 150% more yield losses are expected by 2090. Winter wheat yield response to many parameters were analyzed in the Netherlands including excess precipitation. Except for one precipitation event in week 31 of the calendar year, they could not find any noticeable yield reductions for winter wheat resulting from excess precipitation [22].

The topic discussed in this manuscript relates to the idea of water use efficiency and water footprint. Water-use efficiency [23] is the ratio of aboveground biomass production to the water evapotranspired. The biomass is usually determined as dry weight rather than as fresh weight because moisture content of crops is different, which can mislead the interpretation of the water-use efficiency results. The results are usually expressed in kg L^{-1} or t m^{-3}. In the context of water-use efficiency, the reductions in crop yield during excess precipitation will present a less water efficient scenario. Therefore, care should be taken when interpreting the water-use efficiency results.

Water footprint [24,25] is the inverse of the water-use efficiency described above. The typical units are L kg^{-1} (L of water required to produce a kg of useful yield) or m^3 t^{-1} (m^3 of water required to produce a metric ton of useful yield). Green water footprint is water from precipitation that is stored in the root zone of the soil and evaporated, transpired, or incorporated by plants [24]. For rainfed crops, the inverse of water-use efficiency is analogous to green water footprint. The reductions in crop yield during excess precipitation will produce a relatively large green water footprint. Therefore, care should be taken when interpreting the water footprint results for crops that underwent an excess precipitation scenario like what is discussed in our study. The simplest way to avoid misleading water-use efficiency and green water footprint results are to use the average values from multiple crop growing years capturing a range of climatic scenarios.

The results of this study and other similar studies have applications in payment of crop insurance claims, parameterization of computer models (estimating crop yield reductions based on aeration stress), policy level decisions on rainfed crop selection, yield forecasting, estimating threats to food production, and water footprint analysis.

5. Conclusions

We collected historical crop yield data for Texas by county for grain sorghum from 1973 to 2000 and the corresponding daily precipitation data from weather stations within the counties. After estimating the crop growing season for sorghum in different parts of Texas, we estimated the growing season total precipitation and maximum 4-day total precipitation for each county growing rainfed grain sorghum. Using the two parameters mentioned above as independent variables, and crop yield of sorghum as the dependent variable, we tried to find out relationships between excess precipitation and decreases in crop yields using both graphical and mathematical relationships. We carried out a multiple linear regression (MLR) analysis with and without the use of a principal component analysis (PCA). Based on the results obtained, we can conclude that:

- Excess precipitation during crop growing season can cause yield reduction in rainfed grain sorghum.
- Total precipitation during the growing season and maximum 4-day total precipitation during the growing season are potential indicators of yield reductions in grain sorghum.
- Yield reductions could be in the range of 18% to 38% for rainfed grain sorghum in Texas because of excess precipitation during the growing season.
- When analyzed spatially, the north-central climate region of Texas appears to be more vulnerable to rainfed sorghum yield reductions because of excess precipitation.

Author Contributions: O.P.S. carried out most parts of the study. N.K. conceptualized the overall study, S.C. designed and carried out the principal component analysis; regression analysis was carried out by O.P.S. under the supervision of S.C. and N.K. B.K.P. analyzed the results of the study. Most of the GIS analysis was carried out by C.M. All the authors contributed to the development of this manuscript.

Funding: Funding for this study is provided by Tarleton State University under the Faculty-Student Research and Creative Activity Internal Grants.

Acknowledgments: The authors acknowledge Tarleton State University for supporting this research.

Conflicts of Interest: As the guest editor of the special issue "Water Management for Sustainable Food Production", Narayanan Kannan has a conflict of interest. Therefore, the assistant editors, and the editor-in-chief (of Water-MDPI), were involved in inviting reviewers, analyzing the peer review report and making the decision on acceptance of the article for publication.

Appendix A

Table A1. List of counties in Texas that have rainfed sorghum yield data is available.

Station Number	County	Latitude	Longitude
2	Atascosa	28.92	−98.74
4	Bailey	34.21	−102.73
6	Bee	28.45	−97.70
9	Bosque	32.01	−97.61
16	Cameron	25.91	−97.42
20	Collin	33.03	−96.48
23	Cooke	33.48	−97.15
24	Coryell	31.27	−97.88
26	Dallam	36.23	−102.24
28	Deaf Smith	34.93	−102.98
36	Fannin	33.43	−96.33
39	Floyd	33.98	−101.33
41	Gillespie	30.18	−99.15
42	Gray	35.55	−100.97
43	Hale	34.18	−101.7
45	Hansford	36.19	−101.18
46	Hunt	33.36	−96.06
47	Jackson	28.96	−96.68

Table A1. *Cont.*

Station Number	County	Latitude	Longitude
48	Jones	32.94	−99.8
50	Matagorda	28.68	−95.97
52	Milam	30.61	−97.2
53	Navarro	31.96	−96.68
54	Nolan	32.44	−100.52
58	Randall	34.95	−102.1
66	Wharton	29.31	−96.08
67	Wise	33.35	−97.39

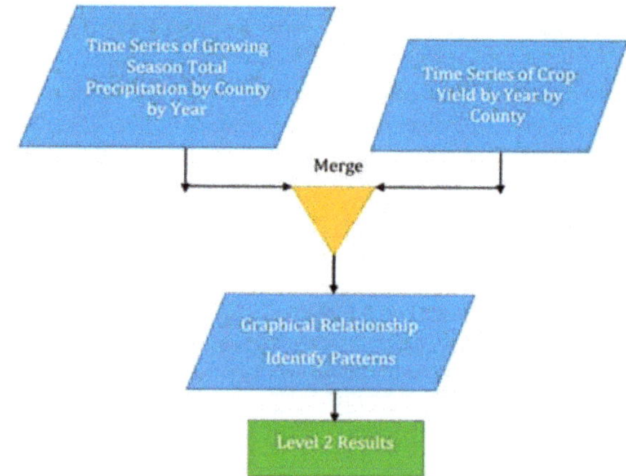

Figure A1. Generation of the graphical relationship between rainfed sorghum yield and growing season total precipitation (level 2 results).

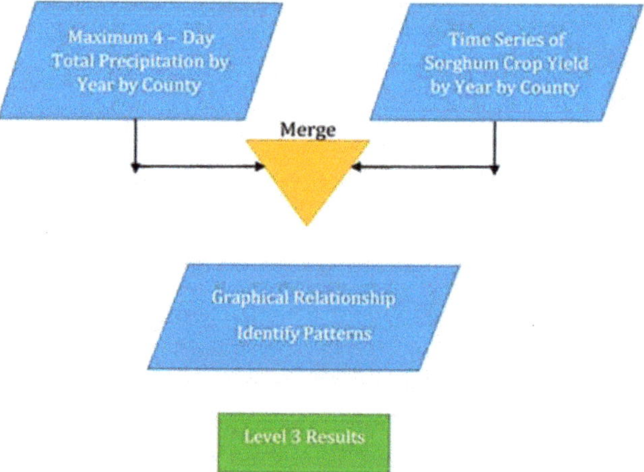

Figure A2. Generation of the graphical relationship between rainfed sorghum yield and maximum 4–day total precipitation (level 3 results).

References

1. Tsuchihashi, N.; Goto, Y. Year-round cultivation of sweet sorghum [*Sorghum bicolor* (L.) Moench] through a combination of seed and ratoon cropping in Indonesian Savanna. *Plant Prod. Sci.* **2008**, *11*, 377–384. [CrossRef]
2. Sorghum Program. Available online: https://www.sorghumcheckoff.com/all-about-sorghum (accessed on 31 July 2019).
3. Eggen, M.; Ozdogan, M.; Zaitchik, B.; Ademe, D.; Foltze, J.; Simane, B. Vulnerability of sorghum production to extreme, sub-seasonal weather under climate change. *Environ. Res. Lett.* **2019**, *14*, 045005. [CrossRef]
4. New, L. *Grain Sorghum Irrigation*; Texas A&M AgriLife Extension: Amarillo, TX, USA, 2004.
5. Arkansas Sorghum Quick Facts. Available online: https://www.uaex.edu/farm-ranch/crops-commercial-horticulture/grain-sorghum/2014-Arkansas-Grain-Sorghum-Quick-Facts.pdf (accessed on 31 July 2019).
6. Promkhambut, A.; Younger, A.; Polthanee, A.; Akkasaeng, C. Morphological and physiological responses of Sorghum (*Sorghum bicolor* L. Moench) to Waterlogging. *Asian J. Plant Sci.* **2010**, *9*, 183–193.
7. Lobell, D.; Burke, M.B.; Tebaldi, C.; Mastrandrea, M.D.; Falcon, W.P.; Naylor, R.L. Prioritizing climate change adaptation needs for food security in 2030. *Science* **2008**, *319*, 607–610. [CrossRef] [PubMed]
8. Grossi, M.C.; Justino, F.; Rodrigues, R.; Andrade, C.L.T. Sensitivity of the sorghum yield to individual changes in climate parameters: Modelling based approach. *Bragantia* **2015**, *74*, 341–349. [CrossRef]
9. Rosenzweig, C.; Tubiello, F.N.; Goldberg, R.; Mills, E.; Bloomfield, J. Increased crop damage in the US from excess precipitation under climate change. *Glob. Environ. Chang.* **2002**, *12*, 197–202. [CrossRef]
10. Balling, R.C., Jr.; Goodrich, G.B. Spatial analysis of variations in precipitation intensity in the USA. *Theor. Appl. Climatol.* **2011**, *104*, 415–421. [CrossRef]
11. Kunkel, K.E.; Karl, T.R.; Brooks, H.; Kossin, J.; Lawrimore, J.H.; Arndt, D.; Bosart, L.; Changnon, D.; Cutter, S.L.; Doesken, N.; et al. Monitoring and understanding trends in extreme storms: State of knowledge. *Bull. Am. Meteorol. Soc.* **2013**, *94*, 499–514. [CrossRef]
12. Texas Natural Resources Information Systems (TNRIS). Available online: https://data.tnris.org/ (accessed on 31 July 2019).
13. Jin, S.; Yang, L.; Danielson, P.; Homer, C.; Fry, J.; Xian, G. A comprehensive change detection method for updating the National Land Cover Database to circa 2011. *Remote Sens. Environ.* **2013**, *132*, 159–175. [CrossRef]
14. United States Department of Agriculture-National Agricultural Statistical Service (USDA-NASS). *Field Crops Usual Planting and Harvesting Dates*; Agricultural Handbook Number 626; USDA-NASS: Washington, DC, USA, 2010.
15. Mishra, A.K.; Singh, V.P. Changes in extreme precipitation in Texas. *J. Geophys. Res. Atmos.* **2010**, *115*. [CrossRef]
16. Nielsen-Gammon, J.W.; Zhang, F.; Odins, A.M.; Myoung, B. Extreme rainfall in Texas: Patterns and predictability. *Phys. Geogr.* **2005**, *26*, 340–364. [CrossRef]
17. Jolliffe, I.T. *Principal Component Analysis*, 2nd ed.; Springer: New York, NJ, USA, 2002.
18. Bluman, A.G. *Elementary Statistics: A Step by Step Approach*, 7th ed.; McGraw Hill Publishers: New York, NJ, USA, 2009.
19. Texas Water Development Board (TWDB). GIS Data. Available online: www.twdb.texas.gov/mapping/gisdata.asp (accessed on 10 June 2019).
20. Tapia-Silva, F.; Itzerott, S.; Foerster, S.; Kuhlmann, B.; Breibich, H. Estimation of flood losses to agricultural crops using remote sensing. *Phys. Chem. Earth* **2011**, *36*, 253–265. [CrossRef]
21. Rosenzweig, C.; Iglesias, A.; Yang, X.B.; Epstein, P.R.; Chivian, E. Climate change and extreme weather events Implications for food production, plant diseases, and pests. *Glob. Chang. Hum. Health* **2001**, *2*, 90–104. [CrossRef]
22. Powell, J.P.; Reinhard, S. Measuring the effects of extreme weather events on yields. *Weather Clim. Extremes* **2016**, *12*, 69–79. [CrossRef]
23. Kirkham, M.B. Water use efficiency. In *Encyclopedia of Soils in the Environment, Reference module in Earth Systems and Environmental Sciences*; Elsevier: Cambridge, MA, USA, 2005; pp. 315–322.

24. Hoekstra, A.Y. *The Water Footprint of Modern Consumer Society*; Routledge: London, UK, 2013.
25. Zhang, Y.; Huang, K.; Yu, Y.; Yang, B. Mapping of water footprint research: A bibliometric analysis during 2006–2015. *J. Clean. Prod.* **2017**, *149*, 70–79. [CrossRef]

 *water*

Article

Analysis of Intra and Interseasonal Rainfall Variability and Its Effects on Pearl Millet Yield in a Semiarid Agroclimate: Significance of Scattered Fields and Tied Ridges

Festo Richard Silungwe [1,2,*], Frieder Graef [1], Sonoko Dorothea Bellingrath-Kimura [1,2], Siza Donald Tumbo [3], Frederick Cassian Kahimba [3] and Marcos Alberto Lana [1,4]

[1] Leibniz Center for Agricultural Landscape Research (ZALF), Eberswalder Straße, 84, 15374 Müncheberg, Germany; graef@zalf.de (F.G.); belks@zalf.de (S.D.B.-K.); marcos.lana@zalf.de (M.A.L.)
[2] Humboldt Universität zu Berlin, Faculty of Life Sciences, Unter den Linden 6, 10099 Berlin, Germany
[3] Department of Engineering Sciences and Technology, Sokoine University of Agriculture, P.O. Box 3003, CHUO KIKUU, Morogoro 3003, Tanzania; siza.tumbo@gmail.com (S.D.T.); fredkahimba@sua.ac.tz (F.C.K.)
[4] Crop. Production Ecology, Swedish University of Agricultural Sciences, Ulls väg 16, 75007 Uppsala, Sweden
* Correspondence: festo.richard@zalf.de; Tel.: +255-767786036

Received: 1 February 2019; Accepted: 17 March 2019; Published: 20 March 2019

Abstract: Establishing food security in sub-Saharan African countries requires a comprehensive and high resolution understanding of the driving factors of crop production. Poor soil and adverse climate conditions are among the major drivers of poor regional crop production. Drought and rainfall variability challenges are not fully being addressed by rainfed producers in semiarid areas. In this study, we analysed the spatiotemporal rainfall variability (STRV) and its effects on pearl millet yield using two seasons of data collected from 38 rain gauge stations scattered randomly in farm plots within a 1500 ha area of semiarid central Tanzania. The STRV effects on pearl millet yield under flat and tied ridge management were analysed. Our results show that seasonal rainfall can vary significantly for neighboring fields at distances of less than 200 m, which impacts yield. The STRV for daily rainfall was found to be more critical than for total seasonal rainfall amounts. Scattering fields can help farmers avoid total harvest loss by obtaining at least some yield from the areas that received adequate rain. The use of tied ridges is recommended to conserve soil moisture and improve yields more than flat cultivation in semiarid areas.

Keywords: spatiotemporal rainfall variability; tied ridges; scattered plots; pearl millet; yield loss

---

1. Introduction

Spatiotemporal rainfall variability (STRV) and drought are among the primary challenges in rainfed agricultural communities [1,2]. STRV and drought both limit crop production and increase crop yield uncertainties among farmers. The situation is particularly severe in semiarid areas in sub-Saharan Africa (SSA) [3,4], exacerbating chronic food insecurity [5–10]. To address such challenges in these areas, the literature accentuates the importance of adopting more water-saving technologies through the efficient storage and use of water [11]. Several studies described STRV on different scales [12–16]; however, these studies rarely demonstrated the potential relationship between STRV and yield variability among farmer fields located within the same agricultural watershed. Rainfall studies in the forms of trend analyses and spatial variability over large areas are numerous, but these studies have limited connections to local agricultural challenges. These studies have rarely prioritized farmer risk management strategies, including crop upgrading strategies (UPS) [17], which are important for understanding the cycle of annual harvest losses, either partially or totally, for farmers in semiarid areas.

The population is increasing annually in the SSA region; therefore, the production of staple food crops has been emphasized to meet the increasing food demand. Pearl millet is an important crop in the region. With drought tolerance characteristics, pearl millet crops provide cultivation opportunities for farmers in drier areas. However, pearl millet production can significantly increase if the water needs of the crop are improved and vice versa. Historical data from the Food and Agriculture Organization Statistical Databases of the United Nations (FAOSTAT) indicate that the production of pearl millet in the SSA region has declined over the last two decades (FAOSTAT was visited on 10 December 2018), which can be directly attributed to poor soil and weather conditions, among other factors. The weather conditions are more severely challenging to most farmers, with spatiotemporal variation in rainfall frequently reported [18–20]. The current practices which are being used to address STRV are limited and the influence of STRV on crop yields at higher resolutions is poorly understood.

Since crop yield can vary even within a single farm due to different individual or combined factors, ranging from soil, weather, topography, and management [21,22], studies are required to provide a comprehensive understanding of the harvest losses at the village and farm levels for pearl millet crops, which would aid in providing practical recommendations to improve crop production. Yield losses in small plots accumulate when there are a considerable number of plots, thus reducing small area losses is advantageous for farmers in dry areas. Eventually, too many farmers with significant annual yield losses results in serious food shortages [23]. In the food shortage context, our research aim was to analyse high-resolution spatiotemporal data on daily rainfall, seasonal rainfall, and pearl millet yield to understand their variability and potential reasons for crop yield variability. Therefore, we specifically aimed to (1) analyze the spatiotemporal rainfall variabilities in neighboring fields, (2) evaluate the significance of rainfall variability on pearl millet yields among farmers, and (3) evaluate the effectiveness of tied ridges and scattering fields in reducing the risks of harvest loss.

2. Materials and Methods

2.1. Study Area

The study was conducted in the Dodoma Region of central Tanzania (Figure 1). The region lies between latitudes 4°7″ and 7°21″ S and between 36°43″ and 35°5″ E. The region has a population of 2.084 million people [24]. Most of the region is semiarid with low and erratic rainfall, which averages less than 600 mm per year. We selected the village of Idifu as our case study area, as more than 70% of the land is annually cultivated for pearl millet crops [23].

Figure 1. Location of the study site.

2.2. Spatiotemporal Rainfall Data Collection

Rainfall data were collected from November 2016 to May 2018, which included two growing seasons. Season one was from November 2016 to May 2017 (SES1) and season two was from November 2017 to May 2018 (SES2). We collected the data from 38 rain gauges randomly located in a rectangular section of Idifu (Figure 2). The rain gauge positions were defined using the K-means clustering algorithm method [25], then displayed in quantum geographic information system (QGIS) and modified onsite depending on the site conditions and features. Distances between rain gauges were limited to a minimum of 150 m between any pair of rain gauges and varied as shown in the rectangular area of 2.5 by 6 km (1500 ha) to cover a portion that contained many of the village farmers. We recorded the daily rainfall using manual rain gauges. Farmers living close to each location were identified, and at least one farmer was trained how to record the daily rainfall at 8:00 a.m. daily with the supervision of an agricultural field officer. The numbers of rainy days (events) were counted as any nonzero readings from an accumulated rainfall measurement recorded each day throughout the season.

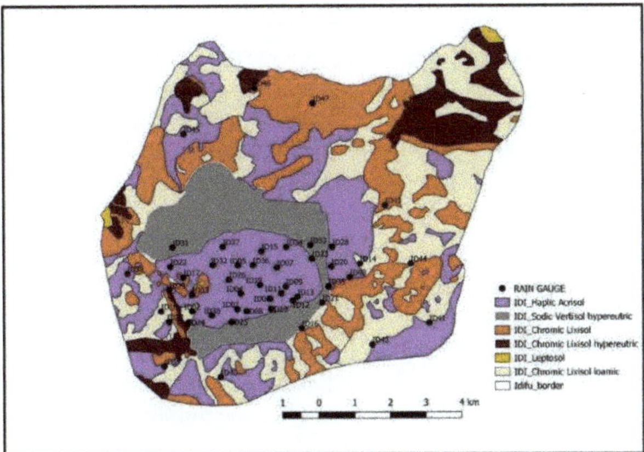

Figure 2. Soil pattern and rain gauge positions at the Idifu study site.

2.3. Pearl Millet Yield Data

To evaluate the variability in pearl millet yield among farmers, we collected yield data from 38 locations representative of the rain gauges shown in Section 2.2. Pearl millet (Okoa variety) was planted in all locations by all farmers under flat cultivation (a common practice by most farmers in the village) and under tied ridges. The tied ridges (in situ) rainwater harvesting practice is among the four soil management strategies recommended as most suitable in semiarid areas [17]. Tied ridges are long, narrow, and elevated strips of land (a ridge) crossed by earthbands within the furrow called ties (Figure 3). The practice is well described in literature [17]. Over 80 farmers across the study area had adopted tied ridge practices at more than 20 spatial rain gauge positions. For each location, we collected yield data from farmers for 2–4 plots with areas of 100 m^2 over two seasons from both flat and tied ridges practices.

Figure 3. Tied ridges prepared by different farmers in semiarid Dodoma.

2.4. Soil Physical and Chemical Properties

Since yield variability is highly influenced by the soil properties, we used a local soil map and underlying data from the physical and chemical properties of the soil [26]. In general, the farmers' soils matched with respect to classification and fertility but were noticeably different in terms of texture with predominantly higher sand content [26]. As shown in the soil map (Figure 2), these soils were chromic lixisol loamic (CLL), chromic lixisol hypereutric (CLH), chromic lixisol (CL), haplic acrisol loamic (HA), and sodic vertisol hypereutric (SVH). The majority of the plots were on HA (71%), followed by CLL (14%), CL (8%), SVH (6%), and CLH (1%) soils.

2.5. Data Analysis

We calculated the rainfall variability in terms of (1) the daily and seasonal amounts, (2) number of rainy days, and (3) total seasonal amounts. We recorded the start and end dates of the rainy season (onset and cessation dates). We used natural neighbor kriging interpolation in QGIS to describe the seasonal rainfall patterns and analyse the spatial rainfall variability. We calculated the variation coefficient of daily and seasonal rainfall amounts and for the number of events.

We determined the probability of an event covering the entire study site (P_{100}) and the probability of covering at least half of the study site (P_{50}) using daily rainfall events for both seasons:

$$P100 = \frac{\text{Number of rainfall events recorded by all 38 rain gauge stations}}{\text{Total number of rainfall events per season}} \tag{1}$$

$$P50 = \frac{\text{Number of rainfall events recorded by at least half of the 38 rain gauge stations}}{\text{Total number of rainfall events per season}} \tag{2}$$

Using *Statgraphics Centurion XVII software* (Statgraphics Technologies, Inc., The Plains, VA, USA), we also performed an analysis of variance (ANOVA) of daily rainfall for both seasons. We used the Kruskal–Wallis test to compare the medians when there were some significant non-normalities in the daily rainfall data [27].

We performed a kriging analysis for each daily rainfall event for both seasons using QGIS. From kriging maps, we performed a variogram cloud analysis using the variogram cloud tool in QGIS for every daily rainfall event to determine their variance related to distances between rain gauges (Appendix A). We modified the approach from [12], who used a defined set of transects from a kriging map of daily rainfall and assigned the mean differences of rainfall along transects to the distances between gauges. The variogram cloud analysis was used to determine the variance, semivariance, and covariance of the rainfall in all directions (360 degrees) by applying the moment of inertia to the data. We performed a regression analysis for the rainfall differences and their distances (Appendix B). Then, we calculated the correlation coefficients for maximum rainfall differences and their associated distances.

We used the Statgraphics Centurion XVII software to map the seasonal yield of pearl millet. Then, we determined the relationships between rainfall variability and pearl millet yield variability among farmers using a simple linear regression model.

We individually tested how both variables (rainfall (mm) and number of events) influence the yield for both seasons. We used the R-squared statistic to indicate how the fitted linear model explains the influence of rainfall and events on pearl millet yield.

We determined the effect of soil type at the study sites on yield variability by performing an ANOVA, comparing the average yields in different soils. We analyzed the effects of tied ridges compared to flat cultivation. We checked the within variation by computing the coefficients of variation (CVs).

3. Results

3.1. Spatiotemporal Rainfall Variability

3.1.1. Average Daily Rainfall and Variability

There were 15 rainfall events in SES1 and 31 events in SES2; the difference in events was significant between the two seasons. This situation involves considerable risks associated with rainfed agriculture for this semiarid area, showing that strategies are required that can absorb these wide variations that occur within a 1500 ha field. The daily intraseasonal average spatial rainfall per event variability (ASREV) was significantly different in both seasons, similar to the interseasonal average spatial distribution of rainfall per rain gauge (Table 1). Generally, for most rain gauge stations, higher ASREVs were recorded during SES2 than SES1 (Figure 4).

In SES1, the probability that the rainfall covered the entire village (P_{100}) was zero, whereas the probability of at least half the village (P_{50}) being covered by rainfall was 42% (Table 2). No rainy days were observed for the entire village in SES1. In SES2, the probability of rain for the entire village was more than 40%, whereas the chance for at least half the village (P_{50}) being rained on during one event was 87% (Table 2). These results show that for every 10 rainfall events, at least four events would cover the entire village and approximately nine events would cover at least half the village.

Table 1. Analysis of average spatial rainfall (mm) per event within and between seasons.

Season	No. of Gauges	Average (mm)	SD (mm)	CV (%)	Minimum (mm)	Maximum (mm)	*p*-Value (within)	*p*-Value (between)
SES1	38	10.79	1.56	14.5	8.0	15.1	0.00 *	
SES2	38	14.11	1.44	10.2	12.3	18.6	0.00 *	0.00 *

Note: CV is the coefficient of variation. Statistically significant at 0.05 level is denoted by a star (*). The average values were calculated by averaging the daily rainfall (mm) for all events in a season to a single value per rain gauge, and then the variations among rainfall averages (mm) for all 38 rain gauges were tested.

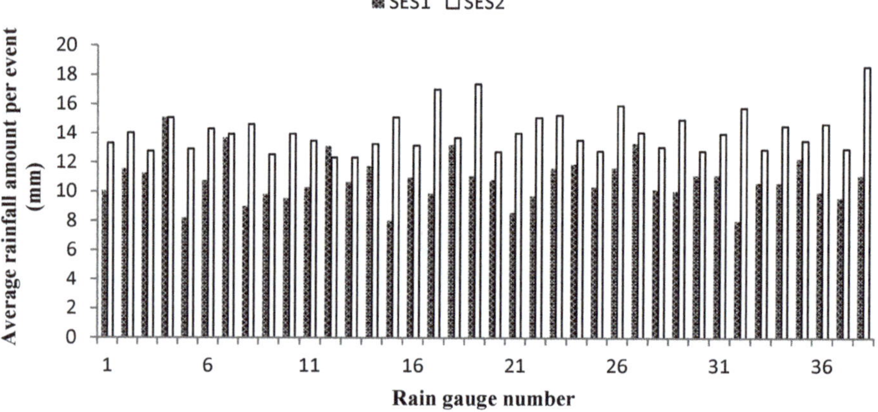

Figure 4. Average rainfall amounts per event for SES1 and SES2.

Table 2. Probabilities of daily rainfall coverage.

Season	No. of Gauges	Average Number of Events	P_{100}	P_{50}
SES1	38	9	0.00	0.42
SES2	38	25	0.56	0.87

Note: The P_{100} and P_{50} values were calculated using daily rainfall data from 38 stations. P_{100} is a probability that for each rainfall event, all 38 rain gauges would record daily rainfall. P_{50} is a probability that for each rainfall event, at least half of the 38 rain gauges recorded daily rainfall. The values were obtained by dividing the maximum possible number of events in a season recorded in any rain gauge out of 38 total gauges by the minimum possible number of events at that gauge.

From the daily rainfall amounts in SES1, the calculated CV was higher (39.6–435.38%) than the seasonal rainfall (14.5%). For SES2, the daily rainfall CVs were between 12.62 and 329.67%, while the seasonal total rainfall was 10.18%, which indicates the significant daily rainfall variability compared to seasonal variability. Figure 5 shows that the daily rainfall in SES1 and SES2 was not normally distributed during many events.

We observed points far outside of the boxes (Figure 5), indicating unusually lower or high rainfall values within the same events, again indicating high spatial variability. However, SES1 shows higher unusual variability than SES2.

Both scenarios, SES1 and SES2, explain the risk of averaging the spatial rainfall per event for fields. We observed significant variations in the gauge station rainfall recorded for all events, and these variations accumulated over the entire season. Consequently, some locations had accumulated deficits resulting in severe shortages in the rainfall amount required to support crop growth, and hence, posing a high risk to crop production. In this study, we found that the accumulated seasonal rainfall amounts recorded over the entire field were significantly different among rain gauges (Table 1). The SES1 rainfall onset varied significantly over five different dates: 14 December 2016; 2 January 2017;

8 January 2017; 15 January 2017; and 30 January 2017. The cessation dates did not vary much as most of the plots (87%) received the least rainfall simultaneously. In contrast, in SES2, we observed that all rain gauges in the field recorded the same rainfall onset and cessation dates, although the rainfall amounts on particular dates varied significantly among gauges ($p < 0.05$).

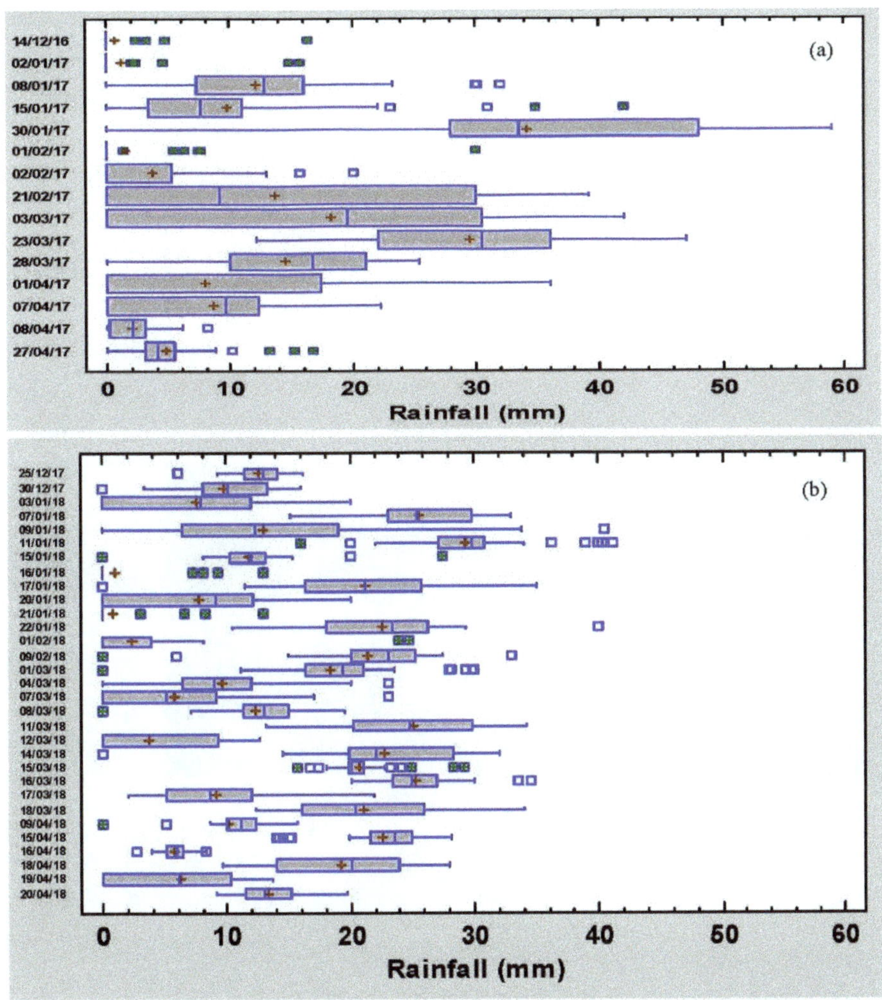

Red plus sign (+) indicates the mean of the dataset. The vertical line is the middle of the box and is the median. The width of the box shows how the data are spread from the mean. The square boxes above the whiskers (lines extending above and below the box) are the outliers. The square boxes above the whiskers with green plus (+) signs indicate the data that are considered to be far outside points (points are three or more times the box width above or below the box).

Figure 5. Distribution of daily rainfall (mm) in (**a**) SES1 and (**b**) SES2.

3.1.2. Seasonal Rainfall Variability

For both seasons, the total seasonal rainfall varied significantly among rain gauges and between seasons within the study site (Figure 6). In SES1 (Figure 6a), the number of total events per rain gauge ranged between 6 and 12, with the total amount of rainfall ranging between 120.1 and 226.6 mm. The average seasonal spatial distribution of rainfall per rain gauge for SES1 was 161.9 mm. For SES2 (Figure 6b), the number of rainy days per rain gauge ranged between 23 and 29, with rainfall ranging between 382 and 576.2 mm. The average seasonal spatial distribution of rainfall per rain gauge was 437.6 mm. The total seasonal rainfall for both seasons, as expected, was highly correlated ($r = 0.97$) with the events (Figure 7). The intraseasonal correlations were far lower than both seasons combined.

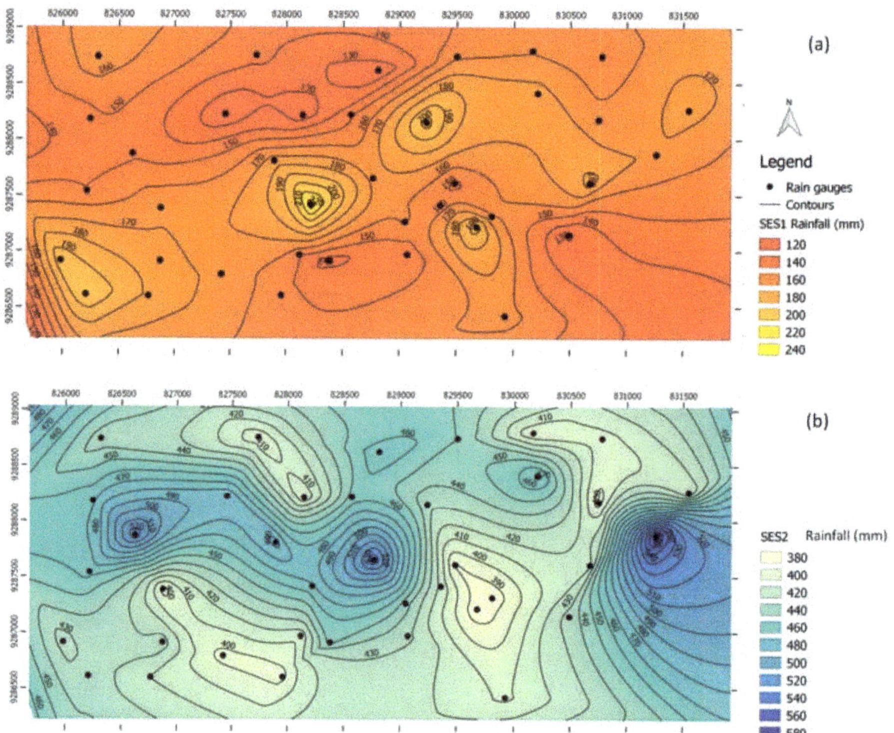

Figure 6. Isohyets of the total seasonal rainfall (mm) during (**a**) SES1 and (**b**) SES2.

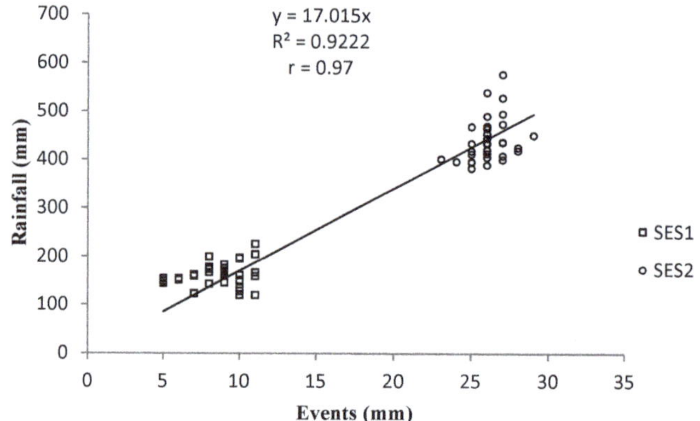

Figure 7. Total rainfall (mm) for two seasons related to the number of events.

3.1.3. Rainfall Variability with Distance between Pairs of Gauges

To understand the relationship between different farmer fields across the study site, we calculated random distances for rain gauge pairs and rainfall variability using a variogram cloud analysis in QGIS. In Appendix A, we present different distances and rainfall variances for pairs of rain gauges randomly picked in different directions (angles). The greatest variance under closely spaced rain gauges occurred at the shortest distance of 164.4 m during SES2. The maximum variance between rain gauges in the area was 27,192.0 mm^2 during SES1. This difference shows that it is possible for a high rainfall season to have a significant variation between a closely (164.4 m) spaced pair of rain gauges. However, higher variation occurred during SES1 with low rainfall (27,192.0 mm^2). In general, the average variance was higher in SES1 (3110.4 mm^2) than in SES2 (901.8 mm^2), which implies that, even for total seasonal rainfall, the variation was high under low rainfall in SES1 and low under comparatively high rainfall in SES2.

3.2. Effects of Spatiotemporal Rainfall Variability on Pearl Millet Grain Yield

The average yields of pearl millet were 360.53 and 637.66 $kgDWha^{-1}$ for SES1 and SES2, respectively (Table 3). In both seasons, the spatial intraseasonal yields were significantly different among farmers. Higher variability was observed in SES2 than SES1 (Figure 8), with higher yields also recorded in SES2 than in SES1. The maximum grain yields for individual locations were 912 $kgDWha^{-1}$ and 1633 $kgDWha^{-1}$ for SES1 and SES2, respectively (Table 4). The rainfall pattern observed in Figure 6 is correlated with the yield pattern in Figure 8, indicating that, for the two seasons, the pearl millet yield was correlated with the recorded amount of seasonal rainfall.

Table 3. Standard deviation, mean, CV, and *p*-values for pearl millet yield ($kgDWha^{-1}$).

Season	No. of Plots	Average ($kgDWha^{-1}$)	SD ($kgDWha^{-1}$)	CV (%)	Minimum ($kgDWha^{-1}$)	Maximum ($kgDWha^{-1}$)	*p*-Value (within Season)	*p*-Value (between Seasons)
SES1	98	360.53	170.6	47.32	105	912	0.00 *	0.00 *
SES2	101	637.66	381.26	59.79	239	1633	0.00 *	

Note: Statistically significant at 0.05 level is denoted by a star (*). For each of the 38 rain gauge positions, we collected samples from a minimum of two plots to a maximum of four plots with flat cultivation and with tied ridges cultivation.

From the correlation analysis, we found that rainfall was moderately weakly but positively correlated with yield in terms of both rainfall amount and rainfall events (Figure 9). However, the rainfall events were more correlated with yield than the total seasonal rainfall amounts in both seasons. In low rainfall SES1, the yield was found to have a small but positive correlation with the rainfall events ($r = 0.37$). A moderately low but positively correlated coefficient ($r = 0.34$) was found between the yield and rainfall amount in SES1. In the wetter SES2, the yield was found to have a low but positive correlation to both events ($r = 0.03$) and seasonal rainfall amount ($r = 0.02$), which means that if the rainfall (during crop growth) is well-distributed, a considerable amount of rainfall can be used by the crops to enhance the yields. We observed a yield increase with better rainfall distribution in SES1; however, the trend appeared negligible or nonsignificant in SES2, which is attributed to a more uniform spatiotemporal seasonal rainfall and event distribution than SES1. Although the variability in seasonal rainfall during SES2 was significant, the rainfall amount was enough to meet most of the pearl millet crop water requirement. The crop water requirement was estimated to be approximately 366.2 mm in Dodoma, which is less than most of the recorded seasonal rainfall amounts. The seasonal rainfall amounts and events were moderately weakly but positively correlated with the pearl millet yield ($r = 0.43$ and 0.44, respectively) (Figure 9). The regression lines for combined seasons showed much stronger correlations than individual seasonal correlations. Thus, apart from variability in rainfall amount and timing, factors other than rainfall may contribute to yield variability.

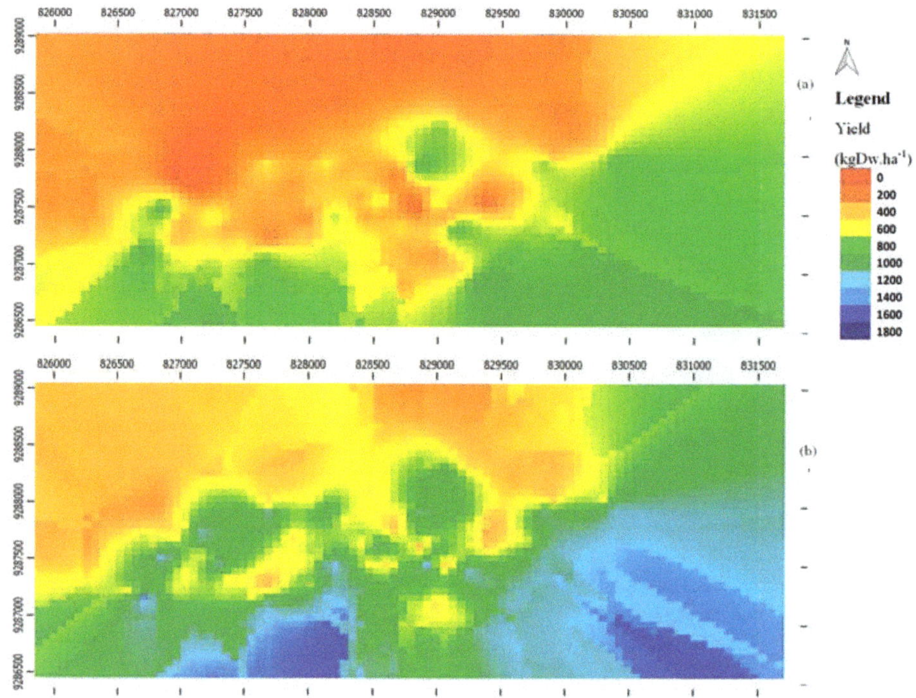

Figure 8. Spatial distribution of pearl millet yield (kgDWha^{-1}) for (**a**) SES1 and (**b**) SES2.

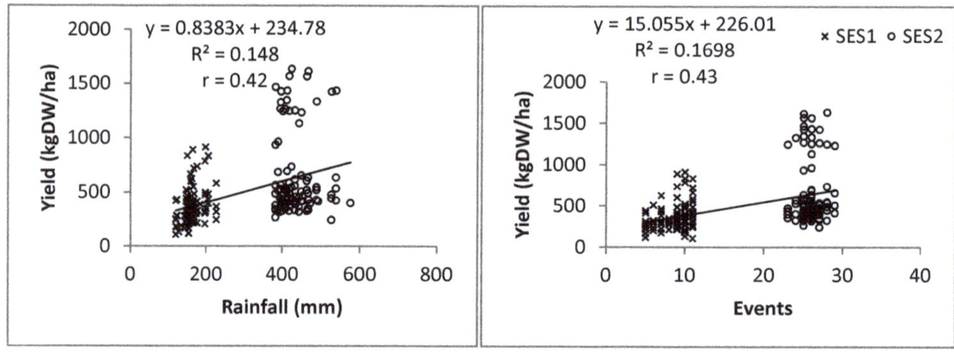

Figure 9. Combined relationships between two seasons of yields (kgDWha^{-1}), seasonal rainfall (mm) amounts, and number of events.

3.3. Yield Variability by Soil and the Influence of the Tied Ridge Management Strategy

3.3.1. Yield Variations among Soil Types

The yield data were collected from plots with either flat or tied ridge management strategies spatially scattered over different soil types in the village. There were differences among the average yields for different soils. The average yield from plots with CLL soils was slightly higher (573.1 kgDWha^{-1}), followed by CL (497.9 kgDWha^{-1}), HA loamic (477.9 kgDWha^{-1}), and SVH (415.6 kgDWha^{-1}). However, from a single-factor ANOVA comparison, we found that the yields among the soils were not significantly different (Table 4). In contrast, the pearl millet yield variability within individual soils varied with CV between 60.3 and 72.5%, and these differences were statistically significant (Table 4). In this context, the pearl millet yield at the study site was not influenced by the soil type.

3.3.2. Yield Variations between Flat and Tied Ridge Cultivations

Tied ridges increased the yields more than flat cultivation and nearly doubled the yields in both seasons (Table 5). The reason for this difference could be due to the tied ridge's ability to prolong soil moisture during the crop growth period, improving the efficiency of rain water usage. However, tied ridge cultivation also increased the yield variations among farmers. We found high yield spatial variability for both SES1 (CV = 40.2%) and SES2 (CV = 44.7%) for tied ridges, which differed from SES1 (CV = 30.59%) and SES2 (CV = 18.7%) under flat cultivation (Table 5). The variations in both seasons were significantly different under tied ridges, but the variations did not differ under flat cultivation (Table 5). This difference may imply that for different soils, tied ridges have a variable advantage in terms of improving pearl millet yield. This difference may also indicate that the ability of the tied ridges to prolong soil moisture affected other factors, such as soil fertility level variations, field slopes, previous crops, and organic matter in the scattered plots, all of which support crop growth differently.

Table 4. Comparison of yield variation according to soils.

Soil Type	No. of Plots	Average (kgDWha^{-1})	SD (kgDWha^{-1})	CV (%)	Minimum Yield (kgDWha^{-1})	Maximum Yield (kgDWha^{-1})	Range (kgDWha^{-1})	p-Value (within Soil)	p-Value (among Soils)
CL	16	497.9	360.85	72.47	216	1424	1208	0.00 *	
CLL	28	573.1	345.76	60.33	214	1633	1419	0.00 *	
HA	141	477.9	320.88	64.75	105	1612	1507	0.00 *	0.38
SVH	14	415.6	266.21	64.05	130	1247	1117	0.00 *	

Note: Statistically significant at 0.05 level is denoted by a star (*). CL—chromic lixisol, CLL—chromic lixisol loamic, HA—haplic acrisol loamic, and SVH—sodic vertisol hypereutric.

Table 5. Overall comparison of yield variations for flat cultivation and tied ridges.

Cultivation Practice	Number of Plots	Average (kgDWha^{-1})	SD (kgDWha^{-1})	CV (%)	Minimum (kgDWha^{-1})	Maximum (kgDWha^{-1})	Range (kgDWha^{-1})	p-Value (within Treatment)	p-Value (between Treatments)	Season
Flat	58	288.5	88.26	30.5	105.0	474	369	0.37		
Tied Ridges	40	470.72	189.29	40.2	163.0	912	749	0.04 *	0.00 *	SES1
Flat	59	418.0	78.15	18.7	239.0	567	328	0.18		
Tied Ridges	42	946.24	423	44.7	343.0	1633	1290	0.00 *	0.00 *	SES2

Note: Statistically significant at 0.05 level is denoted by a star (*).

4. Discussion

Poor and erratic rainfall is challenging rainfed agricultural production in semiarid areas, such that farmers may experience total harvest loss [12]. From our analysis, we found that rainfall can vary significantly in both space and time within a small area between neighboring fields, which agrees with the results reported by other studies [12,28,29]. The variations can be significantly different within a small area (1500 ha study area) in terms of events, rainfall amount per event, total seasonal rainfall amount, and onset and cessation dates. When the total number of seasonal rainfall events is low, the chance of having a lower seasonal rainfall amount and poor distribution is high and vice versa, as we observed during both seasons. Other studies indicated the potential effects of extreme floods and drought events [18]. For instance, few high rainfall events may result in high total seasonal rainfall amounts with poor distribution (during the seasonal crop growth period). In this study, the numbers of events were highly correlated with the seasonal rainfall amount, and the two seasons of data showed significant temporal variability. This situation is common in semiarid areas [18,20]. For both seasons, the spatial distribution of rainfall per event varied significantly within the area. The variation increased with poor total seasonal rainfall, and increased with a nonsignificant linear trend with a distance among rain gauges in the area. Other studies, such as Gao et al. [30], found that the rainfall spatial variation was obvious during the winter dry season. Graef and Haigis [12] reported that the variations along two different transects in Sahel were nearly equal, and the mean differences in the variations increased with the distance between gauges (from ±1.8 mm at 1 km, ±3.5 mm at 2 km, to ±5.7 mm at 3.2 km). The variability increase with distance may be inconsistent when larger areas are considered due to the inherently high local spatial variability behavior of rainfall [30]. In our findings, the correlation coefficient between spatial rainfall differences was found to be weak, which justifies the tendency for examination on larger scales. Buytaert et al. also found that rain gauges separated at a distance of less than 4 km were highly correlated despite having high spatial variability in average rainfall [29].

The rainfall variability directly impacted the farmer's seasonal pearl millet yield. The collected yield discrepancies from different spatial plots within the study area indicate that field scattering is an effective strategy for reducing the probability of total seasonal harvest loss. Previous case studies from the Sahel region show that scattered fields reduce the yield disparity while enhancing the stability of pearl millet yield between households [12,31]. A similar conclusion can be drawn in this study. Thus, a farmer with scattered fields across the study area has a good chance of stable seasonal crop harvest than the one who has all fields concentrated in the same area. The strategy promotes the spatially efficient use of rainfall. For areas with high variations in soil properties, the choice of locations of the scattered fields should consider the quality of soil to reduce the risk associated with soil. Although, in this study the scattered fields in the area were mostly located on soils with similar properties spatially (HA soils), this is not expected to be the case for many areas. There are findings suggesting that yields are poor on gravelly soils and two to three times greater on clay soils [22]. Another study recorded higher yields on under clay soils than sandy soils [32]. However, from the study that checked spatial variability pattern of yields and soils in a 1 ha field, the authors found that soil variables explained 30% of the total yield variation of pearl millet [33]. Thus, to produce higher yields, proper management is required especially in sandy soils. The soil analysis in the Idifu area indicated that the soil has a higher sand content [26] which means creates a risk of lower yields. In addition to the careful selection of soil for the scattered fields practice, overall good crop management is recommended to improve yields of pearl millet.

The yield was consistently correlated with both the rainfall amounts and the number of events in a season. However, if the crop water requirement is met in timing and amount, other factors, such as soil and crop management, may be the risk sources. In most cases, yield would vary depending on soil properties [34–36]; however, we found no statistical evidence of yield variability for different soils in this study site, possibly due to the insignificant effect of soil interactions with other climatic variables. As established by a previous study, the farmers' soils in Idifu matched in terms of classification and fertility but were noticeably different in terms of texture with predominantly higher sand content [26]. Other studies found yield discrepancies even at the within-field scale, as some parts of the field may produce more of a crop relative to the rest of the field, indicating microscale interactions between climate, soil, topography, and management [21]. Generally, soils with higher content of swelling clay and silt better retain and release soil moisture; therefore, under adverse limited rainfall conditions, these properties provide a buffer to crop production [37]. Previous studies suggest that the effect of tied ridges is much more pronounced under limited rainfall in high clay content soils than in sandy soils [37]. In wetter seasons, tied ridges have limited advantages in crop production under clay soils especially when rainfall exceeds its retention capacity. The provision of drainage is important under clay soils. Conversely, sandy soils possess good drainage properties, which make tied ridges useful in dry season and less destructive in wetter seasons.

In contrast, tied ridge cultivation increased the pearl millet yield significantly more than the flat cultivation by prolonging soil moisture from harvested rainfall. Therefore, farmers should use this in situ rainwater harvesting (tied ridges) method in their scattered fields to reduce harvest losses and to manage the high rainfall variability. Elsewhere in semiarid areas, the practice has been successful for other cereals, such as maize [38–40] and sorghum [2,39]. However, in our study, we found that tied ridges increased spatial yield variability. While yield varied in both flat and tied ridge management situations, the CV of the tied ridges was higher. Further studies on the interactions among plants, soil, terrain, and climatic factors should focus on combining management strategies, such as the use of field scattering with tied ridges and fertilization, to increase yields and reduce spatial yield variability. Field scattering according to different soil types may also be a solution to manage yield variability but is often limited by the number and distance of farmers' fields or the types and features of existing soils. Farmers can retain the advantage of not losing the entire harvest if these farmers scatter their fields randomly or purposely within their area. We consider the exploration of the effects of certain factors, such as variable planting dates, to be important. For instance, the variable onset and secession dates have implications for a farmer's decision about when to plant [18]. A modelling study performed in the region on maize production suggested that farmers should better fine-tune the dates that are more likely to enhance crop yield [41]. Similar studies for pearl millet may assist in identifying the best planting dates for achieving the best potential yield.

5. Conclusions

Rainfall in SSA can vary significantly among neighboring fields within a small area, with a distance of less than 200 m between fields, impacting crop yields. The variability in daily rainfall amounts in space is more determinant on crop yield than the total seasonal rainfall variability. The rainfall spatiotemporal variability over such a small distance can result in significant yield variability among farmers. Scattering fields can help farmers to avoid the risk of losing an entire season's harvest by enabling farmers to obtain at least some seasonal yield from locations that received sufficient rain. The use of tied ridges as an infield rain water harvesting system helps to improve yields more than flat cultivation. We recommend this technique as one of the strategies to help farmers reduce yield losses in semiarid areas.

Author Contributions: F.R.S. was involved in conceptualizing the idea, data collection and analysis, experimentation, writing method, implementation, data presentation, writing the original draft paper, and organizing co-authors views. F.G. streamlined the concept, helped in fund acquisition, revised the design of the paper, contributed to its writing, provided additional literature, supervised the whole writing process and administration of the project that financed the work. S.D.B.-K. contributed during conceptualization of the idea, funding acquisition, supported data collection by providing additional resources (moisture measuring tools), and supervision. S.D.T. contributed by revising the idea, assisted in supervising data collection, and provided the reviews to the original draft. F.C.K. contributed during data collection, revised methodology, provided the resource during field work, and supervised the field work. M.A.L. assisted with the conceptualization of the work, supervision of data collection and analysis, revised the method, and completed an overall review of the paper.

Funding: The research conducted in this study was funded the German Federal Ministry of Agriculture and Food (BMEL) in the form of the PhD program of the Federal Office of Agriculture and Food (BLE). The preparation and publication of this article is funded by the Leibniz Center for Agricultural Landscape Research (ZALF).

Acknowledgments: The authors are grateful to the agricultural field officer Fadhili Mbaga for the assistance during data collection and all famers in Idifu village. We also thank the three anonymous reviewers for their valuable comments on an earlier version of the manuscript.

Conflicts of Interest: The authors declare no conflict of interest. The funders had no role in the design of the study; in the collection, analyses, or interpretation of data; in the writing of the manuscript, or in the decision to publish the results.

Appendix A

Table A1. Variogram cloud analysis from QGIS for annual rainfall variance in SES2.

				SES1		SES2	
Distance (m)	Direction Angle (Degrees)	Difference	Variance	Covariance	Variance	Covariance
164.45	52.34	0.2	0	4915.2	1346.9	68.6
231.71	212.44	29.3	858.5	2182.3	46.9	42.9
267.29	102.11	40.8	1664.6	−43.3	262.4	165.9
292.97	175.98	11.8	139.2	161.4	324	−43.6
329.77	198.87	15.6	243.4	549.7	22.6	307.9
347.02	63.96	26.2	686.4	278.2	46.9	−11.4
374.59	122.58	35.9	1288.8	2408.7	1802	−152.7
377.54	143.34	11.5	132.3	449.1	4.8	1662.1
386.54	353.24	73.1	5343.6	−1093.8	228	15.8
397.86	203.91	16.1	259.2	2216.1	14.4	23.6
423.08	89.43	53	2809	−81.8	691.7	305.2
437.56	132.71	6.4	41	4453.1	158.8	−19.3
440.94	154.28	6.6	43.6	4465.8	2430.5	−416.3
457.18	102.79	35.7	1274.5	2401.9	33.1	−7.1
460.1	12.72	55.5	3080.3	−627.1	5700.3	−486.7
470.28	210.79	15.2	231	26.7	858.5	331.4
473.97	0.31	39.6	1568.2	955.5	22.6	162.3
477.42	211.69	93.5	8742.3	−2163.9	2672.9	−584.2
482.79	309.53	59.4	3528.4	1233.1	5.5	127.3
483.99	324.16	95.1	9044	−705.4	27	23.3
488.99	216.32	164.9	27,192	−5072.3	81	153.6
495.36	21.24	16.1	259.2	603.9	3058.1	−760.3

The positive and negative covariance values show that rainfall may increase or decrease with distance in either direction. Distance, randomly chosen distances (m) between two rain gauges, which are compared within a 0–500 m range.

Appendix B

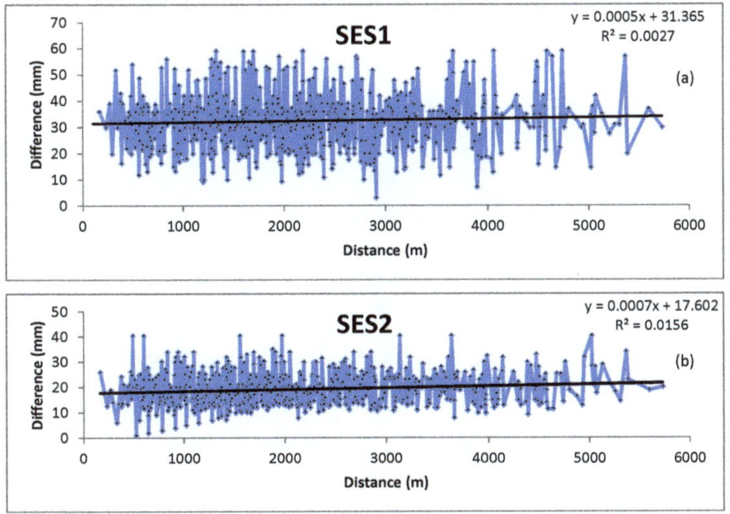

Figure A1. Rainfall variation in space with distance for (**a**) SES1 and (**b**) SES2.

References

1. Granados, R.; Soria, J.; Cortina, M. Rainfall variability, rainfed agriculture and degree of human marginality in North Guanajuato, Mexico. *Singap. J. Trop. Geogr.* **2017**, *38*, 153–166. [CrossRef]
2. Bayu, W.; Rethman, N.F.G.; Hammes, P.S. Effects of tied-ridge, nitrogen fertilizer and cultivar on the yield and nitrogen use efficiency of sorghum in semi-arid Ethiopia. *Arch. Agron. Soil Sci.* **2012**, *58*, 547–560. [CrossRef]
3. van Ittersum, M.K.; Cassman, K.G.; Grassini, P.; Wolf, J.; Tittonell, P.; Hochman, Z. Yield gap analysis with local to global relevance—A review. *Field Crops Res.* **2013**, *143*, 4–17. [CrossRef]
4. Hoffmann, M.P.; Haakana, M.; Asseng, S.; Höhn, J.G.; Palosuo, T.; Ruiz-Ramos, M.; Fronzek, S.; Ewert, F.; Gaiser, T.; Kassie, B.T.; et al. How does inter-annual variability of attainable yield affect the magnitude of yield gaps for wheat and maize? An analysis at ten sites. *Agric. Syst.* **2017**. [CrossRef]
5. Rockstrom, J. Making the best of climatic variability: Options for upgrading rainfed farming in water scarce regions. *Water Sci. Technol.* **2004**, *49*, 151–156. [CrossRef] [PubMed]
6. Vanacker, V.; Linderman, M.; Lupo, F.; Flasse, S.; Lambin, E. Impact of short-term rainfall fluctuation on interannual land cover change in sub-Saharan Africa. *Glob. Ecol. Biogeogr.* **2005**, *14*, 123–135. [CrossRef]
7. Cooper, P.J.M.; Dimes, J.; Rao, K.P.C.; Shapiro, B.; Shiferaw, B.; Twomlow, S. Coping better with current climatic variability in the rain-fed farming systems of sub-Saharan Africa: An essential first step in adapting to future climate change? *Agric. Ecosyst. Environ.* **2008**, *126*, 24–35. [CrossRef]
8. Waongo, M.; Laux, P.; Kunstmann, H. Adaptation to climate change: The impacts of optimized planting dates on attainable maize yields under rainfed conditions in Burkina Faso. *Agric. For. Meteorol.* **2015**, *205*, 23–39. [CrossRef]
9. Sithole, N.J.; Magwaza, L.S.; Mafongoya, P.L. Conservation agriculture and its impact on soil quality and maize yield: A South African perspective. *Soil Tillage Res.* **2016**, *162*, 55–67. [CrossRef]
10. Bekele, D.; Alamirew, T.; Kebede, A.; Zeleke, G.; Melese, A.M. Analysis of rainfall trend and variability for agricultural water management in Awash River Basin, Ethiopia. *J. Water Clim. Chang.* **2017**, *8*, 127–141. [CrossRef]
11. Wang, X.C.; Qadir, M.; Rasul, F.; Yang, G.T.; Hu, Y.G. Response of Soil Water and Wheat Yield to Rainfall and Temperature Change on the Loess Plateau, China. *Agronomy* **2018**, *8*, 101. [CrossRef]
12. Graef, F.; Haigis, J. Spatial and temporal rainfall variability in the Sahel and its effects on farmers' management strategies. *J. Arid Environ.* **2001**, *48*, 221–231. [CrossRef]
13. Satyanarayana, P.; Srinivas, V.V. Regional frequency analysis of precipitation using large-scale atmospheric variables. *J. Geophys. Res.* **2008**, *113*. [CrossRef]
14. Batisani, N.; Yarnal, B. Rainfall variability and trends in semi-arid Botswana: Implications for climate change adaptation policy. *Appl. Geogr.* **2010**, *30*, 483–489. [CrossRef]
15. Nasseri, M.; Zahraie, B. Application of simple clustering on space-time mapping of mean monthly rainfall pattern. *Int. J. Climatol.* **2011**, *31*, 732–741. [CrossRef]
16. Liu, Y.; Liu, L. Rainfall feature extraction using cluster analysis and its application on displacement prediction for a cleavage-parallel landslide in the Three-Gorges Reservoir area. *Nat. Hazards Earth Syst. Sci. Discuss.* **2016**, 1–15. [CrossRef]
17. Silungwe, F.R.; Graef, F.; Bellingrath-Kimura, S.D.; Tumbo, S.D.; Kahimba, F.C.; Lana, M.A. Crop Upgrading Strategies and Modelling for Rainfed Cereals in a Semi-Arid Climate—A Review. *Water* **2018**, *10*, 356. [CrossRef]
18. Munishi, P.K.T. *Analysis of Climate Change and its Impacts on Productive Sectors, Particularly Agriculture in Tanzania*; Sokoine University of Agriculture: Morogoro, Tanzania, 2009; p. 72.
19. Kassile, T. Trend Analysis of Monthly Rainfall Data in Central Zone. *J. Math. Stat.* **2013**, *9*, 1–11. [CrossRef]
20. Mkonda, M.Y.; He, X.H. Are Rainfall and Temperature Really Changing? Farmer's Perceptions, Meteorological Data, and Policy Implications in the Tanzanian Semi-Arid Zone. *Sustainability* **2017**, *9*, 1412. [CrossRef]
21. Maestrini, B.; Basso, B. Drivers of within-field spatial and temporal variability of crop yield across the US Midwest. *Sci. Rep.* **2018**, *8*, 14833. [CrossRef]

22. Falconnier, G.N.; Descheemaeker, K.; Van Mourik, T.A.; Giller, K.E. Unravelling the causes of variability in crop yields and treatment responses for better tailoring of options for sustainable intensification in southern Mali. *Field Crops Res.* **2016**, *187*, 113–126. [CrossRef]

23. Kahimba, F.C.; Mbaga, S.; Mkoko, B.; Swai, E.; Kimaro, A.A.; Mpanda, M.; Liingilie, A.; Germer, J. *Analysing the Current Situation Regarding Biophysical Conditions and Rainfed Crop, Livestock- and Agroforestry Systems: A Baseline Report*; 031A249A; Trans-SEC Consortium: Müncheberg, Germany, 2015; p. 42.

24. URT. *National Sample Census of Agriculture 2007/2008 Small Holder Agriculture: Regional Report—Dodoma Region "[Volume Va]," United Republic of Tanzania*; Regional Administration and Local Governments: Dodoma, Tanzania, 2012; p. 319.

25. Li, Y.; Wu, H. A Clustering Method Based on K-Means Algorithm. *Phys. Procedia* **2012**, *25*, 1104–1109. [CrossRef]

26. Reinhardt, N.; Herrmann, L. Fusion of indigenous knowledge and gamma spectrometry for soil mapping to support knowledge-based extension in Tanzania. *Food Secur.* **2017**, *9*, 1271–1284. [CrossRef]

27. Chan, Y.; Walmsley, R.P. Learning and Understanding the Kruskal-Wallis One-Way Analysis of Variance-by-Ranks Test for Differences Among Three or More Independent Groups. *Phys. Ther.* **1997**, *77*, 1755–1762. [CrossRef] [PubMed]

28. Sucozhañay, A.; Célleri, R. Impact of Rain Gauges Distribution on the Runoff Simulation of a Small Mountain Catchment in Southern Ecuador. *Water* **2018**, *10*, 1169.

29. Buytaert, W.; Celleri, R.; Willems, P.; De Bievre, B.; Wyseure, G. Spatial and temporal rainfall variability in mountainous areas: A case study from the south Ecuadorian Andes. *J. Hydrol.* **2006**, *329*, 413–421. [CrossRef]

30. Gao, R.Z.; Li, F.L.; Wang, X.X.; Liu, T.X.; Du, D.D.; Bai, Y. Spatiotemporal variations in precipitation across the Chinese Mongolian plateau over the past half century. *Atmos. Res.* **2017**, *193*, 204–215. [CrossRef]

31. Akponikpe, P.B.; Minet, J.; Gérard, B.; Defourny, P.; Bielders, C.L. Spatial fields' dispersion as a farmer strategy to reduce agro-climatic risk at the household level in pearl millet-based systems in the Sahel: A modeling perspective. *Agric. For. Meteorol.* **2011**, *151*, 215–227. [CrossRef]

32. Bagayoko, M.; Maman, N.; Palé, S.; Sirifi, S.; Taonda, S.; Traore, S.; Mason, S. Microdose and N and P fertilizer application rates for pearl millet in West Africa. *Afr. J. Agric. Res.* **2011**, *6*, 1141–1150.

33. Stein, A.; Brouwer, J.; Bouma, J. Methods for comparing spatial variability patterns of millet yield and soil data. *Soil Sci. Soc. Am. J.* **1997**, *61*, 861–870. [CrossRef]

34. Usowicz, B.; Lipiec, J. Spatial variability of soil properties and cereal yield in a cultivated field on sandy soil. *Soil Tillage Res.* **2017**, *174*, 241–250. [CrossRef]

35. Miller, M.P.; Singer, M.J.; Nielsen, D.R. Spatial variability of wheat yield and soil properties on complex hills. *Soil Sci. Soc. Am. J.* **1988**, *52*, 1133–1141. [CrossRef]

36. Cox, M.; Gerard, P.; Wardlaw, M.; Abshire, M. Variability of selected soil properties and their relationships with soybean yield. *Soil Sci. Soc. Am. J.* **2003**, *67*, 1296–1302. [CrossRef]

37. Belay, A.; Gebrekidan, H.; Uloro, Y. Effect of tied ridges on grain yield response of Maize (*Zea mays* L.) to application of crop residue and residual N and P on two soil types at Alemaya, Ethiopia. *S. Afr. J. Plant Soil* **1998**, *15*, 123–129. [CrossRef]

38. Wright, J.P.; Posner, J.L.; Doll, J.D. The effect of tied ridge cultivation on the yield of maize and a maize cowpea relay in the gambia. *Exp. Agric.* **1991**, *27*, 269–279. [CrossRef]

39. Mesfin, T.; Tesfahunegn, G.B.; Wortmann, C.S.; Nikus, O.; Mamo, M. Tied-ridging and fertilizer use for sorghum production in semi-arid Ethiopia. *Nutr. Cycl. Agroecosyst.* **2009**, *85*, 87–94. [CrossRef]

40. Kiboi, M.N.; Ngetich, K.F.; Diels, J.; Mucheru-Muna, M.; Mugwe, J.; Mugendi, D.N. Minimum tillage, tied ridging and mulching for better maize yield and yield stability in the Central Highlands of Kenya. *Soil Tillage Res.* **2017**, *170*, 157–166. [CrossRef]

41. Lana, M.A.; Vasconcelos, A.C.F.; Gornott, C.; Schaffert, A.; Bonatti, M.; Volk, J.; Graef, F.; Kersebaum, K.C.; Sieber, S. Is dry soil planting an adaptation strategy for maize cultivation in semi-arid Tanzania? *Food Secur.* **2018**, *10*, 897–910. [CrossRef]

Article

Simulation of Crop Growth and Water-Saving Irrigation Scenarios for Lettuce: A Monsoon-Climate Case Study in Kampong Chhnang, Cambodia

Pinnara Ket [1,2,*] , Sarah Garré [3], Chantha Oeurng [1], Lyda Hok [4] and Aurore Degré [2]

1   Faculty of Hydrology and Water Resources Engineering, Institute of Technology of Cambodia, Russian Federation Bd, P.O. Box 86, Phnom Penh 12156, Cambodia; oeurng_chantha@yahoo.com
2   BIOSE, Gembloux Agro-Bio Tech, Liège University, Passage des Déportés 2, Gembloux 5030, Belgium; aurore.degre@uliege.be
3   TERRA, Gembloux Agro-Bio Tech, Liège University, Passage des Déportés 2, Gembloux 5030, Belgium; sarah.garre@uliege.be
4   Department of Soil Science, Faculty of Agronomy, Royal University of Agriculture, P.O. Box 2696, Phnom Penh 12401, Cambodia; hoklyda@rua.edu.kh
*   Correspondence: ket.pinnara@gmail.com; Tel.: +855-(0)-78-900-477

Received: 19 March 2018; Accepted: 16 May 2018; Published: 21 May 2018

Abstract: Setting up water-saving irrigation strategies is a major challenge farmers face, in order to adapt to climate change and to improve water-use efficiency in crop productions. Currently, the production of vegetables, such as lettuce, poses a greater challenge in managing effective water irrigation, due to their sensitivity to water shortage. Crop growth models, such as AquaCrop, play an important role in exploring and providing effective irrigation strategies under various environmental conditions. The objectives of this study were (i) to parameterise the AquaCrop model for lettuce (*Lactuca sativa* var. *crispa L.*) using data from farmers' fields in Cambodia, and (ii) to assess the impact of two distinct full and deficit irrigation scenarios in silico, using AquaCrop, under two contrasting soil types in the Cambodian climate. Field observations of biomass and canopy cover during the growing season of 2017 were used to adjust the crop growth parameters of the model. The results confirmed the ability of AquaCrop to correctly simulate lettuce growth. The irrigation scenario analysis suggested that deficit irrigation is a "silver bullet" water saving strategy that can save 20–60% of water compared to full irrigation scenarios in the conditions of this study.

Keywords: crop growth; lettuce; AquaCrop; water saving; water productivity; deficit irrigation

---

1. Introduction

Humanity's environmental footprint is unsustainable within the Earth's limited natural resources and assimilative capacity [1]. Climate change and growth in the global population are increasing pressure on these scarce environmental resources, notably water [2–4]. Particularly, increasing relative evapotranspiration from flow regulation and irrigation over the past century raises the global human water consumption and footprint [5]. Improving food production with less water and benchmarking efficiency of resource use is therefore a great challenge of our time, and urgently needed to ensure food security [1,6,7].

Cambodia is considered to be the country most vulnerable to climate change in Southeast Asia [8]. In recent decades, extreme events, such as floods and droughts, have negatively affected the livelihoods of farmers, especially in terms of the loss of crop production [9]. Cambodian farmers are generally conscious of these changes and challenges [9]. Guidelines for agricultural adaptation to improve crop productivity and the sustainability of the farming system and to minimise vulnerability to climate change, are therefore crucial [8,10]. Currently, the production of vegetables, like lettuce, poses more challenges in term of managing irrigation water efficiently, due to the crop's sensitivity to water shortage [11–13]. Lettuce, the most widely consumed leaf vegetable, is also one of the most widely cultivated vegetables in the world [14]. It is also an important to local vegetable production in Cambodia [15,16]. Improving strategies for vegetable farming productivity, including lettuce, for Cambodian farmers, is being increasingly considered [17].

Many irrigation strategies have been investigated for improving irrigation water productivity (IWP) during recent decades, with IWP defined as the ratio of agricultural output to the amount of irrigation water use [18]. Full irrigation via water application with the crop evapotranspiration requirements (ETc) method is an effective irrigation practice for crop production [19–22]. In traditional irrigation scheduling, a technique to meet full irrigation, as well, the soil moisture in the root zone is allowed to fluctuate between an upper limit approximating "field capacity" and the lower limit of the readily accessible water (RAW), referred to as "the threshold", somewhat above where a crop begins to experience water stress [23,24]. These methods have been applied to improve crop water productivity in various regions of the world, including Asian regions [25–30]. Nevertheless, deficit irrigation, as an adaptation strategy for regions with limited water resources or prone to drought, has been proven to be worth considering [31,32].

Deficit irrigation is an irrigation practice whereby a crop is irrigated with an amount of water below the full requirement for optimal plant growth, thereby saving water and minimising the economic impact on the harvest [18,19]. By limiting water applications to drought-sensitive growth stages such as, the vegetative stages and the late ripening period, the aims of this approach is to maximise water productivity and to stabilise, rather than maximise yields [33]. Water deficit can be defined at five levels: severe deficit (with soil moisture (SM) less than 50% of field capacity (FC)), moderate deficit (SM < 50–60% of FC), mild deficit (SM < 60–70% of FC), no deficit or full irrigation (SM > 70% of FC), and overirrigation (application above water requirements) [34]. Crops under deficit irrigation will experience some level of water stress, and often have lower yields than fully irrigated plants [35]. Deficit irrigation can allow irrigation water savings of up to 20–40% at yield reductions below 10% [36], and has been widely investigated in dry regions [36]. Deficit irrigation can be based on applying irrigation water under crop evapotranspiration. Patanè et al. [37] found that deficit irrigation at 50% of ETc for tomato plants resulted in no biomass (B) loss and high irrigation water-use efficiency. Experimental results obtained by Abd El-Wahed et al. [38] suggested that deficit irrigation at 85% of ETc is favourable to save 15% of water provided, with no reduction in the bean crop. The study results of Samperio et al. [39] offered deficit irrigation at 20% and 60% of ETc during stage II and postharvest, respectively, to "Angeleno" Japanese plum as a water-saving strategy, without negatively affecting crop yield. Results from Yang et al. [40] confirmed that the yield loss for cotton was less than 10% under deficit irrigation of 70% of ETc and 85% of ETc. Meanwhile, crop sensitivity to water deficit can be affected by many factors, including climatic conditions, crop species and cultivars, and agronomic management practices, amongst others [34]. Payero et al. [41] suggested that deficit irrigation is not a good strategy for improving the crop water productivity of maize in a semi-arid climate. A study on deficit irrigation treatment on lettuce showed that water stress caused by deficit irrigation at 20% and 40% of ETc significantly reduced leaf number, leaf area index, and dry matter accumulation [42]. Final fresh weight was reduced by 20% to 30% when compared with full irrigation. Kuslu et al. [43] concluded that for lettuce grown in semi-arid regions, full irrigation should be used under no water shortage, and deficit irrigation by 60% of ETc could be used for 40% water saving with a 35.8% yield loss where irrigation water supplies are limited.

Elaborating irrigation strategies merely on the basis of field research is difficult and time consuming [44]. Crop models are effective decision-support tools to investigate irrigation scenarios and to develop improved irrigation strategies [7,45,46]. They can provide a rapid and reasonable accurate prediction of the response of agriculture over a range of environmental conditions [47]. The model AquaCrop, developed by the Food and Agricultural Organisation of the United Nations

(FAO), is a water-driven crop model that simulates daily crop growth (e.g., canopy cover and biomass production) and final crop yield, with a balance between accuracy, simplicity, and robustness in incorporating various agronomy practices [48,49]. It is considered as a valuable tool for improving irrigation water productivity in crop production planning [6,50]. AquaCrop has been calibrated and parameterised to various crops under various environmental and irrigation conditions, including barley [51], soybean [52], sunflower [53], cotton [54,55], corn [56], sugar beet [57], wheat [58,59], potato [60,61], cabbage [62], and rice [63]. However, this has not yet been done in the case of lettuce. Most of these studies proved that the model is capable of accurately simulating crop growth and yield. However, some case studies still report some flaws in simulation of crop evolution and yield, especially under severe deficit irrigation and heat stress conditions. Adeboye et al. [64] found that biomass of soybean simulated by AquaCrop was overestimated under deficit irrigation conditions. Zeleke et al. [65] found that AquaCrop simulated the canopy cover and biomass growth of canola well, but the model was less satisfactory under severe water stress conditions in a semi-arid region. Similarly, a reduction in model reliability in biomass and canopy cover prediction for maize under the severe stress conditions of deficit irrigation in a tropical environment was indicated in a study of Greaves et al. [66]. AquaCrop performed well in biomass simulation of potato in the experiment under deficit irrigation at 120, 100, 80, and 60% of ETc [67]. However, the potato yield simulation was overestimated due to the heat stress, with the authors suggesting the incorporation of a temperature stress coefficient into AquaCrop when a crop is affected by high temperatures. Further research is therefore required to improve the performance of AquaCrop. Furthermore, its performance simulating lettuce growth in Cambodian conditions has not yet been tested. The main objective of this study is to improve the water productivity of lettuce under limited irrigations in the Cambodian climate. More specific objectives are (i) to parameterise the crop model AquaCrop using data from farmer fields, since lettuce is not yet available in the AquaCrop catalogue; and (ii) to assess the impact of water-saving scenarios in full and deficit irrigation in silico using this calibrated model.

2. Materials and Methods

2.1. Experimental Sites

The field experiments were conducted with lettuce plants (*Lactuca sativa* var. *crispa* L.) which are widely used in the study area, during a period from August to September 2017 in two experimental sites located in the villages of Chea Rov (site S1) (104°38′54.442″ E 12°9′15.482″ N) and Ou Roung (site S2) (104°37′16.24″ E 12°11′52.518″ N) in the province of Kampong Chhnang, Cambodia (Figure 1).

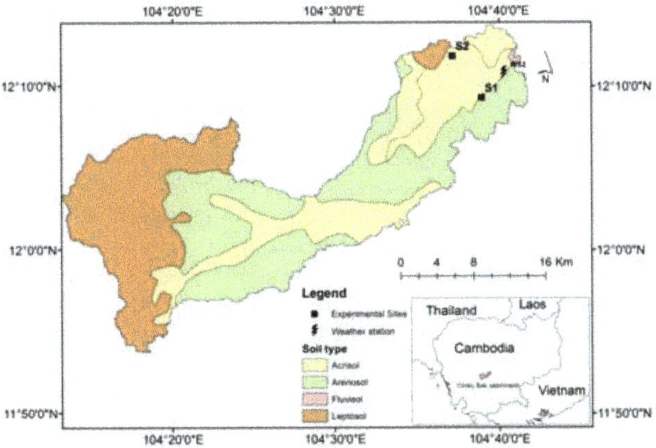

Figure 1. Experimental sites at Chearov (S1) and Ou Roung (S2), located in the Chrey Bak Catchment.

The total land area of the plots was 400 m^2. Lettuce seeds were sown in standard trays (with 123 holes). After 15 days, seedlings were transplanted into raised bed rows (0.30 m in height and with bed tops 0.50 cm wide) and covered with plastic mulch with a planting density of 12 plants m^{-2}. The compost was basally applied at the rate of 20 ton ha^{-1} before transplantation.

Irrigation was carried out using a drip system, with emitters of constructor maximum discharge of 3 L h^{-1} spaced 0.10 m apart. A plastic cover was used to protect the crops from heavy rainfall. Nevertheless, due to the intense rain which flowed between the crop rows, water ponding at 20 cm below the top bed row level was observed between the lettuce rows at both sites during almost the entire growing period. This ponding kept the soil wet during the growing period, and had to be factored into the calibration of the lettuce growing curve. At site S2, irrigation was not applied after a week after planting, due to the benefit of water ponding. At site S1, even though there was also water ponding in the field, the irrigation was applied every other day. The irrigation was determined by checking soil moisture (SM) using the feel and appearance method of Klocke et al. [68]. The irrigation was done when the SM was depleted below field capacity in the root zone at 5 cm, as lettuce have a root depth between 5–10 cm.

2.2. Data Collection and Measurement

2.2.1. Climate Data

Weather data for the experimental sites were collected from a local meteorological station (104°40′21.767″ E; 12°10′45.965″ N) (Figure 1). Daily maximum and minimum temperature, relative humidity, wind speed, rainfall, and solar radiation were recorded automatically at a five minute time step. The daily reference evapotranspiration (ETo) for the growing season, used as input data in AquaCrop, was calculated using the ETo calculator based on the FAO's Penman–Monteith method [69] (Figure 2).

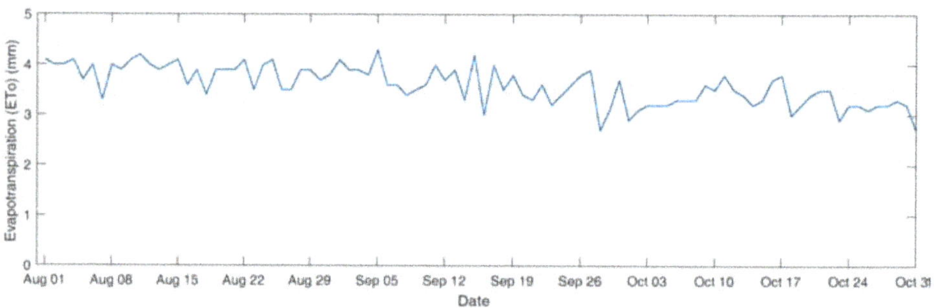

Figure 2. Daily potential evapotranspiration (ETo) during the growing season 2017.

2.2.2. Soil Data

The physical and chemical soil characteristics which were measured are listed in Tables 1 and 2. The soil texture was measured using the pipette method [70]. The bulk density was measured by the core method [71]. Field capacity, wilting point, and saturated hydraulic conductivity were derived from tension and soil moisture monitoring, using MPS2 and 10HS probes and using inverse modelling as presented in [72].

Table 1. Measured physical soil characteristics.

Parameters	Experimental Sites	
	Chea Rov (S1)	**Ou Roung (S2)**
Texture	Sand	Loam
Clay (%)	4.39	7.80
Silt (%)	9.56	41.15
Sand (%)	86.03	51.04
Bulk density (g cm^{-3})	1.5	1.5
Field capacity (m^3 m^{-3}) (sand: at -10 kPa, Loam: at -33 kPa)	0.11	0.14
Wilting point (m^3 m^{-3}) (at 150 kPa)	0.05	0.06
Soil saturation (m^3 m^{-3})	0.27	0.43
Available water content (AWC) (mm m^{-1})	62.48	81.43

Table 2. Measured chemical soil characteristics.

Site	Sampling Time	pH-H$_2$O	EC (uS cm^{-1})	OM (%)	N (%)	P (ppm)	K (meg 100 g^{-1})	CEC (cmol kg^{-1})
S1	Before transplanting	6.28	108	20.31	0.098	13.29	0.77	2.80
	At harvest	6.84	97.4	20.85	0.126	17.08	0.4	4.40
S2	Before transplanting	6.7	223	19.51	0.238	24.07	2.31	7.60
	At harvest	6.8	218	19.78	0.126	15.91	1.45	5.40

Note: EC is electrical conductivity; OM is organic matter content; N is total nitrogen; P is available phosphorous; K is exchangeable potassium; CEC is cation exchange capacity.

2.2.3. Crop Data

Canopy cover was measured at three-day intervals during the growing stage. Four pictures of 1 m^2 were taken randomly using a digital compact camera (Nikon Coolpix p600, Tokyo, Japan) at a fixed height of 1 m above ground level. The canopy cover was analysed using image processing with ImageJ® software (https://imagej.nih.gov). Aboveground dry biomass was determined by harvesting 10 heads at the surface level of each site, oven-drying plant samples at 70 °C for 48 h, and weighing them [73].

2.3. AquaCrop Model

The AquaCrop model is a crop water-driven productivity model developed by the FAO in 2009. A detailed description is presented in [49]. Water is the key limiting factor for crop production in this model [74]. Inputs for the AquaCrop model consist of weather data, crop, and soil characteristics (soil profile and groundwater), and field management practice or irrigation management practices [49].

Canopy cover is a crucial feature of AquaCrop [49]. Under unstressed condition, the exponential growth equation to simulate canopy development for the vegetative stage is

$$CC = CC_o e^{CGC \times t} \tag{1}$$

where CC is the canopy cover at time t and is expressed as fraction of ground covered, CC_o is initial canopy cover size (at t = 0) as a fraction (%), and CGC is the canopy growth coefficient in fraction per growing degree day (GDD), a constant for a crop under optimal conditions, but modulated by stresses.

In the condition of water stress, the CGC is multiplied by a water stress coefficient of expansive growth (Ks_{exp}) (Equation (2)).

$$CGC_{adj} = Ks_{exp}.CGC \tag{2}$$

where Ks_{exp} ranges from 1 to 0, canopy growth begins to slow down below the maximum rate when soil water depletion reaches the upper threshold, and stops completely when the depletion reaches the lower threshold.

Crop transpiration is proportional to the canopy cover and given by

$$Tr = Ks_{sto}Kc_{Tr}ET_o \tag{3}$$

Ks_{sto} is the stress coefficient for stomatal closure. Kc_{Tr} is the crop transpiration coefficient (determined by canopy cover and $Kc_{Tr,x}$), $Kc_{Tr,x}$ is the coefficient for maximum crop transpiration, and ET_o is reference evapotranspiration (mm).

Biomass production is computed from crop transpiration and crop water productivity normalised for ET_o and CO_2 (Equation (4)). The extreme effect of low temperature on crop phenology, biomass accumulation, and harvest index, is considered with adjustment factors [67,75].

$$B = K_{s_b}.f_{WP}.WP^*.\frac{Tr}{ETo} \tag{4}$$

where B is biomass, Tr is crop transpiration (mm day^{-1}), ETo is reference evapotranspiration (mm day^{-1}), and K_{s_b} is the stress coefficient for low-temperature effects on biomass production. f_{WP} is the adjustment factor to account for differences, if any exist, in the chemical composition of the vegetative biomass and harvestable organs. WP^* is normalised crop water productivity, defined as the ratio of biomass produced to water transpired, normalised for the evaporative demand and CO_2 concentration of the atmosphere.

The AquaCrop stress indicators include water storage (not enough water), waterlogging (too much water), air temperature (too high or too low), and soil salinity stress (too high).

2.4. Model Parameterisation

The process of parameterisation is illustrated in Figure 3. The vegetative stage of lettuce refers to the growing period of lettuce growth after germination until harvest. A growing period during the vegetative stage of 59 days after transplanting was simulated in this study.

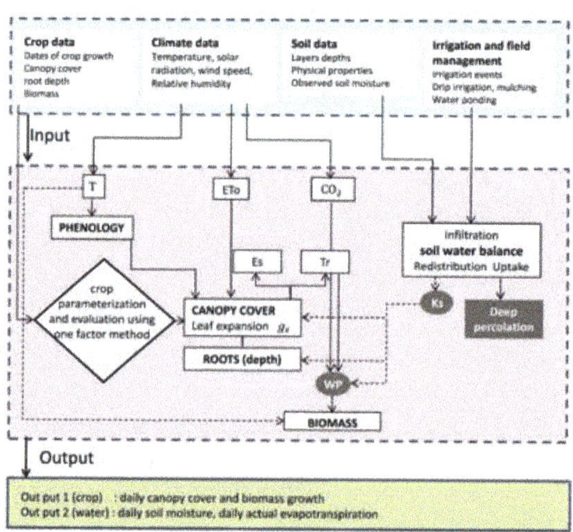

Figure 3. Flow chart of parameterisation of AquaCrop in this study (adjusted from [76]). T is temperature, ETo is potential evapotranspiration, gs is stomatal conductance, WP is water productivity coefficient, Ks is stress coefficient, Es is soil temperature, Tr is crop transpiration.

As lettuce is a crop which is not yet parameterised in AquaCrop, calibration of the model involved adjusting the model parameters to make them match the observed data [54,77].

The primary variables of lettuce growth, e.g., canopy cover and aboveground biomass were parameterised. For the calibration of the curves, the measured data in two experimental fields at the Chearov site (S1) (having sand soil) and Ourong site (S2) (with loam soil) were used, during the growing season in 2017. The AquaCrop model does not allow the use of observed data to build the canopy cover and biomass curves, but allows the data to be used to calibrate the canopy cover and biomass curves [78].

Canopy cover curves are a plot of the development of leaf expansion response to growing time per day, based on Equation (1). Biomass curves are a relationship plot of the growth of lettuce biomass response to growing time per day, based on Equation (4). The calibration of simulated canopy and biomass curves is based on one-at-a-time (OAT) methods (i.e., changing one parameter at a time while holding others constant) [79] and adjusting the parameters by trial and error, by comparing simulated and observed field data, and minimising the function of root mean square error.

We parameterised the canopy cover curve, which is important to the model for transpiration and evaporation [78]. The main parameters of Equation (1), e.g., CCo and CGC for canopy cover curve determination, were adjusted to match the observed canopy cover data. In addition, adjusting the maximum canopy cover (CCx), time to reach maximum canopy cover, and time to recover, is crucial in order to obtain correct simulations of canopy cover growth. Subsequently, the focus was on adjusting the biomass curve of Equation (4). WP* and $Kc_{Tr,x}$ (coefficient for maximum crop transpiration) are the main parameters for regulating biomass curves in AquaCrop [74]. As lettuce is a C3 crop type [80], the recommended values for WP* lie between 15 and 20 g m^{-2}. All calibrated crop parameters are shown in Table 3.

Table 3. Calibrated parameters of lettuce growth.

No	Calibration Step	Calibrated Parameters
1	Canopy cover calibration	Time to recover of transplant, Time to reach the maximum canopy cover, Initial canopy cover (CCo), Canopy growth coefficient (CGC), Maximum canopy cover growth coefficient (CCx)
2	Biomass calibration	Coefficient for maximum crop transpiration ($Kc_{Tr,x}$), Normalised biomass water productivity (WP*)

The model performance for canopy cover and biomass simulation was evaluated using statistic indicators, including root mean square error (RMSE), Nash–Sutcliffe coefficient (N), and coefficient of determination (R^2), defined as below.

$$\text{RMSE} = \sqrt{\frac{\sum_{i=1}^{n}(O_i - S_i)^2}{n}} \tag{5}$$

$$N = 1 - \frac{\sum_{i=1}^{n}(O_i - S_i)^2}{\sum_{i=1}^{n}(O_i - \overline{O})^2} \tag{6}$$

$$R^2 = \left(\frac{\sum_{i=1}^{n}(O_i - \overline{O})(O_i - \overline{S})}{\sqrt{\sum_{i=1}^{n}(O_i - \overline{O})^2 \sum_{i=1}^{n}(O_i - \overline{S})^2}} \right)^2 \tag{7}$$

where O and S are the observed and simulated values at time i, respectively, and n is the total amount of the data. When N and R^2 are close to 1, it is considered to be satisfactory [81]. RMSE should be close to 0.

AquaCrop requires the selection of inputs related to the irrigation method, such as sprinkler, drip, or surface. These methods determine the fraction of the soil surface made wet by irrigation [82] and the impact on irrigation efficiency [83].

Default AquaCrop settings for field management include mulching, and use an adjusted factor for the effect of mulches on soil evaporation. It varied between 0.5 for mulches derived from plant material, and 1.0 for plastic mulch [75].

The drip irrigation method with plastic mulch was applied as the input for field management in the model during the parameterisation, as this is the actual practice of the experiment in this study.

The soil water balance calculation, including soil moisture simulation in AquaCrop, is based on the storage capacity of the soil layers, described in Raes et al. [84], and previously in the BUDGET model [85].

During the experimental period, water ponding at 15 cm and 20 cm below the bed soil at site S1 and S2 respectively, which was observed during the experiment, was taken into account as a boundary condition during the parameterisation of the model. This water ponding resulted in wet soil during the growing period. The values of physical soil available data in the Section 2.2.2 were adopted to simulate soil moisture in this study.

. It was noted that the plantation experiment was during the rainy season when irrigation was not needed. The crop parameters obtained after parameterisation are important for the investigation of the irrigation scenarios for water saving when irrigation is necessary, especially during the dry season.

2.5. Irrigation Scenarios

In the current study, AquaCrop was used to simulate the full and deficit irrigation scenarios described below (and in Table 4), in order to identify the optimal water use efficiency for lettuce.

Table 4. Irrigation Scenarios.

Scenario Code		Short Description
S1 (Sand)	**S2 (Loam)**	
Varied readily available water (RAW) threshold irrigation scenarios		
S0RAW	L0RAW	irrigate at 0% of RAW and refill to field capacity (FC)
S50RAW	L50RAW	irrigate at 50% of RAW and refill to FC
S80RAW	L80RAW	irrigate at 80% of RAW and refill to FC
S100RAW	L100RAW	irrigate at 100% of RAW and refill to FC
S120RAW	L120RAW	irrigate at 120% of RAW and refill to FC
S130RAW	L130RAW	irrigate at 130% of RAW and refill to FC
S150RAW	L150RAW	irrigate at 150% of RAW and refill to FC
S180RAW	L180RAW	irrigate at 180% of RAW and refill to FC
S200RAW	L200RAW	irrigate at 200% of RAW and refill to FC
Varied field capacity threshold irrigation scenarios		
S100FC	L100FC	full irrigation-daily irrigation at 100% of field capacity (FC)
S70FC	L70FC	deficit irrigation at 70% of FC
S60FC	L60FC	deficit irrigation at 60% of FC
S50FC	L50FC	deficit irrigation at 50% of FC
S40FC	L40FC	deficit irrigation at 40% of FC

2.5.1. Varied RAW Threshold Irrigation Scenarios

Figure 4 presents the calculation process of varied RAW threshold irrigation scenarios.

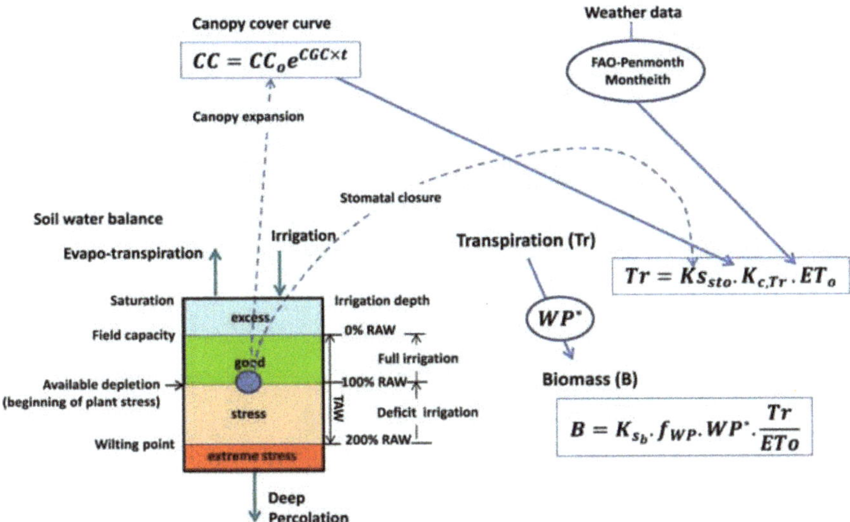

Figure 4. Schematic representation of the crop response to varied RAW threshold irrigation scenarios simulated by AquaCrop (adjusted from [77]). RAW is readily available water content, TAW is total available water content, CC is the simulated canopy cover, CC_o is initial canopy cover size, CGC is canopy growth coefficient in fraction per growing degree day (GDD), Ks_{sto} is the water stress for stomatal closure, Kc_{Tr} is the crop transpiration coefficient (determined by CC and $Kc_{Tr,x}$ at maximum canopy cover), ETo is the reference evapotranspiration, Ks_b is the stress coefficient for low-temperature effects on biomass production, f_{WP} is the adjustment factor to account for differences in chemical composition of the vegetative biomass and harvestable organs, WP* is the normalised water productivity.

These irrigation scenarios applied irrigation scheduling based on soil moisture depletion [86] by applying readily available water depletion in the default option in AquaCrop. The time and irrigation dose were calculated with the criteria below:

1. Soil water content depleted until a fixed lower threshold (RAW) and refill to field capacity (time criteria).
2. Irrigation dose can be determined by the following Equation (8) [87].

$$ID = AD \times RAW \tag{8}$$

where ID is irrigation depth (mm), $RAW = p\ TAW = p\ 1000(FC - PWP)Z_r$, p is soil water depletion threshold, set to 0.3 for lettuce recommended by [69], and Z_r is root depth (m). TAW is the amount of water that a crop can extract from its root zone [88]. FC is field capacity, that is, the amount of water well-drained soil should hold against gravitational forces ($m^3\ m^{-3}$) [88]. PWP is permanent wilting point, referring to soil water content when a plant fails to recover its turgidity on watering ($m^3\ m^{-3}$) [88]. RAW is readily available soil water, referring to the fraction of TAW that a crop can extract from the root zone without suffering water stress [88]. AD is allowable depletion, defined as the percentage of RAW that can be depleted before irrigation water has to be applied.

Full irrigation scenarios with varied RAW thresholds were simulated by selecting allowable depletion levels at 0, 50, 80, 100% in AquaCrop, that avoid drought stress during the growth stage [41]. The irrigation schedule is generated by selecting a so-called "time" and "depth" criterion, with "back to field capacity" and "allowable depletion", respectively. In other words, the different full irrigation scenarios result in decreasing irrigation frequency.

Deficit irrigation scenarios with varied RAW thresholds were similar to the full irrigation scenario criteria, but applied allowable depletion levels at 120, 130, 150, 180, and 200%. These levels result in drought stress during the growing stage, since soil moisture can decrease to a level below RAW before an irrigation event is triggered [41].

2.5.2. Varied Field Capacity Threshold Irrigation Scenarios

Figure 5 illustrated concept of the varied field capacity threshold irrigation scenarios.

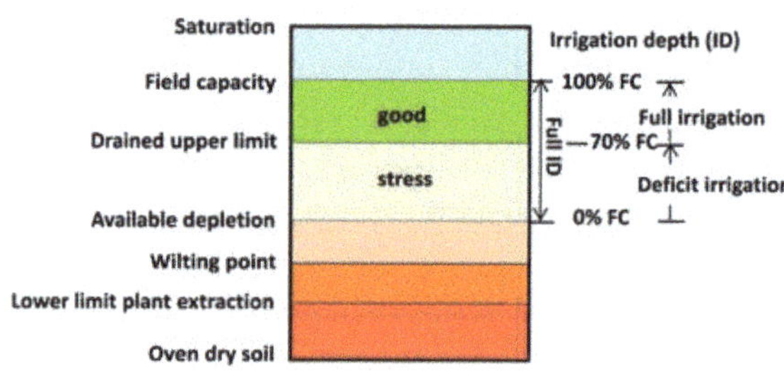

Figure 5. Schematic illustration of the soil water reservoir concepts of varied irrigation depth under field capacity irrigation scenarios (adjusted from [89]). FC is field capacity, full ID is full irrigation depth.

The full irrigation scenario, based on a fixed irrigation frequency maintained the soil moisture in the root zone at field capacity on a daily basis, since the literature claims this is the optimal status to maximise lettuce yield [90]. The irrigation schedule was generated with a fixed time interval (daily) (time criteria) and refill to field capacity (depth criteria).

Deficit irrigation scenarios with varied field capacity threshold reduce the irrigation dose below the dose at field capacity but keeping the same irrigation frequency, as in full irrigation scenario. Daily generated irrigation doses obtained in full irrigation scenario were reduced by 70, 60, 50, and 40%.

Irrigation water productivity (IWP) was used to evaluate the irrigation scenarios for efficient irrigation water use [31,91]. IWP is the ratio between the yield and the irrigation water use [31].

$$IWP = \frac{Y}{I} \tag{9}$$

where IWP is irrigation water productivity (kg m^{-3}), Y is simulated yield (kg ha^{-1}) and interest yield in this study is biomass, and I is irrigation water use (mm).

The adjusted crop parameters obtained from the parameterisation process were used in the scenario simulation under the same weather conditions, using no soil surface cover in model field management, and no ground water at bottom soil profile boundary condition.

3. Results

3.1. Plant Growth and Soil Moisture Status

Figure 6 shows both the lettuce growth measurement and simulation by AquaCrop. Biomass accumulated at a very low rate during the first two weeks of the growing season, and increased sharply in the final week. This trend accords with results obtained by Gallardo et al. [73].

The measured canopy cover and biomass yields were 34% and 0.11 ton ha^{-1}, respectively, at site S1 with sand soil, and 18.5% and 0.11 ton ha^{-1}, respectively, at site S2, which has loam soil. The measured results are comparable with Fazilah et al. [92], who found observed canopy cover of 33% and biomass yields of 0.22 ton ha^{-1} for lettuce under similar tropical conditions. Zhang et al. [93] found higher measured biomass for lettuce with a range of 0.33 to 0.63 ton ha^{-1} under lower temperatures of 20–25 °C. Thus, high day temperatures above 23 °C often limit lettuce production [94]. Optimum growth for lettuce occurs between 15–20 °C [12]. Unfavourable weather conditions, of high average temperature 33/25 °C (day/night) during the experiment, can be the reason of the low measured biomass yields for this study.

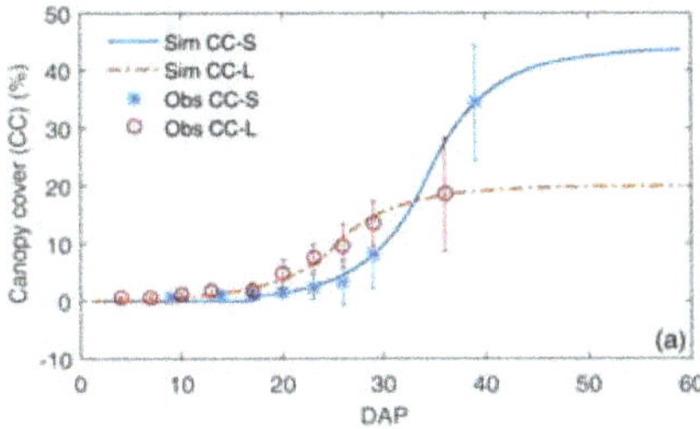
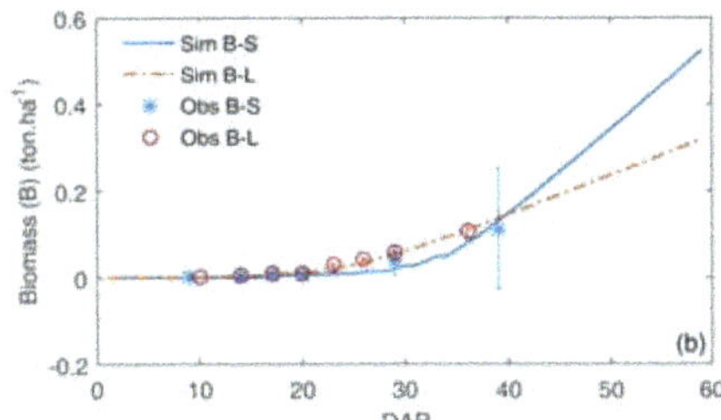

Figure 6. Observed (Obs) data and simulation (Sim) of lettuce growth of AquaCrop: (**a**) canopy cover at site S1 (CC-S) (sand soil) and site S2 (CC-L) (loam soil); (**b**) aboveground biomass at site S1 (B-S) and site S2 (B-S). The error bars were based on 10 biomass samples, except the last observed, which was based on 60 samples at harvest time. Sim CC-S is simulated canopy cover at site S1, Sim CC-L is simulated canopy cover at site S2, Obs CC-S is observed canopy cover at site S1, Obs CC-L is observed canopy cover at site S2, Sim B-S is simulated biomass at site S1, Sim B-L is simulated biomass at site S2, Obs B-S is observed biomass at site S1, Obs B-L is observed biomass at site S2.

3.2. Model Parameterisation and Evaluation

The primary crop variables calibrated for daily lettuce growth were canopy cover and biomass, with the daily soil moisture simulated by AquaCrop, by adapting available physical soil data.

Table 5 presents the adjusted model parameters for canopy cover and biomass curve simulation of lettuce growth. The time to recovery of transplant, the time to reach the maximum canopy cover, the initial canopy cover (CCo), the maximum canopy cover growth coefficient (CCx), the coefficient for maximum crop transpiration (Kc$_{\text{Tr,x}}$), and the normalised biomass water productivity (WP*) were mainly calibrated.

Table 5. AquaCrop variables parameterised.

Parameters	Symbol and Unit	Value S1 Initial	S1 Calibrated	S2 Initial	S2 Calibrated	Sources
Crop Phenology						
Time to recovered transplant (C)	(GDD)	52	280	52	147	Default
Time to maximum canopy cover (C)	(GDD)	563	859	563	727	Default
Crop Growth						
Plant density (NC)	dp (plants m^{-2})	12	-	12	-	Measure
Initial canopy cover (NC)	CCo (%)	0.72	0.84	0.5	0.6	Default
Maximum effective rooting depth	Zr (m)	0.1	-	0.1	-	Measure
Maximum canopy cover (C)	CCx (%)	34	44	18	20	Measure
Canopy growth coefficient	CGC	22.7	18.5		16.8	Default
Base temperature (C)	Tbase (°C)	4	-	4	-	[95]
Upper temperature(C)	Tupper (°C)	28	-	28	-	[96]
Canopy size of transplanted seeding (C)	CC (cm^2 plant^{-1})	6	-	5	-	Measure
Coefficient for maximum crop transpiration (NC)	Kc$_{\text{Tr,x}}$	1.25	0.65	1.25	0.5	Default
Water productivity, (C)	WP* (g m^{-2})	15	16	15	16	Default

Note: C = conservative, NC = non-conservative.

WP* was adjusted at 16 gm^{-2} for both sites, within the recommended range. Kc$_{\text{Tr,x}}$ was adjusted at 0.65 and 0.5 for site S1 and S2, respectively. These adjusted Kc$_{\text{Tr,x}}$ are lower than crop coefficient for the mid-season ($K_{cb,mid} = 1$) proposed by FAO-56. The difference between the values proposed by FAO-56 and the adjusted $Kc_{Tr,x}$ values is due to the fact that the FAO crop coefficients were obtained for specific agroclimatic conditions, which are different from the conditions of this study [78].

In addition, $Kc_{Tr,x}$ is a major requisite for estimating crop transpiration and biomass. The low adjusted value of this parameter resulted in low simulated biomass yields to fit to measured values.

High temperature stress observed during the experiment could be the reason for the low observed lettuce biomass production [12]. This observation leads to a recommendation for further development of a heat stress factor in relation to canopy cover and biomass simulations for lettuce.

The minimum root depth cannot be adjusted under 0.1 m, while the root development of lettuce was under this limit. Thus, root development in the model requires further modification [91].

The crop growth simulation of canopy cover and biomass fitted the observed data well (Figure 6). The statistical values for model evaluation in Table 6 were satisfactory, resulting in $R^2 = 0.99$, RMSE < 0.8%, N < 4.6 for canopy cover, and $R^2 > 0.98$, RMSE < 0.01 ton ha^{-1}, N < −0.07 for biomass. Thus, the model has ability to simulate well the growth of lettuce in both soil types at the two experimental sites.

Table 6. Statistical evaluation of model simulation.

Statistical Criteria	Sites	Canopy Cover (%)	Biomass (ton ha^{-1})
RMSE	S1	0.69	0.012
	S2	0.84	0.01
R^2	S1	0.99	0.98
	S2	0.99	0.99
N	S1	1.1	−0.015
	S2	4.6	−0.07

The measured and simulated soil moisture, at both soil depths of 5 and 15 cm in both sites, also matched well (Figure 7). The soil moisture simulation resulted in good accuracy with low RMSE of 0.18 and 0.14 m^3 m^{-3} at depths of 5 and 15 cm, respectively, at site S1, and 0.05 and 0.06 m^3 m^{-3} at depths of 5 and 15 cm, respectively, at site S2.

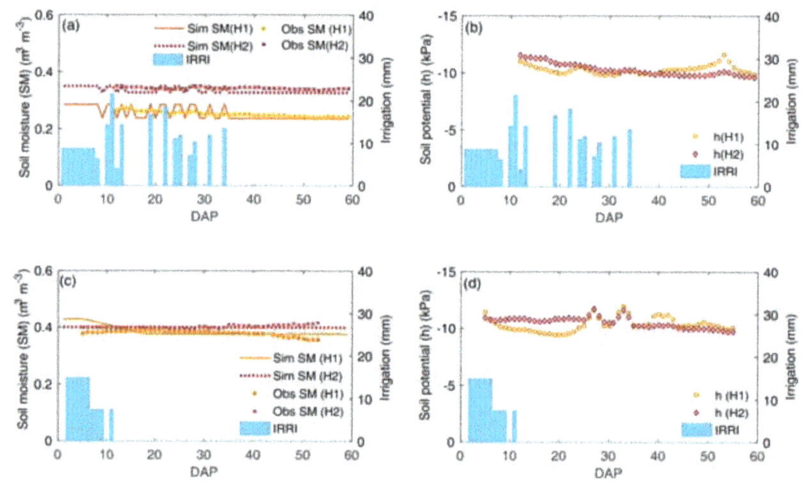

Figure 7. Simulated soil moisture and observed soil moisture data measured at depths of 5 cm (H1) and 15 cm (H2) using soil moisture sensor 10HS and soil potential MPS-2: (**a**) soil moisture at site S1; (**b**) soil potential at site S1; (**c**) soil moisture at site S2; (**d**) soil potential at site S2. DAP is day after planting, Sim SM is simulated soil moisture, Obs SM is observed soil moisture, IRRI is irrigation, h is soil potential.

3.3. Irrigation Scenarios

3.3.1. Irrigation and Soil Moisture Response

The cumulative irrigation in Figure 8, and the fluctuation of the soil moisture depletion in Figure 9, reflect the interaction between irrigation frequency and amount of water applied.

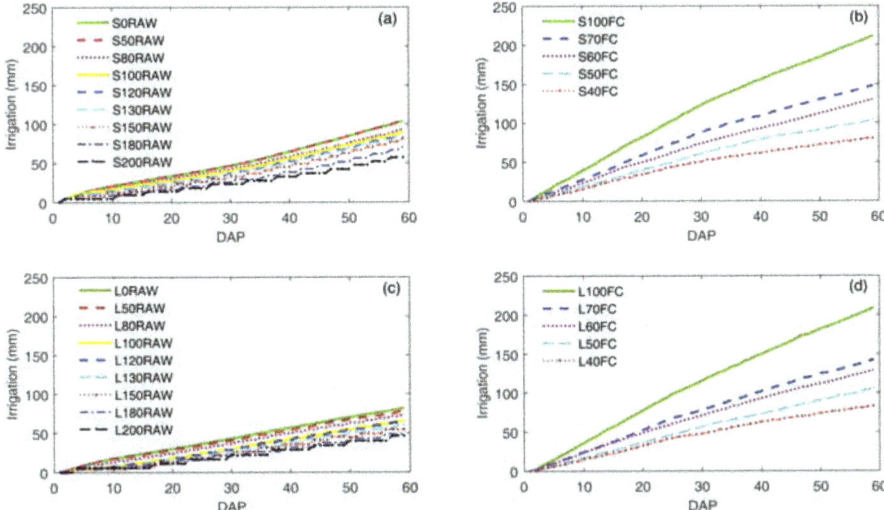

Figure 8. Irrigation accumulation response to different scenarios: (**a**) varied RAW threshold irrigation scenarios at site S1 (sand soil); (**b**) varied field capacity threshold irrigation scenarios at site S1; (**c**) varied RAW threshold irrigation scenarios at site S2 (loam soil); (**d**) varied field capacity threshold irrigation scenarios at site S2. RAW is readily available water content, S0RAW-S200RAW refers to irrigation scenarios with irrigation at 0–200% of RAW for sand soil. L0RAW-L200RAW refers to irrigation scenarios with irrigation at 0–200% of RAW for loam soil. S40FC-S100FC refers to deficit irrigation at 40–100% of field capacity for sand soil. L40FC-L100FC refers to deficit irrigation at 40–100% of field capacity for loam soil.

In both varied RAW and field capacity threshold irrigation scenarios, the irrigation frequency decreased together with decreasing the amount of water applied per irrigation event.

In varied RAW threshold irrigation scenarios, the simulation of irrigation resulted in irrigation depths which ranged from 57 to 104 mm in site S1 (sand soil) and 46–82 mm in site S2 (loam soil) (Figure 8a,c). In varied field capacity threshold irrigation scenarios, irrigation depths ranged from 81–201 mm in site S1 and 83–209 mm in site S2 (Figure 8b,d).

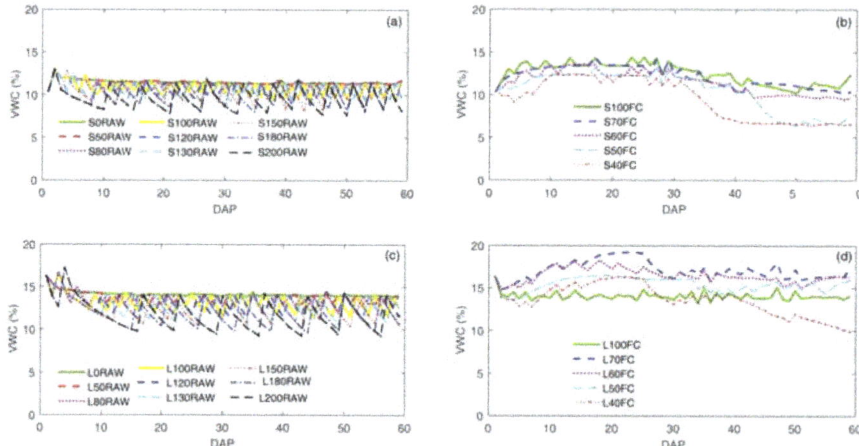

Figure 9. Daily soil moisture (VWC) response to different scenarios: (**a**) varied RAW threshold irrigation scenarios at site S1 (sand soil); (**b**) varied field capacity threshold irrigation scenarios at site S1; (**c**) varied RAW threshold irrigation scenarios at site S2 (loam soil); (**d**) varied field capacity threshold irrigation scenarios at site S2. RAW is readily available water content, S0RAW-S200RAW refers to irrigation scenarios with irrigation at 0–200% of RAW for sand soil. L0RAW-L200RAW refers to irrigation scenarios with irrigation at 0–200% of RAW for loam soil. S40FC-S100FC refers to deficit irrigation at 40–100% of field capacity for sand soil. L40FC-L100FC refers to deficit irrigation at 40–100% of field capacity for loam soil.

3.3.2. Crop Evapotranspiration and Biomass Growth Response

Figures 10 and 11 illustrate the cumulative crop evapotranspiration (ETc) and cumulative biomass of lettuce, respectively, under various irrigation scenarios simulated with AquaCrop calibrated for lettuce.

In varied RAW threshold irrigation scenarios, total simulated ETc ranged from 60 to 100 mm in site S1, and from 53 to 85 mm in site S2 (Figure 10a,c). The main reason for the higher ETc yield in site S1 is the higher adjusted transpiration characteristic of lettuce in sand soil as compared to loam soil. The simulated values of ETc fall within the range reported by Abdullah et al. [97] for lettuce, which varied from 43 mm to 285 mm in response to their different irrigation applications between 0 and 267 mm for open surface soil.

In varied field capacity threshold irrigation scenarios, simulated total crop evapotranspiration ranged from 77 to 205 mm in site S1, and from 83 to 211 mm in site S2 (Figure 10b,d). In both irrigation scenario classes, it was noted that while reducing irrigation events, crop evapotranspiration decreased simultaneously.

Figure 11 shows the response of biomass to the different irrigation scenarios. The varied RAW threshold irrigation scenarios (Figure 11a,c) resulted in biomass yield range from 0.88–1.77 ton ha^{-1} at site S1, and 0.44–0.91 ton ha^{-1} at site S2. By definition, biomass growth is closely related to crop evapotranspiration. Thus, the difference between biomass yields in the two experimental sites is due to the difference in the $Kc_{Tr,x}$ (coefficient for maximum crop transpiration) and CCx (maximum canopy cover) parameters between both sites.

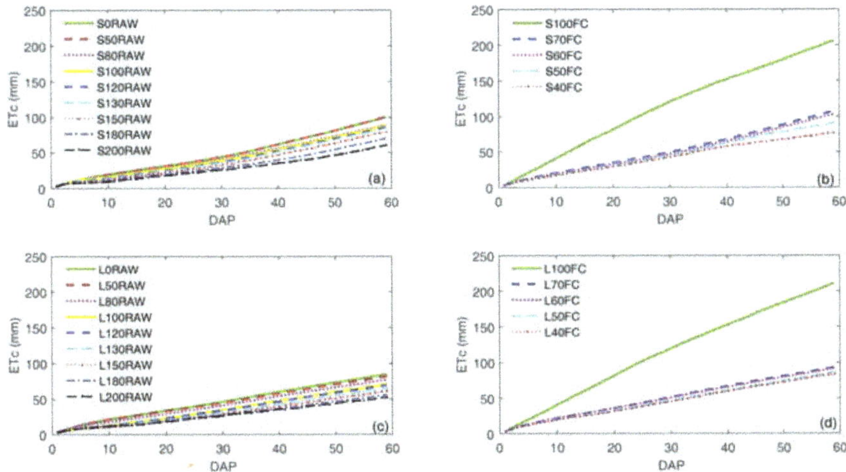

Figure 10. Crop evapotranspiration accumulation responses to different scenarios: (**a**) varied RAW threshold irrigation scenarios at site S1 (sand soil); (**b**) varied field capacity threshold irrigation scenarios at site S1; (**c**) varied RAW threshold irrigation scenarios at site S2 (loam soil); (**d**) varied field capacity threshold irrigation scenarios at site S2. RAW is readily available water content, S0RAW-S200RAW refers to irrigation scenarios with irrigation at 0–200% of RAW for sand soil. L0RAW-L200RAW refers to irrigation scenarios with irrigation at 0–200% of RAW for loam soil. S40FC-S100FC refers to deficit irrigation at 40–100% of field capacity for sand soil. L40FC-L100FC refers to deficit irrigation at 40–100% of field capacity for loam soil.

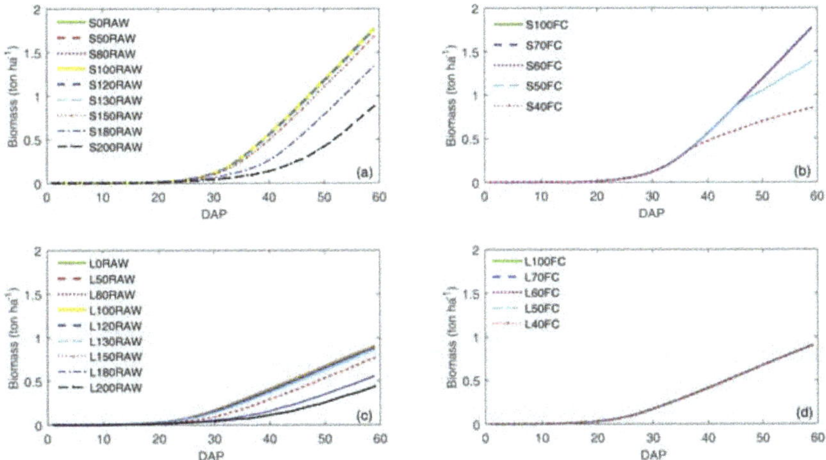

Figure 11. Biomass accumulation responses to different scenarios: (**a**) varied RAW threshold irrigation scenarios at site S1 (sand soil); (**b**) varied field capacity threshold irrigation scenarios at site S1; (**c**) varied RAW threshold irrigation scenarios at site S2 (loam soil); (**d**) varied field capacity threshold irrigation scenarios at site S2. RAW is readily available water content, S0RAW-S200RAW refers to irrigation scenarios with irrigation at 0–200% of RAW for sand soil. L0RAW-L200RAW refers to irrigation scenarios with irrigation at 0–200% of RAW for loam soil. S40FC-S100FC refers to deficit irrigation at 40–100% of field capacity for sand soil. L40FC-L100FC refers to deficit irrigation at 40–100% of field capacity for loam soil.

As expected, in varied RAW threshold irrigation scenarios, the simulations maintained biomass yield at 1.77 ton ha^{-1} at site S1 and 0.90 ton ha^{-1} at site S2 in the full irrigation scenarios with allowable depletion from 0–100% of RAW (e.g., S0RAW to S100RAW for site S1 and L0RAW to L100RAW for site S2), that is due to no-water stress condition. As the water stress started below the RAW line [41], with available depletion from 120–200% of RAW thresholds, the biomass yields decreased up to 50% in the S200RAW (200% of RAW threshold) scenario at site S1 and 52% in L200RAW scenario at site S2.

In varied field capacity threshold irrigation scenarios (Figure 11b,d), biomass yields ranged from 0.85 to 1.77 ton ha^{-1} at site S1, and 0.89 to 0.90 ton ha^{-1} at site S2. At site S1, reducing deficit irrigation at 50% of field capacity (S50FC scenario), the biomass yield started to decrease with 22% and deficit irrigation at 40% of field capacity (S40FC scenario), biomass yields decreased up to 51% compared to full irrigation scenario (S100FC). For site 2, deficit irrigation up to 40% of field capacity (L40FC) did not affect biomass yield.

3.3.3. Relationship between Water Productivity and Irrigation Scenarios

The responses of biomass yield and irrigation water productivity to irrigation depths in various scenarios are presented in Figure 12. Simulated water productivity of varied RAW threshold irrigation scenarios ranged from 1.5 to 2.1 kg m^{-3} for site S1 and 0.9 to 1.4 kg m^{-3} for site S2. In varied field capacity irrigation scenarios, simulated irrigation water productivity (IWP) ranged from 0.8 to 1.36 kg m^{-3} for site S1 and 0.43–1.08 kg m^{-3} for site S2. The simulated irrigation water productivity results are comparable with other studies found in the literature. For instance, Gallardo et al. [98] found a measured IWP for lettuce dry matter of 1.86 kg m^{-3}.

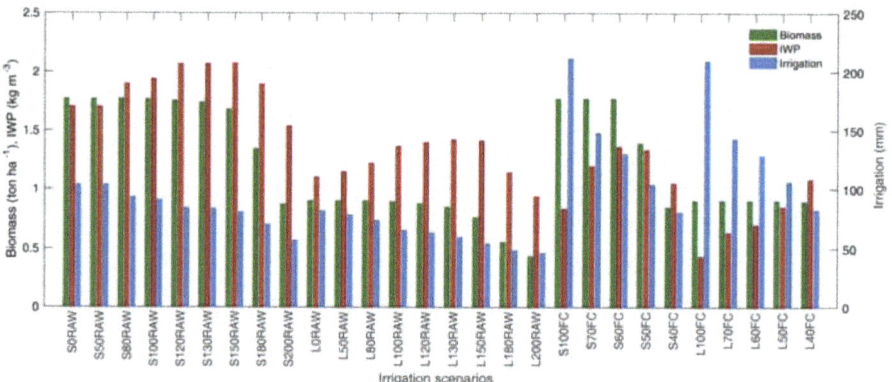

Figure 12. Comparison of biomass and water productivity response (IWP) to different irrigation scenarios. RAW is readily available water content. S0RAW-S200RAW refers to irrigation at 0–200% of RAW threshold irrigation scenarios for sand soil. L0RAW-L200RAW refers to refers to irrigation at 0–200% of RAW threshold irrigation scenarios for loam soil. S40FC-S100FC refers to deficit irrigation at 40–100% of field capacity for sand soil. L40FC-L100FC refers to deficit irrigation at 40–100% of field capacity for loam soil.

Figure 13 shows the relationship curves of biomass yield and irrigation water productivity response to irrigation scenarios. As expected, irrigation water productivity curve response to irrigation depths had parabolic relationships for both soil types in varied RAW threshold irrigation scenarios. Increasing water use efficiency can be enhanced by decreasing the irrigation to an optimum point. The optimum point, which resulted in 22% water saving for site S1, was found at the scenario with depletion of 150% of RAW (S150RAW), resulting in the irrigation water productivity = 2.07 kg m^{-3}, irrigation depth = 81 mm, and biomass yield = 1.68 ton ha^{-1}. For site S2, the optimum irrigation water productivity was at 130% of RAW scenario (L130RAW), resulting in irrigation water productivity = 1.42 kg m^{-3}, irrigation depth = 60 mm, and biomass yield = 0.85 ton ha^{-1}.

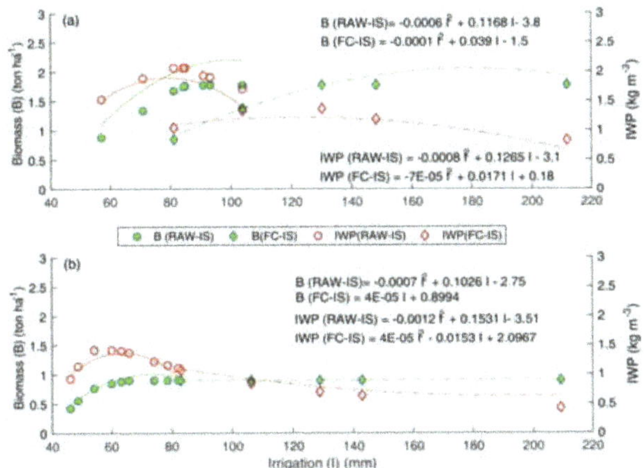

Figure 13. Relationship between biomass and irrigation water productivity responses to different scenarios: (**a**) at site S1 (sand soil) and (**b**) at site S2 (loam soil). I is irrigation, B is biomass, IWP is irrigation water productivity, RAW-IS is varied readily available water content threshold irrigation scenarios, FC-IS is varied field capacity threshold irrigation scenarios.

In varied field capacity threshold irrigation scenarios, for site S1, the optimum irrigation water productivity with 39% water saving was found at deficit irrigation at 60% of field capacity (S60FC) with irrigation water productivity = 1.36 kg m^{-3}, irrigation depth = 130 mm, and biomass yield = 1.77 ton ha^{-1}. For site S2, the optimum water productivity resulted in 60% water saving, which was found at deficit irrigation at 40% of field capacity (L40FC scenario) with irrigation water productivity = 1.08 kg m^{-3}, irrigation depth = 83 mm, and biomass yield = 0.89 ton ha^{-1}.

The varied RAW threshold irrigation scenarios resulted in higher simulated higher irrigation water productivity than the varied field capacity threshold scenarios in this study. Overall, deficit irrigation simulation scenarios in both irrigation scenario classes can provide a remarkable improvement in irrigation water productivity for water saving strategies.

3.3.4. Limitation

Crop models, like AquaCrop, are potentially valuable tools for answering questions primarily relating to research understanding, assessing crop management, and policy decision-making [49,99]. However, it is essential to test the models in diverse field environments, such as those with varied temperatures, elevation transects, or amidst latitudinal variations [99]. Particularly, AquaCrop has some limitations in terms of predicting crop yields only at the single growth cycle, single field scale, and only factoring in vertical water balance. The results of this study, obtained using climate data and field observation data relating to lettuce from a single growth cycle experiment at farm scale, allowed important information to be obtained in terms of calibrating lettuce crop parameters for sand and loam soil, and assessing limited water irrigation scenarios in the Cambodian context. However, it remains limited and the uncertainty on parameters has to be kept in mind. This study should be repeated in a contrasting range of diverse environments. Climate conditions and different cultural practices are the variables that differentiate the scenarios between different sites [99,100]. It has been emphasised that uncertainty model simulation results are themselves uncertain, due to known inadequacies of the model (residual errors in measurement) and due to unknown inadequacies of the model (by inputting new cultivars or different types of management, the model may be wrong in unsuspected ways) [101]. Despite such limitations, AquaCrop has already proven its usefulness in practical applications, and should still be tested widely in broader crop management applications, in diverse field environments [99,100].

4. Conclusions

An AquaCrop model was parameterised to simulate the canopy cover and aboveground biomass growth of lettuce under drip irrigation and plastic mulching for both sand and loam soil in the tropical monsoon climate of Cambodia. The model simulated canopy cover (RMSE < 0.8%) and aboveground biomass (RMSE < 0.01 ton ha^{-1}) in a satisfactory way after adjusting several key parameters, as mentioned in Farahani et al. [54].

Additionally, the results suggested that the incorporation of a heat stress factor affecting canopy cover and biomass growth is necessary to meet the conditions encountered in a tropical climate context.

Shortage of water in Cambodian agriculture has increased due to climate change, and this is a significant challenge facing farmers in their crop production. In this study, the AquaCrop model has helped to develop the simulation process for limited irrigation management strategies to maximise irrigation water productivity. To test the impact of different irrigation scheduling and water saving strategies, two scenario classes were explored: (i) varied readily available water (RAW) threshold irrigation and (ii) varied field capacity threshold irrigation scenarios. The irrigation scenario analysis proposed optimal irrigation strategies for lettuce.

For varied RAW threshold irrigation scenarios, the analysis proposed optimal simulated irrigation water productivity at scenarios of 150% of RAW (irrigation water productivity = 2.1 kg m^{-3}) for sand and 130% of RAW (irrigation water productivity = 1.4 kg m^{-3}) for loam soil. This can save 22% of water, and resulted in a biomass yield reduction of 5 and 2%, respectively, for sand and loam soil. For varied field capacity threshold irrigation scenarios, the optimal deficit irrigation depth was found at 60% of field capacity (irrigation water productivity of 1.4 kg m^{-3}) for sand soil, and at 40% of field capacity (irrigation water productivity of 1.0 kg m^{-3}) for loam soil. It can save water up to 39% and 60%, for sand and loam soil, respectively, maintaining biomass yields compared to full irrigation. These results suggest that deficit irrigation is worth considering as a water saving strategy for lettuce in the monsoon climate of Cambodia.

Overall, AquaCrop is a valuable tool to predict lettuce growth and to investigate different scenarios for providing irrigation scheduling strategies for water saving in Cambodia. However, further research is necessary to standardise the model parameters for lettuce in various irrigation management, environmental, and climatic conditions.

Author Contributions: P.K. performed the experiments, analysed the data, and wrote the paper. S.G. advised on the methodologies, gave comments and corrected the manuscript. C.O. supervised the research and gave comments to improve the manuscript. L.H. advised the agronomy practice during the experiments and gave comments on the manuscript. A.D. guided and supervised the research, gave comments, and corrected the manuscript.

Acknowledgments: This study was funded by the Belgian university cooperation programme, ARES-CCD (La Commission Coopération au Développement de l'Académie de Recherche et d'Enseignement supérieur).

Conflicts of Interest: The authors declare no conflict of interest.

References

1. Hoekstra, A.Y.; Wiedmann, T.O. Humanity's Unsustainable Environmental Footprint. *Science* **2014**, *344*, 1114–1117. [CrossRef] [PubMed]
2. Bae, J.; Dall'erba, S. Crop Production, Export of Virtual Water and Water-Saving Strategies in Arizona. *Ecol. Econ.* **2018**, *146*, 148–156. [CrossRef]
3. Rodríguez-Ferrero, N.; Salas-Velasco, M.; Sanchez-Martínez, M.T. Assessment of Productive Efficiency in Irrigated Areas of Andalusia. *Int. J. Water Resour. Dev.* **2010**, *26*, 365–379. [CrossRef]

4.   Chartres, C. Is Water Scarcity a Constraint to Feeding Asia's Growing Population? *Int. J. Water Resour. Dev.* **2014**, *30*, 28–36. [CrossRef]

5.   Jaramillo, F.; Destouni, G. Local Flow Regulation and Irrigation Raise Global Human Water Consumption and Footprint. *Science* **2015**, *350*, 1248–1251. [CrossRef] [PubMed]

6.   Toumi, J.; Er-Raki, S.; Ezzahar, J.; Khabba, S.; Jarlan, L.; Chehbouni, A. Performance Assessment of AquaCrop Model for Estimating Evapotranspiration, Soil Water Content and Grain Yield of Winter Wheat in Tensift Al Haouz (Morocco): Application to Irrigation Management. *Agric. Water Manag.* **2016**, *163*, 219–235. [CrossRef]

7.   Linker, R.; Ioslovich, I. Assimilation of Canopy Cover and Biomass Measurements in the Crop Model AquaCrop. *Biosyst. Eng.* **2017**, *162*, 57–66. [CrossRef]

8.   Touch, V.; Martin, R.J.; Scott, J.F.; Cowie, A.; Liu, D.L. Climate Change Adaptation Options in Rainfed Upland Cropping Systems in the Wet Tropics: A Case Study of Smallholder Farms in North-West Cambodia. *J. Environ. Manag.* **2016**, *182*, 238–246. [CrossRef] [PubMed]

9.   Chhinh, N.; Millington, A. Drought Monitoring for Rice Production in Cambodia. *Climate* **2015**, *3*, 792–811. [CrossRef]

10.   Montgomery, S.C.; Martin, R.J.; Guppy, C.; Wright, G.C.; Tighe, M.K. Farmer Knowledge and Perception of Production Constraints in Northwest Cambodia. *J. Rural Stud.* **2017**, *56*, 12–20. [CrossRef]

11.   Moreira, M.A.; Dos Santos, C.A.P.; Lucas, A.A.T.; Bianchini, F.G.; De Souza, I.M.; Viégas, P.R.A. Lettuce Production according to Different Sources of Organic Matter and Soil Cover. *Agric. Sci.* **2014**, *5*, 99–105. [CrossRef]

12.   Valenzuela, H.R.; Bernard, K.; John, C. *Lettuce Production Guidelines for Hawaii*; University of Hawaii: Honolulu, HI, USA, 1996.

13.   Cahn, M.; Johnson, L. New Approaches to Irrigation Scheduling of Vegetables. *Horticulturae* **2017**, *3*, 1–20. [CrossRef]

14.   Domingues, D.S.; Takahashi, H.W.; Camara, C.A.P.; Nixdorf, S.L. Automated System Developed to Control pH and Concentration of Nutrient Solution Evaluated in Hydroponic Lettuce Production. *Comput. Electron. Agric.* **2012**, *84*, 53–61. [CrossRef]

15.   Sokhen, C.; Kanika, D.; Moustier, P. *Vegetable Market Flows and Chains in Phnom Penh*; CIRAD-AVRDC-French MOFA: Hanoi, Vietnam, 2004.

16.   De Bon, H.; Parrot, L.; Moustier, P. Sustainable Urban Agriculture in Developing Countries. A Review. *Agron. Sustain. Dev.* **2010**, *30*, 21–32. [CrossRef]

17.   Morris, S.; Davies, W.; Baines, R.N. Challenges and Opportunities for Increasing Competitiveness of Vegetable Production in Cambodia. *Acta Hortic.* **2013**, *1006*, 253–260. [CrossRef]

18.   Xue, J.; Huo, Z.; Wang, F.; Kang, S.; Huang, G. Untangling the Effects of Shallow Groundwater and Deficit Irrigation on Irrigation Water Productivity in Arid Region: New Conceptual Model. *Sci. Total Environ.* **2018**, *619–620*, 1170–1182. [CrossRef] [PubMed]

19.   Adu, M.O.; Yawson, D.O.; Armah, F.A.; Asare, P.A.; Frimpong, K.A. Meta-Analysis of Crop Yields of Full, Deficit, and Partial Root-Zone Drying Irrigation. *Agric. Water Manag.* **2018**, *197*, 79–90. [CrossRef]

20.   Liu, Y.; Luo, Y. A Consolidated Evaluation of the FAO-56 Dual Crop Coefficient Approach Using the Lysimeter Data in the North China Plain. *Agric. Water Manag.* **2010**, *97*, 31–40. [CrossRef]

21.   Verstraeten, W.W.; Veroustraete, F.; Feyen, J. Assessment of Evapotranspiration and Soil Moisture Content across Different Scales of Observation. *Sensors* **2008**, *8*, 70–117. [CrossRef] [PubMed]

22.   Hunsaker, D.J.; French, A.N.; Waller, P.M.; Bautista, E.; Thorp, K.R.; Bronson, K.F.; Andrade-Sanchez, P. Comparison of Traditional and ET-Based Irrigation Scheduling of Surface-Irrigated Cotton in the Arid Southwestern USA. *Agric. Water Manag.* **2015**, *159*, 209–224. [CrossRef]

23.   Thompson, R.B.; Gallardo, M.; Valdez, L.C.; Fernández, M.D. Determination of Lower Limits for Irrigation Management Using in Situ Assessments of Apparent Crop Water Uptake Made with Volumetric Soil Water Content Sensors. *Agric. Water Manag.* **2007**, *92*, 13–28. [CrossRef]

24.   Ferreira, M.I.; Conceição, N.; Malheiro, A.C.; Silvestre, J.M.; Silva, R.M. Water Stress Indicators and Stress Functions to Calculate Soil Water Depletion in Deficit Irrigated Grapevine and Kiwi. *Acta Hortic.* **2017**, *1150*, 119–126. [CrossRef]

25.   Li, S.; Kang, S.; Li, F.; Zhang, L. Evapotranspiration and Crop Coefficient of Spring Maize with Plastic Mulch Using Eddy Covariance in Northwest China. *Agric. Water Manag.* **2008**, *95*, 1214–1222. [CrossRef]

26. Kashyap, P.S.; Panda, R.K. Evaluation of Evapotranspiration Estimation Methods and Development of Crop-Coefficients for Potato Crop in a Sub-Humid Region. *Agric. Water Manag.* **2001**, *50*, 9–25. [CrossRef]

27. Inthavong, T.; Tsubo, M.; Fukai, S. A Water Balance Model for Characterization of Length of Growing Period and Water Stress Development for Rainfed Lowland Rice. *Field Crop. Res.* **2011**, *121*, 291–301. [CrossRef]

28. Davis, S.L.; Dukes, M.D. Irrigation Scheduling Performance by Evapotranspiration-Based Controllers. *Agric. Water Manag.* **2010**, *98*, 19–28. [CrossRef]

29. Kukal, S.S.; Hira, G.S.; Sidhu, A.S. Soil Matric Potential-Based Irrigation Scheduling to Rice (*Oryza sativa*). *Irrig. Sci.* **2005**, *23*, 153–159. [CrossRef]

30. Pereira, L.S.; Paredes, P.; Sholpankulov, E.D.; Inchenkova, O.P.; Teodoro, P.R.; Horst, M.G. Irrigation Scheduling Strategies for Cotton to Cope with Water Scarcity in the Fergana Valley, Central Asia. *Agric. Water Manag.* **2009**, *96*, 723–735. [CrossRef]

31. Pereira, L.S.; Cordery, I.; Iacovides, I. Improved Indicators of Water Use Performance and Productivity for Sustainable Water Conservation and Saving. *Agric. Water Manag.* **2012**, *108*, 39–51. [CrossRef]

32. Afzal, M.; Battilani, A.; Solimando, D.; Ragab, R. Improving Water Resources Management Using Different Irrigation Strategies and Water Qualities: Field and Modelling Study. *Agric. Water Manag.* **2016**, *176*, 40–54. [CrossRef]

33. Geerts, S.; Raes, D. Deficit Irrigation as an on-Farm Strategy to Maximize Crop Water Productivity in Dry Areas. *Agric. Water Manag.* **2009**, *96*, 1275–1284. [CrossRef]

34. Chai, Q.; Gan, Y.; Zhao, C.; Xu, H.L.; Waskom, R.M.; Niu, Y.; Siddique, K.H.M. Regulated Deficit Irrigation for Crop Production under Drought Stress. A Review. *Agron. Sustain. Dev.* **2016**, *36*, 1–21. [CrossRef]

35. Lopez, J.R.; Winter, J.M.; Elliott, J.; Ruane, A.C.; Porter, C.; Hoogenboom, G. Integrating Growth Stage Deficit Irrigation into a Process Based Crop Model. *Agric. For. Meteorol.* **2017**, *243*, 84–92. [CrossRef]

36. Kögler, F.; Söffker, D. Water (Stress) Models and Deficit Irrigation: System-Theoretical Description and Causality Mapping. *Ecol. Model.* **2017**, *361*, 135–156. [CrossRef]

37. Patanè, C.; Tringali, S.; Sortino, O. Effects of Deficit Irrigation on Biomass, Yield, Water Productivity and Fruit Quality of Processing Tomato under Semi-Arid Mediterranean Climate Conditions. *Sci. Hortic. (Amsterdam)* **2011**, *129*, 590–596. [CrossRef]

38. Abd El-Wahed, M.H.; Baker, G.A.; Ali, M.M.; Abd El-Fattah, F.A. Effect of Drip Deficit Irrigation and Soil Mulching on Growth of Common Bean Plant, Water Use Efficiency and Soil Salinity. *Sci. Hortic. (Amsterdam)* **2017**, *225*, 235–242. [CrossRef]

39. Samperio, A.; Moñino, M.J.; Vivas, A.; Blanco-Cipollone, F.; Martín, A.G.; Prieto, M.H. Effect of Deficit Irrigation during Stage II and Post-Harvest on Tree Water Status, Vegetative Growth, Yield and Economic Assessment in "Angeleno" Japanese Plum. *Agric. Water Manag.* **2015**, *158*, 69–81. [CrossRef]

40. Yang, C.; Luo, Y.; Sun, L.; Wu, N. Effect of Deficit Irrigation on the Growth, Water Use Characteristics and Yield of Cotton in Arid Northwest China. *Pedosphere* **2015**, *25*, 910–924. [CrossRef]

41. Payero, J.O.; Melvin, S.R.; Irmak, S.; Tarkalson, D. Yield Response of Corn to Deficit Irrigation in a Semiarid Climate. *Agric. Water Manag.* **2006**, *84*, 101–112. [CrossRef]

42. Karam, F.; Mounzer, O.; Sarkis, F.; Lahoud, R. Yield and Nitrogen Recovery of Lettuce under Different Irrigation Regimes. *J. Appl. Hortic.* **2002**, *4*, 70–76.

43. Kuslu, Y.; Dursun, A.; Sahin, U.; Kiziloglu, F.M.; Turan, M. Short Communication. Effect of Deficit Irrigation on Curly Lettuce Grown under Semiarid Conditions. *Span. J. Agric. Res.* **2008**, *6*, 714–719. [CrossRef]

44. Geerts, S.; Raes, D.; Garcia, M. Using AquaCrop to Derive Deficit Irrigation Schedules. *Agric. Water Manag.* **2010**, *98*, 213–216. [CrossRef]

45. Hassanli, M.; Ebrahimian, H.; Mohammadi, E.; Rahimi, A.; Shokouhi, A. Simulating Maize Yields When Irrigating with Saline Water, Using the AquaCrop, SALTMED, and SWAP Models. *Agric. Water Manag.* **2016**, *176*, 91–99. [CrossRef]

46. Abderrahman, W.A.; Mohammed, N.; Al-Harazin, I.M. Computerized and Dynamic Model for Irrigation Water Management of Large Irrigation Schemes in Saudi Arabia. *Int. J. Water Resour. Dev.* **2001**, *17*, 261–270. [CrossRef]

47. Wolf, J.; Evans, L.G.; Semenov, M.A.; Eckersten, H.; Iglesias, A. Comparison of Wheat Simulation Models under Climate Change. I. Model Calibration and Sensitivity Analyses. *Clim. Res.* **1996**, *7*, 253–270. [CrossRef]

48. Ran, H.; Kang, S.; Li, F.; Tong, L.; Ding, R.; Du, T.; Li, S.; Zhang, X. Performance of AquaCrop and SIMDualKc Models in Evapotranspiration Partitioning on Full and Deficit Irrigated Maize for Seed Production under Plastic Film-Mulch in an Arid Region of China. *Agric. Syst.* **2017**, *151*, 20–32. [CrossRef]

49. Steduto, P.; Hsiao, T.C.; Raes, D.; Fereres, E. AquaCrop—The FAO Crop Model to Simulate Yield Response to Water: I. Concepts and Underlying Principles. *Agron. J.* **2009**, *101*, 426–437. [CrossRef]

50. Singh, A.; Saha, S.; Mondal, S. Modelling Irrigated Wheat Production Using the FAO AquaCrop Model in West Bengal, India, for Sustainable Agriculture. *Irrig. Drain.* **2013**, *62*, 50–56. [CrossRef]

51. Tavakoli, A.R.; Mahdavi Moghadam, M.; Sepaskhah, A.R. Evaluation of the AquaCrop Model for Barley Production under Deficit Irrigation and Rainfed Condition in Iran. *Agric. Water Manag.* **2015**, *161*, 136–146. [CrossRef]

52. Paredes, P.; Wei, Z.; Liu, Y.; Xu, D.; Xin, Y.; Zhang, B.; Pereira, L.S. Performance Assessment of the FAO AquaCrop Model for Soil Water, Soil Evaporation, Biomass and Yield of Soybeans in North China Plain. *Agric. Water Manag.* **2015**, *152*, 57–71. [CrossRef]

53. Todorovic, M.; Albrizio, R.; Zivotic, L.; Saab, M.-T.A.; Stöckle, C.; Steduto, P. Assessment of AquaCrop, CropSyst, and WOFOST Models in the Simulation of Sunflower Growth under Different Water Regimes. *Agron. J.* **2009**, *101*, 509–521. [CrossRef]

54. Farahani, H.J.; Izzi, G.; Oweis, T.Y. Parameterization and Evaluation of the AquaCrop Model for Full and Deficit Irrigated Cotton. *Agron. J.* **2009**, *101*, 469–476. [CrossRef]

55. Hussein, F.; Janat, M.; Yakoub, A. Simulating Cotton Yield Response to Deficit Irrigation with the FAO AquaCrop Model. *Span. J. Agric. Res.* **2011**, *9*, 1319–1330. [CrossRef]

56. Hsiao, T.C.; Heng, L.; Steduto, P.; Rojas-Lara, B.; Raes, D.; Fereres, E. AquaCrop—The FAO Crop Model to Simulate Yield Response to Water: III. Parameterization and Testing for Maize. *Agron. J.* **2009**, *101*, 448–459. [CrossRef]

57. Malik, A.; Shakir, A.S.; Ajmal, M.; Khan, M.J.; Khan, T.A. Assessment of AquaCrop Model in Simulating Sugar Beet Canopy Cover, Biomass and Root Yield under Different Irrigation and Field Management Practices in Semi-Arid Regions of Pakistan. *Water Resour. Manag.* **2017**, *31*, 4275–4292. [CrossRef]

58. Andarzian, B.; Bannayan, M.; Steduto, P.; Mazraeh, H.; Barati, M.E.; Barati, M.A.; Rahnama, A. Validation and Testing of the AquaCrop Model under Full and Deficit Irrigated Wheat Production in Iran. *Agric. Water Manag.* **2011**, *100*, 1–8. [CrossRef]

59. Mkhabela, M.S.; Bullock, P.R. Performance of the FAO AquaCrop Model for Wheat Grain Yield and Soil Moisture Simulation in Western Canada. *Agric. Water Manag.* **2012**, *110*, 16–24. [CrossRef]

60. Rankine, D.R.; Cohen, J.E.; Taylor, M.A.; Coy, A.D.; Simpson, L.A.; Stephenson, T.; Lawrence, J.L. Parameterizing the FAO AquaCrop Model for Rainfed and Irrigated Field-Grown Sweet Potato. *Agron. J.* **2015**, *107*, 375–387. [CrossRef]

61. Casa, A. De; Ovando, G.; Bressanini, L.; Martínez, J. Aquacrop Model Calibration in Potato and Its Use to Estimate Yield Variability under Field Conditions. *Atmos. Clim. Sci.* **2013**, *3*, 397–407. [CrossRef]

62. Wellens, J.; Raes, D.; Traore, F.; Denis, A.; Djaby, B.; Tychon, B. Performance Assessment of the FAO AquaCrop Model for Irrigated Cabbage on Farmer Plots in a Semi-Arid Environment. *Agric. Water Manag.* **2013**, *127*, 40–47. [CrossRef]

63. Deb, P.; Tran, D.A.; Udmale, P.D. Assessment of the Impacts of Climate Change and Brackish Irrigation Water on Rice Productivity and Evaluation of Adaptation Measures in Ca Mau Province, Vietnam. *Theor. Appl. Climatol.* **2016**, *125*, 641–656. [CrossRef]

64. Adeboye, O.B.; Schultz, B.; Adekalu, K.O.; Prasad, K. Modelling of Response of the Growth and Yield of Soybean to Full and Deficit Irrigation by Using Aquacrop. *Irrig. Drain.* **2017**, *66*, 192–205. [CrossRef]

65. Zeleke, K.T.; Luckett, D.; Cowley, R. Calibration and Testing of the FAO AquaCrop Model for Canola. *Agron. J.* **2011**, *103*, 1610–1618. [CrossRef]

66. Greaves, G.E.; Wang, Y.M. Assessment of Fao Aquacrop Model for Simulating Maize Growth and Productivity under Deficit Irrigation in a Tropical Environment. *Water* **2016**, *8*, 1–18. [CrossRef]

67. Montoya, F.; Camargo, D.; Ortega, J.F.; Córcoles, J.I.; Domínguez, A. Evaluation of Aquacrop Model for a Potato Crop under Different Irrigation Conditions. *Agric. Water Manag.* **2016**, *164*, 267–280. [CrossRef]

68. Klocke, N.L.; Fischbach, P.E. G84-690 Estimating Soil Moisture by Appearance and Feel. In *Historical Materials from University of Nebraska-Lincoln Extension*; University of Nebraska: Lincoln, NE, USA, 1984; pp. 1–9.

69. Allen, R.G.; Pereira, L.S.; Raes, D.; Smith, M. Crop Evapotranspiration: Guidelines for Computing Crop Water Requirements. *Irrig. Drain.* **1998**, *300*, 300.

70. Pansu, M.; Gautheyrou, J. *Handbook of Soil Analysis: Mineralogical, Organic and Inorganic Methods*; Springer Science & Business Media: Berlin, Germany, 2007.

71. Margesin, R.; Schinner, F. *Manual for Soil Analysis-Monitoring and Assessing Soil Bioremediation*; Springer Science & Business Media: Berlin, Germany, 2005.

72. Ket, P.; Garré, S.; Oeurng, C.; Degré, A. *A Comparison of Soil Water Retention Curves Obtained Using Field, Lab and Modelling Methods in Monsoon Context of Cambodia*; ARES-CCD: Brussels, Belgium, 2018.

73. Gallardo, M.; Jackson, L.E.E.; Schulbach, K.; Snyder, R.L.L.; Thompson, R.B.B.; Wyland, L.J.J. Production and Water Use in Lettuces under Variable Water Supply. *Irrig. Sci.* **1996**, *16*, 125–137. [CrossRef]

74. Razzaghi, F.; Zho, Z.; Andersen, M.; Plauborg, F. Simulation of Potato Yield in Temperate Condition by the AquaCrop Model. *Agric. Water Manag.* **2017**, *191*, 113–123. [CrossRef]

75. Raes, D.; Steduto, P.; Hsiao, T.C.; Fereres, E. AquaCrop—The FAO Crop Model to Simulate Yield Response to Water: II. Main Algorithms and Software Description. *Agron. J.* **2009**, *101*, 438–447. [CrossRef]

76. Steduto, P.; Raes, D.; Hsiao, T.; Fereres, E. AquaCrop: A New Model for Crop Prediction under Water Deficit Conditions. *Options Méditerr.* **2009**, *33*, 285–292.

77. Steduto, P.; Hsiao, T.C.; Fereres, E.; Raes, D. *Crop Yield Response to Water*; The Food and Agriculture Organization (FAO): Rome, Italy, 2012.

78. Paredes, P.; de Melo-Abreu, J.P.; Alves, I.; Pereira, L.S. Assessing the Performance of the FAO AquaCrop Model to Estimate Maize Yields and Water Use under Full and Deficit Irrigation with Focus on Model Parameterization. *Agric. Water Manag.* **2014**, *144*, 81–97. [CrossRef]

79. Morris, M.D. Factorial Plans for Preliminary Computational Experiments. *Technometrics* **1991**, *33*, 161–174. [CrossRef]

80. Stott, L.D. The Influence of Diet on the δ13C of Shell Carbon in the Pulmonate Snail Helix Aspersa. *Earth Planet. Sci. Lett.* **2002**, *195*, 249–259. [CrossRef]

81. Krause, P.; Boyle, D.P. Comparison of Different Efficiency Criteria for Hydrological Model Assessment. *Adv. Geosci.* **2005**, *5*, 89–97. [CrossRef]

82. Wellens, J.; Raes, D.; Tychon, B. On the Use of Decision-Support Tools for Improved Irrigation Management: AquaCrop-Based Applications. In *Current Perspective on Irrigation and Drainage*; INTECH: London, UK, 2017; pp. 53–67.

83. Zhuo, L.; Hoekstra, A. The Effect of Different Agricultural Management Practices on Irrigation Efficiency, Water Use Efficiency and Green and Blue Water Footprint. *Front. Agric. Sci. Eng.* **2017**, *4*, 185–194. [CrossRef]

84. Raes, D.; Steduto, P.; Hsiao, T.C.; Fereres, E. Calculation Procedures. In *AquaCrop-Reference Manual*; The Food and Agriculture Organization (FAO): Rome, Italy, 2017.

85. Raes, D.; Geerts, S.; Kipkorir, E.; Wellens, J.; Sahli, A. Simulation of Yield Decline as a Result of Water Stress with a Robust Soil Water Balance Model. *Agric. Water Manag.* **2006**, *81*, 335–357. [CrossRef]

86. Navarro-Hellín, H.; Martínez-del-Rincon, J.; Domingo-Miguel, R.; Soto-Valles, F.; Torres-Sánchez, R. A Decision Support System for Managing Irrigation in Agriculture. *Comput. Electron. Agric.* **2016**, *124*, 121–131. [CrossRef]

87. Raes, P.D.; Steduto, T.C.; Hsiao, E.F. *FAO Crop.-Water Productivity Model to Simulate Yield Response to Water. Reference Manual*; Food and Agriculture Organization of the United Nations: Rome, Italy, 2017.

88. Allen, R.G.; Pereira, L.S.; Raes, D.; Smith, M.; Ab, W. Crop Evapotranspiration-Guidelines for Computing Crop Water Requirements. In *FAO Irrigation and Drainage Paper 56*; Food and Agriculture Organization: Rome, Italy, 1998; pp. 1–15.

89. Lamn, F.; Ayars, J.; Nakayama, F. Irrigation Scheduling. In *Microirrigation for Crop Production*; Developments in Agricultural Engineering 13; Freddie, R., Lamm James, E., Ayars Francis, S., Nakayama, Eds.; Elsevier: Houston, TX, 2015; pp. 61–128.

90. Sutton, B.; Merit, N. Maintenance of Lettuce Root Zone at Field Capacity Gives Best Yields with Drip Irrigation. *Sci. Hortic. (Amsterdam)* **1993**, *56*, 1–11. [CrossRef]

91. Tan, S.; Wang, Q.; Zhang, J.; Chen, Y.; Shan, Y.; Xu, D. Performance of AquaCrop Model for Cotton Growth Simulation under Film-Mulched Drip Irrigation in Southern Xinjiang, China. *Agric. Water Manag.* **2018**, *196*, 99–113. [CrossRef]

92. Fazilah, W.; Ilahi, F.; Ahmad, D.; Husain, M.C. Effects of Root Zone Cooling on Butterhead Lettuce Grown in Tropical Conditions in a Coir-Perlite Mixture. *Hortic. Environ. Biotechnol.* **2017**, *58*, 1–4. [CrossRef]

93. Zhang, G.; Johkan, M.; Hohjo, M.; Tsukagoshi, S.; Maruo, T. Plant Growth and Photosynthesis Response to Low Potassium Conditions in Three Lettuce (*Lactuca sativa*) Types. *Hortic. J.* **2017**, *86*, 229–237. [CrossRef]

94. Dufault, R.J.; Ward, B.; Hassell, R.L. Dynamic Relationships between Field Temperatures and Romaine Lettuce Yield and Head Quality. *Sci. Hortic. (Amsterdam)* **2009**, *120*, 452–459. [CrossRef]

95. Parker, R.O. *Plant. & Soil Science: Fundamentals and Applications*, 1st ed.; Delmar Cengage Learning: Independence, KY, USA, 2009.

96. Wheeler, T.R.; Hadley, P.; Morison, J.I.; Ellis, R.H. Effects of Temperature on the Growth of Lettuce (*Lactuca sativa* L.) and the Implications for Assessing the Impacts of Potential Climate Change. *Eur. J. Agron.* **1993**, *2*, 305–311. [CrossRef]

97. Abdullah, K.; Ismail, T.G.; Yusuf, U.; Belgin, C. Effects of Mulch and Irrigation Water Amounts on Lettuce's Yield, Evapotranspiration, Transpiration and Soil Evaporation in Isparta Location, Turkey. *J. Biol. Sci.* **2004**, *4*, 751–755.

98. Gallardo, M.; Snyder, R.L.; Schulbach, K.; Jackson, L.E. Crop Growth and Water Use Model for Lettuce. *Irrig. Drain. Eng.* **1996**, *122*, 354–359. [CrossRef]

99. Boote, K.J.; Jones, J.W.; Pickering, N.B. Potential Uses and Limitations of Crop Models. *Agron. J.* **1996**, *88*, 704–716. [CrossRef]

100. Silvestro, P.C.; Pignatti, S.; Yang, H.; Yang, G.; Pascucci, S.; Castaldi, F.; Casa, R. Sensitivity Analysis of the Aquacrop and SAFYE Crop Models for the Assessment of Water Limited Winter Wheat Yield in Regional Scale Applications. *PLoS ONE* **2017**, *12*, 1–30. [CrossRef] [PubMed]

101. Wallach, D.; Keussayan, N.; Brun, F.; Lacroix, B.; Bergez, J. Assessing the Uncertainty When Using a Model to Compare Irrigation Strategies. *Agron. J.* **2008**, *104*, 1274–1283. [CrossRef]

Article

Water and Nitrogen Productivity of Potato Growth in Desert Areas under Low-Discharge Drip Irrigation

Pavel Trifonov, Naftali Lazarovitch and Gilboa Arye *

French Associates Institute for Agriculture and Biotechnology of Drylands, The Jacob Blaustein Institutes for Desert Research, Ben-Gurion University of the Negev, Sede Boqer Campus, Midreshet Ben-Gurion 84990, Israel; paveltri@post.bgu.ac.il (P.T.); lazarovi@bgu.ac.il (N.L.)

* Correspondence: aryeg@bgu.ac.il

Received: 3 July 2018; Accepted: 21 July 2018; Published: 24 July 2018

Abstract: Narrow profit margins, resource conservation issues and environmental concerns are the main driving forces to improve fertilizer uptake, especially for potatoes. Potatoes are a high value crop with a shallow, inefficient root system and high fertilizer rate requirements. Of all essential nutrients, nitrogen (N) is often limiting to potato production. A major concern in potato production is to minimize N leaching from the root zone. Therefore, the main objective of this study was to examine the potato crop characteristics under drip irrigation with low-discharge (0.6 L h^{-1}) and to determine the optimal combination of irrigation (40, 60, 80, and 100%) and fertigation (0, 50, and 100%) doses. In this study, the 80% (438.6 mm) irrigation dose and a 50% (50 mg N L^{-1}) fertigation dose (W80%F50%) showed that these doses are sufficient for optimal potato yield (about 40 ton ha^{-1}) in conjunction with water and fertilizer savings. Moreover, this treatment did not exhibit any qualitative changes in the potato tuber compared to the 100% treatments. When considering water productivity and yield, one may select a harsher irrigation regime if the available agricultural soils are not a limiting factor. Thus, higher yields can be obtained with lower irrigation and fertigation doses and a larger area.

Keywords: water productivity; nitrogen productivity; fertigation; drip irrigation; low-discharge; arid regions

---

HIGHLIGHTS:

- Potatoes grown under low-discharge drip irrigation in desert region
- Tuber yields and quality were similar to the ones from sprinkler irrigation
- Water productivity affected by water dose and nitrogen level

1. Introduction

Potato (*Solanum tuberosum* L.) growth is characterized by a high demand for nitrogen fertilizer due to its necessity for proper plant and tuber development [1–3]. However, due to a shallow (approximately 30 cm) and inefficient root system, applied water and fertilizer is at risk of leaching below the root zone [4,5]. Although potatoes have a relatively high value, fertilization costs might negatively affect their profitability [6,7]. Therefore, one needs to adjust nitrogen and water availability to crop demand in order to maximize yield, tuber quality, and nitrogen productivity [8–10].

The potato growing season in Israel is from autumn (September–November) to late spring (May–June). The growing area is about 16,000 ha with annual production of approximately 650,000 ton/year. Most growing areas are found in the western Negev region (about 75%); about 20% are located in the center and the remainder in the Galilee and Arava Desert. Whilst appropriate areas for growing potatoes are available in the Arava Desert, there is a shortage of irrigation water and the supply needs to be ensured throughout the growing season. Additionally, the quality of the irrigation water available for potato growth in this region is low due to its relatively high salt content (2–4 dS m^{-1}). This presents a real obstacle to the use of sprinkler irrigation—commonly considered as the best irrigation technique for potatoes [11]. Moreover, due to the low irrigation efficiency of sprinkler irrigation (about 75%), there is a loss of water and leaching of mobile nutrients. These inherent inefficiencies clearly lead to very low recoveries of applied fertilizers as well as the possibility of groundwater contamination.

In Israel, potatoes are usually grown on sandy soils, which have low water and nutrient holding capacity, thus increasing the risk of nitrogen loss by leaching under excess irrigation [12,13]. Thus, an alternative practice should be adopted in such areas and especially in arid areas where water resources are limited.

The use of drip irrigation and fertigation may be the solution. The employment of dripper discharge lower than 0.6 L h^{-1} has been recently demonstrated as an efficient method for potato growth in an arid region [14]. Due to low discharge and precise water application, drip irrigation reduces groundwater contamination, water and fertilizer waste, and energy inputs [15–17]. Surface or subsurface drip irrigation enables high water and oxygen availability in the root zone due to partial wetting [18]. Because of high soil-water pressure head and hydraulic conductivity in the root zone, water and nutrient availability is increased [19,20]. By minimizing the seepage losses beneath the shallow potato root zone, water and nitrogen productivity can be increased [21,22]. However, in the long term, one should monitor and prevent the accumulation of salts in the root zone, especially in arid areas where precipitation is almost negligible [23–25].

In areas where fertigation is applied, it is important to take into consideration the dripper discharge, the spacing between drippers, and soil hydraulic properties in order to achieve the optimal water content [26–29] and nitrogen availability [30] in the root zone.

Increasing water productivity (i.e., the volume of irrigation water required for a desired yield) is currently one of the main goals in arid and semi-arid regions [31,32]. Recently, Trifonov et al. [14] suggested that higher water productivity in potato crops could be achieved in the Arava Desert by using low-discharge drip irrigation. In their study, the combination of emitter discharge (0.6 vs. 1.6 L h^{-1}), spacing (25 vs. 50 cm) and irrigation dose (40, 50, 60, 80, 100 and 120%; where 100% was approximately 620 mm) was examined. In terms of tuber yield, it was found that a combination of sprinkler irrigation for germination (approximately 100 mm) and low discharge drip irrigation produced potato yields similar to the ones obtained from sprinkler irrigation, without harming tuber marketability (i.e., size and quality). The results suggested that the 80% irrigation dose is sufficient for optimal tuber yields and that discharge of 0.6 L h^{-1} is applicable for dripper spacing ranging from 20 to 50 cm. In terms of water productivity, it was demonstrated that the lower the irrigation dose the higher the water productivity.

Following the insights from our previous study [14], we further examined the influence of the irrigation water dose in conjunction with three nitrogen levels on potato growth in the Arava Desert. Hence, the main objective of this study was the optimization of potato growth under a low discharge drip irrigation and fertigation regime.

2. Materials and Methods

2.1. Site Description

The experiment was carried out over the winter of 2014–2015 on Kibbutz Yotvata, situated in the Arava Desert region of Israel. The commonest crops in this area are date palms, onions, and potatoes grown mainly in loamy sand (83% sand, 8% silt, and 9% clay) with a bulk density of 1.3 g cm^{-3}. The saturated water content, field capacity and wilting point are 0.36, 0.13, and 0.05 (v v^{-1}), respectively. The saturated hydraulic conductivity of this soil is 0.15 m h^{-1}. This area has a dry desert climate and during the study period the mean precipitation and class-A evaporation pan were approximately 32 mm and 620 mm, respectively. The annual temperature varied between 16 °C and 31 °C (Figure 1). Additional climatic information on the study area and season can be found in [14].

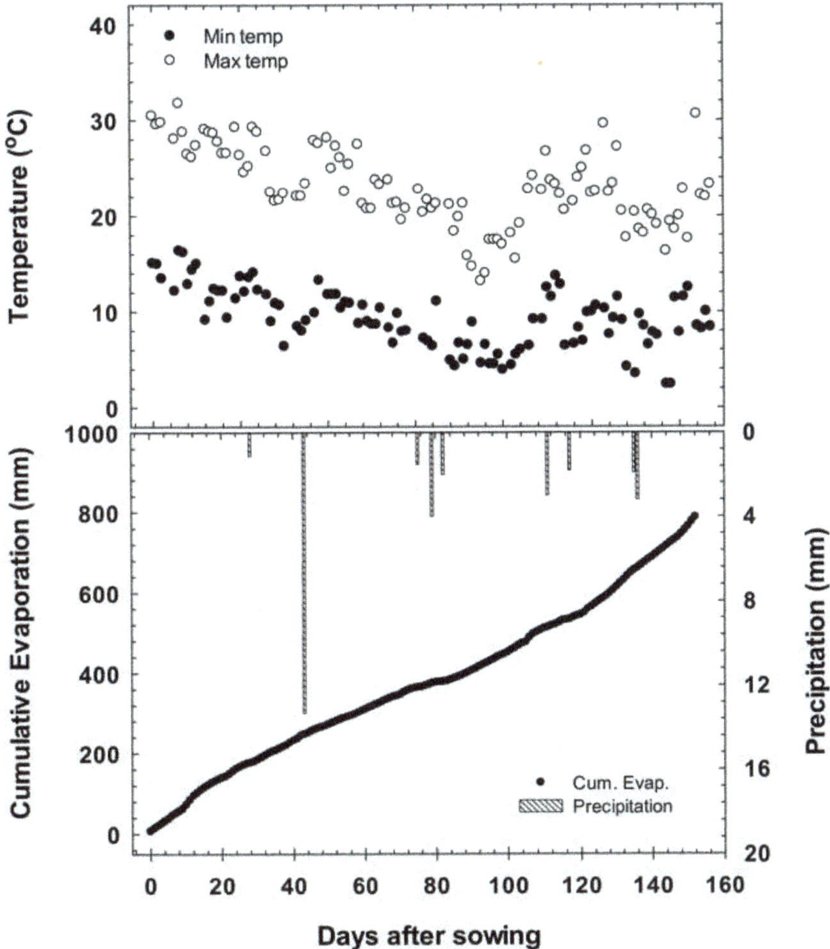

Figure 1. (**a**) Min. and max. Temperature; (**b**) cumulative evaporation (mm), and daily precipitation (mm) during the growing season starting from sowing date.

2.2. Experimental Design

The experimental field was roughly 0.2 ha in size (A = L × W = 180 m × 10.8 m). The experimental design consisted of 12 treatments with four replicates (i.e., 48 plots). Each replicate consisted of three beds, each 15 m by 0.9 m. The drip line had drippers with a 0.6 L h^{-1} discharge and 25 cm spacing between the drippers. As mentioned previously, this setup was selected based on our previous experiments that examined drip discharge and spacing for optimal water productivity [14]. The main findings from the previous study demonstrated that a combination of sprinkler irrigation for germination followed by low discharge drip irrigation could provide potato yields similar to those reported for sprinkler irrigation only, without harming tuber size and quality. During the two years of the previous study, the 80% water dose (i.e., 100% was approximately 620 mm) was sufficient for optimal potato growth in conjunction with water and fertilizer savings (fertigation rate was 100 mg N L^{-1}). Accordingly, the current experiment examined the effect of nitrogen dosage (i.e., 0, 50, and 100%) in conjunction with the water dose (i.e., 40, 60, 80, and 100%) on potato growth and yield. The treatments were marked as following: WX%FX%, where W and F refer to water irrigation and fertigation, respectively, and X is a % value (e.g., W80%F50%: 80% irrigation water dose and 50% fertigation dose). The treatment blocks were randomly distributed over the experimental fields. The representative chemical composition of the irrigation water in Yotvata is shown in Table 1.

Potato tuber (cv. Hermes) sowing took place on 28 October 2014. This date is day zero after sowing (i.e., DAS = 0). Tubers were distributed at a density of five tubers per meter at a depth of 20 cm. The spacing was 1 m between beds and the height of each bed was 40 cm with a trapezoid-like shape—created during placement of the shallow subsurface drip irrigation laterals at 5 cm depth (DAS = 2). The cumulative dose of irrigation water for germination was about 100 mm and was provided by sprinkler irrigation at three-day intervals. The actual cumulative irrigation dose, supplied by drip irrigation for each treatment, is shown in Table 2.

In the Kibbutz Yotvata fields, the main nitrogen sources are fertigation and soil amendment with composted cattle manure, produced from the kibbutz's dairy barns. The representative composition of this composted cattle manure is given in Table 3. It was applied at 40 m^3 ha^{-1} before sowing. Consequently, this was the main nitrogen source for the F0% treatments. The fertigation with nitrogen fertilizer, 100 mg N L^{-1} of "Arava" liquid fertilizer (ICL Ltd., Haifa, Israel), was supplied with the irrigation water. The nitrogen amount for each irrigation regime was proportional to the irrigation dose. However, since a continuous irrigation regime was employed, the available nitrogen concentration in the soil solution was comparable. The Arava fertilizer consists of NH_4NO_3 and KNO_3 as the nitrogen sources. Moreover, it was adjusted to be used with the irrigation water that is common to the Arava region (<1% Cl$^-$ and an acidic pH). It is important to mention that the fertigation started with initiation of drip irrigation (DAS = 26) and halted at the foliage top-kill (DAS = 130). Following the foliage top-kill, irrigation (without fertigation) was continued for three weeks, for the purpose of peel formation, until the harvest (DAS = 152).

2.3. Sampling and Analysis

Each replicate consisted of three rows and, to avoid edge effects, only the middle bed was sampled. During the growing stage, five samples (DAS = 58, 71, 85, 99 and 113) of petiole nitrate-N concentration were taken. The sampling was performed on fully expanded youngest leaves on the main stem, typically fourth or fifth from the top, from three different plants [33]. The sampling was performed at 7 to 14 day intervals [34]. At DAS = 152, the potato yield was assessed on collected tubers from an approximately 2 m^2 area (2.2 m by 0.9 m). The yield from all four replicates was pooled and weighed. In addition to this, two random replicates from each treatment were selected and the following data were collected for about 1600 tubers: Tuber length, tuber width, tuber mass, and their total number. The tuber qualitative evaluation included dextrose concentration and percentage of total solids. The soil, both above (0–5 cm) and beneath (5–20, 20–40 and 40–60 cm) the drippers, was sampled at the beginning and end of the growing season. These samples are referred to later as 5, 20, 40 and 60 cm, respectively. Chemical analyses (EC, Cl$^-$, dissolved organic carbon (DOC), total nitrogen (TN), NO_3^-, and NH_4^+) were performed on 1:1 soil:double distilled water (DDW) extracts.

2.4. Statistical Analysis

Multifactorial analysis of the irrigation and nitrogen treatments showed no significant effect. We performed our statistical analysis based on each treatment. Data were analyzed by analysis of variance (ANOVA) by using JMP software (version 12, 2015, SAS InstituteCary, NC, USA). Means were separated using the Student's-t comparison test at the probability level (p) of 0.05. Differences at the $p < 0.05$ level were considered to be significant. Deviation from the mean is presented in the tables and figures as standard deviation (SD).

Table 1. Representative Yotvata irrigation water composition.

Ca^{2+}	Mg^{+2ss}	Na^+	K^+	PO_4^{-3}	NO_3^-	NH_4^+	SO_4^{-2}	HCO_3^-	EC	pH
				$mg\ L^{-1}$					$dS\ m^{-1}$	-
$192.2 + 3.7$	139.4 ± 5.1	290 ± 5.7	12.8 ± 2.6	0	4 ± 0.3	0	758.5 ± 17.6	226.7	2.5 ± 0.6	7.2 ± 0.1

Table 2. Actual cumulative irrigation (mm) according to the irrigation and fertigation regime.

W40% F0%	W60% F0%	W80% F0%	W100% F0%	W40% F50%	W60% F50%	W80% F50%	W100% F50%	W40% F100%	W60% F100%	W80% F100%	W100% F100%
226.6	331.0	435.2	549.2	224.8	333.3	438.6	547.1	222.6	331.5	436.6	541.5

Table 3. Representative Yotvata composted biosolids composition (source: Dairy barn).

Ca^{+2}	Mg^{+2}	Na^+	K^+	P	OM	N	P	K	C	C/N	EC	pH
	$10^3 \times mg\ L^{-1}$					%wt.				-	$dS\ m^{-1}$	-
120.5	5.8	11.2	21.1	5.5	35.7 ± 0.6	1.6 ± 0.2	1.0 ± 0.0	2.1 ± 0.1	21.0 ± 0.0	12.3 ± 1.1	37.3 ± 4.1	8.1 ± 0.1

3. Results and Discussion

3.1. Water Inputs and Outputs

The impacts of water restriction on potato production will likely increase over the next decades due to climate change. Alternatively, areas that were not previously considered as arid may undergo desertification, thus forcing farmers to adapt to new conditions. Therefore, there is a need for a suitable irrigation regime as well as knowledge of soil properties. Each soil type has its own properties that will indicate the water availability to the crop. Based on the field capacity and wilting point of the soil in our study, the available soil water storage was approximately 8%. Specifically, a bed width of 0.9 m and active root zone of 0.3 m allows for an estimation that the available water (storage) is approximately 22 mm. Assuming that this is the case for all treatments, it makes up about 10% of the low irrigation dose (i.e., 40%) and about 4% of the higher dose (i.e., 100%). Note that the net contribution of rainfall to the water balance was negligible (about 32 mm) as the differences in water storage between the beginning and end of the growing season were small. Specifically, the mean precipitation in the study area is about 30 mm/year, distributed over 5 to 10 rain events (e.g., Figure 1). Considering the potential evaporation in the study region (i.e., approximately 5 mm/day), clearly the rain contribution on a daily basis is negligible. The above precipitation level and distribution is common to the study area [14] and, consequently, the rain contribution to the total water balance. Consequently, the potato growth in this area is entirely dependent on the supplied irrigation and fertigation. The actual irrigation dose corresponded well to the planned doses as can be seen in Table 2, with the average deviation from the planned dose being <1%. The actual 100% irrigation dose was about 545 mm; together with the germination dose (100 mm), the total water applied was 645 mm, comprising about 80% of the cumulative potential evaporation during the growing season (787 mm—Figure 1).

3.2. Tuber Yield

A common yield value for the Hermes cultivar in the Arava region is approximately 40 ton ha^{-1} [14]. The potato yield from this study (3rd year) attained that value and was similar to the results obtained from the study two years earlier. This supports the evidence that the yield under low-discharge drip irrigation is similar to that reported for sprinkler irrigation and ranges from 30 to 50 ton ha^{-1} [11,35,36]. In Figure 2, the results obtained for the total tuber yield as a function of nitrogen (F0, 50 100%) and water doses (W40, 60, 80, 100%) are presented. For a given nitrogen treatment, no significant differences for tuber yield were found between the irrigation doses, excluding the F0% treatment where tuber yield under the W80% irrigation dose was found to be significantly higher than the lower (i.e., W40% and W60%) and higher (i.e., W100%) irrigation doses. A possible explanation for this observation may be attributed to the nitrogen availability. Specifically, since the only source of nitrogen in the F0% treatment is from the applied compost, the rate and extent of its release from the solid compost to the soil solution will govern its availability. In this regard, an irrigation dose lower than W80% may not be sufficient while the higher irrigation dose (i.e., W100%) may leach the available nitrogen from the root zone. Figure 2 suggests that some saving of water and fertilizer are possible. However, additional parameters should be considered in order to conclude the optimal irrigation and fertigation regime.

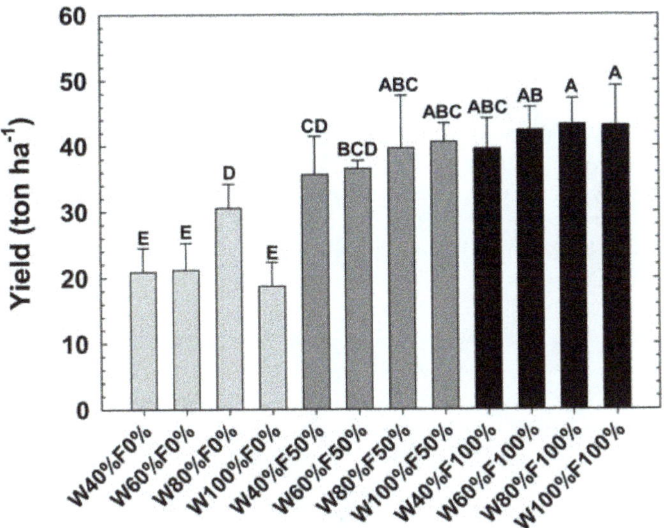

Figure 2. Potato tuber yield (ton ha^{-1}) as a function of irrigation and fertigation regime. Values with the same letter are not significantly different ($p < 0.05$).

3.3. Water Productivity

The amount of irrigation water needed to obtain a given yield (kg m^{-3}) is termed water productivity. The results obtained (Figure 3) clearly demonstrated that, for a given nitrogen treatment, water productivity decreases with increased irrigation dose. These results are in agreement with our previous study on potato growth using low discharge drip irrigation [14].

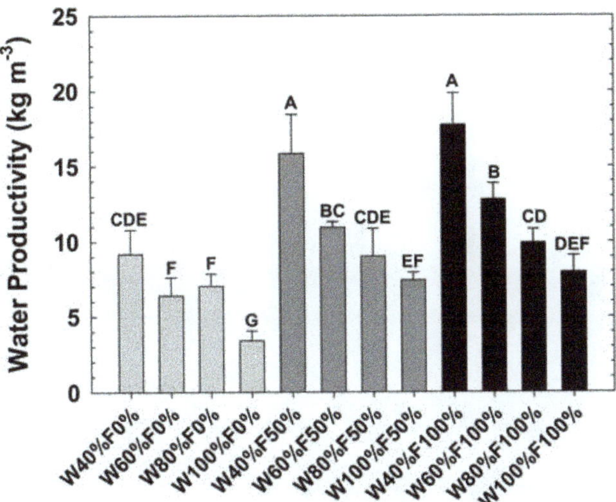

Figure 3. Water productivity (kg m^{-3}) as a function of irrigation and fertigation regime. Values with the same letter are not significantly different ($p < 0.05$).

However, for a given irrigation dose, significant differences in water productivity could only be observed between the F0% treatment relative to the F50% and F100% treatments. The latter did not exhibit a significant difference, implying that for the nitrogen treatments employed in this study the limiting factor of water productivity in potato growth is the dose of irrigation water.

The comparison between the F0% treatment relative to the F50% and F100% treatments implied that both nitrogen and water doses should be considered. Therefore, further reduction in the nitrogen dose should be examined to evaluate the minimum nitrogen treatment from which water productivity is reduced. It was interesting to see that the total tuber yield obtained for the W80%F0% treatment was not significantly different from the one obtained for the W40%F50% and W60%F50% treatments. However, the other F0% treatments exhibited a significantly lower yield than the F50% and F100% treatments, suggesting that at low irrigation doses, nitrogen availability is the limiting factor for tuber yield. Concerning the W100%F0% treatment, we can assume that the additional water caused additional leaching of nutrients such as nitrogen.

If water and fertilizer costs are of a big concern for the farmer, one may apply the W40%F50% regime. However, in order to compensate for a relatively decreased yield, it is necessary to increase the field size in order to produce profitable amounts of potatoes.

3.4. Nitrogen Productivity

Nitrogen productivity refers to the yield per applied N fertilizer (ton kg^{-1}). This value also implicitly depends on the interaction of the given water dose to the fertigation and, as a result, the harvested yield (Figure 4). The main nitrogen addition was due to fertigation and therefore the F0% treatments are not included in Figure 4 due to uncertainty regarding the release dynamics and concentrations of soluble nitrogen from the soil-compost solid phase. Nitrogen productivity decreases with an increasing water dose, which indicates an efficient usage of nitrogen fertilizer. Similar to the results for water productivity, Figure 4 shows that much more nitrogen was applied as compared to the yield with increasing water dose.

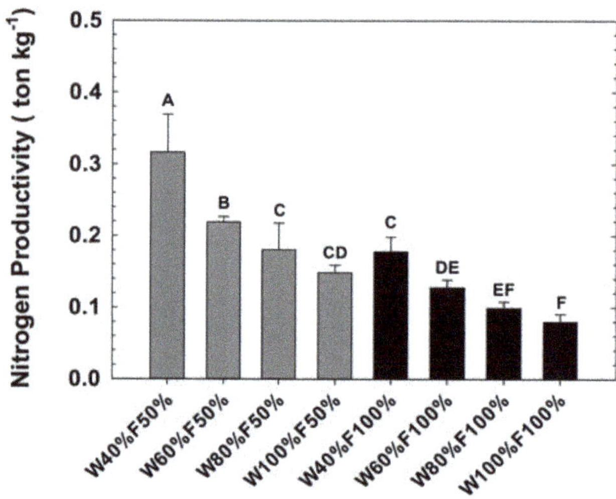

Figure 4. Nitrogen productivity (ton kg^{-1} N) as a function of irrigation and fertigation regime. Values with the same letter are not significantly different ($p < 0.05$).

3.5. Potato Tuber Size Distribution

In addition to the total harvested yield, marketability and commercial value are of considerable importance in directing farmer decision making. In Israel, a major marketability criterion is tuber width. The results of the cumulative width distribution of all treatments are shown in Figure 5. We used the trade standard of selecting potatoes larger than 50 mm width (diameter); vertical dashed lines in Figure 5. The following cumulative width distributions are based on this standard. The results show that the mean value of tuber widths higher than 50 mm for the three fertigation treatments are 45.16 ± 12.76%, 62.06 ±4.06%, 60.55 ± 3.49% for the F0%, F50%, and F100%, respectively (see Figure 4 for detailed values per fertigation treatment). The F50% treatment was similar to F100% and both were very different from the F0%. With regard to the water doses, no significant differences between treatments were found. Tuber length and mass distribution were also recorded (data not shown) and good correlations were found between tuber length/width and tuber mass/width (mass = 0.0007 width$^{2.9273}$, R^2 = 0.98; length = 0.7599 width$^{1.1016}$, R^2 = 0.93). Based on this scaling it is reasonable to use one tuber parameter.

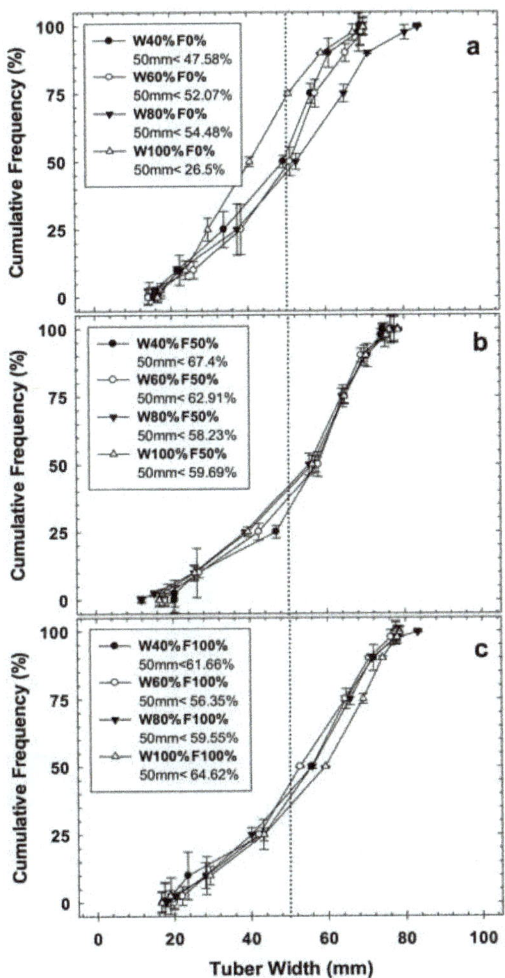

Figure 5. Cumulative frequency of tuber width according to the treatments. (**a**) F0%; (**b**) F50%; (**c**) F100%.

3.6. Potato Tuber Quality

Representative samples of tubers from each treatment were assessed for other commonly used quality parameters. Tuber color was assessed by spectroscope, where a score of 60 (arbitrary units) and above is considered acceptable for market. No significant differences were found between treatments (mean value 67.0 ± 1.94). Total solids percentage (dry mass basis) should range from 19 to 26%. Our results were a low but acceptable mean of 21.3 ± 0.45%. The Hermes cv. is primarily grown as a chipping potato and so low dextrose values are preferred. Higher dextrose values (>0.1 g L^{-1}) are not considered marketable as high dextrose concentrations leads to dark tubers, an unattractive trait for potato chips. While color and total solids were consistent between treatments, the dextrose concentrations showed some variation (mean 0.008 ± 0.005 g L^{-1}), although this was not significant.

Field size may compensate for the total yield. However, the quality of the potato tuber should be a significant parameter in this agricultural practice selection. Considering the large treatment differences in fertigation and irrigation in this study, it is surprising that tuber size and mass distribution was so consistent. Regarding fertigation, the 50% fertigation dose did not result in reduced tuber size. This may be because of the irrigation method used here (i.e., shallow, subsurface and with low discharge rates), which delivers both water and nutrients to the active root zone, and its combination with applied composted manure. Moreover, there were no qualitative differences observed between the treatments.

These insights, in conjunction with the results of water and nitrogen productivity, imply that the additional nitrogen (i.e., the difference between the F50% and F100% treatments), was mainly consumed and allocated to vegetative growth (canopy), and is not reflected in significantly higher tuber yield and/or quality. Moreover, the additional nitrogen also may have been wasted and leached from the root zone, thus leading to economic waste and potential groundwater contamination.

3.7. Nitrogen in the Petioles

The petiole N-NO$_3$ concentration is a common indicator for plant N demand and the available nitrogen from fertilization [33]. We used this metric to better understand nitrogen uptake and demand during the season and thus to monitor the effect of different N regimes. From the multivariate analysis of the petiole nitrogen, we found that, for a given N treatment, there was no significant difference in petiole nitrogen between the water doses. Thus, in Figure 6 the results are the means from the four water treatments at a given nitrogen treatment (i.e., 16 replicates). It can be seen for all the treatments that petiole nitrogen concentration increases with increasing plant size until it reaches a maximum at DAS = 85. After that stage until harvest, there is a decrease in petiole nitrogen concentration, likely due to resource translocation for tuber development. There is a clear difference between the fertigation doses; the higher the fertigation dose the higher the petiole nitrogen concentration in all five sampling events. This is particularly marked for the F100% treatment compared to the F50% and F0% treatments.

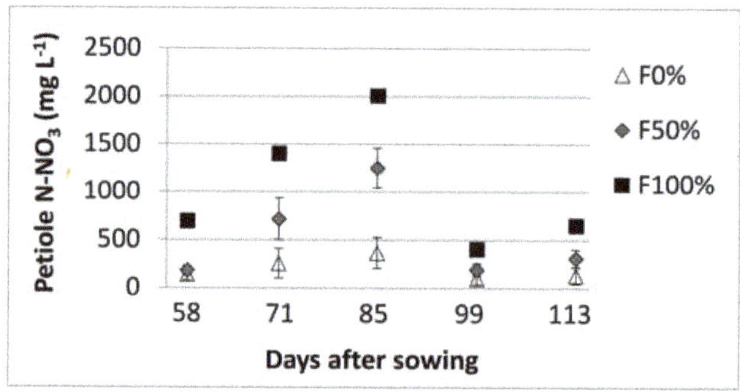

Figure 6. N-NO$_3$ in petioles during the growing season.

3.8. Soil Analysis

Deficient irrigation may cause accumulation of salts, especially when it combines fertigation. In addition, when applying fertilizers that contain highly mobile components such as nitrates, one should consider leaching and possible contamination hazards. As mentioned before, shallow sub-drip irrigation with low discharge can minimize these hazards. However, there is a need to leach the accumulated salts, especially from the top 30 cm of the soil profile (i.e., the potato root zone). Therefore, it is important to examine the soil profile as well as crop parameters.

The soil profile was first sampled after 100 mm sprinkler irrigation was supplied, following germination and initial seedling establishment ('Initial' in Figures 7 and 8). Based on a saturated soil water content of 0.36 (v v^{-1}) this amount of water is equivalent to about 1.4 pore volumes that flow through the top 20 cm of the soil. Theoretically, for non-reactive solute transport, two pore volumes of saturated water flow should fully displace the native salts in this layer.

Even at the lowest irrigation regime (40%, 226.6 mm), approximately 1.6 pore volume of the 20–60 cm layer is displaced and this is enough to leach non-reactive solutes such as Cl$^-$ from the active root zone (Figure 7). Clearly, then, the higher irrigation regimes are able to continuously leach solutes from the soil profile to a depth of 60 cm and deeper. At the end of the growing season, before the harvest and soil sampling, the fertigation was stopped in order to 'kill' the foliage (i.e., plant water uptake and transpiration are negligible), but irrigation with reduced water salinity (due to the lack of fertilizer) continued for a further 21 days, allowing for efficient solute leaching and peel formation. As already mentioned, the dripper was buried to 5 cm and therefore, under continuous fertigation, the top soil accumulates salts continuously due to evaporation and capillary rise. Sprinkler or surface irrigation can achieve removal of these salts between growing seasons when there is no interference of the foliage.

During the harvest, soil samples in proximity to a dripper were taken until a depth of 60 cm. The salinity (EC and Cl$^-$) in the top 5 cm was similar to that at the beginning of the season and behaved similarly in all the treatments. Contrary to the top layer, the rest of the profile was washed.

The EC and Cl$^-$ distributions shown in Figure 7 show that accumulated topsoil salts were displaced by approximately 20 cm and this is in broad agreement with the solute transport/pore volume relationship mentioned above. The top soil layer EC and Cl^{-1} values (3.52 dS m^{-1} and 422 mg L^{-1} respectively) obtained from 1:1 (soil:DDW) extracts indicate a chemical equilibrium was being established in this system.

Figure 7 also shows the DOC concentration in the soil profile. The DOC sources are organic amendments and plant litter. DOC is an available carbon source for microbial activity, which in turn can increase the nitrogen demand and consumption by microorganisms [37]. It can be seen that the DOC concentrations are similar between the treatments. Similar to Cl$^-$, DOC was also washed but with lower impact. This can be explained due to higher retardation of DOC in the soil because of the many DOC functional groups. From the C/N ratio (DOC/TN—data not shown), we observed that with higher fertigation and water dose we have more nitrogen in the soil profile and thus a lower C/N ratio. A low C/N ratio implies high nitrogen concentrations in the soil profile and thus can be interpreted as both a waste of fertilizer and an increased hazard for groundwater contamination.

With regard to the nitrogen, we examined the inorganic and organic fractions. The inorganic nitrogen includes NO$_3$$^-$ and NH$_4$$^+$. The organic fraction is the dissolved organic nitrogen (DON) calculated by subtraction from the total nitrogen (TN): DON = TN − N-NO$_3$$^-$ − N-NH$_4$$^+$.

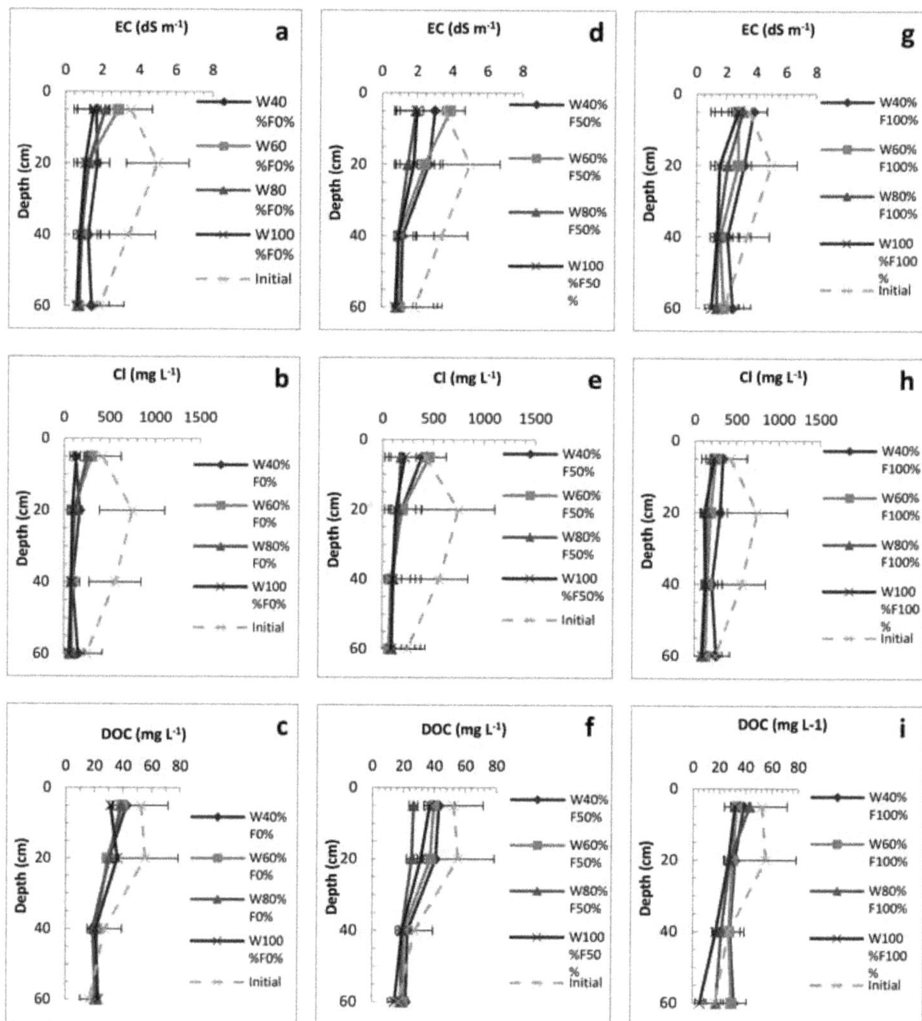

Figure 7. Mean values of soil sample analyses (1:1 soil:water) for EC, Cl⁻, and DOC at four different depths (5, 20, 40, and 60 cm) according to the irrigation regime. The dashed line ('Initial') represents the sampling at the beginning of the season. (**a,b,c**) F0%; (**d,e,f**) F50%; (**g,h,i**) F100%.

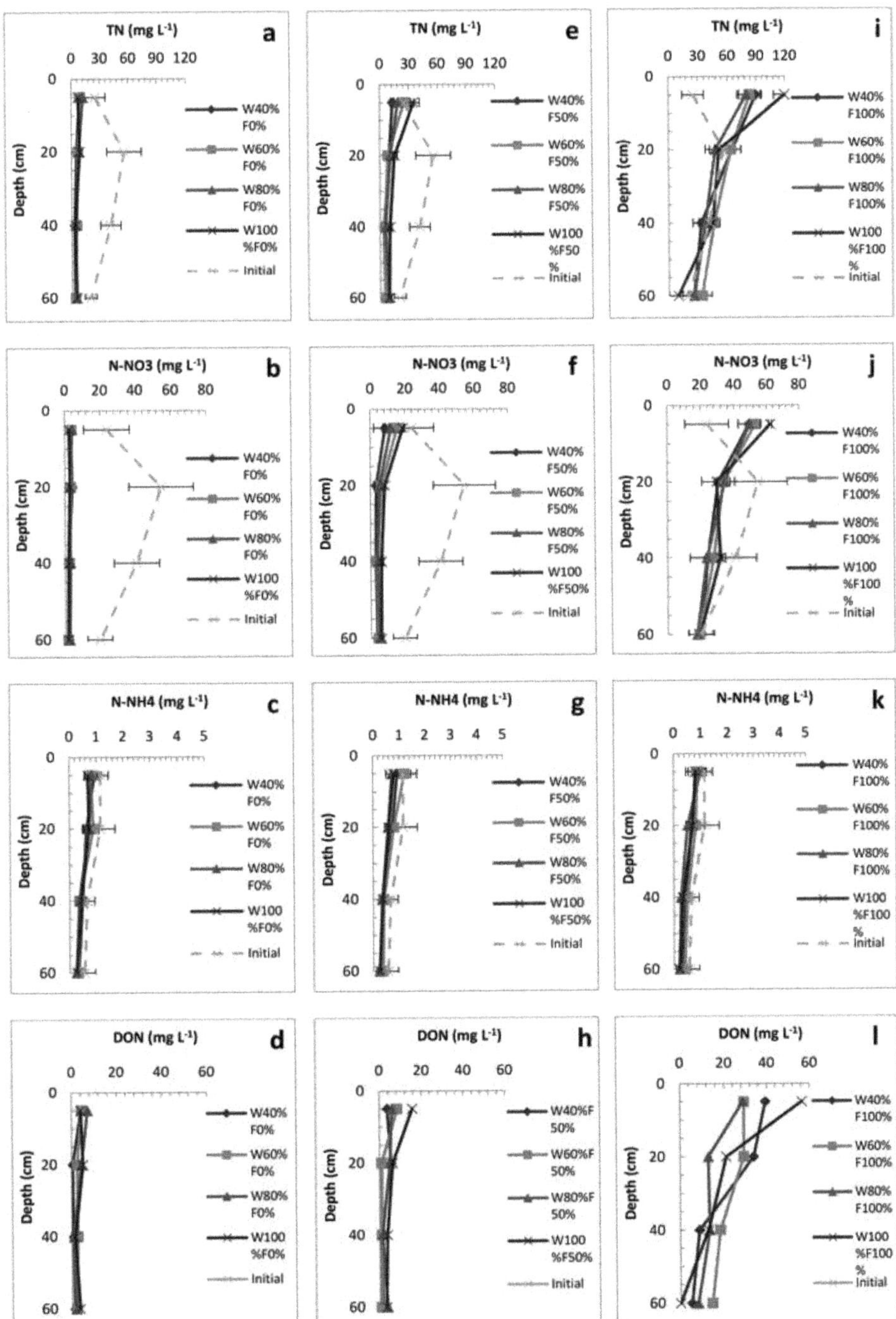

Figure 8. Mean values of soil sample analyses (1:1 soil:water) of TN, NO_3^-, NH_4^+ and DON at four different depths (5, 20, 40, and 60 cm) according to the irrigation regime. The dashed line ('initial') represents the sampling at the beginning of the season. (**a–d**) F0%; (**e–h**) F50%; (**i–l**) F100%.

It can be seen (Figure 8) that the TN was mainly compromised of NO_3^-, except for the F100% treatment that also had a higher fraction of DON. In general, NH_4^+ concentrations were significantly low or negligible for all treatments, both at the beginning and at the end of the season. TN was leached and/or consumed mainly in the F0% and F50% dose treatments, thus minimizing waste and N contamination of groundwater. More importantly, the F50% treatment was able to sustain profitable potato yield and quality. As for the F100% treatments, some accumulation of DON occurred in the top 5 cm. A similar trend was observed for all the irrigation doses with some higher values in the W100%F100% treatment, which showed significantly higher accumulation in this top layer. A possible explanation for this observation is the higher concentration of the mineral nitrogen as reflected by the lower C/N ratio (data not shown). Specifically, since mineral nitrogen was available for both plants and microbial activity, the DON was not consumed and thus accumulated in the root zone.

The soil NO_3^- concentration was most affected by fertigation due to its addition as the main source of nitrogen. The NO_3^- trend at the beginning and end of the season was similar to that of the TN. It can be seen that for the F0% and F50% doses, most of the nitrogen was consumed, while for the 100% treatment nitrate accumulated throughout the soil profile, with higher concentrations found in the top 5 cm.

As mentioned previously, the DON fraction was significantly higher for the F100% dose and exhibited a similar trend to the NO_3^-. As a result, when applying sprinkler irrigation between seasons when using the F100% treatments, the farmer may leach nitrogen components below the root zone, reducing the salinity but with possible groundwater contamination.

4. Summary and Conclusions

In arid regions, suitable irrigation techniques and regimes are highly important in order to acquire a profitable yield. Potato growth is mainly limited by two factors: Water and nitrogen availability. Excess irrigation can reduce tuber yield due to insufficient oxygen in the root zone and increase nutrient loss through leaching, which can also result in ground water contamination. Deficit irrigation may result in decreased tuber yield due to lack of water and a reduction in tuber quality through unavailability of nutrients or salt accumulation. Moreover, improper plant development will result in smaller foliage and thus reduced photosynthesis per unit leaf area.

Our results demonstrate that a sequential practice of sprinkler irrigation for the germination phase followed by low discharge drip irrigation can result in similar potato yields to traditional methods that utilize sprinkler irrigation, without unduly affecting marketable tuber size and quality. The W80%F50% treatment (438 mm water and 50 mg N L^{-1}) showed that this dose is sufficient for optimal potato growth in conjunction with water and fertilizer savings. Furthermore, water productivity is higher under the lower irrigation regime and, with regard to the economics of food production, farmers may benefit from these findings. Greater efficiency can be achieved with even lower irrigation doses (40%) but at the cost of utilizing a greater area. This is feasible if the availability of soils is not limiting but with the condition that this strategy has to take into account other fixed and variable costs, as well the environmental consequences of utilizing saline irrigation water.

Author Contributions: The study was conceived and designed by G.A. and N.L. P.T. performed the field experiment. All authors took a part in the data analysis, interpretation, and writing the paper. All authors have read and approved the final manuscript.

Funding: The study was funded by the Chief Scientist of the Israeli Ministry of Agriculture (857069512), the Goldinger Trust, Jewish Federation of Delaware and the Frances and Elias Margolin Trust.

Acknowledgments: We thank O.M. and his team from Kibbutz Yotvata for providing the agricultural facilities. We also appreciate the technical assistance and support of E.V. (Netafim LTD), Y.M., Y.S. and D.S.

Conflicts of Interest: The authors declare no conflict of interest.

References

1. Ojala, J.C.; Stark, J.C.; Kleinkopf, G.E. Influence of irrigation and nitrogen management on potato yield and quality. *Am. Potato J.* **1990**, *67*, 29–43. [CrossRef]
2. Waddell, J.T.; Gupta, S.C.; Moncrief, J.F.; Rosen, C.J.; Steele, D.D. Irrigation and nitrogen management effects on potato yield, tuber quality, and nitrogen uptake. *Agron. J.* **1999**, *91*, 991–997. [CrossRef]
3. Westermann, D.T.; Kleinkopf, G.E.; Porter, L.K. Nitrogen Fertilizer Efficiencies on Potatoes. *Am. Potato J.* **1988**, *65*, 377–386. [CrossRef]
4. Satchithanantham, S.; Krahn, V.; Sri Ranjan, R.; Sager, S. Shallow groundwater uptake and irrigation water redistribution within the potato root zone. *Agric. Water Manag.* **2014**, *132*, 101–110. [CrossRef]
5. Zotarelli, L.; Rens, L.R.; Cantliffe, D.J.; Stoffella, P.J.; Gergela, D.; Burhans, D. Rate and timing of nitrogen fertilizer application on potato 'FL1867'. Part I: Plant nitrogen uptake and soil nitrogen availability. *Field Crops Res.* **2015**, *183*, 246–256. [CrossRef]
6. Ierna, A.; Pandino, G.; Lombardo, S.; Mauromicale, G. Tuber yield, water and fertilizer productivity in early potato as affected by a combination of irrigation and fertilization. *Agric. Water Manag.* **2011**, *101*, 35–41. [CrossRef]
7. Watkins, K.B.; Lu, Y.; Huang, W. Economic and environmental feasibility of variable rate nitrogen fertilizer application with carry-over effects. *J. Agric. Resour. Econ.* **1998**, *23*, 401–426.
8. Hou, Z.N.; Li, P.F.; Li, B.G.; Gong, J.; Wang, Y.N. Effects of fertigation scheme on N uptake and N use efficiency in cotton. *Plant Soil.* **2007**, *290*, 115–126. [CrossRef]
9. Guo, S.; Wang, J.; Zhang, F.; Wang, Y.; Guo, P. An integrated water-saving and quality-guarantee uncertain programming approach for the optimal irrigation scheduling of seed maize in arid regions. *Water* **2018**, *10*, 908. [CrossRef]
10. Papadopoulos, I. Nitrogen fertigation of trickle-irrigated potato. *Fertil. Res.* **1988**, *16*, 157–167. [CrossRef]
11. Unlu, M.; Kanber, R.; Senyigit, U.; Onaran, H.; Diker, K. Trickle and sprinkler irrigation of potato (*Solanum tuberosum* L.) in the middle Anatolian Region in Turkey. *Agric. Water Manag.* **2006**, *79*, 43–71. [CrossRef]
12. Ju, X.T.; Kou, C.L.; Zhang, F.S.; Christie, P. Nitrogen balance and groundwater nitrate contamination: Comparison among three intensive cropping systems on the North China Plain. *Environ. Pollut.* **2006**, *143*, 117–125. [CrossRef] [PubMed]
13. Levallois, P. Groundwater contamination by nitrate associated with intensive potato culture in Québec. *Sci. Total Environ.* **1998**, *217*, 91–101. [CrossRef]
14. Trifonov, P.; Lazarovitch, N.; Arye, G. Increasing water productivity in arid regions using low-discharge drip irrigation: A case study on potato growth. *Irrig. Sci.* **2017**, *35*, 287–295. [CrossRef]
15. Dahan, O.; Babad, A.; Lazarovitch, N.; Eliani, E.; Kurtzman, D. Nitrate leaching from intensive organic farms to groundwater. *Hydrol. Earth Syst. Sci.* **2014**, *18*, 333–341. [CrossRef]
16. Giménez, L.; Paredes, P.; Pereira, L.S. Water use and yield of soybean under various irrigation regimes and severe water stress. Application of AquaCrop and SIMDualKc models. *Water* **2017**, *9*, 393. [CrossRef]
17. Rajput, T.B.S.; Patel, N. Water and nitrate movement in drip-irrigated onion under fertigation and irrigation treatments. *Agric. Water Manag.* **2006**, *79*, 293–311. [CrossRef]
18. Mmolawa, K.; Or, D. Root zone solute dynamics under drip irrigation: A review. *Plant Soil.* **2000**, *222*, 163–190. [CrossRef]
19. Shenker, M.; Ben-Gal, A.; Shani, U. Sweet corn response to combined nitrogen and salinity environmental stresses. *Plant Soil.* **2003**, *256*, 139–147. [CrossRef]
20. Silber, A.; Xu, G.; Levkovitch, I.; Soriano, S.; Bilu, A.; Wallach, R. High fertigation frequency: The effects on uptake of nutrients, water and plant growth. *Plant Soil* **2003**, *253*, 467–477. [CrossRef]
21. Lesczynski, D.B.; Tanner, C.B. Seasonal variation of root distribution of irrigated, field-grown Russet Burbank potato. *Am. Potato J.* **1976**, *53*, 69–78. [CrossRef]
22. Opena, G.B.; Porter, G.A. Soil management and supplemental irrigation effects on potato. II. Root growth. *Agron. J.* **1999**, *91*, 426–431. [CrossRef]
23. Aragüés, R.; Medina, E.T.; Martínez-Cob, A.; Faci, J. Effects of deficit irrigation strategies on soil salinization and sodification in a semiarid drip-irrigated peach orchard. *Agric. Water Manag.* **2014**, *142*, 1–9. [CrossRef]
24. Rozema, J.; Flowers, T. Crops for a salinized world. *Science* **2008**, *322*, 1478–1480. [CrossRef] [PubMed]

25. Sharma, B.R.; Minhas, P.S. Strategies for managing saline/alkali waters for sustainable agricultural production in South Asia. *Agric. Water Manag.* **2005**, *78*, 136–151. [CrossRef]

26. Ben-Asher, J.; Yano, T.; Shainberg, I. Dripper discharge rates and the hydraulic properties of the soil. *Irrig. Drain. Syst.* **2003**, *17*, 325–339. [CrossRef]

27. Elmaloglou, S.; Diamantopoulos, E. Soil water dynamics under surface trickle irrigation as affected by soil hydraulic properties, discharge rate, dripper spacing and irrigation duration. *Irrig. Drain.* **2010**, *59*, 254–263. [CrossRef]

28. Hinnell, A.C.; Lazarovitch, N.; Furman, A.; Poulton, M.; Warrick, A.W. Neuro-Drip: Estimation of subsurface wetting patterns for drip irrigation using neural networks. *Irrig. Sci.* **2010**, *28*, 535–544. [CrossRef]

29. Lazarovitch, N.; Poulton, M.; Furman, A.; Warrick, A.W. Water distribution under trickle irrigation predicted using artificial neural networks. *J. Eng. Math.* **2009**, *64*, 207–218. [CrossRef]

30. Olfs, H.W.; Blankenau, K.; Brentrup, F.; Jasper, J.; Link, A.; Lammel, J. Soil and plant-based nitrogen fertilizer recommendations in arable farming. *J. Plant Nutr. Soil Sci.* **2005**, *168*, 414–431. [CrossRef]

31. Expósito, A.; Berbel, J. Microeconomics of deficit irrigation and subjective water response function for intensive olive groves. *Water* **2016**, *8*, 254. [CrossRef]

32. Sadras, V.O. Does partial root-zone drying improve irrigation water productivity in the field? A meta-analysis. *Irrig. Sci.* **2009**, *27*, 183–190. [CrossRef]

33. Gardner, B.R.; Jones, J.P. Petiole analysis and the nitrogen fertilization of Russet Burbank potatoes. *Am. J. Potato Res.* **1975**, *52*, 195–200. [CrossRef]

34. Shrestha, R.K.; Cooperband, L.R.; MacGuidwin, A.E. Strategies to reduce nitrate leaching into groundwater in potato grown in sandy soils: Case study from North Central USA. *Am. J. Potato Res.* **2010**, *87*, 229–244. [CrossRef]

35. Matović, G.; Broćić, Z.; Djuričin, S.; Gregorić, E.; Bodroža, D. Profitability assessment of potato production applying different irrigation methods. *Irrig. Drain.* **2016**, *65*, 502–513. [CrossRef]

36. Starr, G.C.; Rowland, D.; Griffin, T.S.; Olanya, O.M. Soil water in relation to irrigation, water uptake and potato yield in a humid climate. *Agric. Water Manag.* **2008**, *95*, 292–300. [CrossRef]

37. Kalbitz, K.; Solinger, S.; Park, J.H.; Michalzik, B.; Matzner, E. Controls on the dynamics of dissolved organic matter in soils: A review. *Soil Sci.* **2000**, *165*, 227–304. [CrossRef]

Article

Potential of Deficit and Supplemental Irrigation under Climate Variability in Northern Togo, West Africa

Agossou Gadédjisso-Tossou [1,2,*] iD, **Tamara Avellán** [1] iD and **Niels Schütze** [2] iD

[1] United Nations University Institute for Integrated Management of Material Fluxes and of Resources (UNU-FLORES), Ammonstrasse 74, 01067 Dresden, Germany; avellan@unu.edu

[2] Institute of Hydrology and Meteorology, Technische Universität Dresden, 01069 Dresden, Germany; niels.schuetze@tu-dresden.de

* Correspondence: Agossou.Gadedjisso-Tossou@tu-dresden.de; Tel.: +49-351-7999-3816

Received: 22 October 2018; Accepted: 4 December 2018; Published: 7 December 2018

Abstract: In the context of a growing population in West Africa and frequent yield losses due to erratic rainfall, it is necessary to improve stability and productivity of agricultural production systems, e.g., by introducing and assessing the potential of alternative irrigation strategies which may be applicable in this region. For this purpose, five irrigation management strategies, ranging from no irrigation (NI) to controlled deficit irrigation (CDI) and full irrigation (FI), were evaluated concerning their impact on the inter-seasonal variability of the expected yields and improvements of the yield potential. The study was conducted on a maize crop (*Zea mays* L.) at a representative site in northern Togo with a hot semi-arid climate and pronounced dry and wet rainfall seasons. The OCCASION (Optimal Climate Change Adaption Strategies in Irrigation) framework was adapted and applied. It consists of: (i) a weather generator for simulating long climate time series; (ii) the AquaCrop model, which was used to simulate the irrigation system during the growing season and the yield response of maize to the considered irrigation management strategies; and (iii) a problem-specific algorithm for optimal irrigation scheduling with limited water supply. We found high variability in rainfall during the wet season which leads to considerable variability in the expected yield for rainfed conditions (NI). This variability was significantly reduced when supplemental irrigation management strategies (CDI or FI) requiring a reasonably low water demand of about 150 mm were introduced. For the dry season, it was shown that both irrigation management strategies (CDI and FI) would increase yield potential for the local variety TZEE-W up to 4.84 Mg/ha and decrease the variability of the expected yield at the same time. However, even with CDI management, more than 400 mm of water is required if irrigation would be introduced during the dry season in northern Togo. Substantial rainwater harvesting and irrigation infrastructures would be needed to achieve that.

Keywords: AquaCrop model; maize; deficit irrigation; crop-water production function; West Africa

1. Introduction

The present world population of 7.3 billion will increase to 9.7 billion by 2050 [1]. Similarly, the medium variant of the UN Population Division [2] predictions disclose that the total population of the West African region would increase from 350 million in 2015 to 450 million in 2030, and nearly 800 million in 2050. FAO [3] estimates that agricultural production will have to rise by 60% by 2050 to meet the world's projected demands for food and feed. In West Africa, Liniger et al. [4] reported that food production should increase by 70% by 2050 to meet the necessary caloric requirements. However, a lack of available water for agricultural production, the energy sector, and other forms of anthropogenic water consumption is already harming several parts of the world. This lack of water is projected to become more severe with the growing population, rising temperatures, and altering precipitation patterns [5]. The variation of the food diets in many developing countries compound this problem and lead to the demand for more processed food and animal proteins by consumers [6].

The World Bank [7] reports that the rate of increase in food demand is projected to be higher in developing than in developed countries. These are also the regions that are subject to a wide yield gap. The world demand (billion tons) of cereals was 1.20 in 1974, 1.84 in 1997 and is expected to be 2.50 in 2020 [8]. In addition, van Ittersum et al. [9] pointed out that Sub-Saharan Africa (SSA) is the region with lowest food security because by 2050 its demand for cereals will almost triple, whereas current levels of cereal consumption already rely on considerable importations.

Lobell and Gourdji [10] pointed out that, in the past several decades, air temperatures have been increasing in most of the main cereal cropping areas around the world. They added that the changes in temperature and the intensity and seasonal volume of rainfall are impacting soil moisture. In turn, soil moisture is of high importance for crop production. In developing countries, particularly, the changes in these climatic variables over time are likely to have a damaging impact on water accessibility, which in turn affects crop yield. Kotir [11] and Druyan [12] stressed the fact that researchers have described Sub-Saharan Africa as the most sensitive region to the impacts of climate variabilities and change because of its dependence on rainfed agriculture and low capacity for adaptation. Moreover, Sarr [13] contended that the West African region has faced decades of severe drought, which have affected agricultural production substantially. The observations already show the late onset and early cessation dates of rainfall and the reduction of length of growing period.

According to the Togolese Ministry of the Environment and Forestry (MERF) [14], in the dry savannah of northern Togo, a West African country, the wet season, which spanned six months in the 1970s, has reduced to five or four months nowadays. Consequently, on the one hand, a substantial amount of rainwater falls within a short period causing flooding, while, on the other hand, frequent dry spells in the wet season lead to crop failure [15]. In addition, there is no rainfed agricultural activity during the dry season in northern Togo because of a lack of rainfall [16].

Researchers and practitioners are putting more focus on producing more with limited resources in agriculture to meet the food demand and at the same time address the adverse effects of climate change [17–19]. Agriculture, which accounts for 38% of Togo's gross domestic product, provides over 20% of export earnings and employs 70% of the active population. Togolese agriculture is predominantly rainfed [20,21]. According to the International Commission on Irrigation and Drainage (ICID) [22], rainfed agriculture is "agriculture without application of irrigation. It may be without, or with a drainage system." A promising practice to overcome water shortage in rainfed cropping systems is supplemental irrigation (SI). The ICID [22] defines SI as: "the addition of small amounts of water to essentially rainfed crops during times when rainfall fails to provide sufficient moisture for normal plant growth, in order to improve and stabilize yields." SI practice increases yields and water productivity in rainfed cropping systems [23]. In addition, conventional irrigation systems can be used to improve crop productivity. The ICID [22] defines conventional irrigation as: "the replenishment of soil water storage in plant root zone through methods other than natural precipitation".

Irrigation scheduling is the procedure of deciding when, where, and how much water to apply [24] for irrigation. Farmers can apply the total crop-water requirements or more in the right period if water is available. This practice is called full irrigation (FI). When water provisions are limited, or irrigation expenses are great, FI may be substituted by deficit irrigation (DI) [25]. This is limited irrigation scheduling in agriculture [26]. DI can be controlled or otherwise. Uncontrolled DI is equivalent to rainfed agriculture. English [27] and English and Raja [28] defined controlled deficit irrigation (CDI) as the concept of intentionally and systematically under-irrigating a crop. English [27] developed an analytical framework to evaluate the profit when optimizing water use. Thus, he included implicitly economic aspects in the definition. Later, Lecler [29] provided a more explicit definition: "CDI is an optimization strategy by which net returns are maximized by lessening the volume of irrigation water applied to a crop to a level that results in some yield loss caused by water stress". Recently, Fereres and Soriano [30] defined CDI as the application of water below full crop-water requirements or evapotranspiration. The objective of applying limited water is to cope with scarce water supplies and improve productivity. Kögler and Söffker [31] reported that CDI practice contributes to saving up to 20–40% irrigation water at yield reductions under 10%. It can contribute to increasing farmers' net income where water is scarce [27]. Thus, CDI is an irrigation management practice that contributes to enhancing food security.

Many studies that applied the simulation-based approach to assess deficit irrigation strategies failed to consider the variability of relevant climate factors—such as precipitation and temperature—and soil properties [32,33]. Semenov [34] and Brumbelow and Georgakakos [35], among others, analyzed possible impacts of climate variability and climate change on agriculture using process-based simulation models. Most of these studies only look at rainfed or non-irrigated sites or assumed full irrigation. Few researchers, including Schütze and Schmitz [36] and Brumbelow and Georgakakos [35], assessed limited irrigation systems and the impact of climate variability on crop-water production functions (CWPF). Brumbelow and Georgakakos [35] derived probability distribution functions of CWPF (CWPF-PDs) using climate change scenarios data of the Intergovernmental Panel on Climate Change (IPCC). Schütze and Schmitz [36] delved into the CWPF concept and suggested a stochastic framework in the form of a decision support tool for Optimal Climate Change Adaption Strategies in Irrigation (OCCASION) for deriving site-specific stochastic CWPFs (SCWPFs). To perform such analyses, one needs to utilize crop models to simulate the potential or expected crop yield for a given soil, climate, and management practice condition.

Several crop simulation models such as DSSAT [37], AquaCrop [38–40], DAISY [41], CropWat [42], APSIM [43], and PILOTE [44] are available in the literature to simulate yield response to water. It is important to recognize that most of these models show substantial complexities and require several data to run. Most of these models require many parameters to run, and many are not readily available in the field and need to be determined experimentally [45]. Exceptionally, the AquaCrop model uses relatively few explicit and mostly intuitive parameters and input variables, requiring simple methods for their derivation [46]. For instance, unlike AquaCrop, the DSSAT model requires input data about crop genetics and pest management [37], while APSIM requires NO_3 and NH_4 content of the soil layers [43].

Few studies have investigated irrigation management strategies on crops in the dry savannah area of northern Togo [20]. Therefore, this study assessed the potential of deficit and supplemental irrigation in northern Togo. Specifically, the study aimed at: (i) characterizing the climate of a water-scarce site in northern Togo, West African region; and (ii) evaluating five irrigation management strategies, ranging from no irrigation (NI) to CDI and FI for a maize crop (*Zea mays* L.) at a representative site in northern Togo with pronounced dry and wet rainfall seasons.

2. Materials and Methods

2.1. Study Area

Togo is a small West African francophone country. It is bordered by the Bight of Benin and Burkina Faso in the south and north, respectively. Togo is bound in the west by Ghana and in the east by Benin. Geographically, it lies between latitudes 6° N and 11° N, and longitudes 0° E and 2° E. It covers a surface of 56,600 km^2 and has a long, narrow profile, stretching more than 550 km from north to south but not exceeding 160 km in width [47]. Its population is estimated to be 6,191,155 [48].

We conducted this study in the Dapaong district, northern Togo (Figure 1). Dapaong belongs to the Southern-Guinea-Savannah agro-ecological zone [49]. The principal rainfed crops grown include maize (*Zea mays*), sorghum (*Sorghum bicolor*), and pearl millet (*Pennisetum glaucum*), mainly for subsistence, while cash crops such as cotton (*Gossypium hirsutum*) are also cultivated. Some vegetables and legumes such as okra (*Abelmoschus esculentus*), cowpea (*Vigna unguiculata*), and soybean (*Glycine max*) are grown in association with the cereals mentioned above. The vegetation type is a woody savannah, with noticeable agricultural farms. The primary tree species are *Parkia biglobosa, Butyrospermum parkii,* and

Acacia sieberiana [50]. The Togolese Institute of Agricultural Research (ITRA) [51] and Didjeira et al. [52] identified maize crop as the staple food in Togo, and it represents 60% of the cereals consumed by the population. On the farms close to the houses, the main cropping system is intercropping (cereal–legume mixtures), while on the farms far from the houses, farmers practice monoculture [53]. Since cotton is grown with a high level of pesticides, intercropping is not possible on cotton farms. Hoes and cutlasses are the primary tools of cultivation.

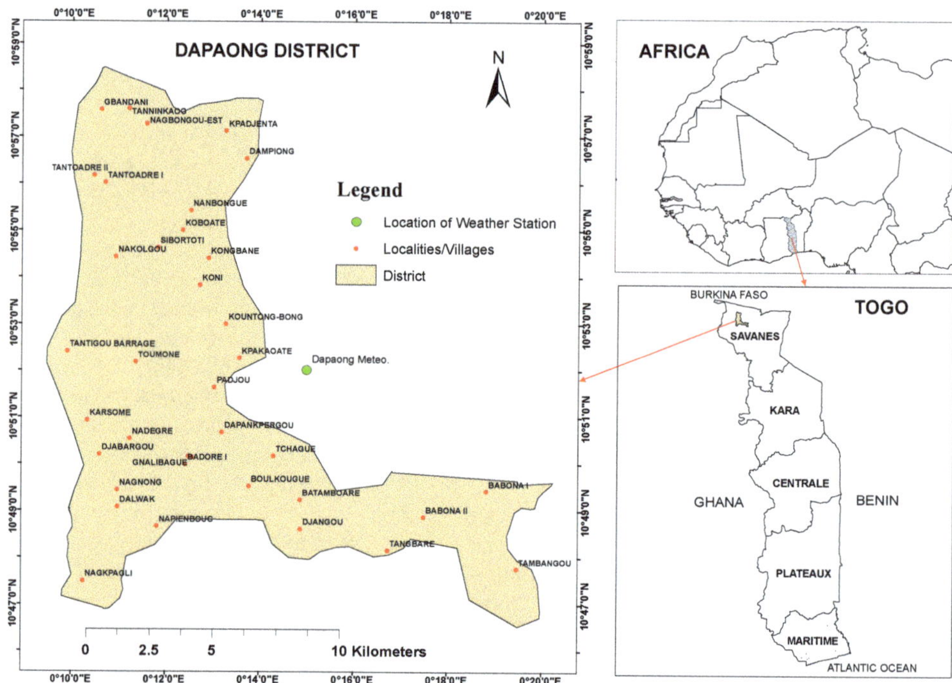

Figure 1. Map of northern Togo indicating the study area (Dapaong district).

According to Köppen–Geiger's climatic classification, the climate of Dapaong district is hot semi-arid (BSh) [54]. The period from mid-April to mid-October is humid, while in the other months dry conditions predominate in Dapaong. The months from June to September show high rainfall (Figure 2). These high annual values of rainfall are sufficient for rainfed cereal crops in northern Togo. The annual rainfall is, however, very unequally dispersed. From November to March (or sometimes April), there is practically no rainfall in the area. From May to October, a substantial amount of rainfall is recorded. Consequently, northern Togo is characterized by a single wet season in a year. This explains why farmers adopt intercropping to obtain the range of crops they need. Introducing irrigated crops in the dry season may help farmers to sustain their production. The mean annual temperature is 28.1 °C, and the annual total precipitation is 1050 mm. The mean daily maximum temperature of the driest month is around 37 °C, whereas the mean daily minimum temperature of the wettest month is 20 °C (Figure 2). In January and February, a robust dusty wind named harmattan, blowing in the northeast direction from the Sahara Desert, increases the dryness of the weather in the area [16].

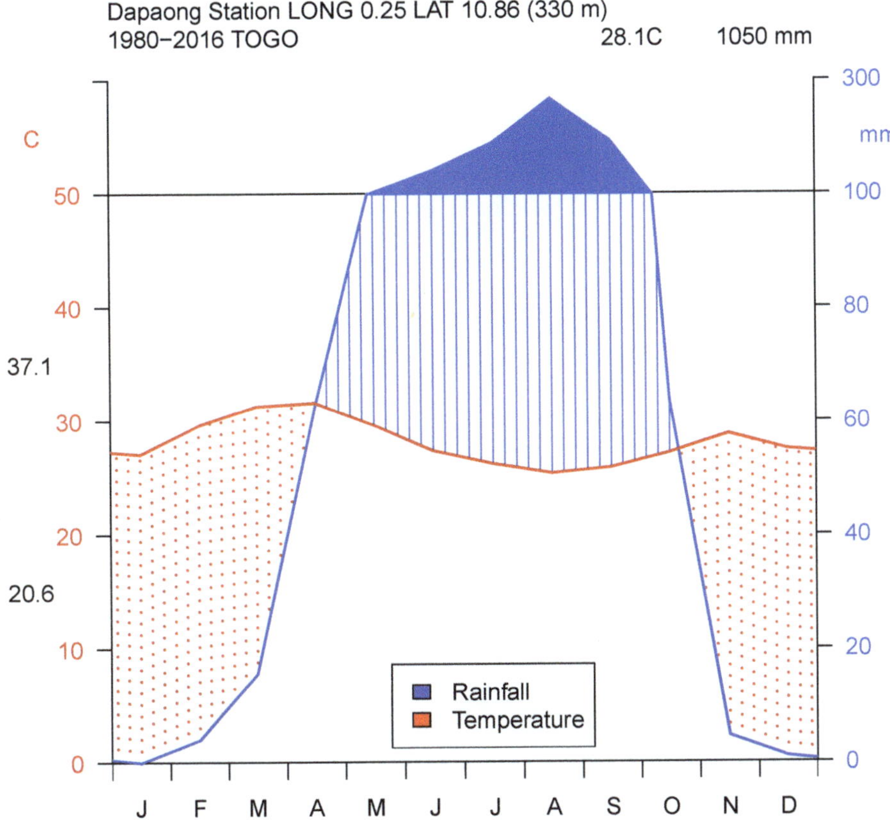

Figure 2. Walter–Lieth [55] climate diagram for northern Togo based on data collected at Dapaong Meteorological Station (Latitude: 10°51′44.10″ N, Longitude: 0°12′27.43″ E, Altitude: 330 m above sea level). Rainfall and temperature data were measured between 1980 and 2016.

With a population density of 96 inhabitants per km^2, over 88% of the population live under the poverty line (US\$ 2/day) [56,57]. Complicated communal land tenure favors men, and encourages farm fragmentation. Women access only marginal lands characterized by reduced soil fertility. Most farmers are smallholders with less than 1.5 ha of land under cultivation [53]. Crop yields are generally low due to erratic rainfall, low soil fertility, low-quality seeds, and inappropriate land preparation tools, among others. Farmers' livelihood depends on small-scale farms with low input, and mixed crop–livestock agriculture. Regarding poultry, most farmers have local hens, cocks, and guinea fowls in their houses. Some families raise local dwarf goats and pigs [53].

2.2. Methods

2.2.1. Adapted Framework for the Evaluation of Irrigation Management Alternatives

In this study, we investigated five irrigation management strategies. These are NI, CDI for supplemental irrigation, CDI for conventional irrigation, FI for supplemental irrigation, and FI for conventional irrigation. The NI is equivalent to the rainfed system, the type of agriculture most farmers are practicing in Dapaong. When rainfall is unevenly distributed throughout the wet season, farmers have the option to apply an optimal amount of irrigation water to supplement the shortage (CDI for SI) or use the fully required amount (FI for SI). On the other hand, in the dry season, farmers can deliberately apply an optimal amount of irrigation water (CDI for conventional irrigation) or fully irrigate the plants (FI for conventional irrigation). When combining these strategies with dry and wet seasons, we obtain the following: (i) NI for the wet season (WS-NI); (ii) CDI for supplemental irrigation system in the wet season (WS-CDI); (iii) full irrigation for supplemental irrigation system in the wet season (WS-FI); (iv) CDI for conventional irrigation system in the dry season (DS-CDI); and (v) full irrigation for conventional irrigation system in the dry season (DS-FI). In this study, one should bear in mind that we only dealt with the physiological and agronomical aspects of DI—crop response to different irrigation regimes—without any economic evaluation. The summary can be seen in Table 1.

Table 1. Irrigation management strategies investigated.

Type of Irrigation System	Irrigation Management Strategies			Application Scenarios	
	Limited Supply		Full Supply		
	Uncontrolled	Controlled	Controlled	Wet Season (WS)	Dry Season (DS)
No irrigation	NI	–	–	x	–
Supplemental irrigation	–	CDI	FI	x	–
Conventional irrigation	–	CDI	FI	–	x

CDI, controlled deficit irrigation; FI, full irrigation; NI, no irrigation.

The OCCASION framework was adapted and used to assess the five irrigation management strategies mentioned above (Figure 3). The adapted framework consists of: (i) a weather generator for simulating long climate time series; (ii) the AquaCrop model, which was used to simulate the irrigation system during the growing season and the yield response of maize to the considered irrigation management strategies (Figure 3, Loop 1); and (iii) a problem-specific algorithm for optimal irrigation scheduling with limited water supply (Figure 3, Loop 2). The latter is named Global Evolutionary Technique for OPTimal Irrigation Scheduling (GET-OPTIS) (For more details, see [33]). A range of given maximum volumes of water is then assigned; a complete CWPF can be derived. The produced CWPF characterizes the maximum yields that can be attained with a given amount of water and is designated the potential CWPF. Then, the crop simulation model was run for a long-term climate time series data yielding a necessary amount of CWPFs. Also, optimized irrigation schedules are obtained. Subsequently, the resulting CWPFs were analyzed, and the SCWPFs obtained through parameters of descriptive statistics such as mean, median, and probability of exceedance, among others. SCWPFs are empirical probability functions where, for every volume of applied irrigation water, the marginal distribution function of the yield related to it can be derived. The probability of exceedance represents the reliability that a specific yield can be achieved [32].

2.2.2. Processing of Climate Data and Set-Up of the LARS Weather Generator

Historical weather observations, including daily maximum temperature, daily minimum temperature, daily rainfall, daily wind speed, daily minimum humidity, and daily maximum humidity were obtained from the nearest meteorological station to the study site—courtesy of the National Weather Service of Togo. These daily weather data available at the station range from 1983 to 2011. In addition, the observed monthly rainfall and maximum and minimum temperatures data from 1980 to 2016 were provided. These monthly data were utilized to characterize the climate of northern Togo with the climate diagram of Walter and Lieth [55]. The Dapaong meteorological station is located at latitude 10°51′44.10″ N, longitude 0°12′27.43″ E, and altitude 330 m above sea level (Figure 1). The solar radiation data, as well as sunshine hours data, were not available at Dapaong weather station. As a substitute, the uncorrected gridded incident solar radiation from the Prediction of Worldwide Energy Resource dataset from the National Aeronautics and Space Administration project NASA-POWER [58] was utilized. Van Wart et al. [59] showed that NASA-POWER is a good source of climate data for crop yields simulation studies. It is publicly accessible, shows acceptable general agreement with ground data for incident solar radiation, and has been used by similar previous studies (See Section 2.2.4).

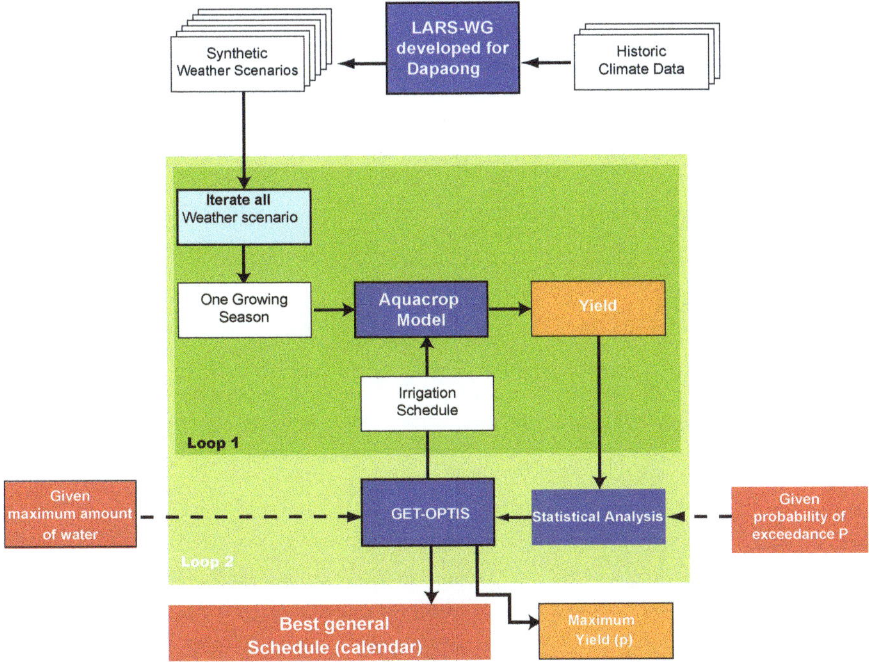

Figure 3. General framework for generating stochastic crop water production functions (adapted from Schütze and Schmitz [36]).

Since the 29-year period (1983–2011) of weather data is not long enough to be used in the assessment of climate variability effect on crop yield, the Long Ashton Research Station Weather Generator version 4.0 (LARS-WG)—a stochastic weather generator—was used to generate a 100-year period of near future climate data. In this study, out of the existing weather generators, LARS-WG was used for two reasons. Firstly, it uses more complex distributions for weather variables and has been tested for diverse climates and found to be better than some other weather generators such as WGEN [60] (Appendix A). Secondly, Semenov [61] recently tested LARS-WG at different locations across the world and revealed its ability to model rainfall extremes with acceptable performance. Similarly, Mehan et al. [62] provided insights into the suitability of LARS-WG for use with water resource applications. Guo et al. [63] suggested performing more than a single realization when generating weather data using LARS-WG for hydrologic and environmental applications. We assessed the performance of the LARS-WG in simulating weather data of Dapaong by comparing the observed and the simulated data with the Kolmogorov–Smirnov test (KS-test). We used the KS-test for the comparison of the probability distributions for each month. The KS-test is a non-parametric and distribution-free test that tries to determine if two datasets are extensively different and come from different distributions. It is an alternative to the Chi-square goodness of fit test. The KS-test compares the two empirical distribution functions as in Equation (1) [64].

$$D = |E_1(i) - E_2(i)| \tag{1}$$

where E_1 and E_2 represent the empirical distribution functions of the two distributions, and D is the absolute difference between them.

The KS-test examines for changes in distributions coming from the generated and observed weather. The KS-test calculates a test statistic and an equivalent *p*-value [65]. It shows how likely it is that the generated and observed data originate from the same distribution. If the *p*-value is very low and below the significance level, set to 0.01 or 0.05, the simulated climate is unlikely to be the same as the "true" climate. Although a *p*-value of 0.05 is the standard significance level employed in most statistics, the authors of the LARS-WG model recommended that a *p*-value of 0.01 should be considered as the satisfactory significance level.

The calibrated LARS-WG for Dapaong was then used to forecast the 100-year daily rainfall and temperature data mentioned above for a near future. For this, the outputs of the General Circulation Models (GCMs) HADCM3 (Hadley Centre Coupled Model version 3) of the IPCC Special Report on Emission Scenarios (SRES) A2 were inputted into LARS-WG. The HADCM3 is the product of the UK Meteorological Office, gridded as $2.5° \times 3.75°$. These long-term data were used to run the AquaCrop model to assess the five irrigation management strategies.

2.2.3. Description and Set-Up of the Crop Simulation Model

AquaCrop, a water-driven crop simulation model, was developed in 2009 by the Food and Agriculture Organization (FAO) of the United Nations [38–40]. The development of the AquaCrop model is based on the algorithm of yield response to water in FAO Irrigation and Drainage Paper No. 33 [66]. AquaCrop evolves from the previous Doorenbos and Kassam [66] K_y approach (Equation (2)), where relative evapotranspiration (ET) is pivotal in calculating yield.

$$\frac{(Y_x - Y_a)}{Y_x} = K_y \left[\frac{(ET_x - ET_a)}{ET_x} \right] \tag{2}$$

where Y_x and Y_a are the maximum and actual yield, respectively; ET_x and ET_a are the maximum and actual evapotranspirations, respectively; and K_y is the proportionality factor between relative yield loss and relative reduction in evapotranspiration.

AquaCrop simulates crop yield in four steps: crop development, crop transpiration, biomass formation, and yield formation [40]. Four water stress response coefficients are considered in the model. These are related to canopy expansion, stomatal conductance, canopy senescence, and harvest index [67].

2.2.4. Soil Data and Calibration of the Crop Simulation Model

We retrieved the physical characteristics data of soils in Dapaong from Poss [68]. These measured soil physical characteristics were used as input into the Soil Water Hydraulic Properties Calculator (http://hydrolab.arsusda.gov/soilwater/Index.htm) to compute various soil hydraulic parameters required to run AquaCrop. We used this soil water hydraulic properties calculator because it has been employed in previous studies in the West African region (e.g., Akumaga et al. [69]). These include volumetric soil water content at field capacity, permanent wilting point, saturation, and saturated hydraulic conductivity (Table 2). Poss [68] classified the soil of Dapaong as sandy loam. According to the World Reference Base for Soil Resources, the soil in northern Togo is characterized Dystric-Ferric Luvisols [70,71].

Table 2. The soil description and properties of Dapaong (See Poss [68]).

Soil Depth (cm)	Texture			OM (%)	dB (g/cm)	SAT (Vol.%)	FC (Vol.%)	PWP (Vol.%)	Ksat (mm/da)	Textural Class
	Sand (%)	Silt (%)	Clay (%)							
0–20	72.5	20.5	7.0	1.5	1.5	42.7	13.3	5.3	1252.6	Sandy Loam
20–50	72.0	19.0	9.0	0.9	1.6	40.8	13.5	5.9	503.0	Sandy Loam
50–110	66.5	18.0	15.5	0.7	1.6	39.9	18.3	10.0	239.5	Sandy Loam

FC, field capacity; PWP, permanent wilting point; SAT, saturation (SAT); Ksat, saturated hydraulic conductivity; dB, soil bulk density; OM, organic matter content in the soil.

Regarding the crop parameters, some of them were assumed to be conservative. The values of conservative parameters used in our study are the same as values proposed by FAO [72] (not presented here). The others, non-conservative or crop-specific, were estimated using measured data retrieved from the ITRA [51], Didjeira et al. [52], and Worou and Saragoni [73] studies conducted in northern Togo (Table 3). These data were used to fine-tune the maize parameters to the local agronomic and management conditions of the study area before running the simulations in AquaCrop. These parameters include information about sowing, canopy cover, canopy senescence, flowering, rooting depth, harvest index, soil management, and the maize cultivar used. Regarding the calibration of the canopy cover, we used the options in AquaCrop to estimate the initial canopy cover (CCo) from sowing rate, seed weight, seed number and estimated germination rate. Subsequently, the canopy expansion rates were automatically estimated by AquaCrop after we entered the phenological dates such as dates of emergence, maximum canopy cover, senescence and maturity. The AquaCrop model simulations were run in growing degree day (GDD) calculated from temperature data used as climate input. Geerts et al. [74], Salemi et al. [75], and Silvestro et al. [76] reported on the most sensitive parameters in AquaCrop obtained through sensitivity analysis testing. The essential crop-specific parameters used to calibrate the AquaCrop model for simulating maize growth and productivity for the study area are presented in Table 3. It should be noted that the calibration of AquaCrop model in this study is preliminary, thus the conclusions that emanated from the simulations are qualitative. The main idea was to compare the irrigation management strategies assessed in this study qualitatively.

Table 3. Non-conservative parameters adjusted and agronomic information for Dapaong, Togo.

Parameter Description	Value	Units or Meaning
Time from sowing to emergence	7 (135)	DAP(GDD)
Time to maximum canopy cover	60 (1109)	DAP(GDD)
Time from sowing to maximum rooting depth	67 (1257)	DAP(GDD)
Time from sowing to start of canopy senescence	76 (1408)	DAP(GDD)
Time from sowing to maturity	100 (1898)	DAP(GDD)
Time from sowing to flowering	54 (1018)	DAP(GDD)
Duration of flowering	10 (183)	DAP(GDD)
Length of building up HI	42 (778)	DAP(GDD)
Maximum effective rooting depth, Z	1	meter
Minimum effective rooting depth, Zn	0.3	meter
Reference harvest index, HI	50	%
Cultivar (TZEE-W)	–	TZEE-W
Planting method	–	Direct sowing
Planting density	62,500	Plants/ha
Soil fertility	65	Moderate (%)
Surface mulches	0	%
Curve number, CN	66	–
Readily Evaporable water, REW	2	mm

DAP, days after planting; GDD, growing degree days; HI, harvest index.

Table 4 summarizes the potential and selected sources of the input data used in this study and reasons for selecting these specific sources.

Table 4. Input data sources.

Type of Data	Possible Sources	Selected Sources for the Study	Reasons of Selecting Specific Sources for the Study
Temperature, rainfall, wind speed, and humidity	-Local meteorological station -Observed data online (NOAA, etc.) -Satellite data (NASA, etc.)	Local meteorological station	Observed data with no missing values
Solar radiation and sunshine hours	-Observed data online (NOAA, etc.) -Satellite data (NASA, etc.)	Satellite data (NASA-POWER project)	Publicly accessible, shows acceptable general agreement with ground data
Soil data	-Poss [68] -National soil survey -FAO Harmonized World Soil Database -ISRIC Soil Geographic Databases	Poss [68]	Publicly accessible and with good resolution (field)
Crop data: conservative parameters	AquaCrop manual	AquaCrop manual	In line with AquaCrop model
Crop data: non-conservative parameters	-AquaCrop manual -ITRA [51], Didjeira et al. [52], and Worou and Saragoni [73]	ITRA [51], Didjeira et al. [52], and Worou and Saragoni [73]	Specific to the maize variety used in the study

2.2.5. Optimal Irrigation Scheduling with Limited Water Supply

Matlab, AquaCrop interface, and Plugin-ACsaV40 (version 4; http://www.fao.org/aquacrop/en/) were used to simulate multiple projects for successive years. The soil and crop phenological data described in Tables 2 and 3, respectively, were used to calibrate AquaCrop. First, AquaCrop was run for a given amount of irrigation water for the maize crop under a specific climate scenario during the dry season of the Dapaong area. GET-OPTIS was employed as irrigation scheduling optimizer and crop yield maximizer. Then, we iterated over a range of given water volumes. As a result, a complete crop-water production function (CWPF) was derived. The 100-year maize crop simulations were run for the wet season as well as the dry season to assess the irrigation management strategies described above, in northern Togo.

3. Results and Discussion

3.1. Traits of the Climate in Dapaong

The temperature is high during the dry season reaching 37 °C and 26 °C maximum and minimum temperatures, respectively, while, in the wet season, the maximum temperature is 30 °C and the minimum temperature is close to 26 °C (Figure 4a). Due to these high temperatures, especially in the dry season, it is likely that the evapotranspiration is relatively high in the area. This argument is corroborated by Djaman and Ganyo [77] who found that the potential annual reference evapotranspiration—computed using the FAO-56 Penman–Monteith method—in northern Togo is higher than 1800 mm on average.

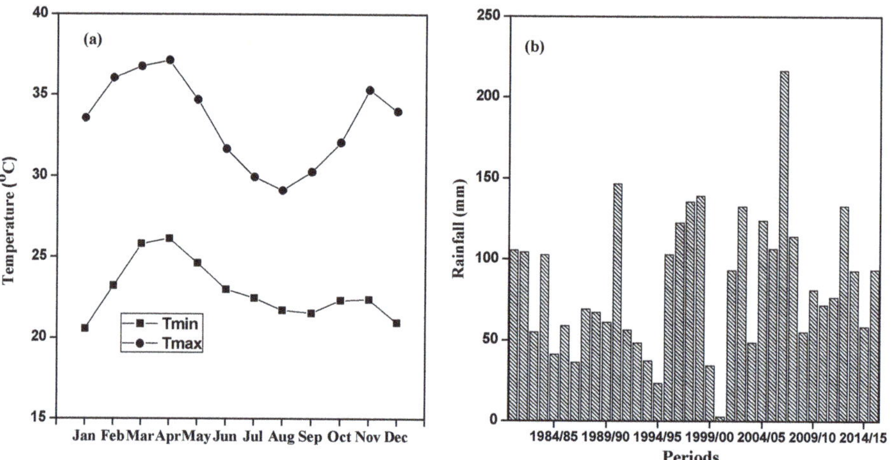

Figure 4. (**a**) Monthly mean temperature; and (**b**) mean total rainfall from November to April in Dapaong, Togo (1980–2016).

Figure 4b depicts the mean total rainfall during the dry season (November–April) in Dapaong district. The rainfall recorded during the dry season varies significantly from year to year. On average, the total rain that falls within this period is lower than 85 mm. In some years, the volume of rain which falls in the same period is up to 100 mm. The highest amount was reached in 2006/2007 (216 mm). Globally, this rainfall occurs on an average of five days only. Thus, none of the main cereals grown in the area such as maize, millet, and sorghum can survive under the dry season climatic conditions without an additional water supply. These findings prove again the fact that farmers only grow crops during the wet season. Overall, the climate of Dapaong in northern Togo is unfavorable to agricultural activities throughout the year because of its vagaries and uncertainties compromising crop yield. These results are in agreement with studies by Ogounde and Abotchi [16].

3.2. Validation and Application of the LARS Weather Generator

The LARS-WG model showed robust compliance between observed and simulated data for the maximum as well as minimum temperatures (Table 5). These findings showed no significant differences between the observed and simulated temperatures for all months. All *p*-values were close to one. It means that the observed and simulated data were from the same distribution. Therefore, based on these results, we conclude that the performance of the LARS-WG model in simulating the climatic variables such as minimum and maximum temperatures of Dapaong district is satisfactory. Similar results were obtained by Semenov et al. [60] at 18 sites in the USA and Europe. However, the standard deviations of the monthly mean simulated values are less than half of the standard deviations of observed values for all months. This means that the extreme temperature values in the minimum and maximum temperatures simulated are smaller than in the observed data.

The observed and simulated rainfall values for most of the months do not correlate significantly (Table 5). This result agrees with studies by Osman et al. [78] in Iraq. However, there are significant differences between December and January, when LARS-WG was incapable of reproducing the observed rainfall, partly because these periods are the driest during the dry season. The standard deviations of the monthly mean rainfall of observed and predicted values are similar for January, February, and April (Table 5). These results imply that there are fewer extreme rainfall values in the dry months, which are of our interest in this study. Overall, the performance of LARS-WG in predicting the rainfall of the Dapaong area is at an acceptable level. It means that the quality of the long-term data that were generated based on these calibration results is not affected.

3.3. Evaluation of Irrigation Management Strategies

3.3.1. Wet Season—Rainfed and Supplemental Irrigation Systems

➢ *Maize Crop under Rainfed Conditions (WS-NI)*

While Figure 5a shows the results of the expected maize crop yields that can be achieved during the rainfed cropping system, Figure 5b portrays the rainfall statistics within the same period. The volume of rainwater that falls within the cropping period of the wet season in Dapaong ranges from 450 mm to 1100 mm approximately. The frequency of the rainfall is high, between 600 mm and 900 mm (Figure 5b). The distribution of the expected rainfed yields is moderately skewed left with a higher coefficient in absolute values (1.91) (Figure 5a). The standard deviation of the expected yields obtained under rainfed conditions is higher than in the case of irrigated maize, regardless of the volume of water used, in northern Togo (See Section 3.3.2). These results show that the variability, as well as the uncertainty, in the yields, are higher under the rainfed conditions (WS-NI) than under the dry season CDI and FI. The high variability under rainfed conditions is likely due to inadequate rainfall distribution and dry spells in the wet season [79]. On average, the expected maize crop yield achieved in the wet season is 3.5 Mg/ha (Figure 5a). These results agree with the findings by Didjeira et al. [52] who indicated the range of 3.5–5 Mg/ha as the expected yield for the maize variety used in this study. Similarly, these results are in line with that of Fosu-Mensah [80] who reported that, in sub-humid Ghana under projected climate change (2030–2050) for scenario A1B of IPCC, the rainfed maize grain yield varies from 3.16 Mg/ha to 4.09 Mg/ha. Therefore, the calibrated AquaCrop model in this study performs well. These results can be improved if data on more site-specific parameters are made available. Akumaga et al. [69] suggested that the AquaCrop model can be utilized as a tool in the study and modeling of maize productivity in West African region.

Table 5. Kolmogorov–Smirnov test statistics for rainfall, maximum and minimum temperatures in Dapaong.

Month	RAINFALL				MAXIMUM TEMPERATURE				MINIMUM TEMPERATURE			
	SD of Observed Data	SD of Simulated Data	K-S	*p*-Value	SD of Observed Data	SD of Simulated Data	K-S	*p*-Value	SD of Observed Data	SD of Simulated Data	K-S	*p*-Value
January	0.11	0.17	0.57	0.00	1.34	0.46	0.11	1.00	1.60	0.52	0.05	1.00
February	14.89	13.89	0.17	0.84	1.24	0.41	0.16	0.91	1.66	0.48	0.11	1.00
March	19.53	31.90	0.15	0.94	0.73	0.28	0.11	1.00	1.04	0.36	0.16	0.91
April	44.99	43.27	0.11	1.00	0.95	0.43	0.11	1.00	0.84	0.43	0.11	1.00
May	38.39	44.09	0.05	1.00	1.25	0.40	0.11	1.00	0.83	0.38	0.11	1.00
June	54.58	53.28	0.03	1.00	0.89	0.32	0.05	1.00	0.72	0.30	0.05	1.00
July	69.68	85.13	0.05	1.00	0.72	0.37	0.05	1.00	0.57	0.26	0.05	1.00
August	85.44	99.96	0.06	1.00	0.60	0.30	0.05	1.00	0.57	0.26	0.11	1.00
September	61.77	68.39	0.08	1.00	0.58	0.37	0.05	1.00	0.58	0.25	0.05	1.00
October	43.20	57.00	0.01	1.00	1.04	0.40	0.11	1.00	0.83	0.27	0.05	1.00
November	12.00	15.73	0.13	0.98	0.83	0.26	0.11	1.00	1.34	0.34	0.11	1.00
December	5.66	10.98	0.26	0.36	1.04	0.43	0.05	1.00	1.38	0.44	0.05	1.00

SD, standard deviation; K-S, Kolmogorov-Smirnov test coefficient.

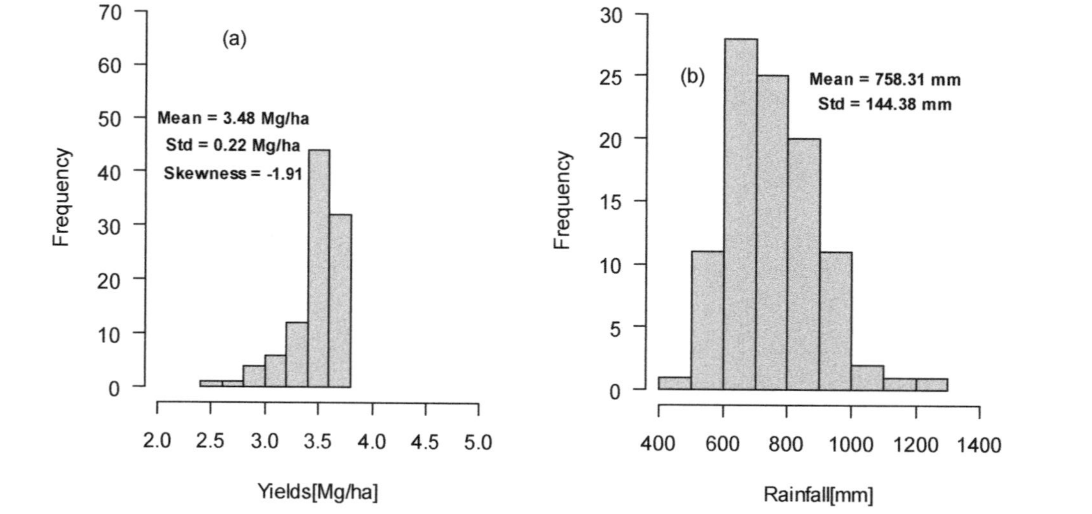

Figure 5. Histogram of distributions of: (**a**) expected yield of maize grown in a rainfed system (WS-NI); and (**b**) the rainfall during the wet season in Dapaong.

> *Maize under Supplemental Irrigation (WS-CDI and WS-FI)*

To improve yield while reducing its variability at the same time, one may apply supplemental irrigation during the rainfed cropping system whenever the crops are experiencing severe water stress, and rainfall is not occurring. The stochastic crop-water production functions for supplemental irrigation conditions are shown in Figure 6a. It can be hypothesized that, when more than 150 mm supplemental irrigation water is applied, the variation in the resulting expected crop yield is likely due to the variation of temperature and radiation in the area. These assumptions are supported by the nearly symmetric distributions of the corresponding expected crop yields (Figure 6a). Besides, at volumes of supplemental water lower than 150 mm, the variation in the expected crop yield can result from the combined effects of the uneven distribution of rainfall and the climate parameters mentioned above. The 90% of SCWPF exceedance probability of yield achievement seems to be the best option for enhancing food security in northern Togo. This might be because it is the only option which helps to achieve the highest level of crop yield improvement (15% or more) (Figure 6a). Applying supplemental irrigation in northern Togo for maize crop cultivation will not only contribute to improving crop grain yield and enhancing food security [81–83] but also help to improve farmers' livelihood. Nevertheless, supplemental irrigation alone cannot improve the rainfed yields significantly; it needs to be combined with other field management aspects such as soil preparation and fertility, pests and diseases management, and the choice of suitable crop varieties. It can be concluded that CWPF is a useful planning tool to assess water requirement for crops, especially in water-scarce regions. Heng et al. [84] and Stricevic et al. [85] reported that, due to its sufficient degree of simulation accuracy, the AquaCrop model is a valuable tool for estimating crop productivity under rainfed conditions, deficit and supplemental irrigation, and on-farm water management strategies for improving the efficiency of water use in agriculture.

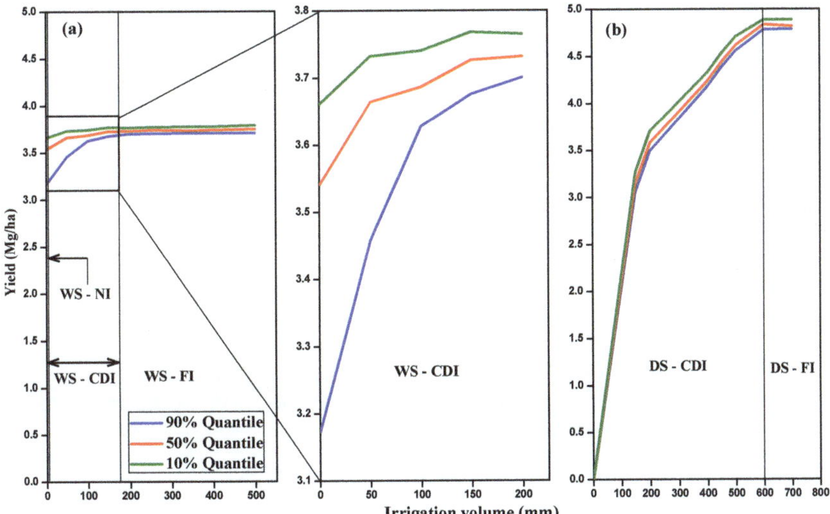

Figure 6. Stochastic crop-water production function for: (**a**) rainfed and supplemental irrigated systems in the wet season; and (**b**) optimized conventional irrigation system in the dry season for maize in Dapaong.

Figure 7 shows the detailed results of the expected yields at various amounts of supplemental irrigation water. With supplemental irrigation (WS-CDI), the rainfed yield increased from 3.48 Mg/ha to 3.74 Mg/ha. The yield becomes constant when the volume of water applied is equal to or greater than 150 mm. Then, the variability in the yields as well as the skewness decreases in absolute value.

These results imply that supplemental irrigation is beneficial up to 150 mm. Above this value, the advantages of supplemental irrigation (WS-FI) become insignificant. Therefore, rainfed maize crop yields may be improved in northern Togo by applying supplemental irrigation assuming that water is available.

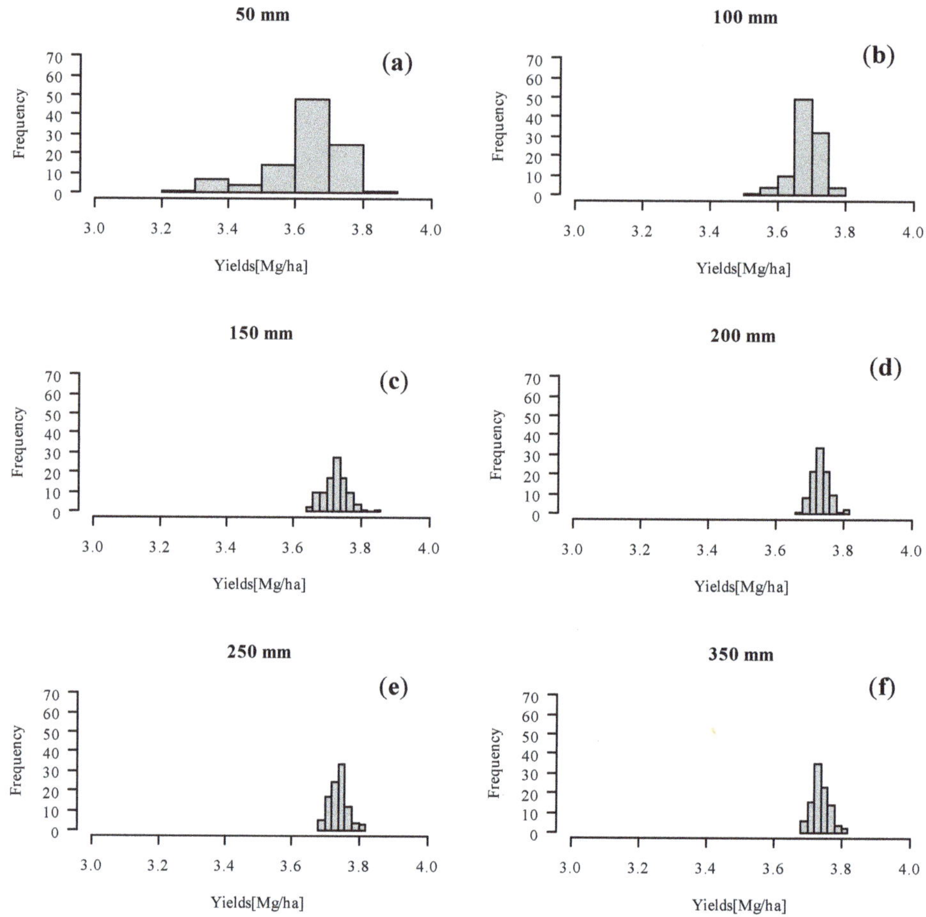

Figure 7. Histogram of distributions of expected yield using water for supplemental irrigation of maize in the wet season in Dapaong: (**a**) 50 mm; (**b**) 100 mm; and (**c**) 150 mm (WS-CDI); and (**d**) 200 mm; (**e**) 250 mm; and (**f**) 350 mm (WS-FI).

3.3.2. Dry Season—Conventional Irrigation System (DS-CDI and DS-FI)

Figure 6b shows the stochastic crop-water production functions (SCWPF) for optimized irrigated maize crop in the dry season in northern Togo. The quantile percentage represents the probability of exceedance. Since rainfall can be ruled out, it is believed that, when the optimal full irrigation conditions are met, the variation of temperature and radiation can explain the variability in the expected crop yield. These assumptions are corroborated by the nearly symmetric distributions of the expected crop yields at full irrigation (Figure 8). These findings are supported by the results presented by Schütze and Schmitz [36]. These two parameters are part of the yield defining factors, as highlighted in the papers explaining the principles of ecology production [86]. In addition, for volumes of water lower than full irrigation, the variation in the expected crop yield can result from the combined effects of drought stress on crops and the climate parameters mentioned above. The maximum expected yields were 4.79 Mg/ha (90% quantile) and 4.89 Mg/ha (10% quantile) at near full irrigation (600 mm) (Figure 6b). The controlled deficit irrigation ranges from 0 to 600 mm for maize in northern Togo. The DS-CDI strategy seems to save water with an insignificant reduction in the grain yield relative to full irrigation [87–92]. Overall, growing maize crop in the dry season in northern Togo may be feasible under CDI if water is available. Irrigation is vital for improving crop yield and stabilizing crop production [93] amidst the threats of climate change [94].

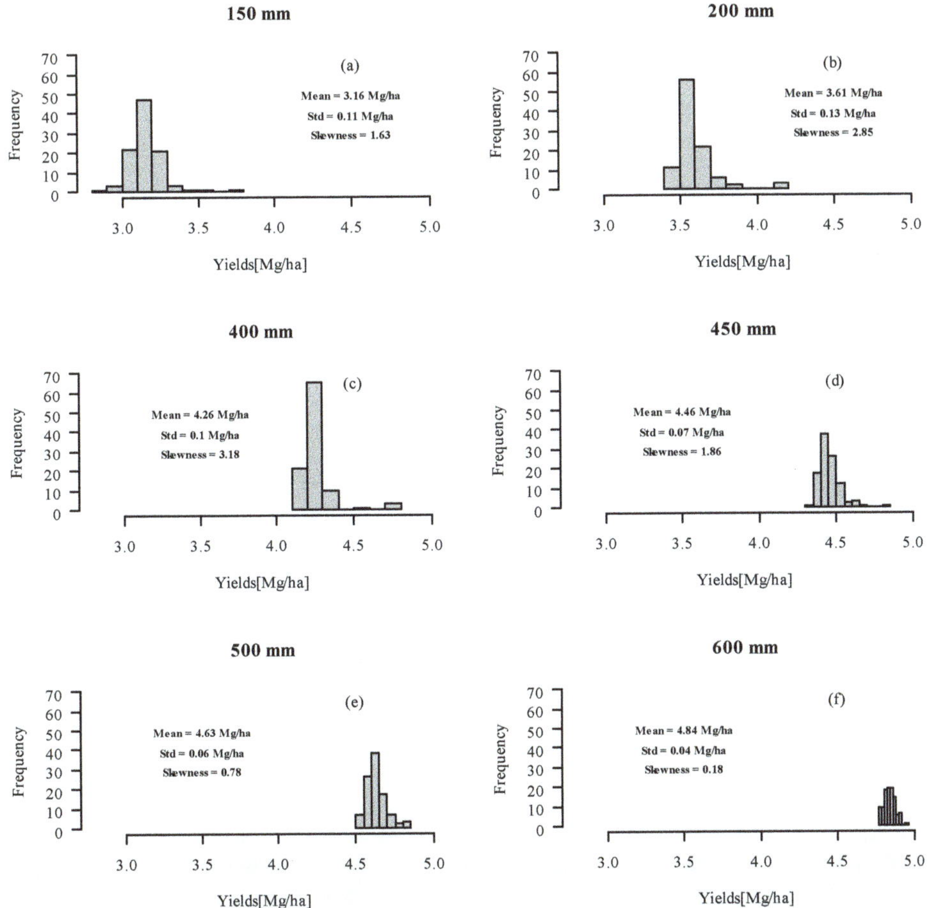

Figure 8. Histogram of distributions of expected yield using water for irrigation of maize in the dry season in Dapaong: (**a**) 150 mm; (**b**) 200 mm; (**c**) 400 mm; (**d**) 450 mm; and (**e**) 500 mm (DS-CDI); and (**f**) 600 mm (DS-FI).

In Figure 8, detailed results of the expected yields at various amounts of irrigation water are given. There is a change in the histogram distribution among the various volumes of irrigation water. The average expected yields concerning the amount of irrigation water used range from 3.16 Mg/ha to 4.84 Mg/ha at 150 mm and 600 mm, respectively. With the increasing application of irrigation water (DS-CDI), the yield increases to a level at which additional water supply fails to raise the crop yield any further (around 600 mm). Thus, the latter volume of water is assumed to be near full irrigation. The frequency distribution shows a positive sign for all the histograms. The coefficients of skewness of the expected yields for 150 mm, 200 mm, and 400 mm water volumes are 1.63, 2.85, and 3.18, respectively. On the contrary, at 600 mm volume of water (DS-FI), the distributions of the expected yields are symmetrical. In addition, the standard deviation is relatively low for the yields at these volumes of water. Abedinpour et al. [95] reported that the AquaCrop model can predict maize yield with acceptable accuracy under variable irrigation in a semi-arid environment.

3.4. Summary of the Discussion

The variability in rainfall during the wet season (WS-NI) was high, inducing a considerable variability in the expected yield for rainfed conditions. The variability in the expected yield would decrease significantly if supplemental irrigation (WS-CDI or WS-FI) were applied. At the same time, supplemental irrigation would improve the expected yields and contribute to avoiding crop failure. The dry season irrigation management strategies (DS-CDI and DS-FI) would increase yield potential and decrease the variability of expected yield at the same time. Thus, the application of supplemental or dry season irrigation management strategies investigated in this study would help to enhance food availability in the West African region.

There are a few caveats that readers should keep in mind when interpreting the results of this study: The AquaCrop model in this study was calibrated with crop and soil data retrieved from previous studies conducted in the area. Thus, the conclusions derived from the outputs of the model simulation are qualitative—ranking of the irrigation management strategies assessed in the study. There are several uncertainties in the general circulation model outputs as well as crop model simulations. The uncertainties related to crop yield exist because AquaCrop assumes a disease- and pest-free environment and considers no effect of weed or extreme climate events such as flooding. Another point worth considering is that, by concluding that there is potential for deficit and supplemental irrigation for maize crop in northern Togo, we assumed that a proper soil fertility management is guaranteed, and water is available for irrigation management. Finally, it is important to note that substantial investments in irrigation infrastructure, as well as extension services to farmers, would be necessary to enhance food security in northern Togo. The calibrated crop model needs to be validated with experimental data to improve the accuracy of the resulting simulations.

4. Conclusions

The AquaCrop model was used to assess the potential of deficit and supplemental irrigation in the dry savannah area of northern Togo under climate variability. For this, the climate of the study area was characterized. The performance of the weather generator used to produce the long-term time series climate data for the crop simulation was also evaluated. In summary, the climate of northern Togo is unimodal with the dry season ranging from November to April. According to Köppen–Geiger's classification, the climate is hot semi-arid in northern Togo. During the dry season the mean maximum and minimum temperatures are 35 °C and 25 °C, respectively, and the mean total rainfall is 85 mm. In short, the performance of the LARS Weather Generator in predicting the climate of northern Togo was found satisfactory. Overall, we found that the deficit irrigation water requirement ranges from 0 to 600 mm. The maximum expected maize grain yield that can be reached under irrigated conditions is 4.84 Mg/ha with TZEE-W local variety. The rainfed yield can be improved from 3.48 to 3.74 Mg/ha with 150 mm of supplemental irrigation water. At the same time, the variability in the yield was significantly reduced. Irrigation practice in agriculture helps to lower crop yield variability as well as crop failure.

Thus, growing maize crop in the dry season in northern Togo may be feasible. In general, irrigation can help to alleviate food insecurity, while supplemental irrigation is a climate-related management practice for crop yield improvement. The latter also contributes to improving farmers' livelihood. Further maize crop genetic improvements would be needed to fine-tune the seeds to the dry season climate. Irrigation infrastructures would be needed to implement in northern Togo the irrigation management strategies investigated in this study. In addition, realistic irrigation water pricing and cost recovery policies should be enforced and followed by all stakeholders to maintain the irrigation infrastructures and ensure the viability of the system. Institutional reforms relevant to the development and management of irrigation systems should be made. The complicated land tenure issue in northern Togo needs to be addressed to incentivize investment in, and management of, irrigation systems. Moreover, the institutional arrangement—market and connectivity among farmers and other agents—should be improved.

To develop regional water management strategies, the adapted framework used in this study may be applied to other sites in the West African region. Field experiments are needed to validate the results of this study before the implementation of its recommendations. In addition, the framework can be extended by adding a soil variability dimension to it. The analysis can be made more comprehensive by considering farmers' socioeconomic characteristics.

Author Contributions: A.G.-T. and N.S. developed the concept and design of the numerical experiment. A.G.-T. carried out the simulations, analyzed the data, and wrote the manuscript; T.A. and N.S. critically reviewed the manuscript. All authors revised and approved the final manuscript.

Funding: This research was supported by a grant to A.G.-T. PhD scholarship under the Merit Scholarship Programme (MSP) 2015/2016 of the Islamic Development Bank (IsDB).

Acknowledgments: This research received logistical assistance from the United Nations University Institute for Integrated Management of Material Fluxes and Resources (UNU-FLORES) and Technische Universität Dresden (TU Dresden), Germany. We extend our thanks to the administration of the national meteorological service of Togo for providing us with the climate data. Our gratitude goes to the editor and anonymous reviewers whose comments and suggestions expressively contributed to the improvement of this paper. Our thanks also go to Atiqah Fairuz Salleh for her editorial input to the manuscript.

Conflicts of Interest: The authors have no competing interests to declare.

Appendix A

List of Abbreviations

APSI	Agricultural Production Systems Simulator
MCDI	Controlled Deficit Irrigation
CWPF	Crop-Water Production Functions
DAISY	Danish simulation model for transformation and transport of energy and matter in the soil plant atmosphere system
DAP	Days After Planting
DI	Deficit Irrigation
DS	Dry Season
DSSAT	Decision Support System for Agrotechnology Transfer
FI	Full Irrigation
GET-OPTIS	Global Evolutionary Technique for OPTimal Irrigation Scheduling
GDD	Growing Degree Days
HI	Harvest index
ICID	International Commission on Irrigation and Drainage
IPCC	Intergovernmental Panel on Climate Change
ISRIC	International Soil Reference and Information Centre
ITRA	Togolese Institute of Agricultural Research
LARS-WG	Long Ashton Research Station Weather Generator
MERF	Togolese Ministry of the Environment and Forestry
NI	No Irrigation
OCCASION	Optimal Climate Change Adaption Strategies in Irrigation
PILOTE	An operative crop model for soil water balance and yieldestimations under conventional tillage
REW	Readily Evaporable Water
SCWPF	Stochastic Crop-Water Production Functions
SI	Supplemental Irrigation
SSA	Sub-Saharan Africa
WGEN	Weather Generator
WS	Wet Season

References

1. UN DESA. *World Population Prospects: The 2015 Revision, Key Findings and Advance Tables*; UN DESA: New York, NY, USA, 2015.

2.   UN DESA. *World Population Prospects: The 2017 Revision, Key Findings and Advance Tables*; UN DESA: New York, NY, USA, 2017.

3.   FAO. *Climate-Smart Agriculture Sourcebook*; FAO: Rome, Italy, 2013; ISBN 978-92-5-107720-7.

4.   Liniger, H.; Mekdaschi Studer, R.; Hauert, C.; Gurtner, M. *Sustainable Land Management in Practice: Guidelines and Best Practices for Sub-Saharan Africa*; TerrAfrica, World Overview of Conservation Approaches and Technologies (WOCAT) and Food and Agriculture Organization of the United Nations (FAO): Rome, Italy, 2011; ISBN 9789250000000.

5.   Elliott, J.; Deryng, D.; Müller, C.; Frieler, K.; Konzmann, M.; Gerten, D.; Glotter, M.; Flörke, M.; Wada, Y.; Best, N.; et al. Constraints and potentials of future irrigation water availability on agricultural production under climate change. *Proc. Natl. Acad. Sci. USA* **2014**, *111*, 3239–3244. [CrossRef] [PubMed]

6.   Edgerton, M.D. Increasing Crop Productivity to Meet Global Needs for Feed, Food, and Fuel. *Plant Physiol.* **2009**, *149*, 7–13. [CrossRef] [PubMed]

7.   World Bank. *World Development Report 2008: Agriculture for Development*; World Bank: Washington, DC, USA, 2008.

8.   Rosegrant, M.W.; Paisner, M.S.; Siet, M.; Witcover, J. *2020 Global Food Outlook*; International Food Policy Research Institution: Washington, DC, USA, 2001; pp. 1–24.

9.   van Ittersum, M.K.; van Bussel, L.G.J.; Wolf, J.; Grassini, P.; van Wart, J.; Guilpart, N.; Claessens, L.; de Groot, H.; Wiebe, K.; Mason-D'Croz, D.; et al. Can sub-Saharan Africa feed itself? *Proc. Natl. Acad. Sci. USA* **2016**, *113*, 14964–14969. [CrossRef] [PubMed]

10.   Lobell, D.B.; Gourdji, S.M. The Influence of Climate Change on Global Crop Productivity. *Plant Physiol.* **2012**, *160*, 1686–1697. [CrossRef] [PubMed]

11.   Kotir, J.H. Climate change and variability in Sub-Saharan Africa: A review of current and future trends and impacts on agriculture and food security. *Environ. Dev. Sustain.* **2011**, *13*, 587–605. [CrossRef]

12.   Druyan, L.M. Studies of 21st-century precipitation trends over West Africa. *Int. J. Climatol.* **2011**, *31*, 1415–1424. [CrossRef]

13.   Sarr, B. Present and future climate change in the semi-arid region of West Africa: A crucial input for practical adaptation in agriculture. *Atmos. Sci. Lett.* **2012**, *13*, 108–112. [CrossRef]

14.   Ministère de l'Environnement et des Ressources Forestières (MERF). *Plan d'Action National d'Adaptation aux Changements Climatiques (PANA)*; MERF: Lome, Togo, 2009. (In French)

15.   Mcsweeney, C.; New, M.; Lizcano, G. *UNDP Climate Change Country Profiles, Togo*; School of Geography and Environment, Oxford University: Oxford, UK, 2009.

16.   Ogounde, L.; Abotchi, T. *Quelques contraintes à la croissance Agricole dans la région des Savanes du Nord-Togo. Bulletin de la société Neuchâteloise de Geographie*; Société Neuchâteloise de Geographie: Neuchâtel, Switzerland, 2003. (In French)

17.   Dobermann, A.; Nelson, R.; Beever, D.; Bergvinson, D.; Crowley, E.; Denning, G.; Griller, K.; d'Arros Hughes, J.; Jahn, M.; Lynam, J.; et al. *Solutions for Sustainable Agriculture and Food Systems—Technical Report for the Post-2015 Development Agenda*; The United Nations Sustainable Development Solutions Network (UNSDSN): New York, NY, USA, 2013.

18.   Rockström, J.; Williams, J.; Daily, G.; Noble, A.; Matthews, N.; Gordon, L.; Wetterstrand, H.; De Clerck, F.; Shah, M.; Steduto, P.; et al. Sustainable intensification of agriculture for human prosperity and global sustainability. *Ambio* **2017**, *46*, 4–17. [CrossRef]

19.   Godfray, H.C.J.; Garnett, T. Food security and sustainable intensification. *Philos. Trans. R. Soc. Lond. B. Biol. Sci.* **2014**, *369*, 1–10. [CrossRef]

20.   Bolor, J.K. Analyse de l'état actuel de développement de l'irrigation au Togo. In *Irrigation in West Africa: Current Status and a View to the Future*; Namara, R.E., Sally, H., Eds.; International Water Management Institute (IWMI), Colombo, Sri Lanka: Ouagadougou, Burkina Faso, 2010; pp. 305–312.

21.   Jalloh, A.; Nelson, G.C.; Thomas, T.S.; Zougmoré, R.; Roy-Macauley, H. *West African Agriculture and Climate Change: A Comprehensive Analysis*; IFPRI Research Monograph; International Food Policy Research: Washington, DC, USA, 2013.

22.   International Commission on Irrigation and Drainage (ICID). Basic Introduction: Irrigation. Available online: http://www.icid.org/res_irrigation.html (accessed on 10 September 2018).

23.   Rockström, J.; Hatibu, N.; Oweis, T.; Wani, S.; Barron, J.; Bruggeman, A.; Qiang, Z.; Farahani, J.; Karlberg, L. Managing Water in Rainfed Agriculture. In *Water for Food, Water for Life: A Comprehensive Assessment of Water Management in Agriculture*; Molden, D., Ed.; Earthscan: London, UK, 2007; pp. 315–352.

24. Pereira, L.S. Higher performance through combined improvements in irrigation methods and scheduling: A discussion. *Agric. Water Manag.* **1999**, *40*, 153–169. [CrossRef]

25. English, M.J.; Nuss, G.S. Designing for Deficit Irrigation. *J. Irrig. Drain. Div.* **1982**, *108*, 91–106.

26. Djaman, K.; Irmak, S.; Rathje, W.R.; Martin, D.L.; Eisenhauer, D.E. Maize evapotranspiration, yield production functions, biomass, grain yield, harvest index, and yield response factors under full and limited irrigation. *Am. Soc. Agric. Biol. Eng.* **2013**, *56*, 273–293.

27. English, M. Deficit Irrigation. I: Analytical Framework. *J. Irrig. Drain. Eng.* **1990**, *116*, 399–412. [CrossRef]

28. English, M.; Raja, S.N. Perspectives on deficit irrigation. *Agric. Water Manag.* **1996**, *32*, 1–14. [CrossRef]

29. Lecler, N.L. Integrated methods and models for deficit irrigation planning. In *Agricultural Systems Modeling and Simulation*; Lecler, N.L., Peart, R.M., Eds.; Marcel Dekker Inc.: New York, NY, USA, 1998; pp. 283–299.

30. Fereres, E.; Soriano, M.A. Deficit irrigation for reducing agricultural water use. *J. Exp. Bot.* **2006**, *58*, 147–159. [CrossRef] [PubMed]

31. Kögler, F.; Söffker, D. Water (stress) models and deficit irrigation: System-theoretical description and causality mapping. *Ecol. Model.* **2017**, *361*, 135–156. [CrossRef]

32. Kloss, S.; Pushpalatha, R.; Kamoyo, K.J.; Schütze, N. Evaluation of Crop Models for Simulating and Optimizing Deficit Irrigation Systems in Arid and Semi-arid Countries Under Climate Variability. *Water Resour. Manag.* **2012**, *26*, 997–1014. [CrossRef]

33. Schütze, N.; De Paly, M.; Shamir, U. Novel simulation-based algorithms for optimal open-loop and closed-loop scheduling of deficit irrigation systems. *J. Hydroinformatics* **2012**, *14*, 136–151. [CrossRef]

34. Semenov, M.A. Development of high-resolution UKCIP02-based climate change scenarios in the UK. *Agric. For. Meteorol.* **2007**, *144*, 127–138. [CrossRef]

35. Brumbelow, K.; Georgakakos, A. Consideration of Climate Variability and Change in Agricultural Water Resources Planning. *J. Water Resour. Plan. Manag.* **2007**, *133*, 275–285. [CrossRef]

36. Schütze, N.; Schmitz, G.H. OCCASION: New Planning Tool for Optimal Climate Change Adaption Strategies in Irrigation. *J. Irrig. Drain. Eng.* **2010**, *136*, 836–846. [CrossRef]

37. Jones, J.W.; Hoogenboom, G.; Porter, C.H.; Boote, K.J.; Batchelor, W.D.; Hunt, L.A.; Wilkens, P.W.; Singh, U.; Gijsman, A.J.; Ritchie, J.T. The DSSAT cropping system model J.W. *Eur. J. Agron.* **2003**, *18*, 235–263. [CrossRef]

38. Hsiao, T.C.; Heng, L.; Steduto, P.; Rojas-Lara, B.; Raes, D.; Fereres, E. Aquacrop-The FAO crop model to simulate yield response to water: III. Parameterization and testing for maize. *Agron. J.* **2009**, *101*, 448–459. [CrossRef]

39. Raes, D.; Steduto, P.; Hsiao, T.C.; Fereres, E. Aquacrop-The FAO crop model to simulate yield response to water: II. main algorithms and software description. *Agron. J.* **2009**, *101*, 438–447. [CrossRef]

40. Steduto, P.; Hsiao, T.C.; Raes, D.; Fereres, E. Aquacrop-the FAO crop model to simulate yield response to water: I. concepts and underlying principles. *Agron. J.* **2009**, *101*, 426–437. [CrossRef]

41. Hansen, S.; Jensen, H.E.; Nielsen, N.E.; Svendsen, H. *DAISY: A Soil Plant System Model. Danish simulation Model for Transformation and Transport of Energy and Matter in the Soil Plant Atmosphere System*; The National Agency for Environmental Protection: Copenhagen, Denmark, 1990.

42. Smith, M. *CROPWAT: A Computer Program for Irrigation Planning and Management*; Food and Agriculture Organization of the United Nations, Ed.; FAO irrigation and drainage paper 46; ISBN1 9251031061. Food and Agriculture Organization of the United Nations: Rome, Italy, 1992; ISBN2 9251031061.

43. Keating, B.A.; Carberry, P.S.; Hammer, G.L.; Probert, M.E.; Robertson, M.J.; Holzworth, D.; Huth, N.I.; Hargreaves, J.N.G.; Meinke, H.; Hochman, Z.; et al. An overview of APSIM, a model designed for farming systems simulation. *Eur. J. Agron.* **2003**, *18*, 267–288. [CrossRef]

44. Mailhol, J.C.; Olufayo, A.A.; Ruelle, P. Sorghum and sunflower evapotranspiration and yield from simulated leaf area index. *Agric. Water Manag.* **1997**, *35*, 167–182. [CrossRef]

45. Iqbal, M.A.; Shen, Y.; Stricevic, R.; Pei, H.; Sun, H.; Amiri, E.; Penas, A.; del Rio, S. Evaluation of the FAO AquaCrop model for winter wheat on the North China Plain under deficit irrigation from field experiment to regional yield simulation. *Agric. Water Manag.* **2014**, *135*, 61–72. [CrossRef]

46. Vanuytrecht, E.; Raes, D.; Steduto, P.; Hsiao, T.C.; Fereres, E.; Heng, L.K.; Garcia Vila, M.; Mejias Moreno, P. AquaCrop: FAO's crop water productivity and yield response model. *Environ. Model. Softw.* **2014**, *62*, 351–360. [CrossRef]

47. Department of Immigration and Citizenship (DIC). *Togolese Community Profile*; Department of Immigration and Citizenship, Commonwealth of Australia: Lomé, Togo, 2007.

48. RGPH. *Recensement Générale de la population et de l'habitat. Direction Générale de la Statistique et de la Comptabilité Nationale*; RGPH: Lomé, Togo, 2010.

49. Ali, E. A review of agricultural policies in independent Togo. *Int. J. Agric. Policy Res.* **2017**, *5*, 104–116. [CrossRef]

50. Poch, R.M.; Ubalde, J.M. Diagnostic of degradation processes of soils from northern Togo (West Africa) as a tool for soil and water management. In *Proceedings of the Workshop for Alumni of the M.Sc. Programmes in Soil Science, Eremology and Physical Land Resources*; Langouche, D., Van Ranst, E., Eds.; Workshop IC-PLR: Ghent, Belgium, 2006; pp. 187–194.

51. Institut Togolais de Recherche Agronomique (ITRA). *Bien cultiver et conserver le maïs. Collection Brochures et Fiches Techniques*; ITRA: Lomé, Togo, 2008. (In French)

52. Didjeira, A.; Adourahim, A.A.; Sedzro, K. *Situation de référence sur les principales céréales cultivées au Togo: Maïs, Riz, Sorgho, Mil*; ITRA: Lomé, Togo, 2007.

53. Desplat, A.; Rouillon, A. *Diagnostic agraire dans la région des Savanes au Togo: Cantons de Nioukpourma, Naki-Ouest et Tami*; Master de Recherche, Institut des Sciences et Industries du Vivant et de L'environnement, AgroParisTech: Paris, France, 2011.

54. Kottek, M.; Grieser, J.; Beck, C.; Rudolf, B.; Rubel, F. World Map of the Köppen-Geiger climate classification updated. *Meteorol. Z.* **2006**, *15*, 259–263. [CrossRef]

55. Walter, H.; Lieth, H.H.F. *Klimadiagramm-Weltatlas*; G. Fischer Verlag: Jena, Germany, 1967.

56. Institut National de la Statistique et des Etudes Economiques et Démographiques (INSEED). *Profil de pauvreté: Togo*; INSEED: Lomé, Togo, 2016.

57. Institut National de la Statistique et des Etudes Economiques et Démographiques (INSEED). Statistiques Nationales: Togo 2015. Available online: http://togo.opendataforafrica.org/# (accessed on 13 September 2018).

58. NASA POWER Project-Agroclimatology Data. Available online: https://power.larc.nasa.gov/data-access-viewer/ (accessed on 10 September 2017).

59. Van Wart, J.; Grassini, P.; Yang, H.; Claessens, L.; Jarvis, A.; Cassman, K.G. Creating long-term weather data from thin air for crop simulation modeling. *Agric. For. Meteorol.* **2015**, *209–210*, 49–58. [CrossRef]

60. Semenov, M.A.; Brooks, R.J.; Barrow, E.M.; Richardson, C.W. Comparison of the WGEN and LARS-WG stochastic weather generators for diverse climates. *Clim. Res.* **1998**, *10*, 95–107. [CrossRef]

61. Semenov, M.A. Simulation of extreme weather events by a stochastic weather generator. *Clim. Res.* **2008**, *35*, 203–212. [CrossRef]

62. Mehan, S.; Guo, T.; Gitau, M.; Flanagan, D.C.; Mehan, S.; Guo, T.; Gitau, M.W.; Flanagan, D.C. Comparative Study of Different Stochastic Weather Generators for Long-Term Climate Data Simulation. *Climate* **2017**, *5*, 26. [CrossRef]

63. Guo, T.; Mehan, S.; Gitau, M.W.; Wang, Q.; Kuczek, T.; Flanagan, D.C. Impact of number of realizations on the suitability of simulated weather data for hydrologic and environmental applications. *Stoch. Environ. Res. Risk Assess.* **2018**, *32*, 2405–2421. [CrossRef]

64. Chakravarti, I.M.; Laha, R.G.; Roy, J. *Handbook of Methods of Applied Statistics*; Wiley: New York, NY, USA, 1967.

65. Semenov, M.A.; Barrow, E.M. Use of a stochastic weather generator in the development of climate change scenarios. *Clim. Chang.* **1997**, *35*, 397–414. [CrossRef]

66. Doorenbos, J.; Kassam, A.H. *Yield Response to Water*; FAO Irrigation and Drainage Paper No. 33; FAO: Rome, Italy, 1979.

67. Greaves, G.E.; Wang, Y.-M. Assessment of FAO AquaCrop Model for Simulating Maize Growth and Productivity under Deficit Irrigation in a Tropical Environment. *Water* **2016**, *8*, 557. [CrossRef]

68. Poss, R. *Etude Morphopédologique du Nord du Togo à [au] 1/500,000*; Institut français de recherche scientifique pour le développement en coopération (ORSTOM): Lomé, Togo, 1996.

69. Akumaga, U.; Tarhule, A.; Yusuf, A.A. Validation and testing of the FAO AquaCrop model under different levels of nitrogen fertilizer on rainfed maize in Nigeria, West Africa. *Agric. For. Meteorol.* **2017**, *232*, 225–234. [CrossRef]

70. IUSS Working Group WRB. *World Reference Base for Soil Resources 2014, Update 2015 International Soil Classification System for Naming Soils and Creating Legends for Soil Maps*; FAO: Rome, Italy, 2015.

71. Worou, K.S. *Sols Dominants du Togo—Corrélation avec la Base de Référence Mondiale. Quatorzième Réunion du Sous-Comité ouest et Centre Africain de Corrélation des sols. Rapport sur les Ressources en Sols du Monde 98*; Food and Agriculture Organization of the United Nations (FAO): Rome, Italy, 2002. (In French)

72. Raes, D.; Steduto, P.; Hsiao, T.C.; Fereres, E. Chapter 2—Users guide. In *Reference Manual: AquaCrop, Version 4.0*; FAO, Land and Water Division: Rome, Italy, 2012; pp. 1–164.

73. Worou, S.; Saragoni, H. *La Culture du Maïs de Contre Saison Est-elle Possible au Togo Meridional? Premières Conclusions d'une Experimentation sur la Station de Recherche Agronomique d'ativémé*; Institut français de recherche scientifique pour le développement en coopération (ORSTOM): Lomé, Togo, 1988.

74. Geerts, S.; Raes, D.; Garcia, M.; Miranda, R.; Cusicanqui, J.A.; Taboada, C.; Mendoza, J.; Huanca, R.; Mamani, A.; Condori, O.; et al. Simulating Yield Response of Quinoa to Water Availability with AquaCrop. *Agron. J.* **2009**, *101*, 499–508. [CrossRef]

75. Salemi, H.; Amin, M.; Soom, M.; Lee, T.S.; Farhad Mousavi, S.; Ganji, A.; Kamilyusoff, M. Application of AquaCrop model in deficit irrigation management of Winter wheat in arid region. *Afr. J. Agric. Res.* **2011**, *610*, 2204–2215. [CrossRef]

76. Silvestro, P.C.; Pignatti, S.; Yang, H.; Yang, G.; Pascucci, S.; Castaldi, F.; Casa, R. Sensitivity analysis of the Aquacrop and SAFYE crop models for the assessment of water limited winter wheat yield in regional scale applications. *PLoS ONE* **2017**, *12*, 1–30. [CrossRef] [PubMed]

77. Djaman, K.; Ganyo, K. Trend analysis in reference evapotranspiration and aridity index in the context of climate change in Togo. *J. Water Clim. Chang.* **2015**, *6*, 848–864. [CrossRef]

78. Osman, Y.; Abdellatif, M.; Al-Ansari, N.; Knutsson, S.; Jawad, S. Climate Change and Future Precipitation in Arid Environment of Middle East: Case study of Iraq. *J. Environ. Hydrol.* **2017**, *25*, 1–18.

79. Assefa, S.; Biazin, B.; Muluneh, A.; Yimer, F.; Haileslassie, A. Rainwater harvesting for supplemental irrigation of onions in the southern dry lands of Ethiopia. *Agric. Water Manag.* **2016**, *178*, 325–334. [CrossRef]

80. Fosu-Mensah, B.Y. *Modelling the Impact of Climate Change on Maize (Zea mays L.) YieLd under Rainfed Conditions in Sub-Humid Ghana*; United Nations University–Institute for Natural Resources in Africa (UNU-INRA): Accra, Ghana, 2013.

81. Chauhan, C.P.S.; Singh, R.B.; Gupta, S.K. Supplemental irrigation of wheat with saline water. *Agric. Water Manag.* **2008**, *95*, 253–258. [CrossRef]

82. Fox, P.; Rockström, J. Supplemental irrigation for dry-spell mitigation of rainfed agriculture in the Sahel. *Agric. Water Manag.* **2003**, *61*, 29–50. [CrossRef]

83. Wakchaure, G.C.; Minhas, P.S.; Ratnakumar, P.; Choudhary, R.L. Optimising supplemental irrigation for wheat (*Triticum aestivum* L.) and the impact of plant bio-regulators in a semi-arid region of Deccan Plateau in India. *Agric. Water Manag.* **2016**, *172*, 9–17. [CrossRef]

84. Heng, L.K.; Hsiao, T.; Evett, S.; Howell, T.; Steduto, P. Validating the FAO AquaCrop Model for Irrigated and Water Deficient Field Maize. *Agron. J.* **2009**, *101*, 488–498. [CrossRef]

85. Stricevic, R.; Cosic, M.; Djurovic, N.; Pejic, B.; Maksimovic, L. Assessment of the FAO AquaCrop model in the simulation of rainfed and supplementally irrigated maize, sugar beet and sunflower. *Agric. Water Manag.* **2011**, *98*, 1615–1621. [CrossRef]

86. Tittonell, P.; Giller, K.E. When yield gaps are poverty traps: The paradigm of ecological intensification in African smallholder agriculture. *Field Crops Res.* **2013**, *143*, 76–90. [CrossRef]

87. Bell, J.M.; Schwartz, R.; McInnes, K.J.; Howell, T.; Morgan, C.L.S. Deficit irrigation effects on yield and yield components of grain sorghum. *Agric. Water Manag.* **2018**, *203*, 289–296. [CrossRef]

88. Greaves, G.E.; Wang, Y.-M. Effect of regulated deficit irrigation scheduling on water use of corn in southern Taiwan tropical environment. *Agric. Water Manag.* **2017**, *188*, 115–125. [CrossRef]

89. Hergert, G.W.; Margheim, J.F.; Pavlista, A.D.; Martin, D.L.; Isbell, T.A.; Supalla, R.J. Irrigation response and water productivity of deficit to fully irrigated spring camelina. *Agric. Water Manag.* **2016**, *177*, 46–53. [CrossRef]

90. Kifle, M.; Gebretsadikan, T.G. Yield and water use efficiency of furrow irrigated potato under regulated deficit irrigation, Atsibi-Wemberta, North Ethiopia. *Agric. Water Manag.* **2016**, *170*, 133–139. [CrossRef]

91. Li, X.; Kang, S.; Zhang, X.; Li, F.; Lu, H. Deficit irrigation provokes more pronounced responses of maize photosynthesis and water productivity to elevated CO_2. *Agric. Water Manag.* **2018**, *195*, 71–83. [CrossRef]

92. Mustafa, S.M.T.; Vanuytrecht, E.; Huysmans, M. Combined deficit irrigation and soil fertility management on different soil textures to improve wheat yield in drought-prone Bangladesh. *Agric. Water Manag.* **2017**, *191*, 124–137. [CrossRef]

93. Lee, S.O.; Jung, Y. Efficiency of water use and its implications for a water-food nexus in the Aral Sea Basin. *Agric. Water Manag.* **2018**, *207*, 80–90. [CrossRef]

94. Gunn, K.M.; Baule, W.J.; Frankenberger, J.R.; Gamble, D.L.; Allred, B.J.; Andresen, J.A.; Brown, L.C. Modeled climate change impacts on subirrigated maize relative yield in northwest Ohio. *Agric. Water Manag.* **2018**, *206*, 56–66. [CrossRef]

95. Abedinpour, M.; Sarangi, A.; Rajput, T.B.S.; Singh, M.; Pathak, H.; Ahmad, T. Performance evaluation of AquaCrop model for maize crop in a semi-arid environment. *Agric. Water Manag.* **2012**, *110*, 55–66. [CrossRef]

Article

Multi-Crop Production Decisions and Economic Irrigation Water Use Efficiency: The Effects of Water Costs, Pressure Irrigation Adoption, and Climatic Determinants

Yubing Fan [1,*][id], Raymond Massey [2] and Seong C. Park [1]

[1] Texas A&M AgriLife Research, Vernon, TX 76384, USA; scpark@ag.tamu.edu
[2] Division of Applied Social Sciences, University of Missouri, Columbia, MO 65211, USA; masseyr@missouri.edu
*   Correspondence: yubing.fan@ag.tamu.edu; Tel.: +1-940-552-9941

Received: 9 October 2018; Accepted: 9 November 2018; Published: 12 November 2018

Abstract: In an irrigated multi-crop production system, farmers make decisions on the land allocated to each crop, and the subsequent irrigation water application, which determines the crop yield and irrigation water use efficiency. This study analyzes the effects of the multiple factors on farmers' decision making and economic irrigation water use efficiency (*EIWUE*) using a national dataset from the USDA Farm and Ranch Irrigation Survey. To better deal with the farm-level data embedded in each state of the U.S., multilevel models are employed, which permit the incorporation of state-level variables in addition to the farm-level factors. The results show higher costs of surface water are not effective in reducing water use, while groundwater costs show a positive association with water use on both corn and soybean farms. The adoption of pressure irrigation systems reduces the soybean water use and increases the soybean yield. A higher *EIWUE* can be achieved with the adoption of enhanced irrigation systems on both corn and soybean farms. A high temperature promotes more the efficient water use and higher yield, and a high precipitation is associated with lower water application and higher crop yield. Intraclass correlation coefficients (ICC) suggest a moderate variability in water application and *EIWUE* is accounted by the state-level factors with ICC values greater than 0.10.

Keywords: climate variability; water use efficiency; multi-crop production; pressure irrigation systems; water costs; corn; soybeans

---

1. Introduction

In many countries, agricultural production relies heavily on water resources [1]. Most of the cropland is irrigated and some traditionally rain-fed agriculture systems have seen growing irrigation to increase production and mitigate climate risks. Accounting for more than 80–90% of the total water withdrawals, irrigated agriculture needs to contribute an increasing share of food production to meet the growing demands of a rising population [2]. Faced with the dramatic impacts of climate change, many arid and semiarid areas are suffering from severe water shortages, for instance, the Western U.S. [3] and Northwestern China [4]. At the same time, some areas that were not facing water deficiencies are experiencing more, frequent droughts, for instance, the Midwestern U.S. [5,6], thus, increasing the stress on current water resources. In addition, in many areas, the water demand from other sectors is expected to grow faster. Though a large proportion of water demand could be satisfied through new investments in water supply and irrigation systems, and the expansion of water supply could be met with some non-traditional sources, the shrinking water availability increases both economic and environmental costs of developing new water supplies [2,7,8]. Therefore, investments in water systems and developing new water sources to meet growing demands will not be a sufficient solution.

As a more practical path to achieve the sustainability of water resources, water can be saved in current uses through increasing the irrigation water use efficiency (total yield per unit of land divided by irrigation water applied) in agricultural production [9]. The traditional flood (also called furrow or gravity) irrigation systems have been reported to lose 50–70% of the water applied as soil evaporation, seepage, and deep drainage [10,11]. Potential improvements in irrigation water use efficiency can be realized by adopting enhanced pressure irrigation systems.

Most of the studies on irrigation water use efficiency are conducted at the field level based on experiments [12,13]. Two foci of field experiments include the comparison of irrigation water use efficiency at different water application levels and utilizing various irrigation methods, and the interaction and compatibility of improved irrigation systems and other farm-related management practices that are considered the best (e.g., film or straw mulching, irrigation scheduling, and soil testing) [14–17]. Previous studies on irrigation water use efficiency (IWUE) typically used experimental data in one field, collected over multiple years. Because of limited research funding, heterogeneity of experimental fields, and the diversity of cropping systems and farming structures, the available farm-level data are limited. As a result, the evaluation of crop IWUE in multiple fields is very challenging. At the farm level, producers usually plant two or more crops in one growing season. In addition to making adoption decisions regarding different irrigation systems, farmers also need to make decisions on land allocation and irrigation water application for each crop that they choose to plant. These decisions can determine whether the water is used efficiently or not.

The farm-level irrigation and production decisions to improve irrigation efficiency in a multi-crop system are understudied, in particular, across regions with different cropping patterns and climatic conditions [18]. In addition, production decisions in irrigated agriculture may be affected by other factors like water sources, input costs, and the farming area [19]. Analysis of irrigation decisions and crop irrigation water use efficiency, as affected by these and other factors, could help farmers and policymakers adapt to potential climate risks, better manage the irrigation water application, and achieve the sustainable use of limited water resources. Furthermore, given the heterogeneity of farms and states, multi-level models (MLMs) can be readily utilized to deal with the hierarchical nature of the farm-level data and to extract the percentage of variability in each response accounted for by farm- and state-level factors. The multilevel model has been applied in social science research [20,21] and agricultural sciences. To analyze the hierarchically structured data, Neumann et al. [22] adopted the multilevel model to investigate the global irrigation patterns using country-level data, and Giannakis and Bruggeman [23] studied the labor productivity in agricultural system in Europe. However, MLMs have never been used to analyze crop production decisions or farm irrigation efficiency. Given the data structure of the United States Department of Agriculture Farm and Ranch Irrigation Survey (USDA FRIS)—i.e., farms are embedded in states—we explore the applicability of the MLMs to multiple equations relating to production decisions in irrigated multi-crop agriculture.

Therefore, the objective of this study is to better understand the production decisions for irrigated agriculture and economic irrigation water use efficiency of major crops in the U.S., as well as the effects of water costs, the adoption of pressure irrigation systems, and the climatic determinants in a multi-crop production system.

Specifically, this research aims to answer the following fundamental questions:

(1) Are enhanced irrigation systems conserving water and are they more efficient than the traditional systems under diverse farm conditions?

(2) How does climate variability affect production decisions in an irrigated agriculture?

(3) What are the major influential factors and how are the multi-crop production decisions affected by these factors at the farm and state levels?

The layout of the analyses in this paper is presented in Figure 1. Focusing on irrigated farms in a multi-crop production system, four equations on land allocation, water application, crop yield, and economic irrigation water use efficiency are estimated using multilevel models. Intensive and extensive margins of water use to water price and energy costs are calculated. Intraclass correlation coefficients (ICC) as defined later are calculated to find out the proportion of variability in each response is accounted for by each level. Econometric results from the multilevel models are provided to show the effects of exogenous variables on each response variable.

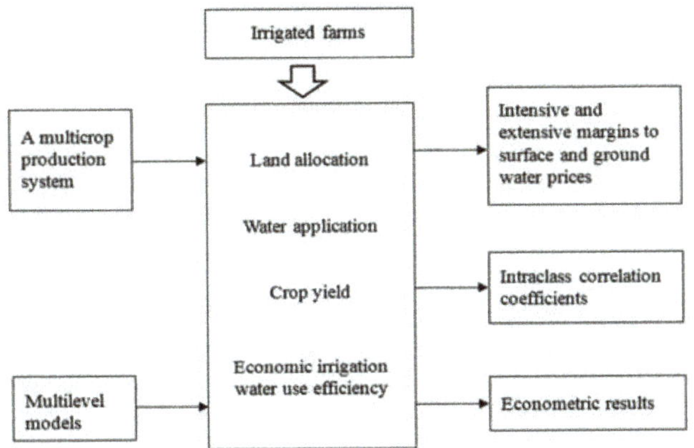

Figure 1. An analytical structure of irrigated multi-crop farming decisions.

2. Literature Review

Crop Water Use Efficiency

In general, water management includes issues relating to five sub-systems existing on most irrigated farms: supply systems, on-farm storage systems, on-farm distribution systems, application systems, and recycling systems [11]. In a report on the Australian cotton industry, Dalton et al. [11] defined water use efficiency at the farm level by focusing on three dimensions: agronomic efficiency, economic efficiency, and volumetric efficiency. The agronomic water use efficiency includes a gross production water use index (yield/total water applied), an irrigation water use index (yield/irrigation water applied), a marginal irrigation water use index (marginal yield due to irrigation/irrigation water applied), and a crop water use index (yield/evapotranspiration). The economic water use efficiency includes a gross production economic water use index (total value/total water applied), an economic irrigation water use index (value/irrigation water applied), a marginal economic irrigation water use index (value due to irrigation/irrigation water applied), and a crop economic water use index (value/evapotranspiration). The volumetric water use efficiency includes the overall project efficiency, conveyance efficiency, distribution efficiency, and field application efficiency, which emphasize irrigation uniformity to avoid over- and under-irrigation issues (reducing the water use efficiency and yield, respectively). Moreover, Pereira [24] discussed various measurements for both distribution uniformity and application efficiency in various irrigation systems.

From a multi-disciplinary perspective, Nair et al. [25] reviewed the efficiency of irrigation water use. Among all the measures of WUE, agronomists defined it as yield per unit area divided by the water used to produce the yield. The yield can be grain yield or the total aboveground biomass depending on the use of the crop produced, and the water can refer to crop evapotranspiration, soil water balance, or precipitation plus irrigation. However, from an economist's perspective, the efficient level of irrigation water occurs "when the marginal revenue (the price of the crop produce in a perfectly competitive market) is equal to the price of water" [25]. The water application level at Stage II in the classical production function was identified as the economically efficient water use amount. Stage II ranges from the point where the marginal physical product (MPP) equals the average physical product (APP), i.e., $w/p = Y/X$ ($MPP = APP$) with w being the water cost, p being the output price, Y being the output quantity, and X being the input quantity, to the yield maximizing point, where $dY/dX = w/p = 0$ (i.e., $MPP = 0$). Other researchers proposed an operating profit water use index to evaluate the water use efficiency, which is defined as $(R - VC - OC)/WU$ with R being the gross return, VC being the variable costs, OC being the overhead costs, and WU being the total amount of water used [26].

Comparing the WUE measures from perspectives of agronomists and economists, a major difference is whether to consider the output price. For example, the economic irrigation water use index (value of crop or grains/irrigation water applied) is the product of the irrigation water use index (yield/irrigation water applied) and the crop price. Because producers are price takers in a competitive market, different farmers growing the same crop will sell it for the same price in the same market. Thus, exogenous variables affecting economic irrigation efficiency and agronomic irrigation efficiency will have the same effects in terms of signs and significance levels, though the magnitude will be different proportionally. To make the analyses easier and follow the mainstream of decision-making on land allocation and water use in order to maximize the expected profit as formulated in the model section below, this study uses the economic measure of irrigation water use efficiency ($EIWUE$) (crop value/irrigation water use), incorporating state-average crop prices in the econometric estimation.

Various approaches have been explored to conserve irrigation water use, such as developing new irrigation techniques [27]; increasing investment in irrigation infrastructure such as canals, wells and drip systems [16]; and designing water conservation policies [28]. Water-conserving irrigation systems have been proposed and applied to various crops in many farming areas around the world. For instance, in eastern Australia [29], arid and semi-arid areas in China [16,30], and southern and southwestern U.S. [15,31,32]. Examples include pressure (or pressurized) irrigation systems (versus gravity irrigation methods), such as linear move, center pivot, sprinkler, and drip irrigation methods. Field experiments with sprinkler and drip irrigation and their comparison with traditional flood or furrow irrigation have been conducted on various crops worldwide [14,33–36]. As a result, a substantial amount of water could be saved using enhanced irrigation systems and crop irrigation water use efficiency can be improved.

3. Hypotheses

In this section, factors affecting farmers' adoption behaviors and irrigation decisions are reviewed, and hypotheses are constructed. Farmers' irrigation decisions are hypothesized to be a function of the expected profit, costs, perceived barriers, information availability, farm and farmer characteristics, and their environmental attitudes and perceptions of climate variability.

Literature reviews on agricultural production and economics show that many changes in socioeconomic, agronomic, technical, and institutional aspects can have considerable positive/negative effects on water use, crop yield, and crop water use efficiencies, and thus diverse effects on the profitability of crop production [37,38]. Farm management practices including controlling the amount and timing of irrigation water, fertilizer/manure use, mulching, and tillage can affect farm returns and profits [39]. Through analyzing various measurements of water use efficiency, Pereira [24] recommended combining improved irrigation methods and scheduling strategies to achieve a higher performance. Pressure irrigation systems are thus expected to decrease water application and increase efficiency.

Based on field-level measurements, Canone et al. [40] assessed the surface irrigation efficiency in Italy. The results from both simulated scenarios and monitored irrigation events highlighted the necessary strategies to improve irrigation efficiencies by reducing the flow rates and increasing the duration of irrigation events. Thus, we hypothesize a higher water availability from various sources and more wells decrease crop water use efficiency.

In addition, the diverse effects of physical factors on farm yield and profits have been reported based on farm-level studies. For instance, with carrot farmer interviews in Pakistan, Ahmad et al. [41] found that farm-level yield and profitability were affected by many factors including expenditures on facility and labor investments regarding the application of fertilizer, irrigation, and weeding. In a similar study, Dahmardeh and Asasi [42] evaluated the effects of the costs of fertilizer, seeds, and water on the profitability of corn farms as well as the effects of income sources. Boyer et al. [19] examined the effects of different energy sources, energy prices, and field sizes on corn production. Thus, the facility expenses and labor payment at the farm level are hypothesized to have positive effects on water application and crop yield, but a mixed effect on water use efficiency.

Farmers face many barriers and challenges when making irrigation and production decisions. Using data on 17 western states from the USDA FRIS, Schaible et al. [43] studied the dynamic adjustment of farmers' irrigation decisions and pointed out some major barriers impacting the adoption of enhanced irrigation technologies. The most important barriers were related to investment cost and financing issues. A greater sharing of costs by government or landlords for installation of advanced irrigation techniques can improve their adoption rates especially for beginning farmers with limited resources and social disadvantages [2]. Moreover, uncertainty about future water availability and farming status could influence farmers' willingness to adopt. Hence, uncertainties regarding potential costs and future benefits will limit the adoption of water conservation practices, and thus discourage farmers to use water more efficiently [44,45].

Information availability and its sources can affect farm irrigation decisions [46]. On the one hand, limited information can be an obstacle to using water efficiently. Rodriguez et al. [47] pointed out that a lack of information on irrigation, crop management, the effectiveness of practices and government programs could be common obstacles for resource-limited farmers when facing the uncertainty of changing to something unknown. On the other hand, effective information can facilitate optimal irrigation decisions by farmers [48]. Frisvold and Deva [49] studied water information used by irrigators and the relationship of information acquisition and irrigation management. Their study indicated that appropriate information use could benefit irrigation management and crop production for farmers with varying acreage. Thus, more information on how to conserve water and use water more efficiently is expected to decrease water use, increase crop yield, and improve irrigation efficiency [37,44].

Regional variables could capture differences in climate, water institutions, and supporting infrastructure [50], as well as farming systems. More generally, which irrigation decisions are appropriate will vary spatially. For example, western states tend to have concentrated irrigation acreage and their irrigation institutions are well established [50]. Eastern and southern states receive moderate amounts of rainfall to support agriculture and do not rely as heavily on irrigation. Thus, we hypothesize that compared with those in the High Plains states, farmers in western states will irrigate more, while farmers in midwestern and southern states will irrigate less.

Furthermore, farmers are also motivated to respond facing varying weather conditions. Climate conditions can influence farm yield and revenue, and irrigation can be considered as a strategy to mitigate the adverse effects and increase profits [51]. Specifically, an awareness of climate change (e.g., drought and heat waves) could motivate farmers to prepare for and take actions to adapt to future risks to production [4,52]. Olen et al. [18] found that farmers were more likely to irrigate crops to mitigate and adapt to various weather and climate impacts including frosts, heats, and droughts. Li et al. [53] reported diverse effects of climate change on corn yield in the United States and China. Therefore, farmers are hypothesized to increase water application rates and decrease irrigation water use efficiency if they perceive or experience less precipitation, higher temperature, or more grain losses due to droughts. This is proxied by changes of weather conditions in 2011, 2012, and 2013.

4. Methods

In this section, we adopt a model of profit maximization [54] and then turn to the maximization of economic irrigation water use efficiency to deal with market failure in water management. In multi-crop irrigated agriculture, producers make decisions on land allocation to each crop, and the amount of water for irrigation [55,56]. Choosing from common crops, a typical producer may plant two or more crops on a farm. Then decisions on land allocation and water supply can be made to maximize the expected total profit [57].

Following a multi-crop production model by Moore et al. [54], the expected profit functions of the multi-crop system and specific crop i can be represented by $\Pi(p, r, b, N; x)$ and $\pi_i(p_i, r, b, n_i; x)$, respectively. p is a vector of crop prices; p_i is the price of crop i, $i = 1, \ldots, m$; r is a vector of variable input prices excluding water prices; b is the water prices; N is the total farming area as a constraint; n_i is the land allocation for crop i; x represents other exogenous variables including land characteristics, water sources, the adoption of various irrigation systems, and climate perceptions. Each crop-specific profit function π_i is assumed to be convex and homogeneous of degree one in output prices, water price, and other prices of variable inputs, nondecreasing in output price and land allocation, and non-increasing in water prices and other variable input prices.

We extend the model of Moore et al. [54] by adding crop irrigation water use efficiency. A single producer makes production and irrigation decisions to maximize profits. While to achieve sustainability of the water resource, the total profit function of the whole society needs to consider the marginal user cost and higher pumping cost externality of extracting water by every farmer. Thus, in addition to the decision-making on conserving water use and increasing crop yield, the way to achieve higher crop irrigation water use efficiency should be explored. Following the discussion on indicators of water use performance and productivity by Pereira et al. [58], the following definition can be used to calculate the farm-level crop-specific economics irrigation water use efficiency.

$$EIWUE = \frac{Crop\ yield \times P}{Total\ amount\ of\ irrigation\ water\ applied} \tag{1}$$

where $EIWUE$ is the economic irrigation water use efficiency, crop yield is the marketable grain yield, P is the crop price, and irrigation water application is measured based on all irrigation water sources, including well, on- and off-farm surface water. The greater the $EIWUE$ value [59], the higher the efficiency due to irrigation water application.

To analyze the effects, $EIWUE$ can be a function of the exogenous variables affecting both yield and water application.

$$EIWUE_i = h_i(p, r, b, N; x)\ i = 1, \ldots, m \tag{2}$$

In addition, the farm-level water application can be decomposed to analyze the role of water price on production decisions regarding each crop [54]. The crop-specific water application can be decomposed into an extensive margin of water use (an indirect effect on water use due to land allocation change) and an intensive margin of water use (a direct effect on water use due to water application).

The farm-level total water application (W) equals the sum of water application for each crop grown on the farm with the optimal land allocation [54,60]:

$$W = \sum_{i=1}^{m} w_i(p_i, r, b, n_i^*(p, r, b, N; x); x)\ i = 1, \ldots, m \tag{3}$$

Taking the derivative of the equation with respect to water price gives

$$\frac{\partial W}{\partial b} = \sum_{i=1}^{m} \left(\frac{\partial w_i}{\partial b} + \frac{\partial w_i}{\partial n_i^*} \times \frac{\partial n_i^*}{\partial b} \right) \tag{4}$$

where $\frac{\partial w_i}{\partial b}$ is the intensive margin, and $\frac{\partial w_i}{\partial n_i^*} \frac{\partial n_i^*}{\partial b}$ is the extensive margin. The total effect can be obtained by summing the effects on all the crops. The intensive margin will decrease in price and $\frac{\partial w_i}{\partial b}$ should have a negative sign for each crop. The sign of the extensive margin depends on $\frac{\partial n_i^*}{\partial b}$. The total farm-level effect on water use should be negative, which indicates a decreasing water application as water price increases. This decomposition of the total marginal effect has been lately employed by Hendricks and Peterson [61], and Pfeiffer and Lin [62].

Multilevel Models

Multilevel models have the advantage of examining individual farms embedded within states and assess the variation at both farm and state levels. The multilevel regression model is commonly viewed as a hierarchical regression model [63]. A multilevel linear modeling technique is utilized to analyze the effects of influential factors on land allocation, water application, crop yield, and *EIWUE*.

For the research questions, we have N individual crop-specific farms ($i = 1, \ldots, N_j$) in J states ($j = 1, \ldots, J$). The X_{ij} represent a set of independent variables at the farm level, and a series of state-level independent variables are represented by Z_j. The model estimation includes two steps. For the first step, a separate regression equation can be specified in each state to predict the effects of independent variables on dependent variables.

$$Y_{ij} = \beta_{0j} + \beta_{1j}X_{1ij} \tag{5}$$

For the second step, the intercepts, β_{0j}'s are considered parameters varying across states as a function of a grand mean (γ_{00}) and a random term (u_{0j}). The β_{1j}'s are also assumed to be varying across states and are presented as a function of fixed parameters (γ_{10}) and a random term (u_{1j}).

$$\beta_{0j} = \gamma_{00} + \gamma_{01}Z_j + u_{0j} \tag{6a}$$

And

$$\beta_{1j} = \gamma_{10} + u_{1j} \tag{6b}$$

Combining Equations (5), (6a) and (6b), we have

$$Y_{ij} = \gamma_{00} + (\gamma_{10} + u_{1j})X_{ij} + \gamma_{01}Z_j + u_{0j} \tag{7}$$

The model is called a random-intercept and random-slope model, as the key features are not only that the intercept parameter in the Level-1 model, β_{0j}, is assumed to vary at Level-2 (state) [64], but that the slope is also random with an error term u_{1j}. The γ_{01} coefficient captures the effects of the state-level variables (Z_j) on the β_{0j}'s, whereas γ_{10} predicts the constant parameter, β_{1j}, (with errors).

To analyze the multi-crop production, four sequential models are estimated for each decision due to their continuous nature, that is, a unconstrained two-level model with random effects for the intercept only and without any predictors (Model 1); random effects for the intercept and fixed effects for level 2 (Model 2); a random intercept as well as a fixed and random level 1 (Model 3); and a random intercept, fixed and random level 1 as well as a fixed level 2 (Model 4) (see Table S1 in the Supplementary Materials for specifications and comparisons of the four models). To determine how much of the variability in the responses is accounted for by factors at the state level, the intraclass correlation coefficient is usually computed from the null model (Model 1) [65] following:

$$ICC = \frac{\tau_{00}}{\tau_{00} + 3.29} \tag{8}$$

where τ_{00} is the covariance parameter estimate for the intercept, and 3.29 is the estimated level-1 error variance [66].

The data were analyzed using the SAS package in the USDA data lab in St. Louis, Missouri, with official permission.

5. Data and Variables

This study uses a national dataset from the 2013 USDA FRIS. Null models for all equations of 17 crops are estimated to calculate the intraclass correlation coefficient. However, only models in the further steps on land allocation [67], water application, crop yield, and *EIWUE* are estimated and presented in this paper focusing on corn and soybeans as they have the most observations but different distribution patterns across the five regions (specified below).

The lower 48 states are grouped into five regions according to the USDA National Agricultural Statistics Services (NASS) [68], including the Western, Plains, Midwestern, Southern, and Atlantic states [69]. The descriptive statistics of the corn and soybean farms [70] at the national level are presented in Table 1. Of the 19,272 irrigated farms, 6030 farms grow corn for grain with an average area of 357 acres, and 3933 farms grow soybeans with an average area of 341 acres [71]. For corn farms, the mean water application is 1.11 acre-feet/acre; the mean yield is 190 bu/acre; and *EIWUE* is 1311 USD/acre-foot on average. For soybean farms, the mean water application, yield, and *EIWUE* are 0.81 acre-foot/acre, 55 bu/acre, and 1221 USD/acre-foot, respectively.

The independent variables are at two levels. At the farm level, the explanatory variables are related to water sources, costs on surface water and energy, expenditures on irrigation equipment, labor payment, farm characteristics including the farming area, number of wells, irrigation systems, barriers for improvements to conserve water, and information sources related to irrigation. Variables related to water sources, federal assistance, barriers, and information sources are dummy variables (Yes = 1, No = 0), and all other independent variables are continuous.

At the state level, in addition to the dummy variables related to the five regions, six explanatory variables on state-wide weather conditions are included using the data from the United States National Oceanic and Atmospheric Administration. The variables are state average precipitation changes in 2011, 2012, and 2013, and the temperature changes in 2011, 2012, and 2013.

Table 1. The summary statistics of crop-specific dependent variables and state-level weather-related independent variables.

Variables	Description (Unit)	N	Mean	Std Dev	CV	Min	Max
Crop-Specific Dependent Variables							
Corn							
Land allocation	Average farming area (acre)	6030	356.84	1426.66	4.00	-	-
Water application	Average water application (acre-foot)	6030	1.11	1.97	1.77	-	-
Crop yield	Average yield of all farms (bu/acre)	6030	190.29	87.19	0.46	-	-
EIWUE	Average economic irrigation water use efficiency ($/acre-foot)	6030	1310.99	3199.15	2.44	-	-
Soybeans							
Land allocation	Average area of all farms (acre)	3933	340.79	1195.10	3.51	-	-
Water application	Average water application of all farms (acre-feet)	3933	0.81	1.13	1.40	-	-
Crop yield	Average yield of all farms (bu/acre)	3933	54.76	27.89	0.51	-	-
EIWUE	Average economic irrigation water use efficiency ($/acre-foot)	3933	1220.55	2352.57	1.93	-	-
State-Wide Average Weather-Related Variables							
PrecipChange2011	Precipitation in 2011—Average precipitation in 1981–2010 (inch)	43	1.51	8.26	5.46	−15.87	17.61
PrecipChange2012	Precipitation in 2012—Average precipitation in 1981–2010 (inch)	43	−3.66	4.74	1.29	−12.21	10.30
PrecipChange2013	Precipitation in 2013—Average precipitation in 1981–2010 (inch)	43	1.74	5.36	3.08	−15.19	14.26
TempChange2011	Temperature in 2011—Average temperature in 1981–2010 (°F)	43	0.54	1.09	2.03	−2.70	2.10
TempChange2012	Temperature in 2012—Average temperature in 1981–2010 (°F)	43	2.47	1.11	0.45	−1.70	4.00
TempChange2013	Temperature in 2013—Average temperature in 1981–2010 (°F)	43	−0.50	0.75	1.48	−2.20	0.90

6. Results

6.1. Descriptive Statistics

The summary statistics of the farm-level independent variables are presented in Table 2. Four water sources are investigated including groundwater only, on- and off-farm surface water [72] only, and two or more water sources (Yes = 1, No = 0). For corn and soybean farms, about 71% and 81% use groundwater only, respectively. Water from on- or off-farm surface sources only account for about 4.5% of soybean farms (about 10.5% of corn farms only use off-farm surface water). About 12% of both farms get water from two or more sources.

Table 2. The summary statistics of farm-level independent variables and region dummies ($: USD).

Variables	Corn (n = 6030)		Soybean (n = 3933)	
	Mean	**Std Dev**	**Mean**	**Std Dev**
Water sources				
Groundwater only (base)	0.713	1.124	0.808	0.926
On-farm surface water only	0.058	0.579	0.045	0.488
Off-farm surface water only	0.105	0.762	0.031	0.406
Two or more water sources	0.124	0.819	0.116	0.752
Costs				
Cost for off-farm surface water ($/acre-foot)	6.891	113.473	4.215	47.154
Energy expenses ($/acre)	47.047	184.994	35.602	62.841
Facility expenses ($/acre)	37.605	367.721	25.131	293.385
Labor payment ($/acre)	5.237	197.95	1.454	25.398
Farm characteristics				
Number of wells used	5.755	23.632	7.365	23.585
Total acre	1879	13497	1665	5238
Percent of owned land	0.497	0.937	0.448	0.852
Pressure irrigation	0.799	0.996	0.708	1.07
Gravity irrigation (base)	0.201	0.996	0.292	1.07
Federal assistance	0.202	0.998	0.219	0.973
Barriers to improvements				
Investigating improvement is not a priority	0.165	0.921	0.14	0.816
Risk of reduced yield or poorer quality crop	0.089	0.708	0.071	0.605
Limitation of physical field or crop conditions	0.11	0.776	0.104	0.718
Not enough to recover implementation costs	0.172	0.937	0.195	0.932
Cannot finance improvements	0.129	0.834	0.114	0.748
Landlords will not share improvement costs	0.119	0.805	0.137	0.808
Uncertainty about future water availability	0.11	0.776	0.08	0.637
Will not be farming long enough	0.075	0.656	0.059	0.554
Will increase management time or cost	0.079	0.671	0.065	0.579
Information sources				
Extension agents	0.33	1.169	0.401	1.153
Private irrigation specialists	0.354	1.188	0.366	1.133
Irrigation equipment dealers	0.31	1.15	0.308	1.086
Local irrigation district employee	0.082	0.683	0.059	0.555
Government specialists	0.153	0.895	0.146	0.831
Media reports	0.118	0.802	0.122	0.769
Neighboring farmers	0.231	1.047	0.231	0.991
E-information services	0.188	0.972	0.191	0.925
Regions				
West	0.139	0.859	0.005	0.171
High Plains (base)	0.554	1.235	0.532	1.174
Midwest	0.16	0.912	0.182	0.908
South	0.113	0.787	0.242	1.008
Atlantic	0.033	0.445	0.038	0.45

All variables have been weighted using weights provided within the FRIS data.

Water costs are measured by the payment for off-farm surface water and energy expenses for pumping groundwater. The average cost for off-farm surface water is 6.89 and 4.22 USD/acre-foot for corn and soybean farms, respectively. The water price measure frees the irrigator from being bind by water institutions [54]. The average energy expenses are 47.05 and 35.60 USD/acre for corn and soybean farms. The energy expenses are a proxy of groundwater price [54]. The average facility expenses and labor payments in 2013 are 37.61 and 5.24 USD/acre for corn, and 25.13 and 1.45 USD/acre for soybeans. The units of costs measure follow the convention by Moore and others [73].

Regarding the farm characteristics, the average number of wells used to irrigate corn and soybeans are 5.76 and 7.37, respectively. The mean areas of the total land are 1879 and 1665 acres/farm for corn and soybeans, and the percentage of owned land is 50% and 45%. For irrigation systems, about 20% of corn farms use gravity systems and 29% of soybean farms use gravity systems, while those using pressure irrigation account for 80% and 71%, respectively. About 20% of the corn farmers received federal assistance to improve irrigation and/or drainage systems, compared to 22% for soybean farmers.

Regarding the barriers to implementing improvements for the reduction of energy costs or water use, nine barriers are investigated in the national survey. The major ones include the following: investigating improvement is not a priority at this time (17% for corn farmers and 14% for soybean farmers), limitation of physical field or crop conditions (11% for corn farmers and 10% for soybean farmers), not enough to recover implementation costs (17% for corn farmers and 20% for soybean farmers), cannot finance improvements (13% for corn farmers and 11% for soybean farmers), and landlords will not share improvement costs (12% for corn farmers and 14% for soybean farmers).

For the eight sources of irrigation information, the top ones are extension agents (33% for corn farmers and 40% for soybean farmers), private irrigation specialists (35% for corn farmers and 37% for soybean farmers), irrigation equipment dealers (31% for both corn and soybean farmers), neighboring farmers (23% for both corn and soybean farmers), e-information services (19% for both), and government specialists (15% for both).

Regarding location, this study includes more irrigated farms in the Plains states, 55% for corn and 53% for soybeans. Farms in the Midwest and South account for 16% and 11% for corn, and 18% and 24% for soybeans, with fewer farms in the Midwest and South.

The state-wide average weather-related variables are presented in Table 1 for the 43 states planting corn. Compared to the 1981–2010 average precipitation, the changes for 2011, 2012, and 2013 are 1.51, −3.66, and 1.74 inches, respectively. Compared with the 1981–2010 average temperature, the changes for 2011, 2012, and 2013 are 0.54, 2.47, and −0.50 °F. While in 2013, the year covered by the survey, it's more favorable for agricultural production as far as the rainfall.

6.2. Decomposition of Farm-Level Water Application

To decompose the effect of water cost on farm-level water application, extensive and intensive margins are provided in Table 3. This paper takes corn and soybeans as examples [74]. The estimated coefficients on crop acreage and water costs in the water application equation suggest a change in water use given a change in land use ($\frac{\partial w_i}{\partial n_i}$), and a marginal change in water use given a change in water cost ($\frac{\partial w_i}{\partial b}$). The estimated coefficients on water cost in the land allocation equation represent a change in land use given a change in water cost ($\frac{\partial n_i}{\partial b}$). The intensive margin can be obtained with $\frac{\partial w_i}{\partial b}$ while adjusting for the estimated probability that the crop is grown. The extensive margin can be calculated using $\frac{\partial w_i}{\partial n_i}\frac{\partial n_i}{\partial b}$. Summing the intensive and extensive margins for each crop gives the total effect of a change in water cost. Further summing the effects on all crops gives the total effect on a typical farm growing both crops.

Margins on both on-surface water costs and energy costs are calculated. Only water from off-farm surface sources is priced and investigated in the survey. Energy expenses on groundwater pumping are considered as the proxy of water price for groundwater. The results show that only $\frac{\partial n_i}{\partial b}$ decreases in energy expenses for soybeans, and other values of $\frac{\partial n_i}{\partial b}$ and $\frac{\partial w_i}{\partial b}$ are positive, which is contradictory to expectations. This indicates more water is used as water prices increases. This is probably true in practice when the adoption of enhanced irrigation systems increase acreage under irrigation and thus increase the amount of irrigation water, as reported in Kansas [75]. There are many debates regarding the empirical changes in water use as a result of changing prices and increasing the adoption of agricultural irrigation technologies [53,61,75]. A numerical illustration can help understand the effects of water prices. A 1 USD increase in groundwater costs (energy expenses) ($\Delta b = \$1$) would lead to a decrease of 0.109 acre-feet of water application per acre of soybeans, and an increase of 0.0737 acre-feet of water per acre of corn. In a multi-crop system, a typical farm growing both corn and soybeans would decrease water application by -10.87 acre-feet as a result of a \$1 increase in energy expenses. These results show water use is highly inelastic in water cost [54]. While this may be different for regions/states with varying availability of water resources, an in-depth analysis of regional or state effect of water costs on water use can be helpful.

Table 3. The crop-specific extensive and intensive margins to surface water cost and energy expenses.

Variables	dw/dn	dn/db	dw/db	Share of Crop-Specific Farms	Extensive Margin	Intensive Margin	Total Effect (Acre-Feet Per Acre)	Total Effect-Farm (Acre-Feet Per Farm)
Surface Water Cost								
Corn	1.0266	0.1766	0.0030	0.3129	0.0567	0.0009	0.0577	20.5769
Soybeans	1.0040	0.0816	0.0006	0.2041	0.0167	0.0001	0.0168	5.7396
Farm total								26.3165
Energy Expenses								
Corn	1.0266	0.2282	0.0012	0.3129	0.0733	0.0004	0.0737	26.2881
Soybeans	1.0040	-0.5334	0.0012	0.2041	-0.1093	0.0002	-0.1090	-37.1606
Farm total								-10.8725

Following the definitions by Moore et al. [54], $\frac{\partial w_i}{\partial n_i}$ is the estimated coefficient of crop acreage in the water application equations, where w_i is the acre-feet of irrigation water on crop i and n_i is the acres of growing crop i. $\frac{\partial n_i}{\partial b}$ is the estimated coefficient of the water price in the land allocation equations, with b being the water price. $\frac{\partial w_i}{\partial b}$ is the estimated coefficients of the water price in the water application equation. The calculation of both intensive and extension margin should be adjusted by the share of the crop planted.

6.3. Intraclass Correlation Coefficients

The first step in conducting a multilevel model is to calculate the ICC which shows how much of the variability in one response variable is accounted for by level 2. The intraclass correlation coefficients for crop-specific multilevel models are presented in Table 4. To better understand these values, for example, the ICC for the water application equation of corn is 0.2102, which suggests about 21% of the variability in water application decisions is accounted for by the factors at the state level, leaving 79% of the variability to be accounted for by the farm-level factors. A moderate variability in water application and *EIWUE* is accounted by the state-level factors, with an ICC value greater than 0.10. However, a higher variability of land allocation and crop yield is accounted for by farm-level factors. In the following sections, results for each estimated equation are presented for corn and soybeans jointly to facilitate the comparison of the effects on the two crops.

Table 4. The intraclass correlation coefficients for null models of each crop-specific multilevel model.

State-Level Variation	Land Allocation	Water Application	Crop Yield	*EIWUE*
Corn	0.0068	0.2102	0.0270	0.1501
Soybeans	0.0291	0.1365	0.0277	0.1763

EIWUE: economics irrigation water use efficiency. The table only presents results from Model 4 (Model 3 + fixed state level) in each equation. More results can be found in Tables S2–S9 of the Supplementary materials.

6.4. Land Allocation

The estimated coefficients from MLMs for land allocation of corn and soybeans are presented in Table 5. The results are shown compared to groundwater use, water uses from on-, off-farm surface only and more sources have a positive effect on land allocation to corn planting. While water from more sources increases the planting of both crops.

Table 5. The results of multilevel models for the land allocation for corn and soybean.

Variables	Corn		Soybeans	
	Estimate	Std Err	Estimate	Std Err
Fixed Effects				
Intercept	−533.020 ***	55.823	12.756	241.4
Water sources				
On-farm surface water only	78.808 ***	23.286	29.948	24.573
Off-farm surface water only	151.370 ***	22.352	49.96	32.727
Two or more water sources	85.086 ***	15.758	71.824 **	20.262
Costs				
Cost for off-farm surface water($/acre-foot)	0.177	0.116	0.082	0.302
Energy expenses ($/acre)	0.228 ***	0.073	−0.533 *	0.253
Facility expenses ($/acre)	0.170 ***	0.036	0.009	0.037
Labor payment ($/acre)	0.034	0.063	0.168	0.428
Farm characteristics				
Number of wells used	38.663 ***	4.081	18.199 ***	2.566
LN(total acre)	81.107 ***	4.816	92.613 ***	10.983
Percent of owned land	−20.374	14.257	−17.387	14.205
Pressure irrigation	21.763	15.337	5.519	22.995
Federal assistance	−26.108	16.153	−25.000 **	12.067
Barriers to improvements				
Investigating improvement is not a priority	−10.970	13.782	23.756 *	13.675
Risk of reduced yield or poorer quality crop	−0.682	19.7	1.521	19.659
Limitation of physical field or crop conditions	−27.543	17.893	−19.917	16.972
Not enough to recover implementation costs	29.853 **	14.16	−2.248	13.184
Cannot finance improvements	−22.241	15.403	−1.889	15.567
Landlords will not share improvement costs	−9.536	16.863	−29.516 **	14.914
Uncertainty about future water availability	−35.165 **	17.261	−28.533	18.987
Will not be farming long enough	3.877	20.056	13.336	20.68
Will increase management time or cost	3.408	20.145	4.782	20.079
Information source				
Extension agents	−29.179 **	11.727	−18.427 *	10.38
Private irrigation specialists	23.782 **	11.218	8.49	10.429
Irrigation equipment dealers	−3.008	11.816	−4.227	11.089
Local irrigation district employee	−2.399	19.916	23.229	21.612
Government specialists	−5.252	15.564	4.839	14.125
Media reports	−0.350	16.675	−0.632	15.178
Neighboring farmers	−27.456 **	12.756	−17.951	11.69
E-information services	12.732	13.76	11.728	12.817
State-level variables				
PrecipChange2011	−2.820	1.936	−8.475	6.35
PrecipChange2012	−3.283	2.405	16.053	9.861
PrecipChange2013	−7.201 **	2.715	30.293 **	11.885
TempChange2011	2.116	16.396	11.812	80.683
TempChange2012	−0.811	12.082	−28.899	73.028
TempChange2013	1.927	18.476	134.110 *	75.466
West	45.578	34.692	−276.610	217.03
Midwest	64.791 *	33.045	−94.133	141.99
South	131.760 ***	41.441	−994.360 ***	163.42
Atlantic	119.730 **	55.951	−343.380	216.11
Error Variance	**Estimate**	**Std Err**	**Estimate**	**Std Err**
Intercept	<0.0001 ***	<0.0001	<0.0001 ***	<0.0001
Residual	858,667 ***	15,746	433,053 ***	9882
Fit Statistics				
N	6030		3933	
−2 Log Likelihood	93,300		58,421	
AIC	93,378		58,513	
AICC	93,379		58,514	
BIC	93,446		58,571	

Significance levels: * 10%; ** 5%; *** 1%.

Surface water price does not affect land allocation, which is consistent with the expectations as the decision on how much land allocated to grow a crop is made mainly depending on the expected crop price and input costs with little consideration of water price, while energy expenses as a proxy of groundwater price increase corn planting and decrease soybean planting. Higher facility expenses increase corn planting as more acres can be irrigated.

Regarding farm characteristics, more wells on a farm increase the planting of both crops. Larger areas of cropland increase the land allocation for both crops. Federal assistance on farm irrigation and drainage management has a negative effect on soybean planting. Unfortunately, land tenure and the adoption of pressure irrigation systems do not have a significant effect on land allocation for both crops.

Regarding barriers to improvements, uncertainties about future water availability have a negative effect on corn planting, and not enough to recover implementation costs has a positive effect. For soybean, landlords not sharing improvements costs has a negative effect on soybean planting, while investigating improvement is not a priority shows a positive effect. While positive effects are unexpected, a comparison of the negative effects on the two crops indicates that corn farmers are more concerned with future uncertainties, and soybean farmers with the share of improvement costs.

Information from extension agents and neighboring farmers decreases the planting of corn and soybean planting is also negatively affected by the information from extension agents, while information from private irrigation specialists increases the planting of corn. These findings indicate the effectiveness of extension programs in promoting the growth of water-conserving crops.

At the state level, the precipitation change in 2013 is negatively associated with corn planting. Both the precipitation change and temperature change are positively associated with soybean acreage. These findings suggest that given climate variability, a lower water available for crop production probably promotes farmers growing more water-conserving crops (in this case, soybeans), and vice versa. Compared with Plains farmers, those in the Midwestern, Southern and Atlantic states are more likely to plant corn, while farmers in the Southern states are less likely to plant soybeans.

6.5. Water Application

The parameter estimates for water application equations of corn and soybeans are presented in Table 6. The results are shown compared to groundwater use only, the water use from two or more sources has a positive effect on water application of corn. High surface water cost, energy expenses, and labor payment are positively associated with water application on corn. The energy expenses are also positively associated with water application on soybeans. The positive effects of water prices and energy expenses are unexpected, but this may indicate the ineffectiveness of a higher water price on water conservation. A positive effect of labor payment may suggest that these factors are complements; more labor use facilitates more irrigation, or producers who need more irrigation to maximize profits use more labor.

Table 6. The results of multilevel models for the mean water application on corn and soybean farms.

Variables	Corn		Soybean	
	Estimate	Std Err	Estimate	Std Err
Fixed Effects				
Intercept	1.041 ***	0.328	1.151 ***	0.227
Water sources				
On-farm surface water only	−0.037	0.069	0.015	0.078
Off-farm surface water only	−0.075	0.083	−0.015	0.055
Two or more water sources	0.106 **	0.044	0.039	0.036
Costs				
Cost for off-farm surface water($/acre-foot)	0.003 **	0.001	0.001	0
Energy expenses ($/acre)	0.001 *	0.001	0.001 **	0.001
Facility expenses ($/acre)	0	0	0	0
Labor payment ($/acre)	0.002 **	0.001	−0.001	0.002

Table 6. *Cont.*

Variables	Corn		Soybean	
	Estimate	Std Err	Estimate	Std Err
Farm characteristics				
Number of wells used	0.002	0.001	0.002 *	0.001
LN(total acre)	0.026 *	0.015	0.004	0.01
Percent of owned land	−0.003	0.049	−0.024	0.035
Pressure irrigation	−0.057	0.107	−0.174 ***	0.044
Federal assistance	0.029	0.033	0.046 *	0.025
Barriers to improvements				
Investigating improvement is not a priority	0.056 ***	0.02	0.006	0.02
Risk of reduced yield or poorer quality crop	0.064 **	0.029	0.079 ***	0.028
Limitation of physical field or crop conditions	−0.094 ***	0.026	0.039	0.025
Not enough to recover implementation costs	0.001	0.021	−0.004	0.019
Cannot finance improvements	0.139 ***	0.023	0.027	0.022
Landlords will not share improvement costs	−0.012	0.025	−0.060 ***	0.022
Uncertainty about future water availability	−0.074 ***	0.026	−0.084 ***	0.028
Will not be farming long enough	0.055 *	0.03	−0.068 **	0.03
Will increase management time or cost	−0.077 ***	0.03	−0.014	0.029
Information source				
Extension agents	−0.062 ***	0.017	−0.041 ***	0.015
Private irrigation specialists	−0.040 **	0.017	−0.042 ***	0.015
Irrigation equipment dealers	0.055 ***	0.018	−0.044 ***	0.016
Local irrigation district employee	0.098 ***	0.03	0.024	0.031
Government specialists	0.038	0.023	0.054 ***	0.02
Media reports	−0.009	0.025	−0.042 *	0.022
Neighboring farmers	−0.069 ***	0.019	−0.037 **	0.017
E-information services	0.060 ***	0.02	0.037 **	0.019
State-level variables				
PrecipChange2011	−0.043 ***	0.015	−0.008	0.006
PrecipChange2012	−0.064 ***	0.019	−0.018 *	0.009
PrecipChange2013	−0.079 ***	0.022	−0.036 ***	0.011
TempChange2011	0.05	0.109	0.111	0.072
TempChange2012	−0.330 ***	0.095	−0.152 **	0.067
TempChange2013	−0.335	0.146	−0.135 *	0.071
West	0.961 **	0.204	0.809 ***	0.192
Midwest	0.180 ***	0.262	−0.160	0.122
South	−0.101	0.276	−0.131	0.149
Atlantic	0.591	0.437	−0.227	0.195
Error Variance	**Estimate**	**Std Err**	**Estimate**	**Std Err**
Intercept	<0.0001 ***	<0.0001	<0.0001 ***	<0.0001
Residual	1.766 ***	0.033	0.886 ***	0.02
Fit Statistics				
N	6030		3933	
−2 Log Likelihood	14,730		6892	
AIC	14,834		6994	
AICC	14,835		6995	
BIC	14,926		7076	

Significance levels: * 10%; ** 5%; *** 1%.

Regarding farm characteristics, the results show that more wells are positively associated with water application on soybean farms, which is consistent with the hypothesis as mentioned above that more wells provide farmers more and easier access to water. A large farming area has a positive association with the average water application on corn farms. The adoption of pressure irrigation systems reduces irrigation water application for soybean farms, which is consistent with the hypothesis that the enhanced pressure irrigation systems reduce water use. Federal assistance increases water use on soybean farms through improved irrigation and drainage.

Barriers showing a negative effect on water application on corn farms include the limitation of physical field or crop conditions, an uncertainty about future water availability, and increase management time or cost. For soybeans, barriers with a negative effect are landlords will not share improvements costs, uncertainty about future water availability, and will not be farming long enough.

These negative effects are in line with the expectations. However, further investigations are needed on variables showing a positive effect.

Information from extension agents, private irrigation specialists, and neighboring farmers have a negative effect on the water use of both corn and soybeans, and irrigation equipment dealers, and media reports also show a negative effect on soybean water use. However, information from E-information services has a positive effect. These findings indicate that certain groups can be more effective in conserving water use.

The state-level variables on climate variability show a very consistent pattern on both corn and soybean water use. Compared to the average precipitation in 1981–2010, more precipitation in 2012 and 2013 leads to less irrigation water application on corn and soybean farms. Compared to the average temperature in 1981–2010, the higher temperature in 2012 and 2013 is negatively associated with the water application of both corn and soybeans in 2013. This indicates that water use is related to both climate variability based on early experience and current water availability. Compared to the farmers in the Plains, those in the West use more water for both crops, which is consistent with the expectations.

6.6. Crop Yield

The MLMs results for crop yield equations of corn and soybeans are presented in Table 7. The results are shown compared to groundwater use only and water from off-farm sources has a positive effect on soybean yield. Unfortunately, none of the cost variables is significantly for both crop yields.

For farm characteristics, more wells used on soybean farms increase the yield. A larger area of farmed land has a positive effect on corn yield, which indicates the economics of scale on corn production. A larger percentage of land owned decreases the yield for both crops. The adoption of pressure irrigation systems shows a positive effect on soybean yield, indicating that soybean yield is increased under enhanced irrigation systems.

Barriers showing a negative effect on yields of both crops include the limitation of physical field or crop conditions, and lack of financing to make improvements. This suggests that crop yield is more related to physical limitation.

Table 7. The results of multilevel models for the mean crop yield of corn and soybean farms.

Variables	Corn		Soybeans	
	Estimate	Std Err	Estimate	Std Err
Fixed Effects				
Intercept	159.780 ***	22.366	51.233 ***	4.905
Water sources				
On-farm surface water only	−6.41	3.823	0.547	1.416
Off-farm surface water only	−2.328	3.64	5.590 ***	1.21
Two or more water sources	1.011	2.681	1.06	0.835
Costs				
Cost for off-farm surface water($/acre-foot)	0.005	0.013	−0.001	0.011
Energy expenses ($/acre)	0	0.025	0	0.014
Facility expenses ($/acre)	0.003	0.007	0	0.001
Labor payment ($/acre)	0.016	0.022	0.035	0.034
Farm characteristics				
Number of wells used	0.02	0.082	0.130 ***	0.032
LN(total acre)	1.980 ***	0.656	−0.262	0.203
Percent of owned land	−5.814 ***	2.129	−2.669 **	1.028
Pressure irrigation	3.956	3.79	2.401 *	1.203
Federal assistance	1.824	1.887	0.81	0.701
Barriers to improvements				
Investigating improvement is not a priority	−1.614	1.151	−0.241	0.484
Risk of reduced yield or poorer quality crop	10.875 ***	1.636	−0.262	0.696
Limitation of physical field or crop conditions	−3.803 ***	1.485	−1.338 **	0.602
Not enough to recover implementation costs	−0.888	1.187	1.401 ***	0.475
Cannot finance improvements	−6.858 ***	1.298	−1.695 ***	0.552
Landlords will not share improvement costs	−0.976	1.39	−0.572	0.53
Uncertainty about future water availability	−2.009	1.441	−1.052	0.677
Will not be farming long enough	0.955	1.682	2.783	0.73
Will increase management time or cost	−0.777	1.679	−0.462 ***	0.711

Table 7. *Cont.*

Variables	Corn		Soybean	
	Estimate	Std Err	Estimate	Std Err
Information sources				
Extension agents	4.107 ***	0.977	1.732 ***	0.368
Private irrigation specialists	3.528 ***	0.94	1.555 ***	0.371
Irrigation equipment dealers	−1.009	0.988	−0.981 **	0.394
Local irrigation district employee	−1.044	1.677	−0.327	0.761
Government specialists	−3.724 ***	1.31	−0.083	0.5
Media reports	2.022	1.381	1.323 **	0.537
Neighboring farmers	−0.522	1.066	0.907 **	0.413
E-information services	2.574 ***	1.147	0.563	0.454
State-level variables				
PrecipChange2011	0.78	0.802	0.045	0.117
PrecipChange2012	−1.145	1.225	0.697 ***	0.171
PrecipChange2013	−1.679	1.234	0.142	0.245
TempChange2011	−11.005	8.57	−3.625 ***	1.388
TempChange2012	3.181	6.313	3.151 **	1.488
TempChange2013	3.242	9.373	5.494 ***	1.439
West	−10.149	18.285	−11.099 ***	4.055
Midwest	9.455	16.155	3.151	2.285
South	25.86	19.538	1.568	2.825
Atlantic	17.813	23.744	−1.05	4.153
Error Variance	**Estimate**	**Std Err**	**Estimate**	**Std Err**
Intercept	326 **	154	<0.0001 ***	<0.0001
Residual	5636 ***	106	534.360 ***	12.309
Fit Statistics				
N	6030		3933	
−2 Log Likelihood	63,275		32,059	
AIC	63,361		32,149	
AICC	63,362		32,150	
BIC	63,440		32,220	

Significance levels: * 10%; ** 5%; *** 1%.

Irrigation information from extension agents and private irrigation specialists show a positive effect on both corn and soybean yield. E-information services only show a positive effect on corn yield, and information from media reports and neighboring farmers have a positive effect on soybean yield. However, information showing a negative effect include government specialists (on corn yield), and irrigation equipment dealers and local irrigation district employees (on soybean yield).

Regarding state-level variables, the precipitation change in 2012 and the temperature changes in 2012 and 2013 show a positive effect on soybean yield. Given the results from the water application regressions, it seems that farmers who have access to more irrigation are able to offset the effects of weather variability. Compared with the Plains States, farms in the West have a lower soybean yield.

6.7. Economic Irrigation Water Use Efficiency

The parameter estimates for *EIWUE* equations of corn and soybeans are presented in Table 8. The results show that irrigation using water from on-farm surfaces only has a positive effect on corn *EIWUE*, compared to groundwater only. Higher water prices decrease *EIWUE* of corn, and higher energy expenses also decrease *EIWUE* of both crops. Combined with the results on water use and yield, these findings suggest that a higher efficiency cannot be achieved through increasing water prices. Higher labor payment also decreases *EIWUE* of corn.

Regarding farm characteristics, the number of wells shows a negative effect on both corn and soybean *EIWUE*. This indicates that fewer wells available on a farm can encourage an efficient use of irrigation water. The adoption of pressure irrigation increases the water use efficiency of both crops, indicating the effectiveness of achieving higher irrigation water use efficiency with the application of enhanced irrigation systems, and this is consistent with the results of water application and crop yield.

Similarly, irrigation efficiency is limited by factors related to the risk of reduced yield or poorer quality crop (on soybeans), limitation of physical field or crop conditions (on soybeans), cannot finance improvements (on corn), and will not be farming long enough (on corn). These findings can be true if water applications are limited by poor water distribution systems and/or farmers are resource-limited.

Effects of information sources are consistent for the two crops. Media reports show a positive effect, and variables showing a negative effect include local irrigation district employees and government specialists.

Regarding the state-level variables on climate variability, for soybean farms, compared with the average precipitation, a higher precipitation in 2011 and 2012 are positively associated with higher irrigation water use efficiency in 2013. The precipitation change in 2013 is positively associated with water use efficiency of both crops. The temperature change in 2011 decreases the *EIWUE* of corn and the temperature changes in 2013 increase *EIWUE* of both crops. These findings suggest that higher temperatures in the growing season lead to farmers using water more efficiently, while perceptions of precipitation are more effective to increase *EIWUE* than perceptions of temperature. Compared to farms in the Plains, both corn and soybean farms in the West have a lower *EIWUE*, while corn farms in the Midwest, South, and Atlantic states have a higher *EIWUE*.

Table 8. The results of multilevel models for the economic irrigation water use efficiency for corn and soybeans.

Variables	Corn		Soybeans	
	Estimate	Std Err	Estimate	Std Err
Fixed Effects				
Intercept	1601.320 ***	341.43	2381.410 ***	692.38
Water sources				
On-farm surface water only	536.290 ***	196.38	22.336	94.799
Off-farm surface water only	561.91	390.29	248.88	289.24
Two or more water sources	−49.571	45.746	−126.34	120.95
Costs				
Cost for off-farm surface water ($/acre-foot)	−11.042 *	6.483	−6.737	4.378
Energy expenses ($/acre)	−3.339 ***	0.986	−3.189 ***	0.859
Facility expenses ($/acre)	−0.461	0.394	0.172	0.55
Labor payment ($/acre)	−0.519 *	0.278	0.147	3.418
Farm characteristics				
Number of wells used	−8.072 ***	2.181	−4.342 *	2.135
LN (total acre)	−17.208	17.449	−18.973	20.326
Percent of owned land	−4.086	62.276	35.36	81.314
Pressure irrigation	141.810 **	64.876	206.590 **	73.067
Federal assistance	14.624	48.547	−12.277	62.92
Barriers to improvements				
Investigating improvement is not a priority	−108.420 ***	38.485	107.210 ***	40.14
Risk of reduced yield or poorer quality crop	−14.715	54.688	−155.590 ***	57.861
Limitation of physical field or crop conditions	−28.084	49.586	−124.660 **	49.947
Not enough to recover implementation costs	−52.497	39.605	−37.805	39.363
Cannot finance improvements	−206.720 ***	43.527	17.588	45.694
Landlords will not share improvement costs	−12.135	46.402	8.707	43.887
Uncertainty about future water availability	111.800 **	48.486	73.29	57.502
Will not be farming long enough	−98.065 *	56.303	64.175	60.414
Will increase management time or cost	188.680 ***	56.305	23.363	59.68
Information sources				
Extension agents	−15.285	32.67	42.179	30.393
Private irrigation specialists	−20.287	31.278	29.727	30.785
Irrigation equipment dealers	−21.818	33.008	6.813	32.874
Local irrigation district employee	−106.130 **	56.215	−119.260 *	64.125
Government specialists	−173.390 ***	43.837	−89.006 **	41.363
Media reports	170.240 ***	46.276	160.840 ***	44.404
Neighboring farmers	54.146	35.616	84.640 **	34.312
E-information services	−35.648	38.361	60.764 *	37.575

Table 8. *Cont.*

Variables	Corn		Soybean	
	Estimate	**Std Err**	**Estimate**	**Std Err**
State-level variables				
PrecipChange2011	7.837	12.116	57.016 ***	14.839
PrecipChange2012	−2.929	17.994	66.779 **	25.763
PrecipChange2013	64.873 ***	18.85	121.720 ***	30.98
TempChange2011	−330.240 ***	117.88	−41.801	195.77
TempChange2012	45.172	96.719	−91.935	215.37
TempChange2013	348.990 **	131.81	375.790 *	198.71
West	−765.200 ***	259.83	−1404.190 **	558.8
Midwest	893.260 ***	224.72	484.59	313.72
South	717.640 **	283.45	−611.14	389.73
Atlantic	1734.600 ***	361.36	−413.26	471.07
Error Variance	**Estimate**	**Std Err**	**Estimate**	**Std Err**
Intercept	<0.0001 ***	<0.0001	98,568 *	68,609
Residual	6,299,184 ***	118,144	3,607,043 ***	84,572
Fit Statistics				
N	6030		3933	
−2 Log Likelihood	105,657		66,842	
AIC	105,759		66,950	
AICC	105,760		66,952	
BIC	105,849		67,037	

Significance levels: * 10%; ** 5%; *** 1%.

7. Discussion

7.1. Balancing Land Allocation

As an important production input, land oftentimes overshadows irrigation water, and the farmers' decision on land input may determine water use and other inputs [76]. A profit maximizing producer might want to optimally allocate land to planting one or more crops while considering the constraints as well as other real and perceived factors. In an irrigated multi-crop production system, water availability is a serious consideration in the production function of farmers. Consistent with Moore and Dinar [73], our results show a better water availability, with more sources and wells for groundwater extraction, as an input for crop production increases land allocation in a multi-crop system. Compared to dryland production, adequate irrigation has been confirmed to increase output and farm income [39]. In particular, for water-consuming crops, farmers' production decisions prefers more water availability and less variation of water supply in growing seasons. Otherwise the yield can be hurt and producers lose incentives to continue farming.

Not only should water sources be considered in agricultural production, but the prices of inputs are important in driving or limiting factors, including the water price and energy price [62]. Water price is integrated with water availability in irrigated farming decisions [54]. While water prices are typically much lower than their real value. Given the inelastic water demand in farm irrigation, a small increase within the low price range may not be effective to conserve water and subsequently may not show a clear influencing pattern on land allocation [54].

In addition to the costs, other farm and farmer characteristics, nonmonetary motivations, or lack thereof, and information availability are equally important in farmers' decision-making [77]. A large farmland may exhibit economies of scale in crop production, thus, increasing the land allocation to more profitable crops. As the cost of per unit input decreases, the cost advantage can be remarkable and larger production returns can be expected. Additionally, farmers' production incentives can be other nonmonetary motivations. In particular, technical assistance and informational support can facilitate scientific farming decisions [78]. Irrigated production incorporates the adoption of agricultural innovations and the best management practices, which are largely adviser-driven and going beyond the farmers' experience. Therefore, the land use decisions are not free from information access and resources for overcoming the obstacles.

Land use decisions can be affected by both weather conditions in the past and farmers' climate perceptions in the coming growing seasons. Climate variability and risks can be a major threat to the farm output if appropriate coping strategies are not in place [18]. Though insurance helps reduce the potential damage, balancing land allocation according to the experienced and perceived weather variability can optimize input combinations and stabilize the expected farm income [4]. In addition, land use decisions are differing with varying geographical conditions at the state and regional levels. Coupled with climatic conditions, farmers growing the same crops may allocate different portions of land to each crop because they are facing different soil types and land slopes, among others [51]. The state and regional boundaries may also represent the implicit effects of water institutions, which influence land allocation decisions through affecting access to different water sources and the priority of water rights.

7.2. Conserving Water Resources

Along with land allocation, water applied for farm irrigation can be managed at the farm level. More sources for water supply may provide better access to water for irrigation purposes and producers may have more flexibility in irrigating crops while considering the real-time soil and crop conditions [78]. According to the conventional production economic theory, a higher input price decreases the amount of input use. In this case, a higher water price should reduce water use [61,62]. The opposite findings from our analysis definitely need close scrutiny, while they might be plausible as an overall effect given the low values of water prices. There are some facts behind these findings: (1) most of the farmers are groundwater users rather than surface water users; (2) some surface water users may have a fee-based surface water delivery system and do not pay a marginal cost for additional water; and (3) surface water use may be highly dependent on the producers' surface water rights, regardless of whether they pay a fee or an additional cost for additional units of surface water [79]. In addition, an increase in water application on per acre basis may be possible if farmers adjust their mix of crops toward more water-consuming crops or varieties, or because yield or revenue can be increased [75].

Meanwhile, the total price effect of surface and groundwater on water use in the multi-crop system is negative and consistent with the previous literature [80]. Pfeiffer and Lin [62] found an increase in the energy price of $1 would decrease groundwater extraction by 5.89 acre-feet per year for an individual farmer in Kansas. Our overall marginal effect is almost doubled, while the area of a typical farm planting multiple crops in their study is less than half of the total area of planting both corn and soybeans in our study. In another study, Hendricks and Peterson [61] found the total elasticity of water demand was −0.10 based on Kansas farm irrigation. As a comparison, Pfeiffer and Lin [62] found an elasticity of −0.26. Therefore, our findings show a modest overall effect of water price on water conservation since we just included two crops, with one being water-intensive and the other being less water-intensive, in our multi-crop production analysis.

Advanced irrigation systems have been promoted in the past decades as a way to conserve irrigation water, while recent studies have reported mixed effects [75,81]. Jevons' Paradox or the rebound effect [82] of an efficient irrigation technology adoption points to an increased water use as a result of crop choices toward more water-intensive crops and an expansion of irrigated acreage [83]. Balanced by both extensive and intensive margins, the rebound effect can be small, moderate, or even larger than 100%. As a typical issue in an irrigated production system with multiple crops, the rebound effect is a serious consideration and it might counteract the water reduction effect of adopting water conservation technologies.

Producers' experienced and perceived climate variability may have a salient influence on water use [5,18]. Similar to the effects of climate risks, a higher variability in rainfall and temperature may ultimately change the real water demand of different crops [51]. To achieve a certain yield goal, farmers routinely want to satisfy the water demand as an attempt to reduce production risks if possible during dry growing seasons, and this is even seen in arid areas [84]. Though the impacts of climate variability on different crops can be different, the effects of the farmers' perceptions may not be proportional to the water demand of different crops [6]. As a result, the effects of climate variability may be mixed and combining those of rainfall, temperature, and others. Additionally, the threshold of climate variability may be of great significance and the effects can depend on the crop-specific and baseline climate conditions [18].

7.3. Improving Crop Yield

The average grain yield on an acre basis has been approved to be higher with adequate irrigation compared with dryland production or with inadequate irrigation [14,42,51]. In a similar vein, a higher water availability by means of either more water sources or more wells facilitates producers to irrigate at a right time, with an appropriate amount of water and at scientific intervals. In the meantime, a large farm may have a higher production efficiency because of the economics of scale and a better ability to mobilize physical and technical resources [19]. Especially, if a large farm owner has the water rights, he can either use as much water as he wants or have a higher priority of withdrawing water for irrigation purposes, even if he grows water-intensive crops. Furthermore, different from the insignificant effect of owned land by Olen et al. [18], our findings show that a larger proportion of owned land decreases average grain yield. This can be true as empirical studies have found farmland rental enhances land productivity [85] and encourages farmers to be more productive and maximize the output within a limited contract period. Leased land may better motivate farmers to utilize machinery and reduce production costs [23], and generally, farmers who rent more land for growing crops specialize in agricultural production [85].

Regarding the barriers to improvements and information sources, their effects can be better understood while jointly looking at the water application and crop yield estimation results. On the one hand, since the barriers are more related to energy reduction or water conservation, their effects are mixed and more indirect. Financial limitation, physical conditions, a short farming horizon, and uncertainty in the future water supply are among the major barriers that push farmers to rely on outdated, conventional irrigation facilities and techniques, which weakens farmers' enthusiasm on water conservation and undermines their ability and effort to maximize crop yield [38,47]. On the other hand, the patterns of information effects are clearer and more direct. Water use can be reduced by extension agents, private specialists, media reports, and neighboring farmers, and these efforts are relatively consistent in enhancing grain yield. As irrigated agriculture becomes increasingly information-dependent, a wide range of scientific and technical information is required for effective decision-making [86]. The information seeking and acquisition behaviors may be influenced by sociodemographic factors as well as the preference of the farmers towards different information sources [87]. Additionally, the information efforts may help overcome the barriers in realizing a lower water consumption and/or higher farm productivity [2,88].

7.4. Enhancing Water Use Efficiency

Farm- and crop-level water use efficiency has been generously reported for different crops under different tillage systems, irrigation levels, and various farm management practices [30,33,39]. By definition, the efficiency is positively correlated with grain yield and negatively correlated with irrigation water application [14]. The effects of the factors on the efficiency can be better understood by comparing their effects on both water applications and crop yield. Water abundance reflected by groundwater use and more wells provide producers easy access and may motivate them to increase the water amount per acre, thus decreasing water use efficiency. In addition, water use efficiency may not have a linear relationship with water application [16,34]. Compared with dryland production or crop growing under drought stresses, a little more water may significantly increase yield, and as a result, the increase in water efficiency can be remarkable [33,34,39], while a higher than usual water use is unlikely to further increase grain yield, and the efficiency change can be reversed [14]. Especially for water-intensive crops like corn, more water usually means no significant yield increase and a declining water efficiency [31].

The costs of variable inputs were previously found to increase water use while having no effect on the grain yield. As a result, water use efficiency decreased. This is largely because more energy and labor were used to provide more water for the water-consuming crop [51], while the adoption of pressure irrigation systems shows a positive effect. This may be because water is saved for water-intensive corn while the yield is not hurt, and both water conservation and grain yield are maintained for less water-intensive soybeans [6].

The weather-related variables show composite effects on water use efficiency by impacting water use and farm yield [6]. On the one hand, more rainfall reduces the supplemental irrigation amount, which results in a higher irrigation efficiency. In particular, the past experiences of ample precipitation may discourage farmers to use more water for a certain yield level [4,51]. On the other hand, a high temperature may have two types of impacts: (1) a hot growing season in previous years, like the drought events in 2011 and 2012, may promote producers to irrigate more to mitigate potential dry conditions in the current year; (2) in a normal year, like 2013, a slightly higher temperature may not lead to notably more irrigation, while the grain yield may be increased as a result of improved photosynthesis [5,51]. Though more fine-scale explorations are necessary to clarify the effects of climate variability, including the direct effects of rainfall and temperature, the evidence here provides insights on the effects of both experienced and perceived weather changes.

8. Conclusions

Using the 2013 USDA FRIS data, this paper analyzes farmers' production decisions relating to irrigated agriculture in a multi-crop production system. To study the role of water costs, the farm-level water application is decomposed into crop-specific application. For each crop, the total effect can be obtained by summing the intensive and extensive margins of water use. With the aggregate effect at the farm level, we can quantify the effect of a one-unit increase in water price. Furthermore, the effects of exogenous variables are analyzed using a multilevel model approach. Four equations regarding land allocation, water application, crop yield, and economic irrigation water use efficiency are formulated using two-level models.

A fundamental finding from the decomposition of farm-level water application illustrates that the higher costs of surface water are not effective to reduce water use for both corn and soybeans through both intensive and extensive margins, while a proxy of groundwater price has a negative effect on soybean water use. This finding is a surprise, but empirically supported by some evidence. Similar to the mixed effects of water price found by Moore et al. [54], water cost is ineffective in conserving water use once producers have made decisions on crop production. Pfeiffer and Lin [75] found farm-level policies to conserve water use may not be effective. In this case, the surface water price is very low and it may not be effective because the water use is inelastic [80,89]. Comparatively, a much higher groundwater price is effective to conserve water use.

In addition, the results from MLMs allow us to make certain of the relative importance of farm- and state-level factors, and the estimation outcomes present the effects of those exogenous variables at both levels. The adoption of pressure irrigation systems reduces the soybean water use and increases the soybean yield. A higher *EIWUE* due to enhanced irrigation methods can also be achieved on both corn and soybean farms.

The findings from MLMs show that the state-level variables on climate variability have consistent effects. A high temperature promotes more efficient water use and higher yield. A high precipitation is correlated with low water application and higher crop yield. Droughts due to less rainfall or high temperature and their perceptions increase farmers' awareness of potential production risks not only during droughts, but in subsequent years [90]. As a result, farmers can be motivated to change land allocation for different crops and irrigate more to mitigate the adverse effects of climate variability. Contrary to Olen et al. [18], we find the irrigation water use is more responsive to precipitation than to temperature. Given the nonlinear impacts of climatic factors, farmers' responses in adapting to climate risks depend on cropping patterns.

This study also leaves some opportunities for future research. The aggregate effect is estimated for a typical farm growing corn and soybeans taking roughly half of the average farming area. Equations on more crops can be estimated to provide a more complete estimate of the water price effect [80], and regional equations can be estimated to account for structural differences across regions. Ideally, the elasticity with respect to water price can be estimated to quantify the price effect from a different and equally important perspective [60]. Though MLMs are supposed to deal with multiple estimation problems, more empirical and methodological investigations are needed, especially on potential endogeneity problems.

Supplementary Materials: The additional Tables S1–S9 are available online at http://www.mdpi.com/2073-4441/10/11/1637/s1.

Author Contributions: Y.F. designed the research, analyzed the data and wrote the paper. Y.F., R.M., and S.C.P. reviewed and commented on the manuscript.

Funding: The research was supported by the USDA National Integrated Water Quality Grant Program number 110.C (Award 2012-03652), the USDA Multi-state Grant W-3190 Management and Policy Challenges in a Water-Scarce World, and the USDA Agricultural Research Service Initiative-Ogallala Aquifer Program (FY2016–2017; FY2017–2018).

Acknowledgments: We thank Brad Parks for his support when the first author analyzed data at the USDA NASS data lab in St. Louis, Missouri. We also appreciate helpful comments by Laura McCann, Hua Qin, and Corinne Valdivia on an earlier version of the paper. The authors are grateful to participants at the 2017 Agricultural & Applied Economics Association Annual Meetings. This research was conducted while Yubing Fan was a doctoral candidate at the University of Missouri-Columbia.

Conflicts of Interest: The authors declare no conflict of interest.

References and Notes

1. Fan, Y.; Massey, R.; Park, S.C. Multicrop production decisions and economic irrigation water use efficiency: Effects of water costs, pressure irrigation adoption and climatic determinants. In Proceedings of the Annual Meeting 2017, Chicago, IL, USA, 30 July–1 August 2017; Agricultural and Applied Economics Association: Milwaukee, WI, USA, 2017; pp. 1–60.

2. Schaible, G.; Aillery, M. *Water Conservation in Irrigated Agriculture: Trends and Challenges in the Face of Emerging Demands*; USDA-ERS Economic Information Bulletin No. 99; USDA-ERS: Washington, DC, USA, 2012.

3. EPA (United States Environmental Protection Agency). Climate Change Indicators in the United States—Weather and Climate, 2014. Available online: https://www3.epa.gov/climatechange/pdfs/climateindicators-full-2014.pdf (accessed on 15 July 2016).

4. Jin, J.; Gao, Y.; Wang, X.; Nam, P.K. Farmers' risk preferences and their climate change adaptation strategies in the Yongqiao District, China. *Land Use Policy* **2015**, *47*, 365–372.

5. Zhang, T.; Lin, X.; Sassenrath, G.F. Current irrigation practices in the central United States reduce drought and extreme heat impacts for maize and soybean, but not for wheat. *Sci. Total Environ.* **2015**, *508*, 331–342. [CrossRef] [PubMed]

6. Zhang, T.; Lin, X. Assessing future drought impacts on yields based on historical irrigation reaction to drought for four major crops in Kansas. *Sci. Total Environ.* **2016**, *550*, 851–860. [CrossRef] [PubMed]

7. Wanders, N.; Wada, Y. Human and climate impacts on the 21st century hydrological drought. *J. Hydrol.* **2015**, *526*, 208–220. [CrossRef]

8. Murray, S.J.; Foster, P.N.; Prentice, I.C. Future global water resources with respect to climate change and water withdrawals as estimated by a dynamic global vegetation model. *J. Hydrol.* **2012**, *448–449*, 14–29. [CrossRef]

9. George, B.A.; Shende, S.A.; Raghuwanshi, N.S. Development and testing of an irrigation scheduling model. *Agric. Water Manag.* **2000**, *46*, 121–136. [CrossRef]

10. Batchelor, C.; Lovell, C.; Murata, M. Simple microirrigation techniques for improving irrigation efficiency on vegetable gardens. *Agric. Water Manag.* **1996**, *32*, 37–48. [CrossRef]

11. Dalton, P.; Raine, S.; Broadfoot, K. Best Management Practice for Maximising Whole Farm Irrigation Efficiency in the Australian Cotton Industry. Final Report for CRDC Project NEC2C. 2001. Available online: http://www.insidecotton.com/xmlui/handle/1/3535 (accessed on 12 November 2016).

12. Qin, W.; Assinck, F.B.T.; Heinen, M.; Oenema, O. Water and nitrogen use efficiencies in citrus production: A meta-analysis. *Agric. Ecosyst. Environ.* **2016**, *222*, 103–111. [CrossRef]

13. Gheysari, M.; Loescher, H.W.; Sadeghi, S.H.; Mirlatifi, S.M.; Zareian, M.J.; Hoogenboom, G. Chapter Three-Water-yield relations and water use efficiency of maize under nitrogen fertigation for semiarid environments: Experiment and synthesis. *Adv. Agron.* **2015**, *130*, 175–229.

14. Ibragimov, N.; Evett, S.R.; Esanbekov, Y.; Kamilov, B.S.; Mirzaev, L.; Lamers, J.P.A. Water use efficiency of irrigated cotton in Uzbekistan under drip and furrow irrigation. *Agric. Water Manag.* **2007**, *90*, 112–120. [CrossRef]

15. Schneider, A.D.; Howell, T.A. Scheduling deficit wheat irrigation with data from an evapotranspiration network. *Trans. ASAE* **2001**, *44*, 1617–1623. [CrossRef]

16. Kang, Y.; Wang, R.; Wan, S.; Hu, W.; Jiang, S.; Liu, S. Effects of different water levels on cotton growth and water use through drip irrigation in an arid region with saline ground water of Northwest China. *Agric. Water Manag.* **2012**, *109*, 117–126. [CrossRef]

17. Nijbroek, R.; Hoogenboom, G.; Jones, J.W. Optimizing irrigation management for a spatially variable soybean field. *Agric. Syst.* **2003**, *76*, 359–377. [CrossRef]

18. Olen, B.; Wu, J.; Langpap, C. Irrigation decisions for major West Coast crops: Water scarcity and climatic determinants. *Am. J. Agric. Econ.* **2016**, *98*, 254–275. [CrossRef]

19. Boyer, C.N.; Larson, J.A.; Roberts, R.K.; McClure, A.T.; Tyler, D.D. The impact of field size and energy cost on the profitability of supplemental corn irrigation. *Agric. Syst.* **2014**, *127*, 61–69. [CrossRef]

20. Dolisca, F.; McDaniel, J.M.; Shannon, D.A.; Jolly, C.M. A multilevel analysis of the determinants of forest conservation behavior among farmers in Haiti. *Soc. Nat. Resour.* **2009**, *22*, 433–447. [CrossRef]

21. Guerin, D.; Crete, J.; Mercier, J. A multilevel analysis of the determinants of recycling behavior in the European countries. *Soc. Sci. Res.* **2001**, *30*, 195–218. [CrossRef]

22. Neumann, K.; Stehfest, E.; Verburg, P.H.; Siebert, S.; Müller, C.; Veldkamp, T. Exploring global irrigation patterns: A multilevel modelling approach. *Agric. Syst.* **2011**, *104*, 703–713. [CrossRef]

23. Giannakis, E.; Bruggeman, A. Exploring the labour productivity of agricultural systems across European regions: A multilevel approach. *Land Use Policy* **2018**, *77*, 94–106. [CrossRef]

24. Pereira, L.S. Higher performance through combined improvements in irrigation methods and scheduling: A discussion. *Agric. Water Manag.* **1999**, *40*, 153–169. [CrossRef]

25. Nair, S.; Johnson, J.; Wang, C. Efficiency of irrigation water use: A review from the perspectives of multiple disciplines. *Agron. J.* **2013**, *105*, 351–363. [CrossRef]

26. Harris, G. *Water Use Efficiency: What Is It, and How to Measure, Spotlight on Cotton Research & Development*; Cotton Research & Development Corporation: Narrabri, Australia, 2007; p. 8. Available online: http://era.daf.qld.gov.au/id/eprint/2986/1/50904_CottonCRC_Final_Report_Harris.pdf (accessed on 5 June 2015).

27. Tanwar, S.; Rao, S.; Regar, P.; Datt, S.; Jodha, B.; Santra, P.; Kumar, R.; Ram, R. Improving water and land use efficiency of fallow-wheat system in shallow Lithic Calciorthid soils of arid region: Introduction of bed planting and rainy season sorghum–legume intercropping. *Soil Tillage Res.* **2014**, *138*, 44–55. [CrossRef]

28. Bozzola, M.; Swanson, T. Policy implications of climate variability on agriculture: Water management in the Po river basin, Italy. *Environ. Sci. Policy* **2014**, *43*, 26–38. [CrossRef]

29. Sadras, V.; Rodriguez, D. Modelling the nitrogen-driven trade-off between nitrogen utilisation efficiency and water use efficiency of wheat in eastern Australia. *Field Crops Res.* **2010**, *118*, 297–305. [CrossRef]

30. Fan, Y.; Wang, C.; Nan, Z. Comparative evaluation of crop water use efficiency, economic analysis and net household profit simulation in arid Northwest China. *Agric. Water Manag.* **2014**, *146*, 335–345. [CrossRef]

31. Salazar, M.R.; Hook, J.E.; Garcia y Garcia, A.; Paz, J.O.; Chaves, B.; Hoogenboom, G. Estimating irrigation water use for maize in the Southeastern USA: A modeling approach. *Agric. Water Manag.* **2012**, *107*, 104–111. [CrossRef]

32. Vories, E.D.; Tacker, P.L.; Lancaster, S.W.; Glover, R.E. Subsurface drip irrigation of corn in the United States Mid-South. *Agric. Water Manag.* **2009**, *96*, 912–916. [CrossRef]

33. Dağdelen, N.; Başal, H.; Yılmaz, E.; Gürbüz, T.; Akçay, S. Different drip irrigation regimes affect cotton yield, water use efficiency and fiber quality in western Turkey. *Agric. Water Manag.* **2009**, *96*, 111–120. [CrossRef]

34. Liu, Y.; Li, S.; Chen, F.; Yang, S.; Chen, X. Soil water dynamics and water use efficiency in spring maize (*Zea mays* L.) fields subjected to different water management practices on the Loess Plateau, China. *Agric. Water Manag.* **2010**, *97*, 769–775. [CrossRef]

35. Salvador, R.; Latorre, B.; Paniagua, P.; Playán, E. Farmers' scheduling patterns in on-demand pressurized irrigation. *Agric. Water Manag.* **2011**, *102*, 86–96. [CrossRef]

36. Usman, M.; Arshad, M.; Ahmad, A.; Ahmad, N.; Zia-Ul-Haq, M.; Wajid, A.; Khaliq, T.; Nasim, W.; Ali, H.; Ahmad, S. Lower and upper baselines for crop water stress index and yield of *Gossypium hirsutum* L. under variable irrigation regimes in irrigated semiarid environment. *Pak. J. Bot.* **2010**, *42*, 2541–2550.

37. Pannell, D.J.; Marshall, G.R.; Barr, N.; Curtis, A.; Vanclay, F.; Wilkinson, R. Understanding and promoting adoption of conservation practices by rural landholders. *Aust. J. Exp. Agric.* **2006**, *46*, 1407–1424. [CrossRef]

38. Knowler, D.; Bradshaw, B. Farmers' adoption of conservation agriculture: A review and synthesis of recent research. *Food Policy* **2007**, *32*, 25–48. [CrossRef]

39. Abd El-Wahed, M.H.; Ali, E.A. Effect of irrigation systems, amounts of irrigation water and mulching on corn yield, water use efficiency and net profit. *Agric. Water Manag.* **2013**, *120*, 64–71. [CrossRef]

40. Canone, D.; Previati, M.; Bevilacqua, I.; Salvai, L.; Ferraris, S. Field measurements based model for surface irrigation efficiency assessment. *Agric. Water Manag.* **2015**, *156*, 30–42. [CrossRef]

41. Ahmad, B.; Hassan, S.; Bakhsh, K. Factors affecting yield and profitability of carrot in two districts of Punjab. *Int. J. Agric. Biol.* **2005**, *7*, 794–798.

42. Dahmardeh, N.; Asasi, H. Determined factors on water use efficiency and profitability in agricultural sector. *Ind. J. Sci. Res.* **2014**, *4*, 48–53.

43. Schaible, G.D.; Kim, C.S.; Aillery, M.P. Dynamic adjustment of irrigation technology/water management in western US agriculture: Toward a sustainable future. *Can. J. Agric. Econ.* **2010**, *58*, 433–461. [CrossRef]

44. Rogers, E.M. *Diffusion of Innovations*, 5th ed.; Free Press: New York, NJ, USA, 2003.

45. Sunding, D.; Zilberman, D. The agricultural innovation process: Research and technology adoption in a changing agricultural sector. *Handb. Agric. Econ.* **2001**, *1*, 207–261.

46. Prokopy, L.; Floress, K.; Klotthor-Weinkauf, D.; Baumgart-Getz, A. Determinants of agricultural best management practice adoption: Evidence from the literature. *J. Soil Water Conserv.* **2008**, *63*, 300–311. [CrossRef]

47. Rodriguez, J.M.; Molnar, J.J.; Fazio, R.A.; Sydnor, E.; Lowe, M.J. Barriers to adoption of sustainable agriculture practices: Change agent perspectives. *Renew. Agric. Food Syst.* **2009**, *24*, 60–71. [CrossRef]

48. Hunecke, C.; Engler, A.; Jara-Rojas, R.; Poortvliet, P.M. Understanding the role of social capital in adoption decisions: An application to irrigation technology. *Agric. Syst.* **2017**, *153*, 221–231. [CrossRef]

49. Frisvold, G.B.; Deva, S. Farm size, irrigation practices, and conservation program participation in the US Southwest. *Irrig. Drain.* **2012**, *61*, 569–582. [CrossRef]

50. Negri, D.H.; Gollehon, N.R.; Aillery, M.P. The effects of climatic variability on US irrigation adoption. *Clim. Chang.* **2005**, *69*, 299–323. [CrossRef]

51. Kresovic, B.; Matovic, G.; Gregoric, E.; Djuricin, S.; Bodroza, D. Irrigation as a climate change impact mitigation measure: An agronomic and economic assessment of maize production in Serbia. *Agric. Water Manag.* **2014**, *139*, 7–16. [CrossRef]

52. Li, C.; Ting, Z.; Rasaily, R.G. Farmer's adaptation to climate risk in the context of China: A research on Jianghan Plain of Yangtze River Basin. *Agric. Agric. Sci. Proc.* **2010**, *1*, 116–125.

53. Li, X.; Takahashi, T.; Suzuki, N.; Kaiser, H.M. The impact of climate change on maize yields in the United States and China. *Agric. Syst.* **2011**, *104*, 348–353. [CrossRef]

54. Moore, M.R.; Gollehon, N.R.; Carey, M.B. Multicrop production decisions in western irrigated agriculture: The role of water price. *Am. J. Agric. Econ.* **1994**, *76*, 859–874. [CrossRef]

55. Producers also need to choose which type of irrigation system(s) to adopt, and this has been examined by much research, for instance, Olen et al. as cited in this paper.

56. Just, R.E.; Zilberman, D.; Hochman, E. Estimation of multicrop production functions. *Am. J. Agric. Econ.* **1983**, *65*, 770–780. [CrossRef]

57. Just, R.E.; Zilberman, D.; Hochman, E.; Bar-Shira, Z. Input allocation in multicrop systems. *Am. J. Agric. Econ.* **1990**, *72*, 200–209. [CrossRef]

58. Pereira, L.S.; Cordery, I.; Iacovides, I. Improved indicators of water use performance and productivity for sustainable water conservation and saving. *Agric. Water Manag.* **2012**, *108*, 39–51. [CrossRef]

59. The calculation of EIWUE (and IWUE) just considers irrigation water applied, while excluding rainfall amounts. The measure of water efficiency is restricted due to data paucity of climate-related variables at the farm level. To exam their effects, the state-level variation is controlled in the multilevel models presented below.

60. Moore, M.R.; Gollehon, N.R.; Carey, M.B. Alternative models of input allocation in multicrop systems: Irrigation water in the Central Plains, United States. *Agric. Econ.* **1994**, *11*, 143–158. [CrossRef]

61. Hendricks, N.P.; Peterson, J.M. Fixed effects estimation of the intensive and extensive margins of irrigation water demand. *J. Agric. Res. Econ.* **2012**, *37*, 1–19.

62. Pfeiffer, L.; Lin, C.-Y.C. The effects of energy prices on agricultural groundwater extraction from the High Plains Aquifer. *Am. J. Agric. Econ.* **2014**, *96*, 1349–1362. [CrossRef]

63. Hox, J.J. *Applied Multilevel Analysis*; TT-Publikaties: Amsterdam, The Netherlands, 1995.

64. Raudenbush, S.W.; Bryk, A.S. *Hierarchical Linear Models: Applications and Data Analysis Methods*, 2nd ed.; Sage: Thousand Oaks, CA, USA, 2002; Volume 1.

65. Ene, M.; Leighton, E.A.; Blue, G.L.; Bell, B.A. Multilevel Models for Categorical data Using SAS PROC GLIMMIX: The Basics. SAS Global Forum 2015. Available online: https://pdfs.semanticscholar.org/a216/864a2a2de19eb194c6523fb8566e601ffa32.pdf (accessed on 15 June 2015).

66. Snijders, T.A.; Bosker, R.J. *Multilevel Analysis: An Introduction to Basic and Advanced Multilevel Modeling*; Sage: Thousand Oaks, CA, USA, 1999.

67. In the empirical analysis below, land allocation refers to harvested acres from the FRIS data.

68. A Map Can Be Found on the USDA NASS Website. Available online: https://www.nass.usda.gov/Charts_and_Maps/Farm_Production_Expenditures/reg_map_c.php (accessed on 16 July 2016).

69. Ideally, analyses on all the production decisions (i.e., 4 equations regarding all crops (17 crops) can be conducted at the region level (i.e., 5 regions). Given the huge amount of work and the focus of this paper, such analyses are beyond the scope.

70. The crop-specific analyses just focus on farms that are at least partially irrigated, while excluding non-irrigated farms.

71. The USDA FRIS targeted at the irrigated farms. The corn and soybean farms included in this analysis are at least partially irrigated. This study only analyzes the harvested acres and excludes the acres that were planted while not harvested due to crop failure or other reasons.

72. According to the USDA FRIS, the on-farm surface water includes recycled water of surface or groundwater that was previous used for irrigation, and reclaimed water from on-farm livestock wastewater after being treated. The off-farm surface water is surface water from off-farm sources, municipal water, rural water supply, as well as reclaimed water from off-farm sources such as municipal reclaimed water, industrial, off-farm livestock operations, and other off-farm sources.

73. Moore, M.R.; Dinar, A. Water and land as quantity-rationed inputs in California agriculture: Empirical tests and water policy implications. *Land Econ.* **1995**, *71*, 445–461. [CrossRef]

74. Ideally, equations on water application and land allocation for each crop can be estimated to obtain both extensive and intensive margins, and then the aggregate effect can be calculated for a typical farm growing all crops. Equations on production decisions can also be estimated for each region to calculate the aggregate effect for a typical farm growing all crops in each specific region. Given the focus of this paper, such analyses are beyond the scope.

75. Pfeiffer, L.; Lin, C.-Y.C. Does efficient irrigation technology lead to reduced groundwater extraction? Empirical evidence. *J. Environ. Econ. Manag.* **2014**, *67*, 189–208. [CrossRef]

76. Moore, M.R.; Gollehon, N.R.; Negri, D.H. Alternative forms for production functions of irrigated crops. *J. Agric. Econ. Res.* **1992**, *44*, 16–32.

77. Koontz, T.M. Money talks? But to whom? Financial versus nonmonetary motivations in land use decisions. *Soc. Nat. Resour.* **2001**, *14*, 51–65.

78. Fan, Y.; McCann, L.M. Farmers' Adoption of Pressure Irrigation Systems and Scientific Scheduling Practices: An Application of Multilevel Models. In Proceedings of the Annual Meeting 2017, Chicago, IL, USA, 30 July–1 August 2017; Agricultural and Applied Economics Association: Milwaukee, WI, USA, 2017; pp. 1–38.

79. We acknowledge the thoughtful ideas from one anonymous reviewer. This issue would be better addressed if there are more detailed data on observations using fee-based systems and the role of water rights in different regions of the United States.

80. Wang, T.; Park, S.C.; Jin, H. Will farmers save water? A theoretical analysis of groundwater conservation policies. *Water Resour. Econ.* **2015**, *12*, 27–39. [CrossRef]

81. Ward, F.A.; Pulido-Velazquez, M. Water conservation in irrigation can increase water use. *Proc. Natl. Acad. Sci. USA* **2008**, *105*, 18215–18220. [CrossRef] [PubMed]

82. While this is not investigated in our study, we realize the potential effect of Jevons' Paradox, which might offset the effect of pressure irrigation in reducing water application.

83. Li, H.; Zhao, J. Rebound effects of new irrigation technologies: The role of water rights. *Am. J. Agric. Econ.* **2018**, *100*, 786–808. [CrossRef]

84. Fan, Y.; Park, S.; Nan, Z. Participatory water management and adoption of micro-irrigation systems: Smallholder farmers in arid north-western China. *Int. J. Water Res. Dev.* **2017**, *34*, 434–452. [CrossRef]

85. Liu, Y.; Wang, C.; Tang, Z.; Nan, Z. Farmland Rental and Productivity of Wheat and Maize: An Empirical Study in Gansu, China. *Sustainability* **2017**, *9*, 1678. [CrossRef]

86. Ali, J.; Kumar, S. Information and communication technologies (ICTs) and farmers' decision-making across the agricultural supply chain. *Int. J. Inf. Manag.* **2011**, *31*, 149–159. [CrossRef]

87. Ma, W.; Renwick, A.; Nie, P.; Tang, J.; Cai, R. Off-farm work, smartphone use and household income: Evidence from rural China. *China Econ. Rev.* **2018**. [CrossRef]

88. Solano, C.; Leon, H.; Perez, E.; Herrero, M. The role of personal information sources on the decision-making process of Costa Rican dairy farmers. *Agric. Syst.* **2003**, *76*, 3–18. [CrossRef]

89. Wang, C.; Segarra, E. The economics of commonly owned groundwater when user demand is perfectly inelastic. *J. Agric. Res. Econ.* **2011**, *36*, 95–120.

90. Peck, D.E.; Adams, R.M. Farm-level impacts of prolonged drought: Is a multiyear event more than the sum of its parts? *Aust. J. Agric. Res. Econ.* **2010**, *54*, 43–60. [CrossRef]

Article

Sustainable Water Use for International Agricultural Trade: The Case of Pakistan

Tariq Ali [1], Abdul M. Nadeem [2], Muhammad F. Riaz [2] and Wei Xie [3,*]

[1]   School of Economics and Management, North China University of Technology, No. 5 Jinyuanzhuang Road, Beijing 100144, China; agri45@gmail.com
[2]   Department of Economics, Government College University Faisalabad, Allama Iqbal Road, Faisalabad 38000, Pakistan; majeednadeem@gcuf.edu.pk (A.M.N.); faraz.riaz@gcuf.edu.pk (M.F.R.)
[3]   China Center for Agricultural Policy, School of Advanced Agricultural Sciences, Wangkezhen Building, Peking University, No.52 Haidian Road Haidian District, Beijing 100871, China
*   Correspondence: xiewei.ccap@pku.edu.cn; Tel.: +86-1521-098-0869

Received: 14 June 2019; Accepted: 4 October 2019; Published: 28 October 2019

Abstract: Sustainable use of resources is critical, not only for people but for the whole planet. This is especially so for freshwater, which in many ways determines the food security and long-term development of nations. Here, we use virtual water trade to analyze the sustainability of water used by Pakistan in the international trade of 15 major agricultural commodities between 1990 and 2016 and in 2030. Most of the existing country-level studies on virtual water trade focused on net virtual water importers, which are usually water-scarce countries as well. This is the first study to concentrate on a water-stressed net virtual water-exporting country. Our results show that Pakistan has been trading large and ever-increasing volumes of virtual water through agricultural commodities. Despite the overall small net export of total virtual water per year, Pakistan has been a net-exporter of large quantities of blue (fresh) virtual water through its trade, even by fetching a lower value for each unit of blue water exported. Given Pakistan's looming water scarcity, exporting large volumes of blue virtual water may constrain the country's food security and long-term economic development. Improving water use efficiency for the current export commodities, for example, rice and exploring less water-intensive commodities, for example, fruits and vegetables, for export purposes can help Pakistan achieve sustainable water use in the future.

Keywords: sustainability; agriculture; virtual water trade; blue; green

1. Introduction

Sustainable development, as defined by the United Nations, is "development that meets the needs of the present without compromising the ability of future generations to meet their own needs" [1]. As an essential primary natural resource, the role of water in sustainable development is well recognized due to its critical functions in social and economic activities [2]. Unsustainable use of freshwater resources by nations can hamper their food security and long-term economic development [3]. The future water vision emphasizes that by 2050, "global markets and trade flows would be monitored through a global water sensitivity certification scheme that ensures water-intensive products are exported from areas with comparatively little or no water stress. We will recognize the economic value of water and all forms of an economic enterprise will take consideration of the water implications of their actions" [2]. Virtual water trade (VWT) can be used to gauge the water that is virtually transferred from one region to another region through commodities and services, both at global and national levels.

Nations can ensure the sustainable use of domestic freshwater by prioritizing their production and trade decisions based on water availability. Trade in virtual water is the amount of hidden/virtual water that crosses borders with commodities/services. Virtual water content (VWC), defined as the ratio of the total water used for the production of the crop to the total volume of crop produced (m^3/ton), is a crucial concept used in the relevant literature [4]. The virtual water content of primary crops is calculated based on crop water requirements and yields [4]. Several studies have demonstrated the role of virtual water trade in ameliorating regional water shortages. Chapagain et al. [5,6] found that virtual water trade saved 6% of global agricultural water use. More recently, References [7–10] demonstrated the role of virtual water in improving water use efficiency and saving global water resources. Several national-level studies have also shown that water-scarce countries have used VWT to alleviate pressure on domestic water resources by importing water-intensive products [11–18]. In his seminal studies on virtual water, Allan [19,20] discussed how the Middle Eastern countries were able to overcome national water shortages through cereal imports. Many other important studies on Brazil, China, Cyprus, Egypt, the Middle East, Morocco, Spain and Tunisia also show the role of VWT in saving (and losses, if any) of domestic water resources for these countries [11–14,21–29].

Pakistan—with the world's fifth-largest population [30]—is a water-stressed country and its freshwater resources face heightened pressure from increasing population and climate change [31]. With an average annual rainfall of about 250 mm, Pakistan is among the most arid countries of the world [32]. Relying upon the most extensive contiguous irrigation system in the world, Pakistan uses over 94% of national water withdrawal in agriculture [33]. In the past, the country failed to add any large water reservoirs, thus adding to the pressure on dwindling water resources, especially in agriculture [34]. Due to declining per capita availability of water, Pakistan will turn from currently water-stressed to a water-scarce country in the 2040s [31]. Groundwater depletion due to unsustainable use of the aquifers for agriculture is becoming more severe in Pakistan [35]. Moreover, increasing quantities of pesticides and fertilizers used in agriculture are causing large-scale uncontrolled pollution of surface and groundwater [31].

Agriculture, the most water-intensive sector in Pakistan, can be taken as the strongest candidate to overcome the growing national water scarcity. Over 26% of crop-related groundwater depletion in Pakistan is due to crops exported to other countries, the majority of which (82%) is embedded in rice exports [35]. Although several studies have discussed the relevance of virtual water in agricultural trade for Pakistan [6,35–38], most of these studies either fail to distinguish between blue and green water (the blue water includes surface water and groundwater, while the green water is the rainwater stored in the soil as soil moisture [37]) or are related to domestic production/consumption of blue/green water. Chapagain et al. [6] report net savings of Pakistan's water resources through the international trade of all major crops and livestock products. However, the study is limited with respect to reporting only point analysis (average for 1997–2001) and reporting only total water savings. Dalin et al. [35] show that Pakistan is facing groundwater depletion through the international food trade. Although an influential study, it is limited to groundwater and presents its analysis only for two years, that is 2000 and 2010. The study by Fraiture et al. [36] shows net water savings for Pakistan. The study only covers total water savings through the cereal trade in 1995. Chapagain and Hoekstra [37] report blue, green and greywater impacts of international trade of rice for Pakistan and other countries for the period 2000–2004. This study, however, lacks a temporal analysis and the coverage of other agricultural commodities. The green, blue and grey water footprint of global crop production, including that of Pakistan, are covered in Reference [38]; however, no trade analysis is presented in the study. Pakistan's high dependence on freshwater for agriculture production, dwindling water resources and increasing trade in agriculture should be a strong motivation for conducting a more detailed study on trade in virtual water for Pakistan.

The primary purpose of this study is to assess the evolution of trade and savings/losses of Pakistan's blue and green virtual water through international trade of the agricultural commodities during 1990–2016 and in 2030. By doing so, we can answer the following research questions: What are the major commodities and source/destination regions involved in Pakistan's VWT? How Pakistan's VWT has evolved over the years and which policies have driven the changes in VWT? More importantly, was the trade dominated by blue or green virtual water? Has Pakistan been saving blue/green water through its trade in agricultural commodities at the national and global levels? Has Pakistan's trade in blue and green VW been economically viable over the years? How would the trends evolve in the future? What is the value of the virtual water contents of these agricultural commodities? A detailed analysis of historical and future blue and green virtual water movements through Pakistan's global agricultural trade can help policy planning on water use for the country. To the best of our knowledge, this is also the first study focused on a water-stressed country that exports large quantities of agricultural products and virtual water.

The rest of the paper is organized as follows. Section 2 describes the methods and data used in this study. Section 3 presents our results on virtual water contents of these commodities for Pakistan; and trade and net savings/losses in terms of total, blue and green virtual water over the past years and in the future. Section 4 discusses the implications of our results and presents policy and research recommendations. The Section 5 presents the conclusions.

2. Materials and Methods

2.1. Selection of Agricultural Commodities

We selected 15 agricultural commodities to display a representative profile of international trade of virtual water for Pakistan. The crops include rice, wheat, maize, cotton, fruits, vegetables (See Appendix A for the complete list of fruits and vegetables included for the analysis), tea, tobacco, oilseeds (soybean, sunflower, groundnuts, copra, rapeseed and cottonseed), other cereals (rye, barley, oats, sorghum and millets), sugar and palm oil. The livestock commodities include the meat of bovine animals (beef), the meat of sheep or goat (mutton) and poultry meat. On average, these crops accounted for over 90% of the harvested area in Pakistan during 1990–2016. Beef, mutton and poultry meat provide over 81% of the animal source protein to consumers in Pakistan [39]. In terms of trade value, all of these commodities account for around 80% of export value and 93% of the import value of Pakistan's total annual agricultural exports and imports, respectively [40].

2.2. Projections of Pakistan's Agricultural Trade in 2030

To estimate the future impacts of Pakistan's agricultural trade on the country's water resources and global savings, we make use of the Global Trade Analysis Project (GTAP) model and its latest database (with the base year 2011) to project Pakistan's trade through to 2030. GTAP is a well-known and widely used global general equilibrium economic model [41,42]. The model assumes cost minimization by producers and utility maximization by consumers. In a competitive market setup, prices adjust until supplies and demands of all commodities equalize. The model and database have been extensively used in research areas such as food security policy, energy, climate change, poverty and migration, among others. There are other model choices as well, like ENVISAGE, FARM, GTEM, AIM and MAGNET, GLOBIOM, GCAM and IMPACT. Most of these computable general equilibrium (CGE) models have their roots in the Global Trade Analysis Project database and the CGE optimizing approach [41] and so have similar model specifications. They mainly differ in parameterization choices, which significantly affect the result. Nelson et al. [43] have discussed the effects of model structure and parameter choice on the results of CGE simulations in detail.

For projecting the future trade of Pakistan with other countries, we use a recursive dynamic method wherein the given GDP targets are met by exogenous shocks to factors of production including population, skilled labor, unskilled labor, capital and natural resources. The exogenous macro assumptions and the procedure for implementing these shocks are discussed in detail in References [41,44], respectively. Most of the data used for projections are based on the CEPII EconMap database 2.4 (2016), which contains 1980–2050 and 2100 data for 167 countries and 6 scenarios (one central scenario and the 5 Shared Socioeconomic Pathways) for GDP, savings, investment, total population, labor force, capital stock, total primary energy consumption, human capital, total factor productivity and energy efficiency [45].

For these projections, we retained maximum disaggregation for agricultural commodities and Pakistan's major trading partners. As for projection for future agriculture production in Pakistan, we use current production technology and the availability of irrigation water. The potential decrease in water availability for agricultural production in the coming decade due to climate change, population growth and urbanization could well be countered by the improved crop production technology (like drought-resistant varieties and new irrigation technology).

2.3. Virtual Water Contents

The virtual water trade between the country/region of production to the country/region of consumption through product trade is the volume of water that is being transferred in virtual form. In this study, the virtual water flows related to Pakistan's agricultural trade have been calculated by multiplying commodity trade flows (ton/year) by their associated virtual water content (m³/ton). The virtual water content (VWC) of a commodity is the quantity of water required to produce one ton of crop biomass, estimated as Equation (1).

$$VWC_{i,g,t} = \frac{ETTOTAL_{i,g,t}}{Yield_{i,g,t}} \tag{1}$$

where *ETTOTAL* is the total evapotranspiration during the cropping period (m³/ha); yield is crop yield (ton/ha) and *i*, *g*, *t* denote crop, grid cell and year, respectively. *ETTOTAL* is further formulated as follows:

$$ETTOTAL_{i,g,t} = \sum_{doy=p}^{h} ET_{c,\,g,\,doy}, \tag{2}$$

where ET is daily evapotranspiration (m³ ha⁻¹ day⁻¹) and *doy*, *p* and *h* denote the day of year, the planting date and the harvesting date, respectively. *ETTOTAL* is divided into blue and green water as follows:

$$ETTOTAL_{i,g,t} = ETTOTAL_{BLUE,i,g,t} + ETTOTAL_{GREEN,i,g,t} \tag{3}$$

Based on the type of water used in *ETTOTAL*, *VWC* can be split into green and blue types, where green and blue water in the context of VW are water consumed by crop vegetation that originated from precipitation and irrigation, respectively [46]. Blue and green VWC have substantially different opportunity costs associated with them (for a more detailed discussion of the method see Hanasaki [46]. Pakistan's virtual water export is the volume of water used to produce export commodities. Similarly, the virtual water import of Pakistan is the volume of water used to produce commodities in the trading partners imported by Pakistan.

The data on VWC of rice, wheat, maize, beef, mutton and poultry are taken from Reference [46] throughout 1990–2005. The original data in the study [46] are based on the H08 model and runs from 1986 to 2005 (The data can be downloaded from the link: https://sites.google.com/site/naotahanasaki/english-contents/data/vwc). Here, we note that in the absence of national estimates of virtual water contents, our study uses assessments from global models, which might contain discrepancies in the results due to the differences in modelling assumptions, input data and parameters adopted by local and global models (see Zoumides et al. [47] for a detailed discussion of these discrepancies). For the period after 2005, where VWC data are unavailable, we make use of the finding that VWC (or water productivity) has a strong inverse (or direct) relationship with crop yield [22]. This method has also been used in other studies like [8,48,49]. Specifically, we employ yield data from Reference [39] in Equation (4) to update the country-specific VWC of these crops and livestock as:

$$VWC_{w,i,r,t} = VWC_{w,i,r,2005} \times Y_{w,i,r,2005} / Y_{w,i,r,t} \tag{4}$$

where *w* represents water type (blue and green), *i* denotes commodity, *r* means country/region and t represents the years, that is, 2006 to 2016.

For updating the VWC for livestock sectors, we extrapolated the VWC values from Reference [46] to future years by assuming a 1% improvement in water productivity every five years for livestock production, which is in the spirit of Reference [27]. For fruits and vegetables, we used VWC for apples and tomatoes, respectively, from Reference [10], which is also in the spirit of Reference [50], which used VWC for apples and tomatoes as representative crops for fruits and vegetables, respectively. The VWC for cotton, palm oil, tea, tobacco, other cereals and oilseeds are from Reference [10], who reported country-wise blue and green VWC of these crops for the period 1996–2005. We applied the method described in Equation (3) to update respective VWCs.

For the future VWC values of rice, wheat, maize, cotton, tea, tobacco, oilseeds, other cereals, sugar and palm oil, from 2017 to 2030, we make use of the forecasts on water productivity from Reference [51]. The study uses IMPACT-WATER model to examine water and food policy and investment issues. Based on the assumptions of enhancements in area and yield growth; decrease in water consumption per hectare and improvement in water supply between 1995 and 2025, the study suggests that over the 30 years, the water productivity of non-rice cereals will improve by 66% (from 0.6 to 1.0 kg/m^3) for developing countries and 40% (from 1.0 to 1.4 kg/m^3) for developed countries. The water productivity for irrigated rice is projected to increase by 33% and 10% for developing and developed countries, respectively. We used these figures to estimate the average annual improvement in the water productivities for developing and developed countries (e.g., 66/30 = 2.2% per annum improvement in water productivity for developing countries) and then the annual change in country-specific VWCs for all the crops mentioned above. The future values of VWC for fruits and vegetables were updated by incorporating 0.5% yearly improvement in water productivity; which is based on [27]. Table 1 contains a full summary of the sources and assumptions for the VWC values used in this study for all the commodities over various periods.

Table 1. Summary of sources and assumptions used for virtual water content (VWC) data.

Commodity	Period	Reference	Assumptions/Recalculations
Rice, wheat, maize, beef, mutton and poultry	1990–2005	[46]	Unprocessed data
Rice, wheat and maize	2006–2016	[22,46]	Recalculated VWC based on the inverse relationship between VWC and crop yield
Fruits and vegetables	1996–2005	[10]	Unprocessed values for apples (for fruits) and tomatoes (for vegetables)
Cotton, palm oil, tea, tobacco, other cereals and oilseeds	1996–2005	[10]	Unprocessed data
Cotton, palm oil, tea, tobacco, other cereals, oilseeds, fruits and vegetables	1990–1995 and 2006–2016	[10,22]	Recalculated VWC based on the inverse relationship between VWC and crop yield
Rice, wheat, maize, cotton, tea, tobacco, oilseeds, other cereals sugar and palm oil	2017–2030	[10,46,51]	Recalculated VWC values by estimating the annual improvement in the water productivities for developing and developed countries based on the forecasts on water productivity from Reference [51]
Fruits and vegetables	2017–2030	[10]	Assumed 0.5% annual improvement in water productivity
Beef, mutton and poultry	2006–2030	[46]	Assumed 1% improvement in water productivity every five years for livestock production based on the values from Reference [46]

2.4. Virtual Water Flows

We calculated the virtual water imports ($VWI_{w,i,r,t}$) and virtual water exports ($VWE_{w,i,r,t}$) of Pakistan for both blue and green water by multiplying the blue and green VWC with import and export quantities of the commodities, as follows:

$$VWI_{w,i,r,t} = M_{i,r,t} \times VWC_{w,i,r,t} \tag{5}$$

$$VWI_{w,i,r,t} = M_{i,r,t} \times VWC_{w,i,PAK,,t} \tag{6}$$

where $M_{i,r,t}$ denotes the quantity of commodity i imported by Pakistan from country/region r during year t. $X_{i,r,t}$ represents Pakistan's export of commodity i to destination r in year t. $VWC_{w,i,r,t}$ is the partner countries' virtual water content (blue and green) of the particular commodity in the respective year (unit: m^3/ton). $VWC_{w,i,PAK,t}$ is Pakistan's virtual water content (blue and green) of the particular commodity in the respective year (unit: m^3/ton). Subtracting VWE from VWI gives us net virtual water imports (NVWI) of Pakistan, as under:

$$NVWI_{w,i,PAK,t} = VWI_{w,i,r,t} - VWE_{w,i,r,t} \tag{7}$$

Pakistan's domestic saving of blue/green water ($DSAV_{w,i,PAK,t}$) through imported commodities is the amount of blue/green water needed to produce the same commodities at home. This is estimated by replacing $VWC_{w,i,r,t}$ of the partner country, with $VWC_{w,i,PAK,t}$ of Pakistan in Equation (5) and then using it in Equation (7).

Pakistan can also contribute to global water saving if it saves more domestic water ($DSAV_{w,i,PAK,t}$) than its net virtual water import ($NVWI_{w,i,PAK,t}$). Specifically, the global saving (or loss) of blue and green water is the difference between the amount of water which Pakistan saves domestically through its food trade and the amount of $NVWI_{w,i,PAK,t}$, which it imports from other countries Equation (8).

$$GSAV_{w,i,t} = DSAV_{w,i,PAK,t} - NVWI_{w,i,PAK,t} \tag{8}$$

Pakistan's food trade would save (waste) water at the global level if we get a positive (negative) value from Equation (8). Total national and global water savings are the sum of respective savings from all commodities and all trading partners.

2.5. Value of Virtual Water Flows

The value produced by one unit of water (value of virtual water, i.e., VVW) is an important indicator of the efficiency of water use in different sectors (unit: US\$/m^3) (the average exchange rate over the period 1990–2016 was 1 US\$ = 60.4 Pakistan Rupee [39]). Due to the differences in the opportunity costs of green and blues water, we calculate an average value of blue virtual water of the studied commodities over 1990-2016 by diving the production value (unit: US\$/ton) by the blue VWC (unit: m^3/ton) that is

$$VVW_{i,t} = NPV_{i,t}/BVWC_{i,t} \tag{9}$$

where $VVW_{i,t}$ is value produced by one unit of blue water by commodity i in year t (US\$/m^3), $NPV_{i,t}$ is the net production (it is the value of net production, where, Net production = Production—Feed—Seed [39]) value of each commodity i produced in year t and $BVWC_{i,t}$ is the blue virtual water content of each commodity i produced in year t. The values for $NPV_{i,t}$ for Pakistan are obtained from Reference [39], while $BVWC_{i,t}$ values are based on various sources described earlier in this section.

3. Results

3.1. Virtual Water Contents of the Commodities

We start with a brief background on the importance of each crop in Pakistan's agriculture by comparing its harvested area to the total harvested area (Table 2) (Appendix A contains a more detailed analysis of Pakistan's agricultural trade; Figures A1 and A2). The analysis shows that crop harvested area in Pakistan is dominated by wheat (42%), seed cotton (14%), rice (12%) and sugar crops (5%) during 1990–2016 (Table 2). Table 2 also compares Pakistan's average yields of the agricultural commodities with global average yields during 1990–2016 (columns 2–3). Except for tobacco (where Pakistan produces a higher yield than many other countries) and cotton lint (where yield in Pakistan is equal to global yield), Pakistan's yields of all the commodities are considerably lower than the corresponding global averages. The yield gap is significantly higher for two crop groups, that is, other cereals and oilseed crops, where yields in Pakistan are 11% and 36% of the global average yields over this period. For livestock production, Pakistan's yield (kg/animal) of mutton (goat and sheep meat) and poultry (chicken meat) are relatively close to the global averages, whereas beef (buffalo and cattle meat) yield is also significantly (39%) lower than the worldwide level. The small agricultural yields are mainly due to poor irrigation water management, lack of advanced methods and quality inputs, fragmented land holdings and inadequate institutional support to the farmers. The relative lower yields of most of the commodities show that agricultural productivity in Pakistan can be substantially improved, which, due to an inverse relationship with VWC, can lower VWC for these commodities. In the absence of such productivity improvements, Pakistan will face the challenge of reducing agricultural exports or increased pressure on water resources.

Table 2. Average shares in total harvest area, yields and virtual water content (blue, green, total, ratios of blue/green) of major agricultural commodities produced in Pakistan (1990–2016).

Commodities	Share in Total Harvested Area (%)	Yield [1]		VWC (m³/ton)			Share of Blue VWC in Total VWC (%)
		Pakistan	World	Blue	Green	Total	
Wheat	41.7	2.4	2.8	1603	260	1862	86.1
Cotton lint [2]	14.4	1.9	1.9	2781	4193	6974	39.9
Rice	12.0	3.1	4.0	2370	576	2947	80.4
Sugar unrefined [3]	5.1	5.4	7.3	2309	577	2887	80.0
Maize	4.9	2.6	4.6	1048	1150	2198	47.7
Other cereals	3.9	2.1	19.4	1048	1150	2198	47.7
Fruits	3.6	8.9	11.1	499	599	1098	45.4
Oilseed crops	2.6	4.2	11.4	617	1536	2153	28.7
Vegetables	2.6	12.7	16.6	371	120	491	75.6
Tobacco [4]	0.3	2.0	1.7	1052	1315	1315	80.0
Palm oil	0.0	-	12.2	0	4833	4833	0.0
Tea	0.0	-	1.4	9	8727	8735	0.1
Beef	-	147.2	204.5	11,713	6871	18,584	63.0
Mutton	-	16.5	14.3	5021	2955	7976	63.0
Poultry	-	1.1	1.6	4100	1388	5487	74.7

[1] Crop yields are in tons/ha; meat yields are kg/animal. [2] Share in harvested area is for seed cotton. [3] Share in harvested area is for both sugarcane and sugar beet. [4] Unmanufactured tobacco. Source: Authors' calculations based on data from Reference [39] (trade volumes) and sources mentioned in the Methods section (VWC).

VWC for the studied agricultural commodities changes with the yearly fluctuation in yield of each commodity. Although annual blue and green VWCs were used for estimating the trade in virtual water in respective years, we report the average VWC (over 1990–2016) for all the commodities (Table 2). The total virtual water contents for the crops vary widely from 491 m³/ton for vegetables to as high as 6974 m³/ton for cotton lint (Table 2, column 6). The VWC for livestock sectors is far higher than the ones for crops, ranging between 5487 m³/ton for poultry to 18,584 m³/ton for beef. Except for oilseed crops, palm oil and tea, blue VWC constitutes much higher shares in the total VWC for both crops and livestock sectors. Notably, for the major crops (wheat, rice and sugar) the percentage of blue VWC in the total VWC ranges between 80% and 86%, indicating very high dependence of these crops on irrigation in Pakistan. The analysis shows that the efforts aimed at improving yields and lowering VWC can significantly lower irrigation water requirements for agricultural commodities in Pakistan.

3.2. Total Virtual Water Trade

Pakistan's annual total virtual water import, although with some inter-annual fluctuations, grew at an annual average rate of 5% between 1990 and 2016 (Figure 1, the sum of yearly points along the red lines). The yearly average of total virtual water import was 8216 million m^3 (Mm^3), 11,218 Mm^3 and 12,202 Mm^3 during the periods 1990–1999, 2000–2009 and 2010–2016, respectively. The annual total virtual water import through agricultural commodities by Pakistan dropped significantly after 2008, from 17,799 Mm^3 in 2008 to 10,257 Mm^3 in 2011, after which it started to increase again. The drop in total virtual water import between 2008 and 2011 was mainly due to a decrease in imports of cotton lint by Pakistan.

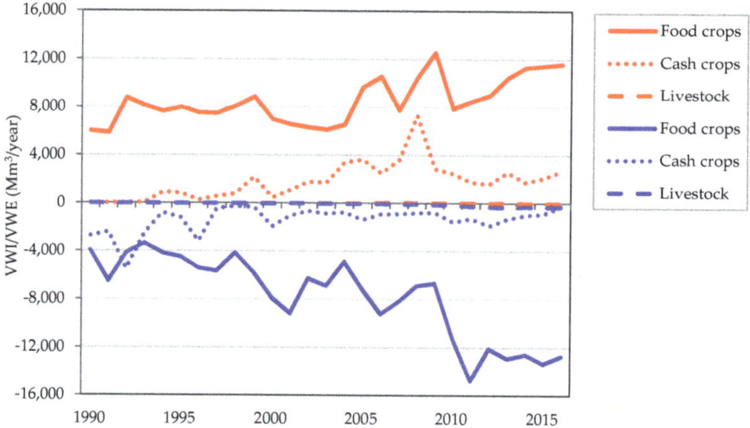

Figure 1. Virtual water import (VWI) and virtual water export (VWE) of different agricultural commodities by Pakistan (1990–2016). Red lines indicate VWI and blue lines indicate VWE. Food crops include rice, wheat, maize, other cereals, oilseed crops, palm oil, sugar, vegetables, fruits and tea; Cash crops include cotton and tobacco; Livestock includes beef, mutton and poultry. Source: Authors' calculations based on data from Reference [39] (trade volumes) and sources mentioned in the Methods section (VWC).

The annual total virtual water exports through the agricultural commodities (Figure 1, the sum of yearly points along the blue lines), however, saw a more rapid and steady increase over the years. Starting at an average of 6663 Mm^3/year during 1990–1999, the total virtual water export moved to an average 8345 Mm^3/year during 2000–2009 and reached 14,205 Mm^3/year during 2010–2016; thus, recording a 113% increase over the period 1990–2016. Food crops (mainly rice) exports were responsible for the rise in Pakistan's total virtual water export.

During the study period, Pakistan's virtual water import (VWI) and virtual water export (VWE) have been dominated by trade in food crops (Figure 1). On average, food crops accounted for over 82% of the total VWI and over 84% of the VWE by the studied commodities. Not only the share of food crops in VWE was higher, but the VWE of food crops has also been increasing rapidly since 2005. VWI of cash crops was low in the early 1990s, after which it started to rise and reached the historical peak (7308 Mm^3) in 2008, before sliding down to a relatively lower level in the later years. During 2008, VWI through cash crops increased mainly due to a sharp rise in imports of cotton by Pakistan. VWI and VWE of livestock have been marginal, with VWE picking up some pace since the early 2000s.

3.3. Pakistan's Net Virtual Water Import

3.3.1. Total Net Virtual Water Import

During 1990–2016, Pakistan's total net virtual water import through the studied commodities has been fluctuating widely from the negative end to the positive end (Figure 2). During 1990–1999, the average annual total net virtual water import remained quite low (1553 Mm^3/year); after that, it increased continuously to reach a historical peak of 10,075 Mm^3 in 2008. Since then, it saw a sharp decline to reach a historic low of −5998 Mm^3 in 2011 and remained negative in most of the later years. Over the years, the evolution of total net virtual water import (with an annual average of mere 1120 Mm^3) seems to indicate that Pakistan has gained some quantities of virtual water through its trade in the major agricultural commodities during 1990–2016. However, further bifurcating total net virtual water import into blue net virtual water import and green net virtual water import would reveal further insights into Pakistan's trade of virtual water, which we present in the next sub-section.

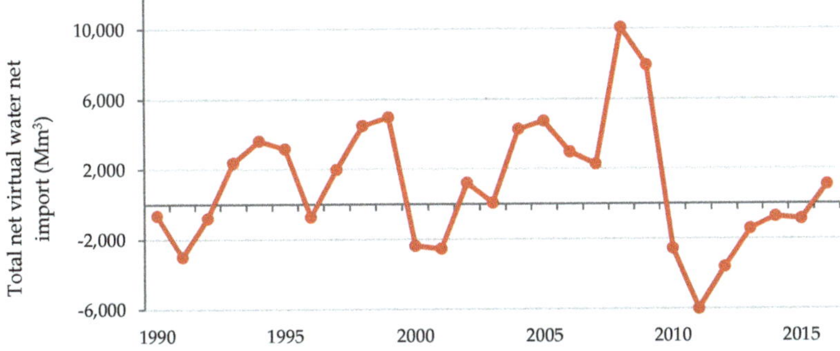

Figure 2. Evolution of Pakistan's total net import of virtual (1990–2016). Source: Authors' calculations based on data from Reference [39] (trade volumes) and sources mentioned in the Methods section (VWC).Cotton also contributed significantly towards Pakistan's blue net virtual water import over the years. For cotton, blue net virtual water import was mostly negative during 1990–2001, after which it turned positive, that is, Pakistan became a net importer of cotton from a net exporter after 2001 (Figure 3a). The reason was increased demand for high-grade cotton by Pakistan's domestic industry, which was not met by the domestic production of cotton, due to lower production in some years and lower quality in others [52,53].

3.3.2. Commodity-Wise Net Import of Blue and Green Virtual Water

The commodity-wise contribution of net VWI of blue and green water varies considerably in Pakistan (Figure 3). Wherein, blue net virtual water import was mostly negative and green net virtual water import was mostly positive. Pakistan's negative blue net virtual water import has been mostly dominated by blue net virtual water import via rice exports, which had an average annual blue net virtual water import of -5680 Mm^3 and accounted for an average share of 94% in the total annual blue net virtual water import over the study period (Figure 3a). Moreover, blue net virtual water import through rice has been rising over the years; from an annual average of −3547 Mm^3 over 1990–1999 it increased to −5271 Mm^3 per annum during 2000–2009 and reached −9314 Mm^3 per year during 2010–2016. The rise in negative net VWI through rice can be attributed to rising domestic production and decreasing share of domestic consumption of rice in the domestic production in Pakistan. Specifically, rice production in Pakistan increased at a rate of 4.3% per year, from 4.89 million tons in 1990 to 10.5 million tons in 2016. Both yield and area contributed to improving production [39]. The combined effect of these factors was a steady year-on-year increase in rice exports from Pakistan. So much so that during 2010, Pakistan exported a staggering 86% of the domestically produced rice to the international market [39].

The contribution of wheat trade towards blue net virtual water import of Pakistan, although relatively small, saw an unusual feature over the study period. During 1990–1999, blue net virtual water import due to wheat had been consistently positive due to the small but sustained net import of wheat by Pakistan. After 1999, however, Pakistan occasionally recorded negative net import of wheat, which lead to sizable negative contributions to the total blue net virtual water import of Pakistan. The reason for this fluctuation in blue net virtual water import through wheat trade over the years was lower production in some years or bumper crops in others and due to weather events and poor stock management [54,55].

The blue net virtual water import of Pakistan due to the rest of the crops like maize, other cereals, oilseed crops, palm oil, sugar, vegetables, fruits, tea and tobacco and livestock such as beef, mutton and poultry has been quite small over the study period (Figure 3a).

Several agricultural commodities have contributed towards green net virtual water import, of which the total contributions due to net imports of palm oil, tea, wheat, oilseeds and cotton have been positive, that is, Pakistan imported more green VW through the trade of these crops than it exported (Figure 3b). Palm oil, with the average yearly contribution of 89% (5465 Mm³/year), was the most dominant commodity. Interestingly, in some years like 1992 and 1999, the contribution of wheat towards green net VWI even surpassed the share of palm oil due to large volumes of wheat imports by Pakistan. Green net virtual water import due to rice trade was negative over the study period. The rest of the crops and livestock had small but predominantly negative green net virtual water imports.

The comparison between both panels of Figure 3 indicates that Pakistan has been exporting ever-increasing quantities of blue virtual water, while at the same time importing almost equal amounts of green virtual water, which in the end seems to have canceled out the effects on Pakistan's water resources. However, the transfer of blue virtual water to other countries through net virtual water export is far more valuable than the green virtual water transfers from other countries into Pakistan through net virtual water import.

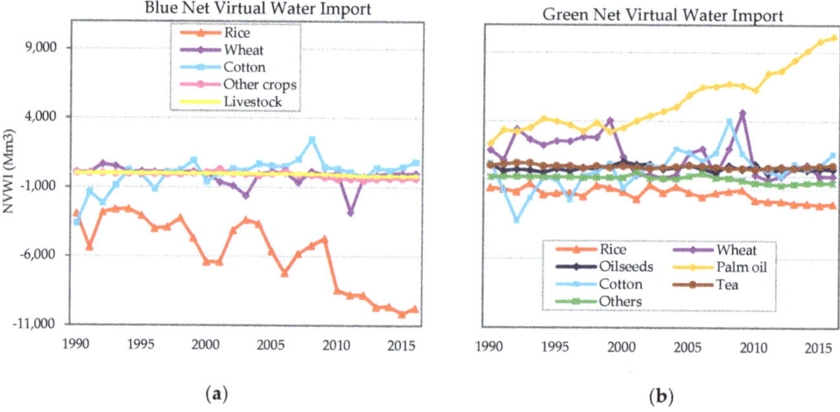

Figure 3. Commodity-wise evolution of (**a**) blue and (**b**) green net virtual water import of Pakistan (1990–2016). Source: Authors' calculations based on data from Reference [39] (trade volumes) and sources mentioned in the Methods section (VWC).

Blue water is far more costly than the green water used for agricultural production; as for the former, there are considerable costs in terms of construction, maintenance and operation of dams and the related irrigation systems, which are usually born by the citizens of exporting countries. However, these costs are not adequately included in the price of blue water or reflected in the price of final products. In addition to economic costs, there are many environmental and social costs of ensuring steady supplies of blue water, which are seldom considered in the literature of virtual water trade.

Exporting commodities heavily laden with blue water also reduces the domestic supply of this precious natural resource, intensifying any water shortage issues.

3.3.3. Region-Wise Net Virtual Water Import

In terms of regional distribution, Pakistan's negative blue net virtual water import (i.e., net blue virtual water export) has been destined mostly to Asian countries—mainly through rice export to Afghanistan, UAE and Iran- and African countries, also primarily via rice export to Kenya and Mozambique. Exports to both the Asian and African regions were responsible for annual blue net virtual water import of −3579 Mm^3 and −2042 Mm^3, respectively, during 1990–2016 (Figure 4a). The share of blue net virtual water import from Asian countries in the total blue net virtual water import decreased from 67% during 1990–2010 to 53% during 2009–2016. For African countries, on the other hand, the share of blue net virtual water imports increased from 27% during 1990–2008 to 42% during 2009–2016. This indicates a gradual shift of Pakistan's export away from destinations in Asia towards African countries. European countries accounted for relatively small but constant share (9%) in the annual blue net virtual water import. North American and South American countries have registered positive (although small) contributions towards Pakistan's blue net virtual water import.

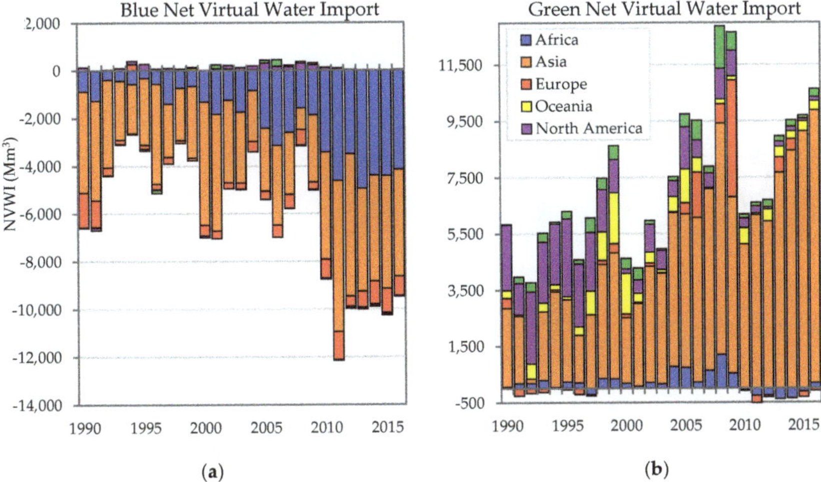

(a)

(b)

Figure 4. Region-wise evolution of (**a**) blue and (**b**) green net virtual water import of Pakistan (1990–2016). Source: Authors' calculations based on data from Reference [39] (trade volumes) and sources mentioned in the Methods section (VWC).In the period 1990–2016, Pakistan's green net VWI related to agricultural commodities was dominated by Asian countries with 4824 Mm^3/year (Figure 4b). About 67% share was due to palm oil imports from Malaysia and Indonesia. On the other hand, trade with countries like Afghanistan, Iran, Saudi Arabia and UAE contributed small but negative shares to green net VWI. We can observe a noticeable shift in sources of green net virtual water imports over the years, with a continuous increase in the share of Asian countries and a gradual decrease in the percentage of North American countries (Figure 4b). Asian countries' share in Pakistan's green net virtual water import increased from 45% (2311 Mm^3/year) during 1990–1997 to 73% (5882 Mm^3/year) during 1998–2016 at the expense of North America's share, which dropped from 42% (2170 Mm^3/year) during 1990–1997 to 8% (603 Mm^3/year) during 1998–2016. Two factors caused the shift, that is, shrinking imports of wheat from North America and increasing imports of palm oil and cotton from Asia (a brief discussion of the water resources and their use in Pakistan and its major trade partners is presented in Table A1 of the Appendix A).

3.4. Water Savings at National and Global Levels

3.4.1. Blue Virtual Water

Our results show that except for a few years, Pakistan has been losing domestic blue virtual water due to its trade in the agricultural commodities (Figure 5a). As Pakistan is a net exporter of blue virtual water, this result is not surprising for the country. The average annual losses during 1990–1999 were −248 Mm^3 and increased to −5956 Mm^3 during 2000–2016. The most substantial losses of blue water were recorded in 2011, with a historic high of −11890 Mm^3. During that particular year, in addition to its traditional export, that is, rice, Pakistan also exported large quantities of wheat, which caused the blue water losses to surge. For comparison, the water loss was equivalent to 7% of the total irrigation water used in Pakistan during 2011. In other words, Pakistan's negative blue net virtual water import is exerting additional pressure on the country's dwindling water resources. There have also been some years when Pakistan saved domestic blue water due to its trade (Figure 5a). The positive contribution of net virtual water import towards Pakistan's domestic blue water saving during the 1990s was due to massive net imports of wheat and during the late 2000s due to sharp but occasional increase in net imports of cotton.

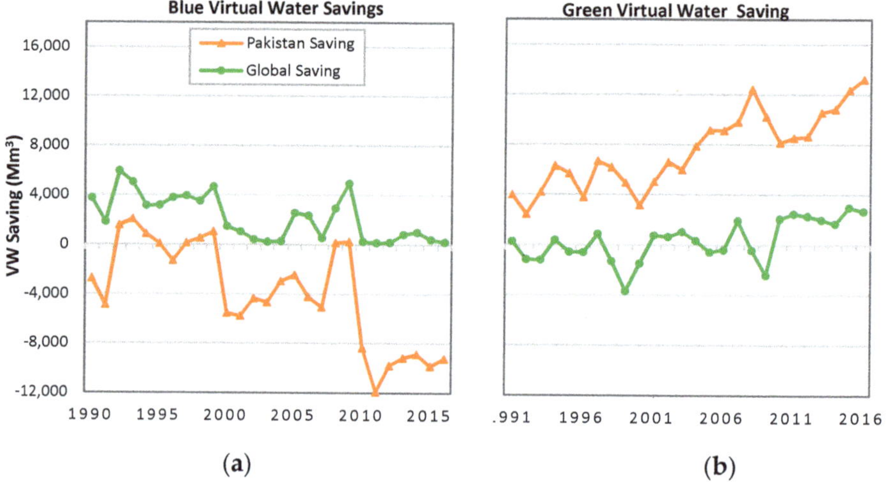

Figure 5. Evolution of (**a**) blue and (**b**) green water-saving for Pakistan and at global level (1990–2016). Source: Authors' calculations based on data from Reference [39] (trade volumes) and sources mentioned in the Methods section (VWC).At the global level, however, the average saving of green virtual water has been positive but small, that is, 307 Mm^3/year (Figure 5b), mainly due to a relatively smaller difference between the green VWC of Pakistan and those of its import partners. The global saving of green virtual water has consistently been positive since 2010.

Pakistan's trade in the agricultural commodities has also contributed to sizable positive savings of blue virtual water at the global level (Figure 5a). This was due to Pakistan's higher blue water productivity in producing its imported commodities, especially those of wheat and cotton. The rate of global blue virtual water savings has been decreasing over the years, with 3896 Mm^3/year blue VW global saving during 1990–1999, 1702 Mm^3/year BVW global saving during 2000–2009 and 459 Mm^3/year blue VW global saving during 2010–2016.

3.4.2. Green Virtual Water

Pakistan's domestic saving of green virtual water has been positive and increasing over the period 1990-2016, with an annual average of 7462 Mm^3 (Figure 5b). Starting with 5191 Mm^3 in 1990, the domestic saving grew by 155% to 13,230 Mm^3 in 2016. The (positive) domestic saving of green virtual water means that Pakistan would have needed an equivalent amount of extra domestic green water to produce the corresponding commodities (mostly palm oil, tea, cotton and wheat) domestically.

3.5. Pakistan's Virtual Water Trade in the Future

By 2030, Pakistan's trade and net imports of blue and green virtual water will increase quite significantly (Table 3). Pakistan's total net import of blue VW will remain negative and increase from −9446 Mm^3 in 2016 to −29528 Mm^3 in 2030, thus recording a 213% change (the figure in the parenthesis in the second row of Table 3) between 2016 and 2030. The net import of green VW during the same period will increase from 10,543 Mm^3 in 2016 to 29,595 Mm^3 in 2030. The ever-increasing quantities of rice exports will be the main factor behind the increase in (negative) net imports of blue VW for Pakistan in 2030. In contrast, cotton imports will contribute significantly to reducing the negative net import of blue VW in 2030. For green VW, on the other hand, the higher net import will be caused not only by the traditional import commodities like palm oil and tea but also by the positive net import of cotton by Pakistan in 2030.

Table 3. Pakistan's virtual water trade in 2030 (million m^3).

Title	Green Water	Blue Water	Total
VW Import	37,012 (178) †	4013 (296)	41,025 (186)
VW Export	7418 (166)	33,541 (221)	40,959 (209)
VW Net Import	29,595 (181)	−29,528 (213)	66 (−94)
National Savings	40,471 (206)	−28,999 (215)	11,472 (185)
Global Savings	10,876 (305)	529 (121)	11,406 (290)

† The figures in the parenthesis are the percentage increase of the respective value in 2030 from the level in 2016.
Source: Authors' calculations

In 2030, the net import of VW through agricultural trade will save substantially more green virtual water but at the same time, cause much higher losses of blue VW for Pakistan (Table 3, row 4). The domestic saving of total VW for Pakistan will be 11,472 Mm^3, a 185% increase from the 2016 savings of 4023 Mm^3. Although the domestic savings of green VW will increase by 206% from 13,230 Mm^3 in 2016 to 40,471 Mm^3, however, the blue water losses will also increase, recording a 215% increase from −9207 Mm^3 to −28,999 Mm^3 over the same period. The trends show that the significant losses of blue VW for Pakistan will further expand in the future.

On the global scale, Pakistan's trade in agricultural commodities will have an overall positive impact in terms of total, blue and green virtual water savings in 2030 (Table 3, row 5). Specifically, Pakistan's agricultural trade in 2030 will save 11,406 Mm^3 of the total virtual water, as compared to 2926 Mm^3 in 2016 (a 290% increase). The contribution of green VW savings towards global VW savings from Pakistan's trade in agricultural commodities will be more than 20 times higher than the contribution from blue VW savings. Between 2016 and 2030, the global savings of green VW will increase 305% (from 2686 Mm^3 to 10,876 Mm^3), while that of blue VW will increase 121% (from 239 Mm^3 to 529 Mm^3). The detailed commodity-wise analysis (not shown in the table for space consideration) shows that with 12,180 Mm^3 savings, palm oil will be the most significant contributor towards global savings of green VW. On the other hand, while the trade in the cotton crop will have a small positive contribution towards global savings of blue VW (398 Mm^3), it will contribute negatively towards global savings of green VW (−1476 Mm^3).

3.6. Value of Virtual Water Flows

The scatter plot between the value of virtual water (VVW) and blue virtual water contents for Pakistan is shown in Figure 6. We can see that commodities like fruits, vegetables and oilseed crops, placed towards the bottom-right corner have high VVW while they use relatively low blue water in their production, which means that these commodities use water more efficiently to produce higher economic value per unit of the blue water. On the other extreme is the beef sector, which uses large quantities of blue water to produce relatively low economic value. In comparison to other sectors, grain crops produced in Pakistan are not only less blue water-intensive but also produce lower economic output per unit of water. Interestingly, although the blue VWC of mutton and poultry are higher than the crops, their VVW is even higher, making them more valuable commodities in terms of water use per unit of production as compared to crops.

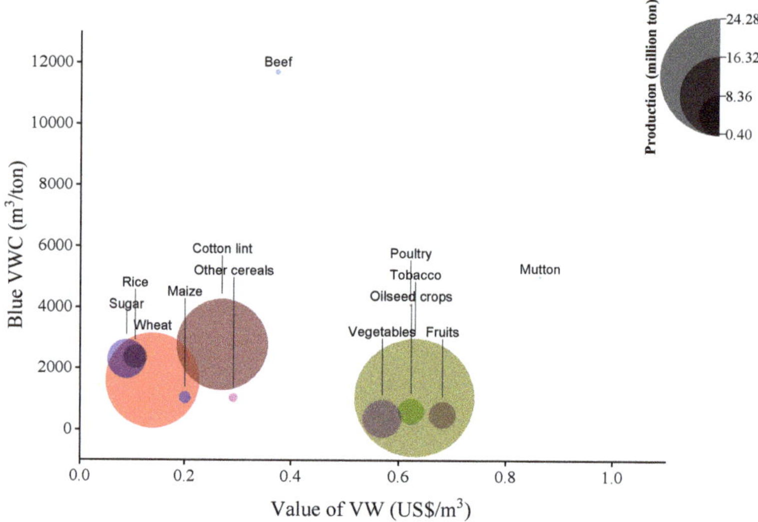

Figure 6. Average blue VWC and per unit water value for different agricultural commodities in Pakistan (1990–2016). 1 US$ = 60.4 Pakistan Rupee over the period 1990–2016. Per unit water value is calculated based on Equation (9). Source: Authors' calculations based on data from Reference [39] (production value) and sources mentioned in the Methods section (VWC).

From the value of virtual water (VVW) estimation, we can argue that to alleviate pressure on water resources and optimize blue water use in country's agriculture sector, Pakistan should promote the production of high-value and low-water-intensive commodities like vegetables, fruits and oilseeds. This type of structural adjustment will increase the value of output per drop of water. The optimization of water use for agricultural production should also consider other environmental aspects, like grey water footprint, especially for the livestock sectors.

4. Discussion and Policy Implications

The United Nations Sustainable Development Goal 12 highlights the need to pursue responsible consumption and production to address global environmental challenges such as water scarcity (Target 6.4 of Reference [56]). Pakistan, with water withdrawals for the agricultural sector of 94%, ranks among the highest in the world, compared to 69% at the global level [33]. Resolving water scarcity in developing countries like Pakistan could be achieved by focused research on the agricultural sector and the food system. In addition to devising localized, sustainable production and consumption systems, the trade in agricultural commodities (through tele-consumption of food) can also play a crucial role in reducing water-scarcity at local and global levels.

Literature is available on the amount of virtual water traded between Pakistan and other countries, but there is a lack of research on Pakistan's blue and green virtual water trade, the relative economic value of blue and green virtual water and the implications for national and global water resources, especially the scarcer blue water. Also, the network properties of a Pakistan-centered virtual water trade network have never been analyzed. Differentiating the analysis into blue and green water components is particularly crucial because blue water has a higher opportunity cost [7] and different environmental effects than green water [57].

Similar to the earlier work in the field [7,58,59], we find that during the 1990–2016 period, Pakistan's total virtual water trade has been increasing. However, the net virtual water import has been small. Our future scenario of Pakistan's trade in agricultural commodities, however, shows that Pakistan will lose a significant amount of virtual water by 2030. At first glance, it appears that Pakistan's historical trade in agricultural commodities neither benefits nor harms the domestic water resources. However, once we separate virtual water trade into blue and green virtual water, we discover a large and increasing net export of blue water from Pakistan, which means that Pakistan indeed exported more expensive and scarce blue water. The future trade scenario shows that the losses of blue virtual water will further increase between 2016 and 2030. Our results clearly show the importance of considering the trade in blue and green water separately.

Our analysis shows that the unit value of blue virtual water export through rice crop (the dominant export crop) 0.10 US$/m^3 has been lower than the corresponding value of green water import through palm oil (the dominant import crop) 0.15 US$/m^3 during last the decade. This further demonstrates that, although unconsciously, Pakistan has not only been exporting its more precious water resource to other countries, it has been doing so by fetching a relatively lower price for the blue water export thus making the water trade even less economically viable. Moreover, the costs of construing and running extensive irrigation infrastructure for the blue water used in agricultural exports further calls into question the actual benefits of exporting blue virtual water embodied in the commodities. One of the main reasons for these inefficient production and trade decisions is that farmers have been using irrigation water quite inefficiently in Pakistan [60], mainly due to subsidized irrigation water supply.

Pakistan's situation becomes even more worrisome when we consider the increasing scarcity of blue water in the country. Pakistan has been a net exporter of blue virtual water despite being a water-deficient country [61]. Other studies have shown that most of the water deficit countries have compensated for their domestic water scarcity by importing increasing imports of cereals [62,63]. The authors assert that upon reaching water availability of roughly 1500 m^3/capita/year, an inverse relationship can be identified between a country's cereal import and its per capita renewable water resources. Pakistan's annual water availability per inhabitant, which had already dropped to 1391 m^3 [33], is expected to reach around 800 m^3 by 2050 [64,65]. Looking ahead, the compounding water shortage in Pakistan would thus make the export of blue virtual water more unsustainable and questionable.

In contrast, the other water-stressed countries have been relying, although partially, on food imports to alleviate some pressure on domestic natural resources. For example, China's food trade saved 215.5 billion m^3 of domestic water in 2015, which was equivalent to about 55% of the country's irrigation water used in the same year [27]. Libya and Israel, having extremely scarce water resources, almost entirely rely on global markets for their cereal supply, with small domestic production [3]. Similarly, for Spain, higher net imports of virtual water during a dry year (8415 million m^3 in 2005) as compared to a wet year (3420 million m^3 in 1997) are consistent with the country's relative water scarcity [66]. These countries seem to have steered their domestic production and trade in agricultural commodities in line with the relative scarcity of water resources, which does not seem quite true for Pakistan.

A silver lining to Pakistan's net exports of domestic blue water resources is that Pakistan has induced a small but positive global saving of blue water over the years. The savings resulted from higher blue water use efficiency in the production of exported commodities (mostly rice) than the water use efficiency of the export destinations. The net imports of green water, on the other hand, saved considerable volumes of domestic green water for Pakistan, while there were quite modest global savings.

Our findings suggest that Pakistan has consistently engaged in exports of much scarcer blue VW, while the VW trade is often determined by factors other than water, like economic and political ones. Virtual water trade alone, therefore, cannot be taken as a yardstick to steer the trade and production policies. It is the combination of agricultural structure adjustment towards high water use-value commodities and active promotion of trade in these commodities that can optimize agricultural water use. By adopting such adjustments, Pakistan can also engage its rural labor force into agriculture, in which the country is better endowed than many other countries in the world [67,68]. Thus, Pakistan can not only make full use of its comparative international advantage in the production of labor-intensive commodities, such as fruits and vegetables [69] but also can save water by improved irrigation technologies, for example, drip irrigation and spray irrigation systems, which are more suitable for horticultural crops [70]. Our suggestion is also in line with the ones found in existing literature where the authors suggest that when water becomes a significant constraint and economic cost factor to agricultural production, improving per unit water value through agricultural structure adjustment is one of the rational options to increase water use efficiency and farmer's income [22].

Importantly, we do not take these results to argue that Pakistan should stop the production or export of blue water-intensive crops like rice. On the contrary, we suggest that the relative scarcity of blue water should be included in the production of agricultural commodities. Such policy measures will not only improve the domestic water use efficiency but will also increase the economic benefits of virtual water exports.

Although the notion of virtual water does not provide unambiguous conclusions about international trade efficiency from a water resources perspective, it might foster cooperation among countries for improving water and land management globally. This is especially relevant when considering adaptation to climate change, together with production and consumption patterns. Virtual water could, therefore, encourage discussions on transboundary water resource management strategies.

A few caveats of our study need to be highlighted here. We used VW values from global model studies, whereas VW data from a national model for Pakistan might be a better choice for future studies on VW trade and footprint. The results for fruits and vegetables should be read with caution because the grouping and reporting for the two sectors are significantly broader than all the other commodities in this study. Future studies should include more detailed analysis for both the groups of commodities, although greywater footprint is also a pertinent topic covered in many studies on VW trade. However, because the environmental side of the virtual water trade (covering greywater) is beyond the scope of this study, future studies should include grey virtual water as well. Due to the unavailability of data on the cost of production for the studied commodities at the national level in Pakistan, we had to use production value for estimating the value of virtual water. Although this is not the ideal measure to account for differences in cost/benefit of various agricultural commodities, the measure still provides the closest proxy for evaluating the economic value of virtual water in agricultural production. Our study did not include the social costs that are often associated with blue water use in exportable commodities. Adding these costs is expected to further support our assertion regarding the inclusion of virtual water trade into domestic water policies. The analysis of water use in agriculture within the country is also an essential topic for prioritizing water use in the country. However, it is out of the scope of this study and should be covered in the future work

5. Conclusions

The quantification of virtual water trade provides an essential perspective on sustainable water use practices. Using Pakistan as a case study, this paper assessed the trade and savings/losses of blue and green virtual water through agricultural commodities, over the period 1990–2016 and in 2030. The results of our study show that in most of the studied commodities, blue VW is the major component in the total water use. Pakistan's export of VW increased more rapidly than the import of VW over 1990–2016. On average, the net VW import (import minus export) was positive but small. Pakistan has been a net exporter of blue VW, mostly through rice export to Asian and African countries. In terms of green VW, Pakistan has been a net importer, mainly through the import of palm oil from Indonesia and Malaysia.

In terms of domestic savings, Pakistan's trade in the agricultural commodities has been saving green VW for the country. However, Pakistan has been losing increasingly higher volumes of blue VW through its agricultural trade over the study period. Putting this into perspective, the blue VW loss was equivalent to about 7% of the total agricultural water withdrawal in the country in 2011. Pakistan's trade in the agricultural commodities has also contributed to positive savings of blue and green virtual water at the global level.

In the future (2030), both Pakistan's domestic savings of green and losses of blue virtual water will increase by more than 200%. The global savings will increase by more than three times for green VW and more than 120% for blue VW. Our results also suggest that Pakistan has been exporting more expensive (with high opportunity cost) blue VW through its agricultural trade to the rest of the world and can thus benefit from improvements in water use efficiency (in the export-oriented crops) and adjustment in its export portfolio of agricultural commodities by promoting the export of commodities with higher value and lower water use intensity.

As discussed in detail, two major shortcomings should be addressed in future research. First, time-series data from the national model(s) should be used in place of the global estimates for VWC. This will enhance the reliability of the results to produce more reliable policy suggestions. Second, time-series data for grey VW for the agricultural commodities are needed to analyze the environmental effects of trade in agricultural commodities, which were not taken into account in this study.

Although numerous studies on VW trade have been published, the studies mainly concerned those countries which are either net-importers and water-stressed or net-exporters and water-abundant. Given that water availability is more crucial for the water-stressed net-exporting countries, the trade in VW for such countries becomes more critical. The analysis of Pakistan's trade in VW through agricultural commodities presented in this study can further promote future work in water-stressed countries.

Author Contributions: Conceptualization, T.A. and W.X.; Formal analysis, T.A. and W.X.; Methodology, T.A., A.M.N. and M.F.R.; Writing—original draft, T.A. and W.X.; Writing—review & editing, T.A., A.M.N. and M.F.R. All authors read and approved the final manuscript.

Funding: The authors acknowledge their respective financial supports from the National Key R&D Program of China (2016YFA0602604) and the National Natural Sciences Foundation of China (71503243; 71873009).

Conflicts of Interest: The authors declare no conflict of interest.

Appendix A

Appendix A.1 Analysis of Pakistan's Agricultural Trade

Despite being an agricultural economy, Pakistan has not only been exporting some agricultural commodities, but it has also been importing significant quantities of some other agricultural commodities. A country's dependence (or independence) on foreign-produced commodities can be gauged by the self-sufficiency ratio (SSR = Production/(Production + Net import) based on data from Reference [39]). Pakistan's average SSR over the period 1990–2013 is shown in Figure A1.

We can observe two extremes in the SSR for Pakistan, that is, on the one hand, Pakistan has been producing almost double the amount of rice it needed for domestic consumption (SSR = 187%), implying that Pakistan exported 47% of its domestically produced rice to other countries over this period. On the other hand, Pakistan's SSR in palm oil (used as edible oil) and tea—the two largest traditional food imports of Pakistan—has been almost zero, indicating that Pakistan has been entirely dependent on other countries for its domestic needs of palm oil and tea. The less than 100% SSR values of other cereals (84%), oilseed crops (89%), wheat (96%) and cotton (96%) show that Pakistan has been importing some amounts of these commodities over the years. The rest of the commodities show relatively high (close to or more than 100%) SSR over the period.

In value terms, Pakistan's exports have been dominated by rice, which—with an average yearly export of 1645 million US$- accounted for over 64% of the total export value of these commodities during 2003–2016 (Figure A2, yellow columns). Fruits (251 million US$), cotton (148 million US$), vegetables (135 million US$) and sugar (108 million US$) have been other major commodities recording sizable export for Pakistan during this period. The five commodities mentioned above were responsible for about 90% share in the total export value of the studied commodities.

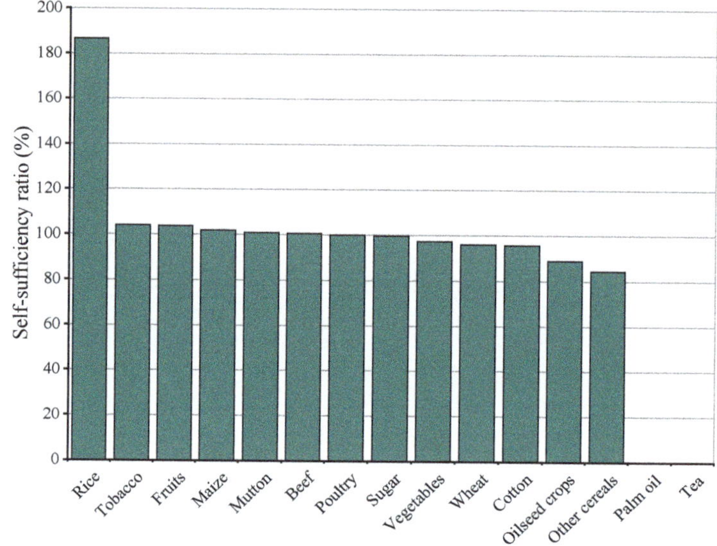

Figure A1. Average self-sufficiency ratios (SSR) of agricultural commodities for Pakistan, 1990–2013. Source: Authors' calculations based on data from Reference [39].

In terms of import value, palm oil takes the lead with an average value of 1434 million US$ (about 37% share) among these agricultural commodities during 2003–2016. Cotton (627 million US$), oilseed crops (544 million US$), vegetables (438 million US$) and tea (293 million US$) also have significant shares in the total import value of the agricultural commodities by Pakistan during the same period. Notice that cotton, vegetables and sugar have substantial shares in both total import and total export values suggesting that over the years the production of these crops in Pakistan have experienced significant ups and downs such that during some years Pakistan exported these crops while it imported the same crops during other years.

Additionally, some quality requirements have also impacted the changing trade situation of some corps, especially in the case of cotton, where Pakistan had to import high-quality cotton from other countries to produce exportable textile goods. It is also noteworthy that although Pakistan's domestic tea production is negligible (there is some domestic production; however, this is close to zero as compared to the total net-import) (Figure A1), the value share of tea import in total imports of the agricultural commodities has been around 8% (Figure A2).

Our future projections under business as usual (BAU) scenarios show that Pakistan will continue to export vast quantities of rice to other countries by 2030, such that its rice exports will increase by 3.3 times from the 2016 level. The exports of sugar and cattle will also increase considerably in percentage terms. However, the corresponding absolute changes will be quite small, bearing little effect on VW exports from Pakistan. At the same time, the exports of some agricultural commodities will be reduced. Notably, both cotton and fruits have relatively high exports in 2016, while their respective exports will drop by 86% and 48% by 2030. In terms of export partners, China will gain increased importance as the export destination, especially for the export of rice, fruits, vegetables and livestock products.

Pakistan's imports of palm oil, tea and cotton will increase considerably by 192%, 31% and 363% during 2016–2030. Palm oil and tea are the most significant traditional agricultural imports by Pakistan. However, our projections show that under the BAU scenario, Pakistan will turn from a net exporter of cotton to a net importer during the coming decade. In 2030, Pakistan's import sources will remain almost the same as they were in 2016.

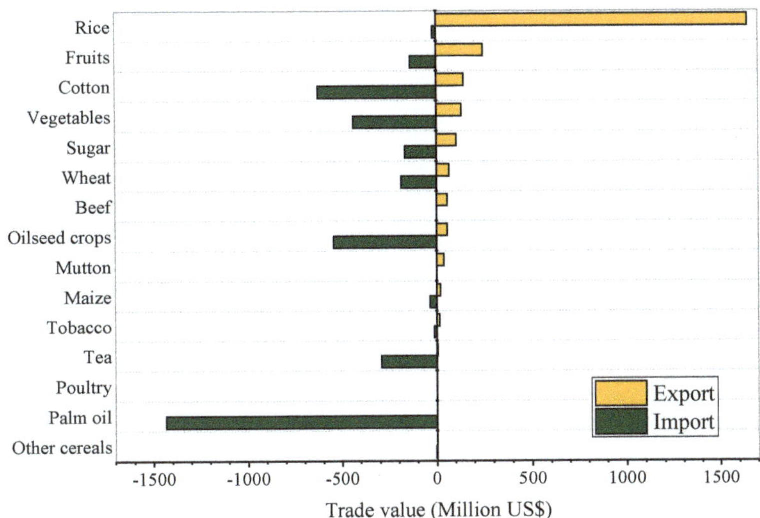

Figure A2. Average trade value of agricultural commodities, 2003–2016. Source: Authors' calculations based on data from Reference [40].

Appendix A.2 Water Resources and Use in Pakistan and its Major Trade Partners

Pakistan's water resources are scarcer than most of its major export partners. At 1711 m^3/capita (By 2014, Pakistan had already become a water-stressed country with water availability 1306 m3/capita.), Pakistan is almost at the edge of entering the water scarcity threshold of 1500–1700 m^3/capita in 2000 [62,71] (Table A1). Apart from Kenya, Saudi Arabia and UAE, Pakistan has much lower per capita renewable water resources than the countries it has negative net virtual water imports of blue VW. Among these three countries, Kenya is almost entirely rainfed, while Saudi Arabia and UAE have desert climates with small agriculture sectors. The per capita renewable water resources in Indonesia and Malaysia, the major import partners of Pakistan in terms of virtual water (most of which is green VW), are around 8 and 20 times that of Pakistan's.

The water use intensity in a country is reflected by the ratio of freshwater withdrawal to total renewable water resources. Again, Pakistan's ratio of 70%, which is much larger than the water criticality threshold suggested by Reference [72], is also higher than most of its major trade partners. The ratio of irrigated area to total crop area in Pakistan is higher than most of its major trading partners, with minimal coverage in Indonesia and Malaysia.

Water stress is indicated by the ratio of total annual freshwater withdrawals to hydrological availability. Pfister (2009) [57] uses the concept of water stress to define water stress index (WSI), which indicates the portion of withdrawals of blue water that deprives other users of freshwater, ranging from zero to one (zero being no stress). We can see from the last column of Table 1A that, except the two desert regions, that is, Saudi Arabia and UAE, Pakistan's WSI is greater than all its trading partners. The higher WSI further indicates that Pakistan is exporting its scarce blue virtual water to relatively water abundant regions of the world.

List of fruits: Almonds shelled, Apples, Apricots, Apricots dry, Avocados, Bananas, Cake copra, Cashew nuts shelled, Cherries, Chestnut, Coconuts, Coconuts desiccated, Copra, Currants, Dates, Figs, Figs dried, Fruit cooked homogenized preparations, Fruit dried nes, Fruit fresh nes, Fruit prepared nes, Fruit tropical fresh nes, Grapefruit (inc. pomelos), Grapes, Hazelnuts shelled, Juice citrus concentrated, Juice citrus single strength, Juice fruit nes, Juice grape, Juice grapefruit, Juice grapefruit concentrated, Juice orange concentrated, Juice orange single strength, Juice pineapple, Juice pineapple concentrated, Kiwi fruit, Lemons and limes, Mangoes, Guavas, Nuts, nes, Nuts prepared (exc. groundnuts), Olives, Olives preserved, Oranges, Papayas, Peaches and nectarines, Pears, Persimmons, Pineapples, Pineapples canned, Pistachios, Plantains, Plums, Plums dried (prunes), Raisins, Strawberries, Tangerines mandarins, satsumas, Walnuts shelled, Walnuts with shell [39].

List of vegetables: Artichokes, Asparagus, Beans dry, Beans green, Broad beans (horse beans dry), Cabbages and other brassicas, Carrots and turnips, Cauliflowers and broccoli, Chickpeas, Chilies and peppers dry, Chilies and peppers green, Cucumbers and gherkins, Eggplants (aubergines), Garlic, Ginger, Juice tomato, Leeks other alliaceous vegetables, Lentils, Lettuce and chicory, Melons other (inc. cantaloupes), Mushrooms and truffles, Mushrooms canned, Onions dry, Onions shallots green, Peas dry, Peas green, Pepper (piper spp.), Potatoes, Potatoes frozen, Pumpkins squash and gourds, Spinach, Sweet potatoes, Tomatoes peeled, Vegetables in vinegar, Vegetables dehydrated, Vegetables fresh nes, Vegetables fresh or dried products nes, Vegetables frozen, Vegetables homogenized preparations, Vegetables preserved nes, Vegetable preserved frozen, Vegetables temporarily preserved, Watermelons [39].

Table A1. Water resources and use in Pakistan and its major trade partners, 2000 [1].

Country	Total Renewable Water Resources (billion m³/year)	Total Water Withdrawal (billion m³/year)	Per Capita Renewable Water Resources (m³/capita/ year)	Per Capita Water Withdrawal (m³/capita/ year)	Freshwater Withdrawal as % of Total Renewable Water Resources (%)	Ratio of Irrigated Area to Total Crop Area (%)	Water Stress Index [2]
Pakistan	246.8	172.6	1711	1196	69.94	55.89	0.971
Major export partners							
Afghanistan	65.33	20.28	3040	943.8	31.04	41.38	0.966
China	2840	549.8	2161	418.4	19.36	42.83	0.478
Iran	137	89.7	2024	1325	65.47	44.73	0.912
Kenya	30.7	2.32	824.1	62.28	7.557	1.899	0.021
Mozambique	217.1	0.8842	11220	45.7	0.4073	2.513	0.197
Saudi Arabia	2.4	23.67	92.01	907.5	943.3	44.08	0.995
UAE	0.15	2.904	44.2	855.6	1556	97.42	0.998
USA	3069	559.3	10639	1939	15.43	15.33	0.499
Major import partners							
Indonesia	2019	113.3	9288	521.2	5.612	13.69	0.043
Malaysia	580	9.305	23769	381.3	1.604	5.101	0.180

Source: [1] This is the latest year for which all the data is available for these countries. [2] is from Reference [57] and the rest of the information is from Reference [39].

References

1. United Nations (UN). Report of the World Commission on Environment and Development: Our Common Future (the Brundtland report). 1987. Available online: http://www.un-documents.net/a42-427.htm (accessed on 10 January 2019).
2. UNESCO. *World Water Development Report. Water for a Sustainable World*; WWAP (United Nations World Water Assessment Programme): Paris, France, 2015.
3. Yang, H.; Zehnder, A.B.J. Water scarcity and food import: A case study from southern Mediterranean countries. *World Dev.* **2002**, *30*, 1413–1430. [CrossRef]
4. Chapagain, A.K.; Hoekstra, A.Y. *Water Footprints of Nations*; Value of Water Research Report Series No. 16; UNESCO-IHE: Delft, The Netherlands, 2004.
5. Chapagain, A.K.; Hoekstra, A.Y.; Savenije, H.H.G. *Saving Water through Global Trade*; Value of Water Research Report Series No. 17; UNESCO-IHE: Delft, The Netherlands, 2005.
6. Chapagain, A.K.; Hoekstra, A.Y.; Savenije, H.H.G. Water saving through international trade of agricultural products, Hydrol. *Earth Syst. Sci.* **2006**, *10*, 455–468. [CrossRef]
7. Hoekstra, A. *The Relation between International Trade and Freshwater Scarcity*; WTO Staff Working Paper, No. ERSD-2010-05; World Trade Organization: Geneva, Switzerland, 2010.
8. Dalin, C.; Konar, M.; Hanasaki, N.; Rinaldo, A.; Rodriguez-Iturbe, I. Evolution of the global virtual water trade network. *Proc. Natl. Acad. Sci. USA* **2012**, *109*, 5989–5994. [CrossRef] [PubMed]
9. Konar, M.; Dalin, C.; Hanasaki, N.; Rinaldo, A.; Rodriguez-Iturbe, I. Temporal dynamics of blue and green virtual water trade networks. *Water Resour. Res.* **2012**, *48*, W07509. [CrossRef]
10. Mekonnen, M.M.; Hoekstra, A.Y. *National Water Footprint Accounts: The Green, Blue and Grey Water Footprint of Production and Consumption*; Value of Water Research Report Series No. 50; UNESCO-IHE: Delft, The Netherlands, 2011.
11. Schyns, J.F.; Hoekstra, A.Y. The added value of Water Footprint Assessment for national water policy: A case study for Morocco. *PLoS ONE* **2014**, *9*, e99705. [CrossRef] [PubMed]
12. Aldaya, M.M.; Martinez-Santos, P.; Llamas, M.R. Incorporating the water footprint and virtual water into policy: Reflections from the Mancha Occidental region, Spain. *Water Resour. Manag.* **2010**, *24*, 941–958. [CrossRef]
13. Chouchane, H.; Krol, M.S.; Hoekstra, A.Y. Virtual water trade patterns in relation to environmental and socioeconomic factors: A case study for Tunisia. *Sci. Total Environ.* **2018**, *613–614*, 287–297. [CrossRef]
14. Zoumides, C.; Bruggeman, A.; Hadjikakou, M.; Zachariadis, T. Policy-relevant indicators for semi-arid nations: The water footprint of crop production and supply utilization of Cyprus. *Ecol. Indic.* **2014**, *43*, 205–214. [CrossRef]
15. Gobin, A.; Kersebaum, K.; Eitzinger, J.; Trnka, M.; Hlavinka, P.; Takáč, J.; Kroes, J.; Ventrella, D.; Dalla Marta, A.; Deelstra, J.; et al. Variability in the water footprint of arable crop production across European regions. *Water* **2017**, *9*, 93. [CrossRef]
16. Allan, J.A. Virtual water: A strategic resource. *Ground Water* **1998**, *36*, 545–547. [CrossRef]
17. Antonelli, M.; Tamea, S. Food-water security and virtual water trade in the Middle East and North Africa. *Int. J. Water Resour. Dev.* **2015**, *31*, 326–342. [CrossRef]
18. Antonelli, M.; Tamea, S.; Yang, H. Intra-EU agricultural trade, virtual water flows and policy implications. *Sci. Total Environ.* **2017**, *587*, 439–448. [CrossRef] [PubMed]
19. Allan, J.A. *Virtual Water: A Long Term Solution for Water Short Middle Eastern Economies?* Occasional Paper, No. 3; Water Issues Study Group, School of Oriental and African Studies, University of London: London, UK, 1997.
20. Allan, J.A. Watersheds and problemsheds: Explaining the absence of armed conflict over water in the Middle East. *Middle East Rev. Int. Aff.* **1998**, *2*, 1.
21. Shuval, H. 'Virtual water' in the water resource management of the arid Middle East. *Water Resour. Middle East* **2007**, *2*, 133–139.
22. Liu, J.; Zehnder, A.J.B.; Yang, H. Historical trends in China's virtual water trade. *Water Int.* **2007**, *32*, 78–90. [CrossRef]
23. Liu, J.; Savenije, H.H.G. Food consumption patterns and their effect on water requirement in China. *Hydrol. Earth Syst. Sci.* **2008**, *12*, 887–898. [CrossRef]

24. El-Sadek, A. Virtual water trade as a solution for water scarcity in Egypt. *Water Resour. Manag.* **2010**, *24*, 2437–2448. [CrossRef]

25. Shi, J.; Liu, J.; Pinter, L. Recent evolution of China's virtual water trade: Analysis of selected crops and considerations for policy. *Hydrol. Earth Syst. Sci.* **2014**, *18*, 1349–1357. [CrossRef]

26. Dalin, C.; Hanasaki, N.; Qiu, H.; Mauzerall, D.L.; Rodriguez-Iturbe, I. Water resources transfers through Chinese interprovincial and foreign food trade. *Proc. Natl. Acad. Sci. USA* **2014**, *111*, 9774–9779. [CrossRef]

27. Ali, T.; Huang, J.; Wang, J.; Xie, W. Global footprints of water and land resources through China's food trade. *Glob. Food Secur.* **2017**, *12*, 139–145. [CrossRef]

28. Huang, H.; Li, X.; Cao, L.; Jia, D.; Zhang, J.; Wang, C.; Han, Y. Inter-Sectoral Linkage and External Trade Analysis for Virtual Water and Embodied Carbon Emissions in China. *Water* **2018**, *10*, 1664. [CrossRef]

29. Da Silva, V.; de Oliveira, S.D.; Hoekstra, A.; Neto, J.D.; Campos, J.; Braga, C.; de Araújo, L.; Aleixo, D.; de Brito, J.; de Souza, M.; et al. Water Footprint and Virtual Water Trade of Brazil. *Water* **2016**, *8*, 517. [CrossRef]

30. UN. *"Total Population—Both Sexes." World Population Prospects, the 2019 Revision*; United Nations Department of Economic and Social Affairs, Population Division, Population Estimates and Projections Section: New York, NY, USA, 2019.

31. World Bank. *Pakistan—Country Water Resources Assistance Strategy: Water Economy Running Dry (English)*; World Bank: Washington, DC, USA, 2005. Available online: http://documents.worldbank.org/curated/en/315851468285362706/Pakistan-Country-water-resources-assistance-strategy-water-economy-running-dry (accessed on 2 July 2018).

32. UNEP. The Environment and Climate Change Outlook of Pakistan. 2008. Available online: https://www.uncclearn.org/sites/default/files/inventory/unep25082015.pdf (accessed on 29 July 2019).

33. FAO. AQUASTAT Main Database, Food and Agriculture Organization of the United Nations (FAO). 2017. Available online: http://www.fao.org/nr/water/aquastat/data/query/index.html?lang=en (accessed on 12 December 2018).

34. Ali, T.; Xie, W. Pakistan needs more reservoirs and fast. *Nature* **2018**, *560*, 431. [CrossRef] [PubMed]

35. Dalin, C.A.C.; Wada, Y.; Kastner, T.; Puma, M.J. Groundwater depletion embedded in international food trade. *Nature* **2017**, *543*, 700–704. [CrossRef] [PubMed]

36. De Fraiture, C.; Cai, X.; Amarasinghe, U.; Rosegrant, M.; Molden, D. *Does International Cereal Trade Save Water? The Impact of Virtual Water Trade on Global Water Use*; Comprehensive Assessment Research Report 4; Comprehensive Assessment Secretariat: Colombo, Sri Lanka, 2004.

37. Chapagain, A.K.; Hoekstra, A.Y. The blue, green and grey water footprint of rice from production and consumption perspectives. *Ecol. Econ.* **2011**, *70*, 749–758. [CrossRef]

38. Mekonnen, M.M.; Hoekstra, A.Y. The green, blue and grey water footprint of crops. *Hydrol. Earth Syst. Sci.* **2011**, *15*, 1577–1600. [CrossRef]

39. FAOSTAT. FAO's Corporate Database, 2017. Available online: http://www.fao.org/faostat/en/#data (accessed on 10 December 2018).

40. DESA/UNSD. (United Nations Comtrade database), 2017. Available online: https://comtrade.un.org/data (accessed on 3 July 2018).

41. Hertel, T.W. *Global Trade Analysis. Modeling and Applications*; Cambridge University Press: New York, NY, USA, 1997.

42. Corong, E.L.; Hertel, T.W.; McDougall, R.; Tsigas, M.E.; van der Mensbrugghe, D. The Standard GTAP Model, Version 7. *J. Glob. Econ. Anal.* **2017**, *2*, 1–119.

43. Nelson, G.C. Climate change effects on agriculture: Economic responses to biophysical shock. *Proc. Natl. Acad. Sci. USA* **2014**, *111*, 3274–3279. [CrossRef]

44. Walmsley, T.L.; Dimaranan, B.V.; McDougall, R. A baseline scenario for the dynamic GTAP model. In *Dynamic Modeling and Applications for Global Economic Analysis*; Ianchovichina, E., Walmsley, T., Eds.; Cambridge University Press: New York, NY, USA, 2006; p. 136.

45. Fouré, J.; Bénassy-Quéré, A.; Fontagné, L. Modelling the world economy at the 2050 horizon. *Econ. Transit.* **2013**, *21*, 617–654.

46. Hanasaki, N. Estimating virtual water contents using a global hydrological model. In *Terrestrial Water Cycle and Climate Change: Natural and Human-Induced Impacts*; Tang, Q., Oki, T., Eds.; John Wiley & Sons, Inc.: Hoboken, NJ, USA, 2016. [CrossRef]

47. Zoumides, C.; Bruggeman, A.; Zachariadis, T. Global versus local crop water footprints: The case of Cyprus. In Proceedings of the Session "Solving the Water Crisis: Common Action toward a Sustainable Water Footprint", Planet under Pressure Conference, London, UK, 26 March 2012.

48. Dalin, C.; Suweis, S.; Konar, M.; Hanasaki, N.; Rodriguez-Iturbe, I. Modeling past and future structure of the global virtual water trade network. *Geophys. Res. Lett.* **2012**, *39*, L24402. [CrossRef]

49. Tuninetti, M.; Tamea, S.; Laio, F.; Ridolfi, L. A Fast Track approach to deal with the temporal dimension of crop water footprint. *Environ. Res. Lett.* **2017**, *12*, 074010. [CrossRef]

50. Liu, J.; Savenije, H.H.G. Time to break the silence around virtual-water imports. *Nature* **2008**, *453*, 587. [CrossRef] [PubMed]

51. Rosegrant, M.W.; Cai, X.; Cline, S.A. *World Water and Food to 2025: Dealing with Scarcity*; International Food Policy Research Institute: Washington, DC, USA, 2002.

52. USDA. *Pakistan: Cotton and Products Annual*; Global Agricultural Information Network (GAIN): Washington, DC, USA, 2010.

53. USDA. *Pakistan: Cotton and Products Annual*; Global Agricultural Information Network (GAIN): Washington, DC, USA, 2018.

54. USDA. *Pakistan: Grain and Feed Update, Wheat Update*; Global Agricultural Information Network (GAIN): Washington, DC, USA, 2010.

55. USDA. *Pakistan Subsidizes Wheat Exports*; Global Agricultural Information Network (GAIN): Washington, DC, USA, 2018.

56. UNDSD. *Transforming our World: The 2030 Agenda for Sustainable Development*; United Nations Division for Sustainable Development: New York, NY, USA, 2015.

57. Pfister, S.; Koehler, A.; Hellweg, S. Assessing the environmental impacts of freshwater consumption in LCA. *Environ. Sci. Technol.* **2009**, *43*, 4098–4104. [CrossRef]

58. Fader, M.; Gerten, D.; Thammer, M.; Heinke, J.; Lotze-Campen, H.; Lucht, W.; Cramer, W. Internal and external green-blue agricultural water footprints of nations and related water and land savings through trade. *Hydrol. Earth Syst. Sci.* **2011**, *15*, 1641–1660. [CrossRef]

59. Konar, M.; Hussein, Z.; Hanasaki, N.; Mauzerall, D.L.; Rodriguez-Iturbe, I. Virtual water trade flows and savings under climate change. *Hydrol. Earth Syst. Sci.* **2013**, *17*, 3219–3234. [CrossRef]

60. Sahibzada, S.A. Pricing Irrigation Water in Pakistan: An Evaluation of Available Options. *Pak. Dev. Rev.* **2002**, *41*, 209–241. [CrossRef]

61. UNEP. *Vital Water Graphics. An Overview of the State of the World's Fresh and Marine Waters*, 2nd ed.; United Nations Environment Programme: Nairobi, Kenya, 2008. Available online: http://www.unep.org/dewa/vitalwater/article77.html (accessed on 10 March 2018).

62. Yang, H.; Reichert, P.; Abbaspour, K.C.; Zehnder, A.J.B. A water resources threshold and its implications for food security. *Environ. Sci. Technol.* **2003**, *37*, 3048–3054. [CrossRef]

63. Yang, H.; Wang, L.; Zehnder, A. Water scarcity and food trade in the Southern and Eastern Mediterranean countries. *Food Policy* **2007**, *32*, 585–605. [CrossRef]

64. Burek, P.; Satoh, Y.; Fischer, G.; Kahil, T.; Jimenez, L.; Scherzer, A.; Tramberend, S.; Wada, Y.; Eisner, S.; Flörke, M.; et al. *Final Report: Water Futures and Solution Fast Track Initiative*; ADA Project Number 2725-00/2014; IIASA: Luxemburg, 2016.

65. Amir, P. *Agricultural Modernization through Better Water Management*; Country Water Resources Assistance Strategy, Background Paper # 12; Pakistan: Country Water Resources Assistance Strategy, Water Economy: Running Dry Report No. 34081-PK; Agriculture and Rural Development Unit, South Asia Region, World Bank: Washington, DC, USA, 2005.

66. Novo, P.; Garrido, A.; Varela-Ortega, C. Are virtual water "flows" in Spanish grain trade consistent with relative water scarcity? *Ecol. Econ.* **2009**, *68*, 1454–1464. [CrossRef]

67. Central Intelligence Agency (CIA). The World Factbook—Central Intelligence Agency. 2018. Available online: https://www.cia.gov/library/publications/the-world-factbook/index.html (accessed on 21 June 2018).

68. Faruqee, R. (Ed.) *Strategic Reforms for Agricultural Growth in Pakistan*; World Bank Institute (WBI) Learning Resources Series; The World Bank: Washington, DC, USA, 1999. Available online: http://documents.worldbank.org/curated/en/205781468780276574/Strategic-reforms-for-agricultural-growth-in-Pakistan (accessed on 21 July 2018). (In English)

69. World Bank. *Revealed Comparative Advantage of Pakistan's Agricultural Exports*; World Bank: Washington, DC, USA, 2010. Available online: http://documents.worldbank.org/curated/en/973131468064727361/Revealed-comparative-advantage-of-Pakistans-agricultural-exports (accessed on 24 February 2018). (In English)

70. Evans, R.G.; Sadler, E.J. Methods and technologies to improve efficiency of water use. *Water Resour. Res.* **2008**, *44*, W00E04. [CrossRef]

71. Falkenmark, M.; Widstrand, C. *Population and Water Resources: A Delicate Balance*; Population Bulletin 47; Population Reference Bureau, UN: Washington, DC, USA, 1992.

72. Alcamo, J.; Henrichs, T. Critical regions: A model-based estimation of world water resources sensitive to global changes. *Aquat. Sci.* **2002**, *64*, 352–362. [CrossRef]

MDPI

St. Alban-Anlage 66

4052 Basel

Switzerland

Tel. +41 61 683 77 34

Fax +41 61 302 89 18

www.mdpi.com

Water Editorial Office

E-mail: water@mdpi.com

www.mdpi.com/journal/water